Air Turbulence and its Methods of Detection

Leonardo Di G. Sigalotti
Department of Basic Sciences
Autonomous Metropolitan University
Azcapotzalco Unit, UAM-A, Mexico

Fidel Cruz Peregrino
Department of Basic Sciences
Autonomous Metropolitan University
Azcapotzalco Unit, UAM-A, Mexico

Alejandro Ramírez-Rojas
Department of Basic Sciences
Autonomous Metropolitan University
Azcapotzalco Unit, UAM-A, Mexico

T0399998

CRC Press
Taylor & Francis Group
Boca Raton   London   New York

CRC Press is an imprint of the
Taylor & Francis Group, an **informa** business

A SCIENCE PUBLISHERS BOOK

First edition published 2023
by CRC Press
6000 Broken Sound Parkway NW, Suite 300, Boca Raton, FL 33487-2742

and by CRC Press
4 Park Square, Milton Park, Abingdon, Oxon, OX14 4RN

Library of Congress Cataloging-in-Publication Data (applied for)

ISBN: 978-0-367-54680-9 (hbk)
ISBN: 978-0-367-54681-6 (pbk)
ISBN: 978-1-003-09014-4 (ebk)

DOI: 10.1201/9781003090144

Typeset in Times New Roman
by Radiant Productions

In memory of our fathers
Girolamo Sigalotti Coassin
and
Airplane Pilot Captain Filadelfo Cruz Zuñiga

A.R.-R. dedicates the book to his
children Gabriela and Alejandro
and
grandchildren Ari and Nicolás

Preface

Clear air turbulence (CAT) – the ubiquitous phenomenon that occurs in cloudless regions and causes the sudden and violent buffeting of aircraft – continues today to elude our understanding and the detection techniques even after several decades of research and technological development. In the literature, CAT has been frequently defined as aircraft turbulence associated with wind shear around the tropopause. In general, it refers to bumps in flight aboard aircraft at cruising altitudes and away from convective clouds. While light and light-to-moderate CAT is almost always encountered in long-haul flights, causing only some discomfort to passengers, moderate-to-severe and severe CAT encounters are much rarer and may result in serious injuries to passengers and crew as well as in damage to the aircraft. Several studies indicate that turbulence is one of the leading causes of weather-related aviation accidents, with CAT being responsible for about 24 percent of these accidents. On the other hand, recent multiple computer climate model simulations are finding that global temperature changes would lead to an increase in the strength and frequency of CAT worldwide in the coming years, leaving some bewilderment and concern among flight attendants and travelers.

Today, convective regions can be accurately detected using radar technology and therefore pilots can easily avoid such regions. In contrast, CAT cannot be detected by current weather radars and some emerging and promising technologies, such as, for example, the Light Detection and Ranging (LIDAR), which would help mitigate the problem, have not become fully operational because of power and cost considerations. Therefore, pilot reports remain one of the most reliable means to avoid CAT. Alternative promising tools rely on the new generation of satellites with higher spatio-temporal resolution and on forecasting and nowcasting systems to improve CAT avoidance. However, current turbulence forecasts suffer from a very limited spatial resolution so that empirical turbulence diagnostics (or indices) are used to resolve aviation-affecting eddies, which work on the principle that the kinetic energy associated with turbulence at the scales of aircraft-sized eddies cascades down from the largest eddies, which can be explicitly resolved by numerical models. Despite a large number of limitations still present in the areas of turbulence observation and detection, nowcasting, forecasting, and verification as well as in simulation and modeling studies, the last two decades have provided growing scientific, technological, and operational advances aimed at improving passenger safety on commercial flights.

In the preparation of this book, the guiding thought has been to develop a logical unified treatment of the subject of CAT and its methods of detection. The book is divided into three main parts. The first part is composed of four chapters, where the first chapter begins with an introductory overview on turbulence with particular emphasis on the physical processes leading to the Kelvin-Helmholtz instability, which is believed to be a primary producer of CAT. Elements on the statistics of turbulence are also briefly revisited followed by Kolmogorov's theory and Richardson's concept of the energy cascade. A brief description of the modern approaches for modeling turbulence is given in Chapter 2, including details of the Reynolds Averaged Navier-Stokes (RANS), the Large Eddy Simulation (LES), and the Direct Numerical Simulation (DNS) approaches. The description of CAT is fully introduced in Chapter 3 along with its causes and effects. This chapter also contains related topics and characteristics associated with CAT, which will clarify many concepts and definitions employed in the rest of the book. The topics include brief descriptions of wake turbulence, gravity waves, wind shear, vortex rolls, jet streams and streaks, lee and mountain waves, temperature gradients, on one hand, and dynamical, aviation, and synoptic meteorology and the Lighthill-Ford theory, on the other hand. The notion of fossil turbulence and the question of whether CAT is increasing close the chapter. Chapter 4 presents an overview of the existing meteorological evidence for CAT, including the meteorological analysis of CAT around the troposphere and its relation with the mesoscale structure of the jet stream region. Details of the probabilistic structure of CAT and turbulence at low altitudes are also given.

The second part of the book comprises Chapters 5 and 6. Chapter 5 presents the different types of forecasts for aviation, while Chapter 6 introduces the reader into the impact of CAT on aviation operations and aircraft, concluding with special topics on the historical records of worldwide aviation accidents and injured people during the last decade.

The third part of the book comprises Chapters 7, 8, 9, 10, and 11. In particular, the primary purpose when writing Chapter 7 was to cover as much as possible the extensive existing material on the earlier methods of CAT detection and warning. This was mandatory because many of the present and emerging technologies for CAT detection and avoidance are based on development work done during the 1980s and 1990s. The chapter covers the basic principles underlying the passive and active optical techniques and various optical sensors and ultrasensitive radars designed and operated during those years. Aspects of the new technologies for reducing aviation weather-induced accidents is provided in Chapter 8 with a focus on the advances made in the area of digital information and communication. Like

Chapter 7, Chapter 9 has been distributed in small type sections to cover as much as possible the material on modern methods and techniques for CAT detection, including emerging technologies and sensor devices such as, the DELICAT, the JAXA, and the AWIATOR airborne LIDARs on one hand, and the KMW and ULTURB forecasting algorithms, on the other hand. The Chapter ends with a brief account of the NCAR/NEXRAD turbulence detection algorithm, the newly employed optical airflow meters for remote sensing, and the emergent small unmanned aircraft systems. The book closes with an overview of the Weather Accident Prevention Project in Chapter 10 along with some miscellaneous issues together with a description of the recent research activities on CAT and major challenges in turbulence detection, prediction, and avoidance, which have been integrated into Chapter 11.

In a subject as extended and complex as the one in this book, it would be impossible to give credit and acknowledge in every case indebtedness to specific books and papers. However, in the various references that appear throughout the text and in the bibliography every effort has been made to indicate the main sources. We apologize to all those authors who have not been cited in the book because of our carelessness or human limitations to review all the literature on the subject. We wish to express our sincere appreciation to Suzanne Kelley of Littleton for allowing us to use her beautiful photograph on the Kelvin-Helmholtz instability in clouds. We also express our thanks to our many colleagues of the Department of Basic Sciences at the Azcapotzalco Unit of the Autonomous Metropolitan University of Mexico for continued encouragement in the preparation of the manuscript. We are particularly indebted to Assistant Editor Vijay Primlani for inviting us to participate in this project and for his infinite patience and continuous assistance during the writing of the book.

Leonardo Di G. Sigalotti
Fidel Cruz Peregrino
Alejandro Ramírez-Rojas

Contents

PART II

PART III

PART I

Chapter 1

The Theory of Turbulence

1.1 Introduction

Turbulence – the intricate phenomenon in which a fluid flow becomes highly irregular, chaotic, and disordered – can be found everywhere in nature and also in many instances of our daily lives. However, not all irregular and chaotic motions are necessarily turbulent. A complete mathematical theory of turbulence is still missing and many efforts have been accomplished from ancient times to understand its very nature. The great physicist and Nobel laureate Richard Feynman remarked in one of his lectures in 1964: *Turbulence remains the last great unsolved problem of classical Physics.* Part of the difficulty lies in the fact that turbulent motions vary on a wide range of length and time scales. In the Cambridge dictionary, turbulence is usually defined as *a state of confusion without any order* or, when the movement of air and water is concerned, it is defined as *strong sudden movements within air or water.* Associated with the definition of turbulence there are terms like *instability* or *disturbance*, where, for example, the former term can be identified with that horrible thing that happens on a plane when flying under a severe storm and that causes it to bounce around violently.

In the language of *Fluid Mechanics* and in contrast to laminar flows, in which fluid parcels, or particles, flow along smooth paths or layers, turbulence is characterized by the fluid parcels experiencing fluctuations, or mixing[1], causing them to suddenly deviate from their paths and acquire a new kind of motion, which can be either rotational or linear in a different direction. Therefore, within a region where turbulent mixing is taking place, the fluid velocity at any point changes

---

[1]In general, mixing can be viewed as the mechanical process by which fluid parcels, or particles, change their position randomly. It is dominated by small-scale random motions of the fluid parcels that bring them into a closer or more distant relationship.

continuously in both magnitude and direction. According to Eckart [264], the stretching and folding of fluid trajectories concern the process of *stirring*, which must be distinguished from true mixing in that the latter tends to erase differences (i.e., gradients) in the physical properties of different portions of the fluid under the subsequent action of molecular diffusion. Following this line of thought, in a recent review on the difference between mixing and stirring, Villermaux [939] defined mixing as the process by which a system evolves under stirring from a state of simplicity to another state of simplicity characterized by complete uniformity.

Over the years, a significant amount of work has been devoted to the development of a theory of turbulence, with much of this effort, however, ending up with clear attempts to develop physically realistic and yet manageable mathematical and numerical models of turbulence. As commented in a Science Briefs of the *National Aeronautics and Space Administration* (NASA) on turbulence by Canuto and Dubovikov [153]: "a reliable description of the atmosphere depends on how well we are able to provide a physically correct description of turbulence." In fact, as they say, "turbulence cannot exist by itself, but rather it requires a continuous supply of energy to survive." For instance, there are at least two important sources of stirring in the atmosphere; one is the occurrence of temperature gradients which give rise to the transport of heat from hot to cold regions by turbulent mixing, and the other is the generation of vortices by wind shear [153].

This chapter deals with a review on some fundamentals of the present theory of turbulence. The first section is an overview on the onset of turbulence and its connection to the Kelvin-Helmholtz instability, while the following sections cover some background material on turbulence as, for instance, its main characteristics, the Kolmogorov's theory of inertial turbulence, some theoretical aspects of isotropic turbulence and the energy spectrum. The chapter ends with a brief description of turbulence intermittency.

Big whirls have little whirls
Which feed on their velocity;
And little whirls have lesser whirls,
And so on to viscosity
  in the molecular sense.

L. F. Richardson
Weather Prediction by Numerical Process. Cambridge University Press,
1922

1.2 The Kelvin-Helmholtz instability and the onset of turbulence

Since the investigations of W. Thomson[2] (Lord Kelvin, 1824–1907) and H. von Helmholtz[3] (1821–1894), the Kelvin-Helmholtz (KH) instability has been the subject of longstanding research because of its role in the dynamics of the Earth system and its prevalence as a precursor of turbulence in many instances of meteorological and oceanographic interest [487, 333, 906, 429, 644, 852, 853, 851]. In particular, it has been frequently pointed out as the basic mechanism leading to clear air turbulence (CAT).

Figure 1.1: Side view of Kelvin-Helmholtz billows in clouds caught at sunset over the Rocky Mountains in Colorado, United States on New Year's Eve, December 31, 2019. The photograph is a courtesy of Suzanne Kelley of Littleton.

The KH instability is a hydrodynamic instability arising at the interface between two immiscible fluid layers in relative (irrotational) motion. It can be easily recognized by the train of characteristic finite-amplitude billows that form, whose shapes resemble those of surface waves breaking on a beach. Figure 1.1 displays a photograph showing an atmospheric example of KH billows as revealed by clouds over the Rocky Mountains in Colorado on December 31, 2019. Starting from an initial wavelike perturbation at the interface, the chain of events that characterize the KH instability begins with a shear flow instability at the interface induced by the discontinuity in the tangential velocity there. Due to advection parallel to the interface, a secondary instability soon arises in the form of small-scale vortices in the shear layer on both sides of the interface. This vorticity in turn induces vertical motions that amplify the original wave, causing an exponential growth of the perturbation. During this process, the perturbed interface evolves toward an unstable vortex sheet. When the growing perturbation

---

[2]William Thomson. 1871. Hydrokinetic solutions and observations. *Philosophical Magazine*, 42:362–377.

[3]Hermann von Helmholtz. 1890. Die energie der wogen und des windes. *Annalen der Physik*, 41:641–662.

Figure 1.2: Schematic representation of the Kelvin-Helmholtz instability. First, waves form by shear flow, which then amplify due to vertical motions and overturn to form billows in a two-dimensional fashion; the process ends with the billows rolling up into spirals. Figure adapted from https://hmf.enseeiht.fr.

achieves a fully nonlinear regime, the KH billows roll up into spirals giving rise to a secondary shear instability. A schematic representation of the steps leading to the KH instability is shown in Fig. 1.2. High resolution simulations that permit solving small-scale motion [851], reveal that a complex combination of these secondary flows with even tertiary instabilities and beyond will ultimately lead to turbulent mixing. When the energy feeding the turbulent motion declines, turbulence is seen to decay leaving behind a much thicker sheared layer owing to irreversible mixing.

The onset of turbulence in atmospheric flows has often been linked to the KH instability and the sequence of events described above. Although the physics of KH billows is fairly well understood, there is as yet no complete theory for the resulting turbulence. In this section, we shall first focus on the physical mechanisms underlying the KH instability as described by linear stability theory for both discontinuous and continuous tangential velocity profiles, where the latter perturbation analysis leads to the well-known Taylor-Goldstein equation and the Richardson number criterion for stability of stratified flows.

1.2.1 The Helmholtz two-layer system

A simple two-dimensional analysis based on energetic considerations is described by Cushman-Roisin and Becker [213], which is used to derive an upper bound to the condition on the mean flow kinetic energy transfer necessary for the KH instability to lead to turbulent fluid mixing. A two-layer stratified fluid with velocity shear is considered for the analysis, as shown on the left of Fig. 1.3. For simplicity both fluids are assumed to be unbounded. This is not a too bad approximation since in most real-world applications of the KH instability, it is often assumed that there is no upper or lower boundary. The intermediate layer of width Δh, which is displayed on the right of Fig. 1.3, represents the thickening of the interface region of average density $\rho = (\rho_A + \rho_B)/2$, where fluid mixing occurs. Therefore, as heavier fluid parcels raise and lighter ones lower, the potential

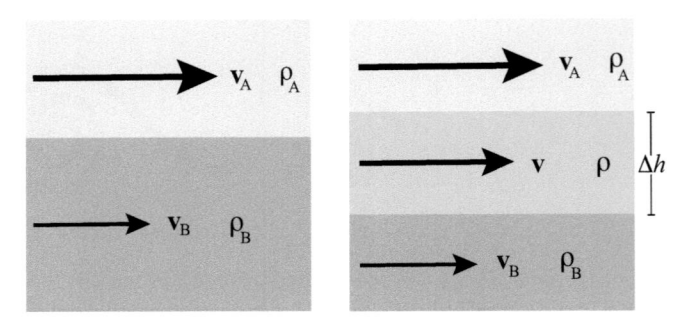

Figure 1.3: Sketch of a two-layer stratified fluid with velocity shear (left) and resulting region of width Δh produced by turbulent mixing (right). Figure adapted from Cushman-Roisin and Beckers [213].

energy after mixing raises at the expense of the kinetic energy carried by these parcels before mixing. This way the gain in potential energy, ΔU, is calculated as

$$
\begin{aligned}
\Delta U &= U_f - U_i = \\
&= \int_0^{\Delta h} \rho g z \, dz - \left(\int_0^{\Delta h/2} \rho_B g z \, dz + \int_{\Delta h/2}^{\Delta h} \rho_A g z \, dz \right) = \\
&= \frac{1}{8} (\rho_B - \rho_A) g (\Delta h)^2,
\end{aligned}
\tag{1.1}
$$

where g is the constant gravitational acceleration. If we appeal to momentum conservation and the Boussinesq approximation[4], the loss of kinetic energy, ΔK, in the absence of external forces is obtained as

$$
\begin{aligned}
\Delta K &= K_f - K_i = \\
&= -\int_0^{\Delta h/2} \frac{1}{2} \rho v_B^2 \, dz - \int_{\Delta h/2}^{\Delta h} \frac{1}{2} \rho v_A^2 \, dz + \int_0^{\Delta h} \frac{1}{2} \rho v^2 \, dz = \\
&= -\frac{1}{8} \rho (v_A - v_B)^2 \Delta h,
\end{aligned}
\tag{1.2}
$$

where $v = (v_A + v_B)/2$ is the average flow velocity characterizing the mixing layer of width Δh. Energy conservation requires that the loss of kinetic energy must be greater than the gain of potential energy in order for mixing to occur, i.e., $|\Delta K| > \Delta U$. This leads to the condition

$$
g \Delta h \frac{(\rho_B - \rho_A)}{\rho} \leq (v_A - v_B)^2.
\tag{1.3}
$$

---

[4]In this approximation the density variations are ignored.

Although this result shows that the road to turbulent mixing requires the conversion of kinetic energy from the shear flow into potential energy, it is rather a weak one in the sense that turbulent mixing cannot be properly described as a two-dimensional phenomenon. We shall return to this argument in the next section where the KH instability is analyzed according to linear theory.

1.2.2 Linear stability theory: discontinuous velocity profiles

As was pointed out previously, the mechanism of development of the KH instability in a stratified flow of two superposed fluids lies on the tangential velocity discontinuity at the separating interface, which induces a shear flow instability. Linear and nonlinear analyses of the KH instability have been carried out by several authors [250, 656, 964, 162] based on a simple model where the flow of a lighter fluid is superposed to that of a heavier fluid, as shown schematically in Fig. 1.4. In contrast to the Rayleigh-Taylor instability, where a heavier fluid is superposed to a lighter one, no gravity and density differences between the two fluids are actually needed for describing the KH instability. In Fig. 1.4, the heavier fluid has a density ρ_2 and a uniform velocity $\mathbf{v}_2$, whereas the density and velocity of the lighter fluid are, respectively, ρ_1 and $\mathbf{v}_1$. The analysis parallels that described by Kundu and Cohen [515], where both fluids are assumed to be inviscid and incompressible.

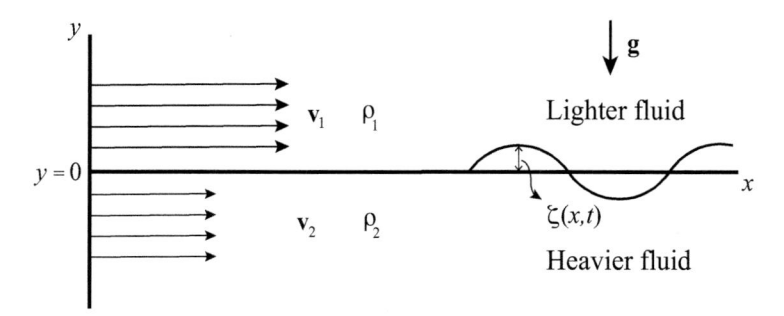

Figure 1.4: Two-layer discontinuous shear flow across a density interface with $\rho_2 > \rho_1$. The unperturbed interface is represented by the line $y = 0$, while the perturbed interface is given by the vertical displacement $y = \zeta(x,t)$.

According to Squire's theorem[5] [863], a two-dimensional description of the KH instability will suffice, even though mixing can only be correctly described in three-dimensions. A schematic diagram in the two-dimensional (x,y)-plane

---

[5]Squire's theorem states that two-dimensional shear flow perturbations are much more unstable than their three-dimensional counterparts. The theorem relies on a coordinate transformation in wavenumber space, where for each unstable mode of a given perturbation in three-space dimensions, there must always exist a corresponding two-dimensional instability at a higher wavenumber.

of the discontinuous shear across the interface (with $\rho_2 > \rho_1$) is shown in Fig. 1.4. The Rossby radius[6] is assumed to be greater than the lateral scale so that the influence exerted by the Coriolis force on the flow can be neglected. Both fluids move with uniform velocities $\mathbf{v}_1 = v_1\mathbf{i}$ and $\mathbf{v}_2 = v_2\mathbf{i}$ along the x-direction, with $v_1 > v_2$. Under the action of gravity, the stratified fluid flow is governed by Euler's equation augmented with the force term $\rho\mathbf{g}$ [521]

$$\rho\frac{\partial\mathbf{v}}{\partial t} + \rho(\mathbf{v}\cdot\nabla\mathbf{v}) = -\nabla p + \rho\mathbf{g}. \tag{1.4}$$

The condition for irrotational flow, $\nabla\times\mathbf{v} = \mathbf{0}$, implies potential flow, $\mathbf{v} = \nabla\phi$, which combined with the condition of incompressibility, $\nabla\cdot\mathbf{v} = 0$, leads to the Laplace equation

$$\nabla^2\phi = 0, \tag{1.5}$$

for the velocity potential. Substitution of $\mathbf{v}$ by $\nabla\phi$ and noting that for irrotational flow $\mathbf{v}\cdot\nabla\mathbf{v} = \nabla(\mathbf{v}\cdot\mathbf{v})/2 = (\nabla v^2)/2$, Eq. (1.4) can be integrated to yield the unsteady Bernoulli equations [515]

$$\frac{\partial\phi_j}{\partial t} + \frac{1}{2}(\nabla\phi_j)^2 + \frac{p_j}{\rho_j} + gy = C_j, \tag{1.6}$$

where the index $j = 1, 2$ is used to denote quantities of the first and second fluid layers, the C_j are constants of integration, and the equality $\mathbf{g} = -\nabla(gy)$ has been used. These equations are subject to the boundary conditions: $\phi_1 \to 0$ as $y \to -\infty$ and $\phi_2 \to 0$ as $y \to \infty$.

At the interface the flow is governed by the kinematic boundary condition, which demands that the normal velocity above and below the interface must be equal (i.e., $\mathbf{n}\cdot\mathbf{v}_1 = \mathbf{n}\cdot\mathbf{v}_2$), and the dynamic boundary condition, which requires that the pressure must be continuous across the interface in the absence of surface tension effects (i.e., $p_1 = p_2$). The disturbance at the interface is characterized by the vertical displacement $y = \zeta(x,t)$ so that the kinematic boundary conditions on either side of the interface obey the equations

$$\frac{\partial\zeta}{\partial t} + v_j\frac{\partial\zeta}{\partial x} = \frac{\partial\phi_j}{\partial y}, \tag{1.7}$$

at $y = 0$ for $j = 1, 2$. Solving for p_j in Eq. (1.6), the dynamic boundary condition takes the form

$$\rho_1\left[\frac{\partial\phi_1}{\partial t} + \frac{1}{2}(\nabla\phi_1)^2 + g\zeta - C_1'\right] = \rho_2\left[\frac{\partial\phi_2}{\partial t} + \frac{1}{2}(\nabla\phi_2)^2 + g\zeta - C_2'\right], \tag{1.8}$$

---

[6]The length scale at which the effects of rotation become comparable to those of buoyancy in the course of a flow perturbation is known as the Rossby radius of deformation. According to the *American Meteorological Society*, an alternative definition of common use in atmospheric dynamics is $\lambda_R = GZ_T/(f_c z_i)$, where G is the geostrophic wind speed, Z_T is the depth of the troposphere, f_c is the Coriolis parameter, and z_i is the depth of the atmospheric boundary layer, which is approximated as $z_i \approx G/N_{BV}$, where N_{BV} is the average Brunt-Väisälä frequency.

at $y = \zeta$, where $C_j = \rho_j C'_j$. Initially there are no transient perturbations (i.e., $y = 0$) and this equality reduces to

$$\rho_1 \left(\frac{1}{2} v_1^2 - C'_1 \right) = \rho_2 \left(\frac{1}{2} v_2^2 - C'_2 \right). \tag{1.9}$$

Assuming that the perturbations are small, the flow can be linearized as

$$\phi_j = v_j x + \phi'_j, \tag{1.10}$$

where $\phi'_j \ll \phi_j$ and the first term represent the uniform main stream flow. Substitution of Eq. (1.10) into Eq. (1.6) gives $\nabla^2 \phi'_j = 0$ for $j = 1, 2$, which must satisfy the boundary conditions $\phi'_1 \to 0$ as $y \to -\infty$ and $\phi'_2 \to 0$ as $y \to \infty$. Using the approximation

$$(\nabla \phi'_j) \approx v_j^2 + 2v_j \frac{\partial \phi'_j}{\partial x}, \tag{1.11}$$

in Eq. (1.8) for the perturbed quantities, where terms of second order are neglected, and summing up the result with Eq. (1.9) to eliminate the constants C'_j, we are left with the equality

$$\rho_1 \left(\frac{\partial \phi'_1}{\partial t} + v_1 \frac{\partial \phi'_1}{\partial x} + g\zeta \right)_{y=0} = \rho_2 \left(\frac{\partial \phi'_2}{\partial t} + v_2 \frac{\partial \phi'_2}{\partial x} + g\zeta \right)_{y=0}. \tag{1.12}$$

The normal mode analysis of Eqs. (1.7) and (1.12) can be performed by assuming wave perturbations of the form

$$[\zeta, \phi_1, \phi_2] = [Z, F_1 \exp(-ky), F_2 \exp(ky)] \exp[i(kx - \omega t)], \tag{1.13}$$

where the wavenumber k is real and the frequency $\omega = \text{Re}(\omega) + i\text{Im}(\omega)$. Substitution of relations (1.13) into Eqs. (1.7) and (1.12) leads to the system of algebraic equations for the amplitudes Z, F_1, and F_2

$$\begin{bmatrix} i(-\omega + kv_1) & k & 0 \\ i(-\omega + kv_2) & 0 & -k \\ g(\rho_1 - \rho_2) & i\rho_1(kv_1 - \omega) & i\rho_2(-kv_2 + \omega) \end{bmatrix} \begin{bmatrix} Z \\ F_1 \\ F_2 \end{bmatrix} = 0, \tag{1.14}$$

whose non-trivial solution (when the determinant of the matrix is zero) leads to the dispersion relation

$$\frac{\omega}{k} = \frac{\rho_1 v_1 + \rho_2 v_2}{\rho_1 + \rho_2} \pm \sqrt{\frac{(\rho_1 v_1 + \rho_2 v_2)^2}{(\rho_1 + \rho_2)^2} - \frac{[\rho_1 v_1^2 + \rho_2 v_2^2 + g(\rho_1 - \rho_2)/k]}{\rho_1 + \rho_2}}. \tag{1.15}$$

The amplification of any instability will then require that ω be a complex number, which is achieved when the radicand in Eq. (1.15) is negative. The condition for this reduces after some rearrangement to the inequality

$$(v_1 - v_2)^2 > \frac{g}{k} \frac{(\rho_2^2 - \rho_1^2)}{\rho_1 \rho_2}. \tag{1.16}$$

If this condition holds, the perturbations will amplify (leading to instability) when $\text{Im}(\omega) > 0$. Otherwise they will decay, leading to a stable solution. Moreover, in the case where $\rho_1 \approx \rho_2 \approx \rho$ then $\sqrt{\rho_1 \rho_2} \to (\rho_1 + \rho_2)/2$ and inequality (1.16) simplifies to

$$(v_1 - v_2)^2 > \frac{2g\,(\rho_2 - \rho_1)}{\rho k}, \tag{1.17}$$

which implies that if there is velocity shear the flow is always unstable to short wavelengths. On the other hand, comparing the conditions given by Eqs. (1.3) and (1.17) yields an estimate for the vertical extent of the mixing interface, namely

$$\Delta h \sim \frac{\rho\,(v_1 - v_2)^2}{g\,(\rho_2 - \rho_1)}, \tag{1.18}$$

which scales as $\lambda_{\max}/\pi$, where $\lambda_{\max}$ is the maximum wavelength of the unstable wave.

If the effects of surface tension are incorporated into the above analysis, a term of the form

$$+\frac{\sigma k}{\rho_1 + \rho_2},$$

must be added to the radicand on the right-hand side of Eq. (1.15), where σ is the surface tension coefficient at the interface. In this case, the condition for instability becomes

$$(v_1 - v_2)^2 > \frac{(\rho_1 + \rho_2)}{\rho_1 \rho_2} \left[\frac{g}{k}\,(\rho_2 - \rho_1) + \sigma \right]. \tag{1.19}$$

A direct comparison with Eq. (1.17) clearly shows that surface tension has a stabilizing effect against shear.

1.2.3 The Taylor-Goldstein equation

According to Eq. (1.17), the kinetic energy available by shear at the interface will always be sufficient to allow the growth of perturbations even when there is not much difference between the density of the two fluid layers. The more general case of parallel flow with a continuous stratification profile is fully developed in the books by Kundu and Cohen [515] and Cushman-Roisin and Becker [213]. Here we parallel the exposition of Cushman-Roisin and Becker [213]. To do so, first consider an observer at rest in a reference frame moving with the mean flow velocity of the system as shown in Fig. 1.5, where the system consists of a two-layer horizontal flow along the x-coordinate, $v_{x,0} = v_{x,0}(y)$ and $v_{y,0} = 0$, and a vertical density stratification, $\rho_0 = \rho_0(y)$, such that the density and the x-velocity component are continuously varying across the interface. As in the discontinuous case of the previous section, we apply Squire's theorem and assume steady-state

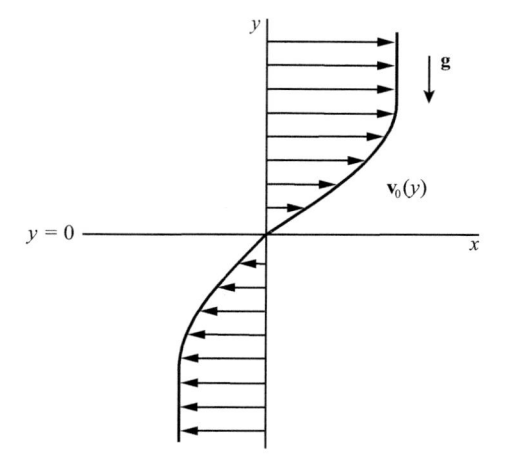

Figure 1.5: Two-layer continuous shear flow across a density stratified medium. Figure adapted from Cushman-Roisin and Beckers [213].

conditions for the background flow accompanied by hydrostatic equilibrium so that

$$\frac{dp_0}{dy} = -g\rho_0(y). \tag{1.20}$$

Under isothermal conditions the flow is governed by the continuity, or mass conservation law

$$\frac{\partial \rho}{\partial t} + \nabla \cdot (\rho \mathbf{v}) = 0, \tag{1.21}$$

and Euler's equation (1.4). The analysis then consists of linearizing these two equations in the (x,y)-plane by just perturbing the background stationary state $(\rho_0, p_0, v_{x,0}, v_{y,0})$ with an infinitesimal disturbance such that $\rho = \rho_0 + \rho'$, $p = p_0 + p'$, $v_x = v_{x,0} + v'_x$, and $v_y = v_{y,0} + v'_y$. Using Cartesian coordinates, Eqs. (1.4) and (1.21) become

$$\begin{aligned}
\frac{\partial v_x}{\partial t} &+ v_x \frac{\partial v_x}{\partial x} + v_y \frac{\partial v_x}{\partial y} = -\frac{1}{\rho}\frac{\partial p}{\partial x}, \\
\frac{\partial v_y}{\partial t} &+ v_x \frac{\partial v_y}{\partial x} + v_y \frac{\partial v_y}{\partial y} = -\frac{1}{\rho}\frac{\partial p}{\partial y} - g, \\
\frac{\partial \rho}{\partial t} &+ v_x \frac{\partial \rho}{\partial x} + v_y \frac{\partial \rho}{\partial y} = 0.
\end{aligned} \tag{1.22}$$

Linearization of these equations leads to the following set of equations for the perturbed state

$$\frac{\partial v'_x}{\partial t} + v_{x,0}\frac{\partial v'_x}{\partial x} + v'_y\frac{\partial v_{x,0}}{\partial y} = -\frac{1}{\rho_0}\frac{\partial p'}{\partial x},$$

$$\frac{\partial v'_y}{\partial t} + v_{x,0}\frac{\partial v'_y}{\partial x} = -\frac{1}{\rho_0}\frac{\partial p'}{\partial y} - \frac{\rho'g}{\rho_0}, \qquad (1.23)$$

$$\frac{\partial \rho'}{\partial t} + v_{x,0}\frac{\partial \rho'}{\partial x} - v'_y\frac{N_{\text{BV}}^2\rho_0}{g} = 0,$$

where N_{BV} is the Brunt-Väisälä frequency. Using the two-dimensional stream function $v'_x = \partial\psi'/\partial y$, $v'_y = -\partial\psi'/\partial x$ into the above equations and assuming normal mode solutions of the form

$$(\rho', p', \psi') = (R, P, \Psi)\exp[i(kx - \omega t)], \qquad (1.24)$$

where the amplitudes $R = R(y)$, $P = P(y)$, and $\Psi = \Psi(y)$, Eqs. (1.23) become

$$\left(v_{x,0} - \frac{\omega}{k}\right)\frac{\partial\Psi}{\partial y} - \Psi\frac{\partial v_{x,0}}{\partial y} = -\frac{1}{\rho_0}P, \qquad (1.25)$$

$$k^2\left(v_{x,0} - \frac{\omega}{k}\right)\Psi = -\frac{g}{\rho_0}R - \frac{1}{\rho_0}\frac{\partial p}{\partial y}, \qquad (1.26)$$

$$\left(v_{x,0} - \frac{\omega}{k}\right)R + \frac{N_{\text{BV}}^2\rho_0}{g}\Psi = 0. \qquad (1.27)$$

Taking the y-derivative of Eq. (1.25), subtracting Eq. (1.26), and using Eq. (1.27) to eliminate the density yields the equation

$$\left(v_{x,0} - \frac{\omega}{k}\right)\left(\frac{\partial^2\Psi}{\partial y^2} - k^2\Psi\right) + \left[\left(v_{x,0} - \frac{\omega}{k}\right)^{-1}N_{\text{BV}}^2 - \frac{\partial^2 v_{x,0}}{\partial y^2}\right]\Psi = 0, \qquad (1.28)$$

subject to the boundary conditions

$$\Psi(0) = \Psi(h) = 0, \qquad (1.29)$$

where a vanishing vertical velocity is imposed at the ground and top of the atmosphere. Equation (1.28) is the Taylor-Goldstein equation. This is a second-order partial differential equation, where the terms proportional to Ψ represent the kinetic energy of the wave perturbation, the effects of buoyancy, and the gradual transfer of kinetic energy from the mean shear to the wave. It expresses the relationship between restoring forces (decaying modes) and destabilizing forces (growing modes) in a parallel flow perturbation.

The extension of the Taylor-Goldstein equation to include the stabilizing effects of molecular viscosity leads to the well-known Orr-Sommerfeld equation, which can be derived from linearization of the Navier-Stokes equations for a homogeneous fluid following a procedure similar to the one employed here to derive Eq. (1.28). Here we shall only write down this equation for completeness and refer the interested reader to Kundu and Cohen [515] for details on its derivation. In terms of the stream function amplitude Ψ, the Orr-Sommerfeld equation reads as follows

$$\left(v_{x,0} - \frac{\omega}{k}\right)\left(\frac{\partial^2 \Psi}{\partial y^2} - k^2 \Psi\right) - \frac{\partial^2 v_{x,0}}{\partial y^2}\Psi$$

$$+ \frac{i}{k\text{Re}}\left(\frac{\partial^2}{\partial y^2} - k^2\right)^2 \Psi = 0, \qquad (1.30)$$

where $\text{Re} = v_{\text{max}}\Delta h/v$ is the Reynolds number[7] for some characteristic depth scale Δh in a stratified fluid flow with kinematic viscosity v. A comparison with Eq. (1.28) clearly shows that viscosity introduces a higher order of complexity, where the damping factor proportional to N_{BV}^2 is replaced with a differential term containing a fourth-order derivative of the structure function $\Psi(y)$. For $\text{Re} \gg 1$, Eq. (1.30) reduces to the Rayleigh stability equation when the viscosity is zero. The same is true for Eq. (1.28) which also reduces to the Rayleigh equation in the absence of gravity, i.e., when $N_{\text{BV}} = 0$. For looking at solutions to Eq. (1.30) applied to parallel shear flow the reader is referred to the many existing papers in the literature, old and new, on this subject.

1.2.4 The Richardson number criterion

The Richardson number, Ri, is a dimensionless number that expresses the importance of buoyancy over flow shear. Values of $\text{Ri} < 1$ indicate that shear dominates over buoyancy, while if $\text{Ri} > 1$, then buoyancy dominates. In aviation, it is often employed as a rough measure of expected air turbulence, where usually $0.1 \lesssim \text{Ri} \lesssim 10$. As before, values less than unity are indicative of shear dominance and therefore of significant turbulence. An analytical derivation of the criterion using the Taylor-Goldstein equation is developed here following the work of Miles [628] and Howard [417].

The first step in the derivation consists of replacing the stream function Ψ in Eq. (1.28) by the potential function Φ, according to the conformal mapping

$$\Psi = \Phi\left(v_{x,0} - \frac{\omega}{k}\right), \qquad (1.31)$$

---

[7]The Reynolds number, defined as $\text{Re} = ul/v$, where u is a characteristic velocity and l a characteristic length of the system, was introduced for the first time by the Irish physicist and mathematician Osborne Reynolds, as a criterion for the transition from laminar to turbulent flow, in his 1883 paper: "An experimental investigation of the circumstances which determine whether the motion of water shall be direct or sinuous, and of the law of resistance in parallel channels, *Philosophical Transactions of the Royal Society of London*, 174, 935–982".

to yield the differential equation

$$\frac{\partial}{\partial y}\left[\left(v_{x,0} - \frac{\omega}{k}\right)\frac{\partial\Phi}{\partial y}\right] - \left\{k^2\left(v_{x,0} - \frac{\omega}{k}\right) + \frac{1}{2}\frac{\partial^2 v_{x,0}}{\partial y^2}\right.$$
$$\left. + \left(v_{x,0} - \frac{\omega}{k}\right)^{-1}\left[\frac{1}{4}\left(\frac{\partial v_{x,0}}{\partial y}\right)^2 - N_{BV}^2\right]\right\}\Phi = 0,$$

$$(1.32)$$

where since the frequency ω takes on complex values[8], the motion potential function Φ is a complex function. Thus multiplying Eq. (1.32) by the complex conjugate potential $\Phi^\star$ and integrating the resulting equation between $y = 0$ and $y = h$, gives

$$\int_0^h |\Phi|^2 \left(v_{x,0} - \frac{\omega}{k}\right)^{-1}\left[N_{BV}^2 - \frac{1}{4}\left(\frac{\partial v_{x,0}}{\partial y}\right)^2\right]dy =$$
$$\int_0^h \left(v_{x,0} - \frac{\omega}{k}\right)\left(\left|\frac{\partial\Phi}{\partial y}\right|^2 + k^2|\Phi|^2\right)dy + \frac{1}{2}\int_0^h |\Phi|^2 \frac{\partial^2 v_{x,0}}{\partial y^2}dy,$$

$$(1.33)$$

where $|\Phi|^2 = \Phi\Phi^\star$ and $|\partial\Phi/\partial y|^2 = (\partial\Phi/\partial y)(\partial\Phi^\star/\partial y)$. The imaginary part of Eq. (1.33) leads to the following relation[9]

$$\mathrm{Im}(\omega)\int_0^h \left[N_{BV}^2 - \frac{1}{4}\left(\frac{\partial v_{x,0}}{\partial y}\right)^2\right]\frac{|\Phi|^2 dy}{|v_{x,0} - \omega/k|^2} = -\mathrm{Im}(\omega)\int_0^h \left(\left|\frac{\partial\Phi}{\partial y}\right|^2 + k^2|\Phi|^2\right)dy.$$

$$(1.34)$$

When the frequency N_{BV} happens to be greater than $(\partial v_{x,0}/\partial y)/2$ everywhere in the flow, Eq. (1.33) has no sense unless $\mathrm{Im}(\omega) = 0$, which will be a condition for critical stability. Therefore, defining the Richardson number as

$$\mathrm{Ri} = \frac{N_{BV}^2}{(\partial v_{x,0}/\partial y)^2} = \left(\frac{g}{\theta}\right)\frac{\partial\theta/\partial y}{(\partial v_{x,0}/\partial y)^2}, \qquad (1.35)$$

where θ is the potential temperature and g is the gravitational acceleration, linear stability would be guaranteed whenever the inequality

$$\mathrm{Ri} > \frac{1}{4}, \qquad (1.36)$$

---

[8]That is, $\omega = \mathrm{Re}(\omega) + i\mathrm{Im}(\omega)$.
[9]In Eq. (1.33) we have made for simplicity

$$\left(v_{x,0} - \frac{\omega}{k}\right)\left(v_{x,0} - \frac{\omega^\star}{k}\right) = v_{x,0}^2 - \frac{2}{k}v_{x,0}\mathrm{Re}(\omega) + \frac{|\omega|^2}{k^2} \rightarrow \left|v_{x,0} - \frac{\omega}{k}\right|^2.$$

holds everywhere in the flow. It must be noticed that this is a necessary but not sufficient condition for flow stability. On the other hand, if Ri happens to be less than $1/4$ somewhere in the flow, this will not necessarily mean that the flow is unstable. Further studies of three-dimensional nonlinear deformations of a planar shear flow by Abarbanel et al. [14] have established that stability of this flow requires that Ri > 1. A critical Richardson number criterion for the air-surface interface of Ri $= 1$ was also derived by Bye [147] for which all the shear energy is converted into potential energy. However, a more recent study by Galperin et al. [322] suggests that turbulence may indeed survive for Ri $\gg 1$, implying that the Richardson number should not be used as a valid criterion for turbulence extinction. Since turbulence is by itself an unstable, stochastic phenomenon, they further argued that the adaptation of the critical Richardson number to turbulent regimes is not straightforward mainly because turbulence lacks a well-defined basic state whose stability can be analyzed. Moreover, the anisotropization of stably stratified turbulence is an additional factor that complicates the use of the critical Richardson number [532].

1.3 Characteristics of turbulence

Turbulence is routinely encountered in many instances in our daily lives. As taught in many fluid mechanics textbooks, a typical example of turbulent flow is that generated by flow past a solid body, where a turbulent wake, i.e., an extended region of turbulent mixing is formed just behind the body [521]. The stability of such steady flows is a hard problem still awaiting solution. However, experimental data for different kinds of flows show that they generally remain stable for small Reynolds numbers (Re) and, as Re increases, a critical value Re_c is eventually reached beyond which the flow becomes unstable against infinitesimal perturbations. Therefore, Re is an important dimensionless number whose critical value for a given flow marks the transition from a laminar to a turbulent state. Routine examples of turbulent boundary layers and wakes are those caused by flowing air around and after bluff bodies such as vehicles, planes, and buildings. Apart from the case where turbulence is generated by the KH instability in shear flow, turbulent flows are also commonly encountered in straight pipes at high Re and pipe elbows and bends even at moderate and low Re, in smoke rising from cigarettes and chimneys, in water flowing through fully open taps, and in jets from a nozzle, among many other examples.

A frequently asked question is: how can we characterize turbulence? or, even more, how can we distinguish turbulent flow from other flows with a complex structure? To answer these questions we have to recognize some properties which are intrinsic to turbulent flows. The first characteristic is their irregularity and randomness, which make the flow approach a rather chaotic appearance. This is one main reason why a quantitative description of turbulence requires the use of statistical or conditional methods. In addition, turbulent flow is characterized by rapid diffusivity of velocity fluctuations in space and time. This is the principal cause of enhanced mixing and high rates of mass, momentum, and energy transfer. Such rates are much more intense than those observed in molecular diffusion. Therefore, in the absence of diffusivity, random and chaotic fluid movements cannot be classified as turbulent. Turbulence is inherently a three-dimensional phenomenon characterized by rotational (vortical) motions at different length scales and very high levels of fluctuating vorticity. Although it is governed by the continuum Navier-Stokes equations, the smallest scales of turbulent motion are always much larger than typical inter-molecular distances. Physically, turbulence is maintained by inertial forces, which dominate over viscous forces. This intuitively explains why high Reynolds numbers are generally required. Turbulence does not last forever since this will need a continuous energy supply owing to internal viscous shear stresses, which convert kinetic energy into heat by viscous dissipation. Hence, when kinetic energy is no longer supplied by external means, turbulence dies out quickly. Therefore, viscous losses are a further characteristics that differentiate turbulent mixing from random fluid motions.

1.3.1 Statistical description of turbulence

In 1926, Richardson [759] advanced the idea that any turbulent motion is composed of eddies[10], arranged in a hierarchy of sizes. In this model, kinetic energy is transferred from the larger to the smaller eddies. This feeding process continues until the energy is converted into heat by viscous dissipation in the smallest eddies. He also argued that small eddies in turbulent flow are statistically homogeneous and isotropic. In fact, that turbulence is the same in all directions is considered to be a property of the turbulent flow structure at small scales.

In statistical mechanics, the concept of ensemble average is based on the existence of independent statistical events. For example, consider an ensemble of turbulent flows, all starting from the same initial flow and subject to the same boundary conditions. Suppose that a random small perturbation is added to each flow in the ensemble, and follow the resulting flows for long enough times until they become uncorrelated with each other. Since turbulent flows vary randomly in space and time, a space average or an average over sufficiently long times will be a good approximation to the ensemble average. Therefore, the velocity of the turbulent flow, say $\mathbf{v}$, can be written as

$$\mathbf{v} = \langle \mathbf{v} \rangle + \mathbf{v}', \tag{1.37}$$

where $\langle \mathbf{v} \rangle$ is the ensemble mean of $\mathbf{v}$ or, simply the mean velocity, and $\mathbf{v}'$ is the perturbed velocity or fluctuating part of $\mathbf{v}$ [521]. By definition, the ensemble mean of $\mathbf{v}'$ is exactly zero. A complete description of $\mathbf{v}$ in space and time is

---

[10]Although the concept of eddy is not well defined, here we shall imagine it as a swirling in the fluid caused by turbulent flow.

given in terms of the probability density function $P(\mathbf{v})$. The quantity $P(\mathbf{v})d\mathbf{v}$ then gives the fraction of the turbulent flows in the ensemble for which the measured velocities lie between $\mathbf{v}$ and $\mathbf{v}+d\mathbf{v}$. Hence the ensemble mean $\langle \mathbf{v} \rangle$ is given by the first moment

$$\langle \mathbf{v} \rangle = \int_{\mathbb{R}^3} \mathbf{v} P(\mathbf{v})d\mathbf{v}, \tag{1.38}$$

and the level of variability of $\mathbf{v}$ over the mean is described by the variance (or second moment) of the fluctuating part

$$\langle \mathbf{v}'^2 \rangle = \int_{\mathbb{R}^3} \mathbf{v}'^2 P(\mathbf{v})d\mathbf{v}. \tag{1.39}$$

This provides a measure of the turbulent kinetic energy per unit mass

$$K_{\text{turb}} = \frac{1}{2}\langle \mathbf{v}'^2 \rangle = \frac{1}{2}\left(\langle v_x'^2 \rangle + \langle v_y'^2 \rangle + \langle v_z'^2 \rangle \right). \tag{1.40}$$

Other two important parameters are the coefficients of *skewness* and *kurtosis*. The former describes the shape of the density function and is defined as the ratio of the third central moment of the fluctuating part over the cube of its standard deviation,

$$\alpha_3 = \frac{\langle \mathbf{v}'^3 \rangle}{(\sigma_{\mathbf{v}'})^3}, \tag{1.41}$$

where $\sigma_{\mathbf{v}'} = (\langle \mathbf{v}'^2 \rangle)^{1/2}$. If $\alpha_3 = 0$, the density probability function will be symmetric about the mean velocity. Otherwise, the function will be skewed to the left or to the right of the mean velocity depending on the sign of α_3. On the other hand, the coefficient of kurtosis is a measure of the degree of peakedness of $P(\mathbf{v})$ and is defined as the ratio of the fourth moment of $\mathbf{v}'$ over its standard deviation to the fourth

$$\alpha_4 = \frac{\langle \mathbf{v}'^4 \rangle}{(\sigma_{\mathbf{v}'})^4}. \tag{1.42}$$

For instance, a time series representing the velocity fluctuations in a point will have a small kurtosis if most measurements are clustered around the mean velocity, and a large kurtosis if, on the other hand, the series is dominated by intermittent values that differ from the mean [714].

Most of our current understanding of turbulence comes from experimental measurements of mean flow velocities, standard deviations of the turbulent velocity fluctuations, and correlation coefficients between the turbulent velocity fluctuation components, which are mostly based on formulations of the joint probability density functions (JPDF) of velocity and scalars for turbulent flows [713, 675, 589, 858]. Such analyses are particularly useful because turbulence involves several random variables which depend on each other. The JPDF of two quantities u and v, say $P(u,v)$, is the probability of finding u and v between the intervals $(u, u+du)$ and $(v, v+dv)$, respectively[11]. $P(u,v)$ is positive definite and its integral over the (u,v)-space is unity by definition. For example, the joint first moment of u and v, denoted $\langle uv \rangle$, is defined by the double integral

$$\langle uv \rangle = \int_{\mathbb{R}^2} uvP(u,v)dudv, \tag{1.43}$$

while the covariance of u and v is given by

$$\text{Cov}(u,v) = \langle uv \rangle - \langle u \rangle \langle v \rangle = \langle u'v' \rangle, \tag{1.44}$$

which is a measure of the asymmetry of the JPDF. The degree of correlation between u and v is measured by the correlation function defined as

$$R(u,v) = \frac{\langle u'v' \rangle}{\sigma_{u'}\sigma_{v'}}, \tag{1.45}$$

where the denominator is the product of the standard deviations of the fluctuating parts of u and v. If u and v are perfectly correlated then $R(u,v) = \pm 1$. Conversely, they are uncorrelated when $\text{Cov}(u,v) = \langle u'v' \rangle = 0$. However, two uncorrelated variables are not statistically independent since this would require that $P(u,v) = P(u)P(v)$. Therefore, independent variables can be also uncorrelated, but the converse is not true. Whereas characterization of the turbulent flow is most often analyzed by means of Eqs. (1.38) and (1.39), as we shall see in Chapter 2, the covariance will be particularly important in the problem of Reynolds averaging.

Based on its statistical analysis, turbulence can be classified as stationary, homogeneous, isotropic, or axisymmetric. For example, stationary turbulence is encountered when its mean quantities (i.e., the first and higher-order moments of turbulent variables) are all time-invariant. If, on the other hand, a stationary variable, say u, is such that its time average is equal to its ensemble average value, $\langle u \rangle$, i.e., if

$$\frac{1}{T}\int_0^T u(t)dt = \langle u \rangle, \tag{1.46}$$

as $T \to \infty$, we then say that u is *ergodic*. In similar fashion, homogeneous turbulence is characterized by the invariance under spatial translations of all its mean quantities, in which case the relation

$$\frac{1}{L}\int_0^L u(x)dx = \langle u \rangle, \tag{1.47}$$

holds as $L \to \infty$. Hence, the spatial average of any turbulent variable, u, is completely equivalent to its ensemble average or mean. On the other hand, isotropic

---

[11]Here u and v may refer to the same field measured at different positions or times or different fields measured at the same place and time. The former measurements are useful to define spatial or temporal autocorrelations.

turbulence occurs when all of its mean quantities are invariant under arbitrary rotations of the coordinates about any of the three spatial directions. In this case, turbulent motions will look the same from all orientations. A special case of this type of invariance happens when all mean quantities remain invariant under rotation about only one specified coordinate axis, in which case we speak of axisymmetric turbulence.

If we place a particle in a turbulent flow, its trajectory can be seen as the sum of a large number of random displacements by small eddies so that its velocity will be the result of an equally large number of random velocity increments. For a number N of independent and uncorrelated increments, the probability density function corresponds to a Gaussian distribution

$$P(u) = \frac{1}{\sqrt{2\pi\sigma^2}} \exp\left(\frac{-u^2}{2\sigma^2}\right), \tag{1.48}$$

where σ^2 is the variance of the incremental changes. However, for turbulent flows the above assumptions are often broken. For instance, probability density functions of turbulent flows are not Gaussian because they are dominated by coherent, spatially, and temporally localized structures. The same is found in intermittent flows, which also displays localized regions of large-magnitude events separated from low-magnitude events by long periods [714].

1.3.2 Spectral analysis of turbulence

Since turbulent motion consists of small eddies superimposed on larger eddies, its dynamics can be described by considering the spatial and temporal distribution of the fluctuations (eddies) among the various scales of motion, together with the interaction, if any, of the motion among the different scales. A usual way to characterize such multiscale flows is through a Fourier spectral analysis, where, for example, the scales of motion are represented using series of trigonometric functions or, more generally, by imaginary exponentials. Attempts have been also made to use functions other than trigonometric or imaginary exponential that are in principle more representative of a typical eddy [577, 283, 799]. However, most analyses have been performed based on trigonometric and imaginary exponential functions. For example, given a sufficiently long time series for the instantaneous fluctuating part of a turbulent velocity component, say $v'(t)$, its Fourier transform to the frequency domain over an arbitrary time interval $(-T/2, T/2)$ is given by

$$\tilde{v}'(\omega_n) = \frac{1}{T} \int_{-T/2}^{T/2} v'(t) \exp\left(\frac{-2i\pi n}{T}\right) dt, \tag{1.49}$$

where $i = \sqrt{-1}$ and $\omega_n = 2\pi n/T$ is the frequency of the nth harmonics for $n = 0, \pm1, \pm2, \dots$. The temporal power spectrum of $v'(t)$ is then obtained as the ensemble average of the quantity $|\tilde{v}'(\omega_n)|^2$. Moreover, if the time interval is such that $T \to \infty$, the power spectrum becomes a continuous function of the frequency, say $\tilde{S}_{v'}(\omega)$, which can be shown to be the Fourier transform of the autocovariance $\langle v'(t)v'(t+T)\rangle$

$$\tilde{S}_{v'}(\omega) = \frac{1}{2\pi} \int_{-\infty}^{\infty} \langle v'(t)v'(t+T)\rangle \exp(-i\omega T)\, dT. \tag{1.50}$$

The function $\tilde{S}_{v'}(\omega)$ calculated this way is a measure of the power content of the turbulent motion. If, on the other hand, this information is known, the autocovariance can be recovered from the inverse Fourier transform

$$\langle v'(t)v'(t+T)\rangle = \int_{-\infty}^{\infty} \tilde{S}_{v'}(\omega) \exp(i\omega T)\, d\omega. \tag{1.51}$$

The power content at any instant t is then obtained by setting $T = 0$ in the above equation. This is just the variance of $v'(t)$, which is equal to the entire area under the power spectral density curve. Since the autocovariance function is both a real and an even function of T, Eqs. (1.50) and (1.51) can be rewritten as the cosine transforms

$$\tilde{S}_{v'}(\omega) = \frac{1}{\pi} \int_{-\infty}^{\infty} \langle v'(t)v'(t+T)\rangle \cos(\omega T)\, dT, \tag{1.52}$$

and

$$\langle v'(t)v'(t+T)\rangle = 2 \int_{-\infty}^{\infty} \tilde{S}_{v'}(\omega) \cos(\omega T)\, d\omega, \tag{1.53}$$

which are referred to as the Wiener-Khintchine relations. From Eq. (1.50) it follows that $\tilde{S}_{v'}(\omega)$ is also real and even.

In the case when we are faced with homogeneous turbulence, it is customary to calculate the three-dimensional Fourier transform of the spatial autocovariance function $\langle v'(\mathbf{r})v'(\mathbf{r}+\mathbf{R})\rangle$ in order to obtain the spatial power spectrum

$$\tilde{S}_{v'}(\mathbf{k}) = \frac{1}{(2\pi)^3} \int_{\mathbb{R}^3} \langle v'(\mathbf{r})v'(\mathbf{r}+\mathbf{R})\rangle \exp(-i\mathbf{k}\cdot\mathbf{R})\, d\mathbf{R}, \tag{1.54}$$

where $\mathbf{k}$ is the wavevector. The quantity $\tilde{S}_{v'}(\mathbf{k})$ provides information of how the spectral contributions to the autocovariance function $\langle v'(\mathbf{r})v'(\mathbf{r}+\mathbf{R})\rangle$ are distributed in wavenumber space. In other words, it represents the decomposition of the autocovariance function (i.e., the two-point velocity correlation) into the eddy sizes contained by the turbulent motion. As for the temporal autocorrelations, the autocovariance function can be recovered from knowledge of the spatial power spectrum by the inverse Fourier transform

$$\langle v'(\mathbf{r})v'(\mathbf{r}+\mathbf{R})\rangle = \int_{\mathbb{R}^3} \tilde{S}_{v'}(\mathbf{k}) \exp(i\mathbf{k}\cdot\mathbf{R})\, d\mathbf{k}. \tag{1.55}$$

Setting $\mathbf{R} = \mathbf{0}$ in the above expression, the variance of $v'(\mathbf{r})$ is recovered as the integral of $\tilde{S}_{v'}(\mathbf{k})$ over all wavevectors. Only when turbulence is isotropic, the Fourier transforms (1.54) and (1.55) will be expressed in terms of the magnitude k of the wavevector, in which case the spatial variance of $v'(\mathbf{r})$ can be written as a single integral over all wavenumbers

$$\langle v'(\mathbf{r})v'(\mathbf{r}) \rangle = 4\pi \int_0^\infty \tilde{S}_{v'}(k)k^2 dk. \tag{1.56}$$

The treatment of two-point correlations for other turbulent quantities is similar to the one described here for the fluctuating part of a turbulent velocity component. From Eq. (1.40) it follows that for homogeneous and isotropic turbulence the partition of the turbulent kinetic energy into its contributions from all wavenumbers defines the energy spectrum $E(k)$ such that

$$K_{\text{turb}} = \frac{1}{2}\langle \mathbf{v}'^2 \rangle = \int_0^\infty E(k)dk. \tag{1.57}$$

At a given wavenumber k, the energy spectrum can be interpreted as being due to eddies whose size l corresponds to the half-wavelength π/k. The analysis of the spectral form of the continuum equations, both averaged and unaveraged, will be discussed in Chapter 2.

1.3.3 Clear air turbulence

Among the many aspects of human life, for instance, turbulence is almost always present during commercial flights, which occasionally may produce sudden changes in the aircraft's speed and altitude. In aviation, a common type of turbulence is called *clear-air turbulence* (CAT) and is caused by the encounter of cold and warm air currents at altitudes between 7 and 12 thousand meters from the ground. However, CAT may well be enhanced by a combination with other factors such as the presence of temperature gradients, jet streams, storms, and mountains during the flight. We shall return to this argument in Chapter 3, where the phenomenon of CAT and its causes will be widely discussed. Here we shall only limit ourselves to outline that in the world of commercial aviation turbulence can be classified as (a) light, (b) moderate, (c) severe, and (d) extreme turbulence. Light turbulence is probably the most common and occurs in nearly every flight. At its lowest levels of intensity, slight changes in the aircraft's altitude or tilt might not even be noticed by the passengers. Under moderate turbulence, the aircraft can bounce around with sudden jumps of up 6–10 meters and the passengers may feel a definite strain against their seatbelts. Among aviation professionals, severe turbulence occurs when the aircraft's altitude changes by up to 20–30 meters. Under these conditions, the pilot could momentarily lose control of the plane and passengers can be injured. On the other hand, extreme turbulence—also colloquially known as violent turbulence—is very rare and potentially dangerous since it can cause serious aircraft damage. Under extreme turbulence, the pilot completely loses control of the plane with the extreme risk of crashing. Although CAT cannot be accurately forecast and is usually difficult to detect on radar, in most cases severe and extreme turbulence can be avoided by just taking into account the weather patterns before a flight.

1.4 Kolmogorov's theory

A crucial step forward to improve our understanding of turbulence was made by A. N. Kolmogorov in 1941, when he published three seminal papers which dealt with the concepts of scale similarity and turbulent energy cascade [497, 498, 496], introducing what is known today as the origin of the modern theory of turbulence. For a brief historical outlook and discussion on Kolmogorov contributions to the theory of turbulence the reader is referred to the paper by Jiménez [440]. Here we will only briefly review the most relevant aspects of the theory.

In an incompressible fluid, turbulent flows are described by the Navier-Stokes equations

$$\frac{\partial \mathbf{v}}{\partial t} + (\mathbf{v} \cdot \nabla \mathbf{v}) = -\frac{1}{\rho} \nabla p + \nu \nabla^2 \mathbf{v} + \mathbf{F}, \tag{1.58}$$

coupled with the continuity equation $\nabla \cdot \mathbf{v} = 0$, where ν is the kinematic viscosity and $\mathbf{F}$ is a force term acting on the fluid, which is assumed to be smooth over large scales. On taking the scalar product between each term of Eq. (1.58) and the flow velocity vector $\mathbf{v}$ yields

$$\frac{\partial K}{\partial t} + \nabla \cdot \left[\mathbf{v} \left(K + \frac{p}{\rho} \right) \right] = \mathbf{v} \cdot \mathbf{F} - \nu |\nabla \mathbf{v}|^2, \tag{1.59}$$

where $K = v^2/2$ is the kinetic energy per unit mass. If the turbulent flow is statistically homogeneous, the energy introduced through the boundary of the domain by the forcing must cancel the viscous dissipation, i.e.,

$$\varepsilon = \nu \langle |\nabla \mathbf{v}|^2 \rangle = \frac{\nu}{V} \int_V |\nabla \mathbf{v}|^2 dV, \tag{1.60}$$

where V is the volume of the domain and the integration is made over a large enough domain. It is clear from relation (1.60) that when $\nu \to 0$, the velocity field is such that $\mathrm{Re} \to \infty$.

1.4.1 The energy cascade

In order to solve the problem of energy dissipation in a turbulent flow, Kolmogorov recovered Richardson's [759] idea that any turbulent motion is composed of a hierarchy of eddies with different sizes, where smaller eddies feed on the larger ones in the hierarchy until a smallest size is reached where viscosity smooths out the flow. In Kolmogorov's view an eddy is conceived as a patch of fluid, localized over a region of size l and with a characteristic velocity difference u_l. The largest eddies are characterized by a length l_0 and a velocity u_{l_0}. This length is comparable with the length, L, of the turbulent flow. Two different characteristic timescales can be defined to describe the evolution of an eddy: the so-called *inertial* timescale, $\tau_l = l/u_l$, during which the eddy deforms and loses its individuality, and the *viscous* timescale, $\tau_v = l^2/v$, during which its internal velocity differences smooth out by viscous effects. In terms of these two timescales, the Reynolds number of an eddy of size l is

$$\text{Re}_l = \frac{\tau_v}{\tau_l} = \frac{u_l l}{v}. \tag{1.61}$$

At scales l comparable to the fundamental scale L, i.e., at the larger scales, $\tau_l \ll \tau_v$ and $\text{Re}_l \gg 1$ so that the eddy deforms so rapidly into smaller eddies that it has no time to dissipate energy. At such large scales, the energy passes to smaller scales once the eddy deforms. Conversely, when $\tau_l \sim \tau_v$ then $\text{Re}_l \sim 1$ and the eddy dissipates its internal velocity differences by viscosity before it has a chance to deform. This occurs at the smallest scale of the hierarchy, the so-called Kolmogorov length. These eddies define the smallest turbulent structures and they are not connected to the external forces that maintain turbulence. At such small scales, turbulence looks statistically isotropic and homogeneous. This is known as Kolmogorov's hypothesis of local isotropy and homogeneity, where isotropic turbulence means that the eddies behave the same way in all directions, while homogeneous turbulence means that the kinetic energy is the same everywhere. Although this is true at the smallest scales, large scale turbulence may still be anisotropic and the characteristic length dividing the anisotropic from the isotropic regime is estimated to be $\sim l_0/6$, where $l_0 \sim L$.

A further important quantity characterizing the turbulent motion is the kinetic energy that flows per unit mass down the cascade (Fig. 1.6), which can be estimated as

$$\varepsilon \sim \frac{u_l^2}{\tau_l} = \frac{u_l^3}{l}, \tag{1.62}$$

from which it follows that

$$u_l \sim (\varepsilon l)^{1/3}. \tag{1.63}$$

The size range for which $l < l_0/6$ is referred to as the *universal equilibrium range* because at these scales the small eddies can maintain dynamic equilibrium with the energy transfer rate imposed by the larger eddies. At the Kolmogorov length scale ($l \sim \eta$) where $\text{Re}_\eta \sim 1$, the characteristic velocity of the viscous eddies can be estimated to be

$$u_\eta \sim (v\varepsilon)^{1/4}, \tag{1.64}$$

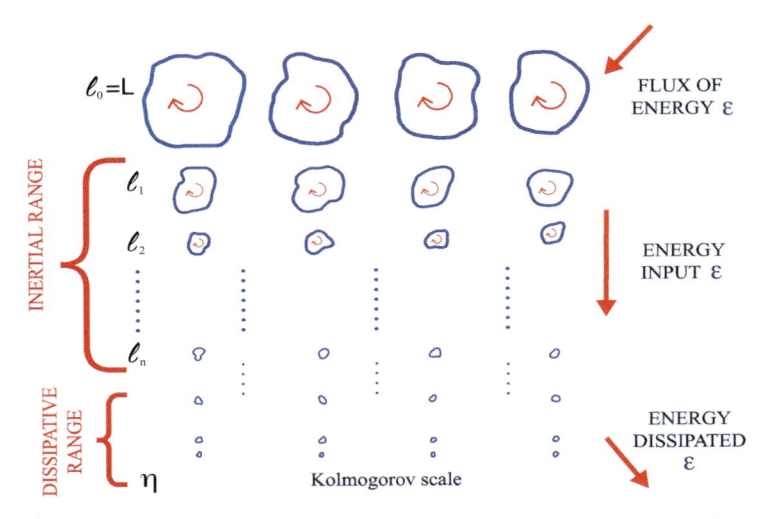

Figure 1.6: Schematic of the Richardson's energy cascade. Figure adapted from Frisch [312].

where $u_\eta \eta \sim \nu$ and relation (1.63) have been used. From this relation and Eq. (1.62) it follows that

$$\eta \sim \left(\frac{\nu^3}{\varepsilon} \right)^{1/4}. \tag{1.65}$$

At this scale the kinetic energy of turbulence is converted into heat. Using Eqs. (1.62) with $l = L$ and Eq. (1.65), the ratio between the sizes of the largest and the smallest eddies is estimated to be

$$\frac{L}{\eta} \sim \left(\frac{u_L L}{\nu} \right)^{3/4} = \mathrm{Re}_L^{3/4}, \tag{1.66}$$

while the characteristic velocity of the largest eddies to the smallest ones scales as

$$\frac{u_L}{u_\eta} \sim \mathrm{Re}_L^{1/4}. \tag{1.67}$$

The effects of increasing the Reynolds number of the turbulent flow are not to decrease the rate of energy dissipation, but rather to lengthen the cascade before reaching the viscous dissipation limit.

1.4.2 Kolmogorov's similarity hypotheses

Given the two parameters ε and ν, the velocity and length scales are uniquely defined by Eqs. (1.64) and (1.65), respectively. In general, this is also true in the universal equilibrium range for which $l < l_0/6$. Based on this observation, Kolmogorov stated the so-called first similarity hypothesis, which reads as follows:

> At sufficiently high Reynolds numbers, the statistics of small scale motions in the universal equilibrium range, being independent of the larger scales, is universally and uniquely determined by the viscosity, v, and the rate of energy dissipation, ε.

From Eqs. (1.64) and (1.65), it follows that the ratio of η over u_η defines the Kolmogorov's time

$$\frac{\eta}{u_\eta} = \left(\frac{v}{\varepsilon}\right)^{1/2}, \tag{1.68}$$

which provides the timescale for eddy evolution in the universal equilibrium range. At scales $l > l_0/6$, i.e., the so-called *inertial* range, the Reynolds number is sufficiently large that the eddies are not affected by viscosity. This observation led Kolmogorov to formulate the second similarity hypothesis, which states the following:

> The statistics of turbulent motion at scales in the inertial range $\eta \ll l \ll l_0$, i.e., at sufficiently high Reynolds number, have a universal form that is independent of v and uniquely determined by ε.

If the turbulent motion is statistically steady, i.e., if the turbulence intensity does not change, then the rate of energy transfer must be the same for all scales. In this steady-state energy cascade it follows that the rate per unit mass and time at which the kinetic energy is supplied to the fluid at the fundamental scale L is equal to that cascading from scale to scale and to the one being ultimately dissipated by viscosity at the Kolmogorov scale η.

1.4.3 The energy spectrum of turbulence

In his 1941 papers [497, 498], Kolmogorov derived a universal scaling for the inertial range in terms of the second-order structure function[12], namely

$$S_2 = C\varepsilon^{2/3}l^{2/3}, \qquad (1.69)$$

where C is a universal constant prefactor. In 1945, Onsager [676] obtained in wavenumber space the corresponding energy spectrum

$$E(k) = C_2\varepsilon^{2/3}k^{-5/3}, \qquad (1.70)$$

where C_2 is a constant and k is the wavenumber. This is the legendary Kolmogorov-Obukhov $-5/3$ law. Kolmogorov's findings were first brought to the fluid dynamics community outside the U.S.S.R. by Batchelor [65], who worked the energy spectrum in real-space, and later in his classical monograph [66], where he worked in wavenumber space. Equation (1.70) was confirmed experimentally almost a decade later. The first convincing data was obtained by Grant et al. [357] from observations of the turbulence generated by strong tidal currents in Seymour Narrows of the Discovery Passage in British Columbia. After that, several other confirmations for different physical systems were obtained in the following years. The results of many of these confirmations are shown in Fig. 6.14 of Ref. [715], which is here reproduced schematically for the reader's convenience as Fig. 1.7. The curve depicts the one-dimensional longitudinal energy spectra for a variety of turbulent flows at different Reynolds numbers, which are indicated by the final numbers in the legends of Fig. 6.14 of Ref. [715]. The reader interested in learning more about the subject is also referred to McComb's book [607]. In the case of isotropic turbulence, Kolmogorov found that the third-order structure function obeys the relation $S_3 = -4\varepsilon l/5$, which is better known as the Kolmogorov 4/5 equation. As was commented by Jiménez [440], this is one of the few equations with no adjustable parameters that have been derived so far in turbulence. It has been found to fit acceptably well the experimental data in the limit when $Re_L \gg 1$.

---

[12]The structure function of order p of the longitudinal velocity difference between two points separated by a distance l along the x-axis is given by

$$S_p = \langle \delta u^2 \rangle = \langle [u(x+l) - u(x)]^p \rangle.$$

For homogeneous and isotropic turbulence, Kolmogorov demonstrated that in the inertial range the structure functions vary with the eddy size as $S_p \sim l^{\zeta_{u(x)}}$, where $\zeta_{u(x)}$ is known to be the scale exponent of the longitudinal structure function. Since in the inertial range, ε remains constant he also assumed a linear dependence between $\zeta_{u(x)}$ and p. The deviations of the experimental results from this linear dependence are known as small-scale intermittency of turbulence. Benzi et al. [84] proposed that the exponents $\zeta_{u(x)}$ can be calculated by plotting the third-order structure function, S_3, on the abscissa against the structure function of any order in the ordinate. They stated that

$$\zeta_{u(x)} = \frac{d|S_p|}{d|S_3|},$$

based on the observation that $|S_p| \sim [|S_3|]^{\zeta_{u(x)}}$, where the structure function is replaced by its modulus.

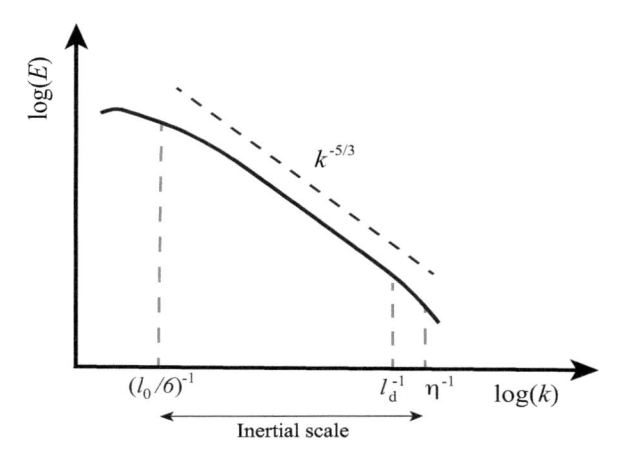

Figure 1.7: Schematic representation of the one-dimensional longitudinal energy spectrum (solid line). The dashed straight line depicts the $-5/3$ slope predicted by Eq. (1.70).

In a 1962 paper [499], Kolmogorov himself provided corrections to Eqs. (1.69) and (1.70) by introducing a length, L_{ext}, which is defined as a characteristic size of the system. In terms of this length the modified second-order structure function is

$$S_2 = C\varepsilon^{2/3}l^{2/3}\left(\frac{L_{ext}}{l}\right)^{-\mu}.$$ (1.71)

Similarly, the modified energy spectrum reads as follows

$$E(k) = C_2\varepsilon^{2/3}k^{-5/3}\left(L_{ext}k\right)^{-\mu}.$$ (1.72)

As was recently demonstrated by McComb and May [608], the presence of L_{ext} in the above expressions destroys the observed asymptotic universality of the energy spectrum at large wavenumbers. Evidently, the form given by Eq. (1.72) poses problems for the spectral scaling shown in Fig. 1.7, and according to McComb and May [608] the corrections in Eqs. (1.71) and (1.72) are unnecessary. They also concluded that the dependence of Eqs. (1.71) and (1.72) on L_{ext} does not show up in the study of wavenumber spectra for real physical systems whose L_{ext} value can vary up to five orders of magnitude, and therefore the modifications given by Eqs. (1.71) and (1.72) may not be correct [608].

1.4.4 The Taylor microscale

We have seen that $l = l_0/6$ divides the entire range of eddy sizes into the so-called *energy containing* range $l_0/6 < l < L$ (fundamental or integral length scale) and the universal equilibrium range $\eta < l < l_0/6$, which in turn can be subdivided into a dissipation range $\eta < l < l_d$ (the Kolmogorov length scale), where $l_d \approx 60\eta$, and an inertial subrange $l_d < l < l_0/6$, which is called the Taylor microscale.

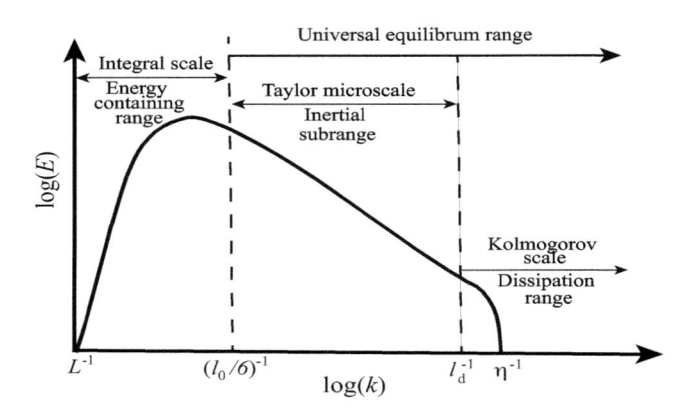

Figure 1.8: Kinetic energy of turbulence for the different scales of motion (or wavelengths).

In this scale, the viscous effects are negligible and the eddies evolve governed by inertial effects. In contrast, in the dissipation range the size of the eddies are already so small that their movement is dominated by strong viscous forces. Based on the Taylor microscale, λ, the Reynolds number is given by

$$\text{Re}_\lambda = \frac{u_\lambda \lambda}{\nu}, \tag{1.73}$$

where u_λ is usually defined as the root mean square of the velocity fluctuations, i.e., $u_\lambda = \langle v' \rangle_{\text{rms}}$. The Taylor microscale is defined through the relation

$$\lambda = \left(\frac{15\nu}{\varepsilon}\right)^{1/2} u_\lambda. \tag{1.74}$$

Although this length scale does not have an easily understood physical significance as the Kolmogorov or integral length scales, it provides a convenient estimate of the fluctuating strain rate field. It was the first length scale that was derived to describe turbulence and therefore sometimes it is also called the *turbulence length scale*. A commonly found relation in terms of the turbulent kinetic energy is

$$\lambda \approx \left(\frac{10\nu K_{\text{turb}}}{\varepsilon}\right)^{1/2}, \tag{1.75}$$

where K_{turb} is as defined by Eq. (1.57). The relation between Re_λ and the Reynolds number of the turbulent flow, Re_L, is

$$\text{Re}_\lambda = \left(\frac{20}{3}\text{Re}_L\right)^{1/2}. \tag{1.76}$$

For a deeper insight into the length scales characterizing turbulent flows the reader is referred to the book by Tennekes and Lumley [902].

1.4.5 Intermittency

The characteristic velocity of eddies can be seen as a representative variable of turbulence, and its variation with scale is commonly referred to as intermittency. In an equilibrium system, Eq. (1.60) expresses energy conservation, i.e., the input kinetic energy that feeds the turbulent flow must be equal to the energy dissipated by viscous effects on the Kolmogorov scale. In the inviscid limit ($v \to 0$), the turbulence becomes singular as Re $\to \infty$. It has been observed experimentally that for high enough Reynolds numbers, the energy input necessary to maintain a turbulent flow becomes almost independent of viscosity [67]. This implies that in the limit of $v \to 0$, the velocity field is such that infinite velocity gradients are developed [438, 439]. In the framework of the cascade model, the similarity approximation (1.63) predicts that at very high Reynolds numbers, the singularities in the flow are uniformly distributed. However, in such limits, the experimental evidence shows that contrarily to the predictions of strict similarity, the singularities are distributed inhomogeneously across the flow, with the inhomogeneous character being developed across the inertial cascade [439]. This way the problem of intermittency reduces to the geometrical characterization of the support of flow singularities in the limit when Re $\to \infty$. However, in the absence of a general solution of the Navier-Stokes equations, most of our present understanding of intermittency in turbulence relies on experimental observations.

Chapter 2

The Numerical Modeling of Turbulence

2.1 Introduction

Modern approaches for modeling turbulence are all based on numerical solutions of the unsteady Navier-Stokes equations. These equations are mathematical statements for the laws of mass and momentum conservation. In many engineering applications, they often appear accompanied by an equation of state, relating the density, pressure, and temperature, and the first law of thermodynamics, embodying the law of energy conservation. It is well known that turbulent flows cannot be modeled as easily as laminar flows due to the enormous difficulty in resolving all relevant length scales (from the largest to the smallest eddies) with the aid of present-day computers. The problem becomes even more demanding under the requirement that the time steps must be small enough to guarantee that motion at the smallest scales is accurately resolved. In particular, solving the motion at all turbulent length scales is the task of the so-called *Direct Numerical Simulation* (DNS) approaches. While their application to simulate complex flows is not likely to happen soon, present research in the numerical simulation of turbulent flows relies primarily on methods that are based on averaging the Navier-Stokes equations. Two promising schemes that have been widely used are the *Large Eddy Simulation* (LES) and the *Reynolds Averaged Navier-Stokes* (RANS) approaches. This chapter is divided into four sections, where in the first section the Navier-Stokes equations are introduced. The other three sections deal with a brief account of the RANS, LES, and DNS approaches.

2.2 The Navier-Stokes equations

The Navier-Stokes momentum equation for an incompressible flow has already been introduced in Chapter 1, Eq. (1.58). However, for the sake of completeness here we write down the set of equations for both compressible and incompressible flows in a form that is suitable for averaging.

2.2.1 Compressible flows

In general, the law of mass conservation in the absence of mass sinks and sources is given by the continuity equation

$$\frac{\partial \rho}{\partial t} + \nabla \cdot (\rho \mathbf{v}) = 0, \tag{2.1}$$

where ρ is the density, $\mathbf{v}$ is the velocity vector, and t is time. In terms of the substantial, or total, derivative

$$\frac{d}{dt} = \frac{\partial}{\partial t} + \mathbf{v} \cdot \nabla \tag{2.2}$$

Eq. (2.1) becomes

$$\frac{d\rho}{dt} + \rho \left(\nabla \cdot \mathbf{v} \right) = 0. \tag{2.3}$$

The most general form of the law of momentum conservation is

$$\frac{d\mathbf{v}}{dt} = \frac{\partial \mathbf{v}}{\partial t} + \mathbf{v} \cdot \nabla \mathbf{v} = -\frac{1}{\rho} \nabla p + \frac{1}{\rho} \nabla \cdot \mathbb{T} + \mathbf{g}, \tag{2.4}$$

where p is the pressure, $\mathbf{g}$ is the gravitational acceleration vector, and $\mathbb{T}$ is the viscous stress tensor defined as

$$\mathbb{T} = \eta \left(\nabla \mathbf{v} + (\nabla \mathbf{v})^t - \frac{2}{3} (\nabla \cdot \mathbf{v}) \mathbb{I} \right) + \zeta (\nabla \cdot \mathbf{v}) \mathbb{I}, \tag{2.5}$$

where η is the dynamic viscosity, ζ is the second viscosity, the superscript t means transposition, and $\mathbb{I}$ is the identity tensor. In compressible flows the dynamic and second viscosities are usually functions of the density and temperature. However, in most applications it is of common use to take ζ as constant and sometimes even as zero. Under the assumption of $\zeta = 0$ and noting that $\nabla \cdot (\nabla \mathbf{v}) = \nabla^2 \mathbf{v}$ and $\nabla \cdot (\nabla \mathbf{v})^t = \nabla (\nabla \cdot \mathbf{v})$, the most general form of the Navier-Stokes momentum equation is sometimes written as

$$\frac{d\mathbf{v}}{dt} = \frac{\partial \mathbf{v}}{\partial t} + \mathbf{v} \cdot \nabla \mathbf{v} = -\frac{1}{\rho} \nabla p + \frac{\eta}{\rho} \left[\nabla^2 \mathbf{v} + \frac{1}{3} \nabla (\nabla \cdot \mathbf{v}) \right] + \mathbf{g}, \tag{2.6}$$

For isothermal flows Eqs. (2.3) and (2.4) are complemented by an equation of state of the form $p = p(\rho)$. Because these equations are highly nonlinear, numerical methods must be invoked for their solution under given boundary conditions. The advent of ever more sophisticated numerical techniques along with increasing computational power allow accurate modeling of fluid motion, while on average real turbulent flows are also very well predicted.

2.2.2 Incompressible flows

For incompressible flows the continuity equation (2.3) becomes

$$\nabla \cdot \mathbf{v} = 0, \tag{2.7}$$

which is also called the condition of incompressibility. This condition implies that $d\rho/dt = 0$, or in other words, that the density remains uniform ($\rho = \rho_0$) during the fluid motion. Hence substitution of Eq. (2.7) into Eq. (2.6) produces the familiar Navier-Stokes momentum equation for incompressible flows

$$\frac{d\mathbf{v}}{dt} = \frac{\partial \mathbf{v}}{\partial t} + \mathbf{v} \cdot \nabla \mathbf{v} = -\nabla w + \nu \nabla^2 \mathbf{v} + \mathbf{g}, \tag{2.8}$$

where $\nu = \eta/\rho_0$ is the kinematic viscosity and $\nabla w = \nabla(p/\rho_0)$ is an internal source term. The scalar w is commonly referred to as the thermodynamic work per unit mass. In general, the incompressible flow assumption is valid when a fluid, whether gas or liquid, moves with a velocity $v \ll c$ [16], where c is the speed of sound in the fluid. However, in more unsteady flows the incompressibility assumption holds when the distance traveled by a sound wave during time t, say ct, is much larger than the typical distance over which significant changes of the fluid velocity occurs. Therefore, over sufficiently small spatial scales and free of external forces all fluids are incompressible with a very good approximation.

2.2.3 Non-isothermal flows: heat transfer

When heat transfer between different parts of the fluid or external sources of heating are involved, the Navier-Stokes equations must be complemented by an energy equation. The first law of thermodynamics applied to a moving viscous fluid leads to the differential equation for the total energy per unit volume E

$$\rho \frac{d}{dt} \left(\frac{E}{\rho} \right) = \frac{\partial Q}{\partial t} - \nabla \cdot (p\mathbf{v}) - \nabla \cdot \mathbf{q} + \nabla \cdot (\mathbb{T} \cdot \mathbf{v}) + \rho \mathbf{g} \cdot \mathbf{v}, \tag{2.9}$$

where

$$\frac{E}{\rho} = e + \frac{1}{2}\mathbf{v} \cdot \mathbf{v} + \text{other forms of energy}, \tag{2.10}$$

and e is the specific internal energy. The first term on the right-hand side of Eq. (2.9) is the rate of heat per unit volume transferred to the fluid by external sources, while $\nabla \cdot \mathbf{q}$ represents the rate of heat conduction between different parts of the fluid, where the heat transfer vector $\mathbf{q}$ is assumed to obey Fourier's law

$$\mathbf{q} = -\kappa \nabla T, \tag{2.11}$$

where T is the temperature and κ is the thermal conductivity. As established by the first law of thermodynamics, Eq. (2.9) describes the time rate of change of the fluid energy due to heat added to (or removed from) the fluid plus the work done on the fluid.

Replacing Eq. (2.10) into the left-hand side of Eq. (2.9) under the assumption that only the internal and kinetic energies are considered to be important in Eq. (2.10), taking the dot product of Eq. (2.4) with the velocity vector, and using the result in Eq. (2.9) it is easy to demonstrate that this equation reduces to a differential equation for the time rate of change of the specific internal energy

$$\rho \frac{de}{dt} = \frac{\partial Q}{\partial t} - p(\nabla \cdot \mathbf{v}) - \nabla \cdot \mathbf{q} + \nabla \cdot (\mathbb{T} \cdot \mathbf{v}) - (\nabla \cdot \mathbb{T}) \cdot \mathbf{v}, \qquad (2.12)$$

where the last two terms on the right-hand side account for the rate of decrease of the mechanical energy of the fluid caused by viscous friction. It describes the irreversible conversion of mechanical energy associated with the fluid ordered motion into energy of disordered motion, and hence ultimately into thermal energy. Sometimes the contribution of these two terms is denoted by Φ, which is customarily called the viscous dissipation function. The energy equation can be expressed in different ways. For example, an equation for the enthalpy, $h = e + p/\rho$, which in some applications is preferred to the internal energy can be derived by combining Eqs. (2.3) and (2.12) to be

$$\rho \frac{dh}{dt} = \frac{dp}{dt} + \frac{\partial Q}{\partial t} - \nabla \cdot \mathbf{q} + \nabla \cdot (\mathbb{T} \cdot \mathbf{v}) - (\nabla \cdot \mathbb{T}) \cdot \mathbf{v}. \qquad (2.13)$$

Equation (2.12) is valid for compressible fluids moving under the presence of external heat sources. For an incompressible flow with no external heat sources, this equation takes the more familiar form

$$\rho \frac{de}{dt} = -\nabla \cdot \mathbf{q} + \nabla \cdot (\mathbb{T} \cdot \mathbf{v}) - (\nabla \cdot \mathbb{T}) \cdot \mathbf{v}. \qquad (2.14)$$

For a thorough account on the Navier-Stokes equations and heat conduction theory the interested reader is referred to the books by Landau and Lifshitz [521] and Tannehill et al. [891], from which part of the formalism presented here has been extracted.

2.3 Reynolds averaged Navier-Stokes (RANS) equations

The RANS equations can be derived from the Navier-Stokes equations by first decomposing the dependent variables into a time-averaged (or mean) part and a fluctuating component as

$$f(\mathbf{x}, t) = \overline{f}(\mathbf{x}) + f'(\mathbf{x}, t), \qquad (2.15)$$

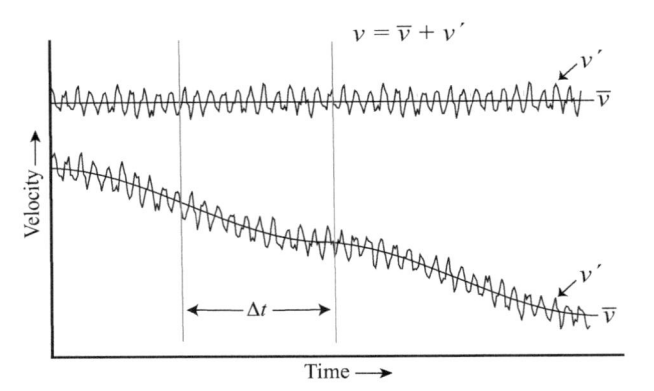

Figure 2.1: Schematic of the Reynolds decomposition of the velocity field in a point for constant and transient flow.

where f may be ρ, $\mathbf{v}$, p, T, e, or h. According to the classical Reynolds averaging, the temporal mean $\overline{f}$ is defined as

$$\overline{f}(\mathbf{x}) = \frac{1}{\Delta t} \int_t^{t+\Delta t} f(\mathbf{x},t)\,dt, \qquad (2.16)$$

where Δt is some finite time interval, which must be sufficiently large compared to the period of the turbulent fluctuations [891]. In what follows the spatial and temporal dependence of variables will be omitted for simplicity. On the other hand, it is easy to show that $\overline{f'} = 0$. For constant and transient flow in a point, Fig. 2.1 shows a schematic of the velocity variation as a function of time. However, in general $\overline{f'f'} = \overline{f'^2} \neq 0$. In particular, if v denotes the flow velocity component in a given direction, then from Eq. (2.15) it follows that $v' = v - \overline{v}$ and from Eq. (2.16)

$$\overline{v'^2} = \frac{1}{\Delta t} \int_t^{t+\Delta t} (v - \overline{v})^2\,dt \neq 0, \qquad (2.17)$$

so that

$$v_{\mathrm{rms}} = \sqrt{\overline{v'^2}}, \qquad (2.18)$$

defines the turbulence intensity in the direction of the flow velocity component v. If f and g are two fluid variables, other useful relations are

$$\overline{\overline{f}g'} = 0, \qquad \overline{\overline{f}g} = \overline{f}\,\overline{g}, \qquad \overline{f+g} = \overline{f}+\overline{g}. \qquad (2.19)$$

2.3.1 Compressible flows

In order to derive the Reynolds averaged form of the continuity equation, it is easier to start from the conservation-law form (2.1) by first decomposing the density and velocity using Eq. (2.15), i.e., $\rho = \overline{\rho} + \rho'$ and $\mathbf{v} = \overline{\mathbf{v}} + \mathbf{v}'$ and then time-averaging the full equation by taking into account relations (2.19). The result is

$$\frac{\partial \bar{\rho}}{\partial t} + \nabla \cdot \left(\bar{\rho}\,\bar{\mathbf{v}} + \overline{\rho'\mathbf{v'}} \right) = 0. \tag{2.20}$$

The Reynolds form of the momentum conservation law is obtained by re-writing Eq. (2.4) as

$$\frac{\partial (\rho \mathbf{v})}{\partial t} + \nabla \cdot \left(\rho \mathbf{v}\mathbf{v} + p\mathbb{I} - \mathbb{T} \right) = \rho \mathbf{g}, \tag{2.21}$$

and following the same steps as before. This yields the Reynolds time-averaged momentum equation

$$\frac{\partial}{\partial t} \left(\bar{\rho}\,\bar{\mathbf{v}} + \overline{\rho'\mathbf{v'}} \right) \quad + \quad \nabla \cdot \left(\bar{\rho}\,\bar{\mathbf{v}}\,\bar{\mathbf{v}} + \bar{\mathbf{v}}\overline{\rho'\mathbf{v'}} \right)$$
$$= \quad -\nabla \bar{p} + \nabla \cdot \left(\overline{\mathbb{T}} - \bar{\rho}\overline{\mathbf{v'}\mathbf{v'}} - \bar{\mathbf{v}}\overline{\rho'\mathbf{v'}} - \overline{\rho'\mathbf{v'}\mathbf{v'}} \right) + \bar{\rho}\mathbf{g}, \tag{2.22}$$

where

$$\overline{\mathbb{T}} = \eta \left[\nabla \bar{\mathbf{v}} + (\nabla \bar{\mathbf{v}})^t - \frac{2}{3} (\nabla \cdot \bar{\mathbf{v}})\,\mathbb{I} \right] + \zeta \,(\nabla \cdot \bar{\mathbf{v}})\,\mathbb{I}. \tag{2.23}$$

The Reynolds form of the energy equation is derived more easily in terms of the total specific enthalpy

$$H = \frac{E}{\rho} + \frac{p}{\rho}, \tag{2.24}$$

where E/ρ is the total energy given by Eq. (2.10). Substitution of relation (2.24) into Eq. (2.9) gives

$$\frac{\partial (\rho H)}{\partial t} + \nabla \cdot \left(\rho H \mathbf{v} + \kappa \nabla T \right) = \frac{\partial Q}{\partial t} + \frac{\partial p}{\partial t} + \nabla \cdot \left(\mathbb{T} \cdot \mathbf{v} \right) + \rho \mathbf{g} \cdot \mathbf{v}. \tag{2.25}$$

After replacing the dependent variables in the above equation with the temporal averages plus the fluctuating components as in Eq. (2.15) for $f = \rho, \mathbf{v}, p, H$, and T and time averaging the equation, it follows that

$$\frac{\partial}{\partial t} \left(\bar{\rho}\bar{H} + \overline{\rho'H'} \right) \quad + \quad \nabla \cdot \left(\bar{\rho}\bar{H}\bar{\mathbf{v}} + \bar{\rho}\overline{H'\mathbf{v'}} + \bar{H}\,\overline{\rho'\mathbf{v'}} + \bar{\mathbf{v}}\overline{\rho'H'} + \overline{\rho'H'\mathbf{v'}} \right)$$
$$= \quad \frac{\partial \bar{Q}}{\partial t} + \frac{\partial \bar{p}}{\partial t} + \nabla \cdot \left(\kappa \nabla \bar{T} \right) + \left(\bar{\rho}\,\bar{\mathbf{v}} + \overline{\rho'\mathbf{v'}} \right) \cdot \mathbf{g} + \overline{\nabla \cdot (\mathbb{T} \cdot \mathbf{v})}, \tag{2.26}$$

where

$$\overline{\nabla \cdot (\mathbb{T} \cdot \mathbf{v})} \quad = \quad \nabla \cdot \left\{ \eta \left[\nabla \bar{\mathbf{v}} \cdot \bar{\mathbf{v}} + \overline{\nabla \mathbf{v'} \cdot \mathbf{v'}} + (\nabla \bar{\mathbf{v}})^t \cdot \bar{\mathbf{v}} + \overline{(\nabla \mathbf{v'})^t \cdot \mathbf{v'}} \right] \right\}$$
$$+ \quad \nabla \cdot \left\{ \left(\zeta - \frac{2}{3}\eta \right) \left[(\nabla \cdot \bar{\mathbf{v}})\bar{\mathbf{v}} + \overline{(\nabla \cdot \mathbf{v'})\mathbf{v'}} \right] \right\}. \tag{2.27}$$

An alternative averaging to the conventional Reynolds averaging was introduced by Favre [284, 285]. In this case, it is of common use in the literature to express the Reynolds decomposition of a fluid variable f as

$$f = \tilde{f} + f'', \tag{2.28}$$

where now the temporal mean corresponds to a mass-weighted average

$$\tilde{f} = \frac{1}{\overline{\rho} \Delta t} \int_t^{t+\Delta t} \rho f \, dt = \frac{\overline{\rho f}}{\overline{\rho}}. \tag{2.29}$$

As before, the overbar denotes Reynolds averaging and from now on the tilde will denote Favre averaging. From relation (2.29), it follows that $\tilde{f} \overline{\rho} = \overline{\rho f}$. When deriving the Favre filtered Navier-Stokes equations, the following rules must be taken into account

$$\overline{\tilde{f}} = \tilde{f}; \quad \overline{\rho \tilde{f}} = \overline{\rho} \tilde{f}; \quad \overline{\rho f''} = 0; \quad \overline{f''} = -\frac{\overline{\rho' f'}}{\overline{\rho}} \neq 0; \quad \widetilde{\rho f} = \overline{\rho} \tilde{f}, \tag{2.30}$$

while

$$\overline{\rho f g} = \overline{\rho} \tilde{f} \tilde{g} + \overline{\rho f'' g''}. \tag{2.31}$$

It is important to mention that only the velocity and energy variables are mass-weighted averaged. Following steps similar to those employed to derive the Reynolds averaged form of the conservation laws and using the definitions in Eqs. (2.28)–(2.30), the Favre filtered compressible continuity equation becomes

$$\frac{\partial \overline{\rho}}{\partial t} + \nabla \cdot (\overline{\rho} \tilde{\mathbf{v}}) = 0. \tag{2.32}$$

Similarly, it can be demonstrated that the Favre filtered momentum and energy equations are, respectively

$$\frac{\partial (\overline{\rho} \tilde{\mathbf{v}})}{\partial t} + \nabla \cdot (\overline{\rho} \tilde{\mathbf{v}} \tilde{\mathbf{v}}) = -\nabla \overline{p} + \nabla \cdot \left(\mathbb{T} - \overline{\rho \mathbf{v}'' \mathbf{v}''} \right), \tag{2.33}$$

and

$$\frac{\partial (\overline{\rho} \tilde{H})}{\partial t} \quad + \quad \nabla \cdot \left(\overline{\rho} \tilde{\mathbf{v}} \tilde{H} + \overline{\rho H'' \mathbf{v}''} \right)$$
$$= \quad \frac{\partial \overline{p}}{\partial t} + \nabla \cdot \left(\kappa \nabla \overline{T} + \mathbb{T} \cdot \tilde{\mathbf{v}} + \overline{\mathbb{T} \cdot \mathbf{v}''} \right), \tag{2.34}$$

where

$$\mathbb{T} \quad = \quad \eta \left[\nabla \tilde{\mathbf{v}} + (\nabla \tilde{\mathbf{v}})^t - \frac{2}{3} (\nabla \cdot \tilde{\mathbf{v}}) \mathbb{I} \right] + \zeta (\nabla \cdot \tilde{\mathbf{v}}) \mathbb{I}$$
$$+ \quad \eta \left[\widetilde{\nabla \mathbf{v}''} \left(\widetilde{\nabla \mathbf{v}''} \right)^t \frac{2}{3} \left(\widetilde{\nabla \cdot \mathbf{v}''} \right) \mathbb{I} \right] + \zeta \left(\widetilde{\nabla \cdot \mathbf{v}''} \right) \mathbb{I}. \tag{2.35}$$

Compared to Eqs. (2.20), (2.22), and (2.26), the Favre filtered Eqs. (2.32)–(2.34) look mathematically much simpler. However, in order to solve these equations some approximations are needed for the fluctuating parts. For compressible flow a commonly used approximation is the so-called Reynolds-stress tensor, which according to the Boussinesq assumption is given by

$$-\overline{\rho \mathbf{v''v''}} = \eta_t \left[\nabla \tilde{\mathbf{v}} + (\nabla \tilde{\mathbf{v}})^t - \frac{2}{3}(\nabla \cdot \tilde{\mathbf{v}})\mathbb{I} \right] - \frac{2}{3}\overline{\rho}k\mathbb{I}, \qquad (2.36)$$

where η_t is the turbulent viscosity and $\overline{\rho}k$ is the kinetic energy of turbulence. The turbulent heat-flux vector is defined as

$$\overline{\rho H''\mathbf{v''}} = -\frac{\eta_t}{\mathrm{Pr}_t}\nabla \tilde{H}, \qquad (2.37)$$

where $\mathrm{Pr}_t = c_p \eta_t / \kappa$ is the turbulent Prandtl number and c_p is the specific heat at constant pressure. The turbulent viscosity is defined as $\eta_t = \rho v_t l$, where l and v_t are characteristic length and velocity scales of turbulence, which must be suitably evaluated. Models based on the Boussinesq assumption are referred to as turbulent viscosity models or first-order models. In contrast, models referred to as stress-equation models or second-order closures do not use the Boussinesq approximation. In this case, exact transport equations are derived for the Reynolds stresses $-\overline{\rho \mathbf{v'v'}}$ [973, 529]. Among this category of models the one known as the *algebraic Reynolds stress model* (ASM), which employs the kinetic energy of turbulence and the dissipation rate as parameters, has shown good performance for isothermal and buoyant thin shear layers. On the other hand, in the category of the so-called *One-Equation Models*, the turbulent viscosity is defined in terms of the kinetic energy equation, which is calculated by solving a transport partial differential equation that is derived from the Navier-Stokes equations. However, the most commonly used closure models are the *Two-Equation Models*, which employ a second transport equation for the dissipation rate ε. These models have given rise to the k-ε closure turbulence models [383, 445, 530]. For a more complete description of the various closure models, the interested reader is referred to the book of Tannehill et al. [891] and references therein.

2.3.2 Incompressible flows

For incompressible flows $\rho' = 0$ and the Favre averaging becomes identical to the conventional Reynolds averaging. Therefore, $\tilde{f} = \overline{f}$ and $f'' = f'$. It is easy to show that the continuity equation becomes

$$\nabla \cdot \overline{\mathbf{v}} = 0, \qquad (2.38)$$

while the momentum equation (2.22) takes the much simpler form

$$\frac{\partial (\rho \overline{\mathbf{v}})}{\partial t} + \nabla \cdot (\rho \overline{\mathbf{v}}\,\overline{\mathbf{v}}) = -\nabla \overline{p} + \nabla \cdot \left(\mathbb{T} - \rho \overline{\mathbf{v'v'}} - \right) + \rho \mathbf{g}. \qquad (2.39)$$

Exactly the same form can be straightforwardly derived from the Favre filtered momentum equation (2.33). In terms of the total enthalpy, the energy equation (2.26) becomes

$$\frac{\partial(\rho\overline{H})}{\partial t} + \nabla \cdot \left(\rho\overline{H}\overline{\mathbf{v}} + \rho\overline{H'\mathbf{v}'}\right) = \frac{\partial\overline{Q}}{\partial t} + \frac{\partial\overline{p}}{\partial t} + \nabla \cdot \left(\kappa\nabla\overline{T}\right) + \rho\mathbf{g}\cdot\overline{\mathbf{v}} + +\overline{\nabla \cdot (\mathbb{T}\cdot\mathbf{v})}, \tag{2.40}$$

which is also identical to the Favre filtered equation for incompressible flows.

2.4 Large eddy simulation (LES) equations

In turbulent flow, the large eddies are more energetic than the small ones. In fact, conserved properties are transported more efficiently at larger scales than at smaller ones. Therefore, it makes sense to treat the large eddies more accurately than the small ones and the LES approach works this way. In general, a simulation based on the LES model is more expensive than one based on any of the RANS approaches. This scheme was originally proposed in 1963 to simulate atmospheric air currents [845] and despite its higher computational cost compared to RANS, it has been widely applied to simulations of atmospheric boundary layer flows [882, 177, 574, 167, 376, 819, 926]. For details on the theory, capabilities, and limitations of LES the reader is referred to the review articles in Refs. [758, 705, 551, 859, 716, 317]. In LES the filtered Navier-Stokes equations describe the evolution of the large eddies which represent the resolved scale field, while the effects of the small (unresolved) eddies are represented using a subgrid-scale stress tensor. Flow variables are then split up into large scale and subgrid-scale variables as [550]

$$f = \overline{f} + f', \tag{2.41}$$

where now the bar and prime quantities have completely different meanings to the variables in Eq. (2.15) for the RANS decomposition. The resolved scale field is the local average of the complete field given by the convolution integral

$$\overline{f(\mathbf{x})} = \int G(\mathbf{x},\mathbf{x}')f(\mathbf{x}')d\mathbf{x}', \tag{2.42}$$

where the integration has been taken over the flow volume. Here $G(\mathbf{x},\mathbf{x}')$ is a localized filter function, which must obey the normalization condition

$$\int G(\mathbf{x},\mathbf{x}')d\mathbf{x}' = 1, \tag{2.43}$$

if the filtering (2.42) has to return the correct value when f is a constant. Among the several filters that can be used, the so-called box filter has been frequently employed [21], which can be written as

$$G(\mathbf{x},\mathbf{x}') = \frac{1}{\Delta}, \tag{2.44}$$

if $\| \mathbf{x} - \mathbf{x}' \| \le \Delta/2$ or zero otherwise [859], where Δ is a length scale such that eddies with length scales smaller than Δ are removed from the filtering. The Navier-Stokes equations can be written in a filtered form suitable for LES by noting that $\overline{\nabla f} = \nabla \overline{f}$ if f vanishes on the fluid boundaries. While the small-scale fluctuations are effectively removed by low-pass filtering of the Navier-Stokes equations, their effect on the modeling of actual turbulent flow fields represents an active area of research.

2.4.1 Incompressible flows

For incompressible flow, the filtered continuity equation is simply

$$\nabla \cdot \overline{\mathbf{v}} = 0, \qquad (2.45)$$

while the LES form of the momentum equation can be obtained by filtering Eq. (2.21) for $\rho = \text{const.}$ and $\nabla \cdot \mathbf{v} = 0$ to give

$$\frac{\partial \overline{\mathbf{v}}}{\partial t} + \nabla \cdot \overline{\mathbf{v}\mathbf{v}} = -\nabla \overline{w} + \nu \nabla^2 \overline{\mathbf{v}} + \mathbf{g}, \qquad (2.46)$$

where the convective flux is decomposed as

$$\overline{\mathbf{v}\mathbf{v}} = \overline{(\overline{\mathbf{v}} + \mathbf{v}')(\overline{\mathbf{v}} + \mathbf{v}')} = \overline{\overline{\mathbf{v}}\,\overline{\mathbf{v}}} + \overline{\overline{\mathbf{v}}\mathbf{v}'} + \overline{\mathbf{v}'\overline{\mathbf{v}}} + \overline{\mathbf{v}'\mathbf{v}'}, \qquad (2.47)$$

so that

$$\overline{\mathbf{v}\mathbf{v}} = \overline{\mathbf{v}}\,\overline{\mathbf{v}} + \tau, \qquad (2.48)$$

where

$$\tau = \left(\overline{\overline{\mathbf{v}}\,\overline{\mathbf{v}}} - \overline{\mathbf{v}}\,\overline{\mathbf{v}} \right) + \left(\overline{\overline{\mathbf{v}}\mathbf{v}'} + \overline{\mathbf{v}'\overline{\mathbf{v}}} \right) + \overline{\mathbf{v}'\mathbf{v}'}, \qquad (2.49)$$

is the subgrid-scale stress tensor. Using relation (2.48) into Eq. (2.46), the filtered momentum equation becomes

$$\frac{\partial \overline{\mathbf{v}}}{\partial t} + \nabla \cdot \overline{\mathbf{v}}\,\overline{\mathbf{v}} = -\nabla \overline{w} + \nu \nabla^2 \overline{\mathbf{v}} - \nabla \cdot \tau + \mathbf{g}. \qquad (2.50)$$

The first term in parentheses on the right-hand side of Eq. (2.49) is the so-called Leonard tensor, while the second and third terms are, the Clark (or cross-term) stress tensor and the Reynolds stress tensor, respectively. It is evident from Eq. (2.49) that time averaging and filtering differs in that in the former case the second averaging vanishes identically, while in the filtering case it yields a different result.

2.4.2 Compressible flows

As for the conventional Reynolds averaging, the LES filtering of the continuity equation is more easily performed by starting from the conservation-law form

(2.1). Application of the Favre filtering (2.29) to solve the low-pass filtering of the convective flow $\rho\mathbf{v}$ produces the LES filtered form

$$\frac{\partial\overline{\rho}}{\partial t}+\nabla\cdot(\overline{\rho}\tilde{\mathbf{v}})=0. \tag{2.51}$$

Although this equation is mathematically identical in form to Eq. (2.32), the averaged density and velocity are different in both cases.

Following similar steps and starting from the momentum equation (2.21) in conservation-law form, the LES filtered momentum equation becomes [946]

$$\frac{\partial(\overline{\rho}\tilde{\mathbf{v}})}{\partial t}+\nabla\cdot(\overline{\rho}\tilde{\mathbf{v}}\tilde{\mathbf{v}})=-\nabla\overline{p}+\nabla\cdot\mathbb{T}-\nabla\cdot(\overline{\rho}\tau^r)+\overline{\rho}\mathbf{g}, \tag{2.52}$$

where $\mathbb{T}$ has the same mathematical form of the shear stress tensor (2.35) and τ^r is the subgrid stress tensor of the Favre filtered momentum field given by

$$\tau^r=\widetilde{\mathbf{v}\mathbf{v}}-\tilde{\mathbf{v}}\tilde{\mathbf{v}}. \tag{2.53}$$

This tensor can be broken up in a manner analogous to Eq. (2.49), where as above the Leonard tensor represents the interaction among large scale eddies, the Clark tensor represents the interaction between large and small eddies, and the Reynolds tensor represents the interaction among the unresolved (sub-filter) scales [361].

2.4.3 Subgrid scale stresses

The simplest subgrid scale model is the so-called *eddy viscosity model* proposed by Smagorinsky [845]. In this formulation, the subgrid stress tensor is defined as

$$\tau=2\nu_t\overline{\mathbb{S}}, \tag{2.54}$$

where ν_t is the turbulent eddy viscosity[1] given by

$$\nu_t=(C_s\Delta)^2\sqrt{2\mathbb{S}:\mathbb{S}}, \tag{2.55}$$

where C_s is the Smagorinsky constant, which takes values in the interval $(0.1, 0.24)$, and Δ is set equal to the minimum grid size allowed, and $\overline{\mathbb{S}}$ is the rate of strain tensor defined as

$$\overline{\mathbb{S}}=\frac{1}{2}\left[\nabla\overline{\mathbf{v}}+(\nabla\overline{\mathbf{v}})^t\right]. \tag{2.56}$$

There exists in the literature more complex models for the subgrid stresses. One of these models is the *Dynamic Smagorinsky Model* introduced by Germano et al. [330]. This model differs from the classical Smagorinsky model in that the

---

[1]Note that ν_t in Eq. (2.54) has units of m^2 s^{-1}.

constant C_s is calculated as part of the solution by utilizing two filters for any turbulent variable f, namely the LES filter, $\overline{f}$, and the so-called test filter, $\hat{f}$. Other approaches as the *Dynamic Localization Model* (DLM) was suggested by Ghosal et al. [334] as an alternative model. Corrections of the DLM model for backscatter representation were suggested in the DLM(k) approach introduced by Carati et al. [154], where the velocity scale $\Delta \parallel \mathbb{S} \parallel$ in the Smagorinsky eddy viscosity is replace by $\sqrt{k}$, where k is a parameter that enters in the definition of the subgrid-scale kinetic energy. Despite many efforts addressed to improve the capabilities of LES, it is well known that it still suffers from errors carried by the subgrid stress models and the numerical discretization scheme. Accurate predictions for turbulence dynamics can be obtained by optimal combinations of the subgrid stress model and the discretization technique [17]. Attempts to revive the Smagorinsky model for LES have also been reported recently based on increased accuracy and computational efficiency when the value of the Smagorinsky constant is correctly chosen, which in turn leads to a correct dissipation in turbulence energy [616]. This finding has recently provided a good impetus to the use of the Smagorinsky model in wind farm aerodynamics and atmospheric applications.

2.5 Direct numerical simulations (DNS)

As was already commented at the beginning of this chapter, DNS involves the direct numerical solution of the unsteady Navier-Stokes equations without the use of turbulence models. Therefore, such models are thought to be capable of resolving the full energy cascade to the smallest eddies and time scales of turbulence within a flow. To some extent, this has been possible thanks to recent increases in computational power. However, the requirement of very fine grids and extremely small time steps has confined the applications of DNS to flow problems with relatively low Reynolds numbers. Therefore, DNS of the fully turbulent flow at high Reynolds numbers must surely wait for further advances in computational hardware. Even with the present limitations, there are examples in the literature where DNS data has been used to evaluate closure models and measurement accuracy, providing further insight into the structure of turbulent boundary layers [594, 737]. DNS results of the dynamics of a stratocumulus-topped boundary layer have shown convergence toward Reynolds number similarity and consistency with LES results and field measurements [617].

Applications of DNS to atmospheric flows have started to appear in the last decade. For instance, Smyth and Thorpe [853] have reported a nice simulation of how a two-layer instability grows to form a train of Kelvin-Helmholtz billows, which then merge and develop convective secondary instabilities in overturned zones of the billow cores; the evolution ending with the growth of a combination of complex tertiary instabilities, leading to a fully developed turbulence state followed by turbulence decay into sharp layers. Figure 2.2 displays the sequence of events showing the transition from a boundary layer instability to Kelvin-

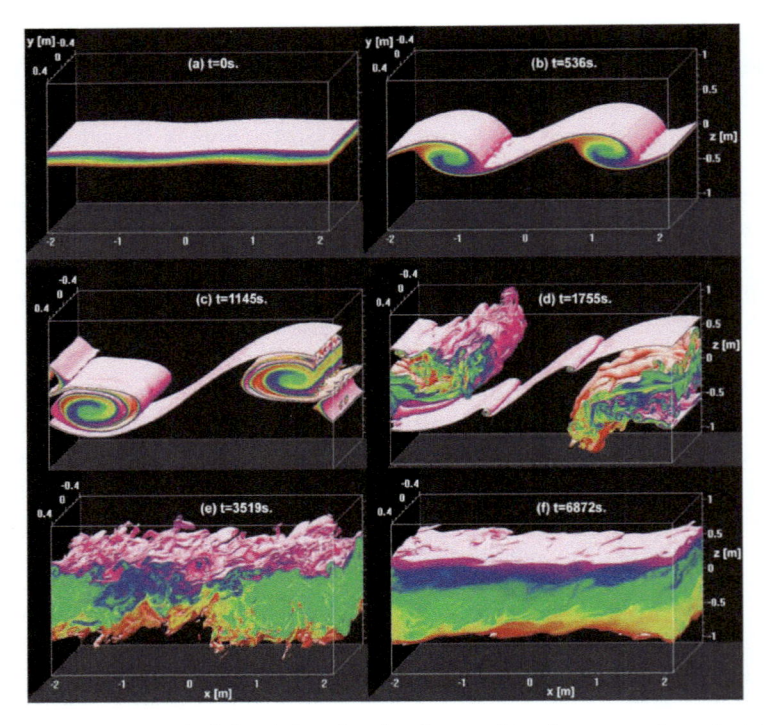

Figure 2.2: Direct numerical simulation showing the transition from a two-layer instability to a fully turbulent state. The upper and lower layers are homogeneous and rendered transparent, while the colors show the density stratification in the transition layer. (a) Initial state where the upper (lighter) layer moves to the right and the lower (denser) to the left. (b) A train of KH billows forms. (c) Convective secondary instabilities develop within the billow rolls. (d) Secondary shear instability amplifies in the braids. (e) Tertiary instabilities lead to a fully turbulent state. (f) Turbulence dacay and thickening of the transition layer by irreversible mixing. Figure taken from Smyth and Thorpe [853].

Helmholtz instability to fully developed three-dimensional turbulence to turbulence decay. For example, the understanding of the development of the atmospheric energy spectrum across different scales is a necessary step toward atmospheric predictability. Recent two-dimensional DNS models have been addressed to investigate energy transfer between the synoptic scale and the mesoscale, the results suggesting the existence of mesoscale feedback on synoptic-scale predictability [266]. Further research on atmospheric boundary-layer based on DNS is moving toward the use of adaptive grids [922], which has been a common strategy employed by computational fluid dynamicists to increase spatial resolution. Given the requirements imposed by DNS of turbulence, this method has a clear potential for numerical simulations of the most challenging atmospheric cases.

Chapter 3

Clear Air Turbulence (CAT)

3.1 Introduction

In the area of aviation safety, the impact of weather phenomena on aircraft operations has long been recognized. As evidence of this, weather is responsible for about 30% of aviation accidents and approximately two-thirds of air carrier delays with a consequent loss of billions of dollars of which a significant part could be avoided [528, 825]. In particular, turbulence associated with convective weather and CAT is frequently encountered in commercial flights, sometimes compromising both the integrity of the passengers and the structure of the aircraft. However, in general CAT is not the cause of aircraft crashes unless it is classified as severe or extreme.

As discussed in Chapter 1, Section 1.3.3, CAT can be defined as erratic air currents often encountered at altitudes between 7 to 12 thousand meters from the ground and caused by rapidly moving air layers in contact with much slower air. Such wind velocity gradients usually occur in regions with a thickness of tens to hundreds of meters. The effects of these erratic currents are even more severe over mountains and close to thunderstorms. In fact, among the most common causes of CAT are the jet streams – the narrow and fast moving air currents close to the tropopause, which are generated by temperature gradients between air masses – and as was mentioned above, by the roughness of the terrain – giving rise to the so-called mountain waves – and by the turbulence in the vicinity of thunderstorms and thunderstorm complexes. The detection of CAT for warning purposes remains a difficult task today and often relies on pilot reports. Atmospheric turbulence has long been a major research topic in the areas of applied fluid mechanics and commercial and military aviation. Indeed, it was not until the World War II that CAT was discovered as a consequence of the danger that it represented to military aviation. However, the scientific investigation of CAT detection began to deepen strongly in 1996 when the *White House Commission on Aviation Safety and Security* was established, which in addition to other aviation topics, also focuses on the study of weather effects upon aviation safety to provide better weather situational awareness. Since then new and modern observation techniques, as well as new methods of CAT detection and warning, have been developed to reduce aviation weather-related accidents.

This chapter will be entirely devoted to a description of CAT as well as its causes and effects. Related topics and characteristics associated with CAT include the phenomena of wake turbulence, atmospheric gravity waves, wind shear, vortex rolls, jet streams and streaks, lee and mountain waves, temperature gradients, on one hand, and dynamical, aviation, and synoptic meteorology as well as applications of the Lighthill-Ford theory of spontaneous imbalance to CAT forecasting, on the other hand. The chapter ends with a brief description of fossil turbulence and its implications on CAT and a tentative answer to the all-important question of whether CAT is indeed increasing.

3.2 Definition of CAT

CAT is a class of turbulence caused by physical and atmospheric conditions that have direct interaction with airplanes during take-off, flight, and landing. In general, it refers to bumps in flight aboard aircraft at high altitudes. Different definitions of CAT can be found in the literature. For example, among the simplest definitions, we find that CAT is a kind of turbulence that occurs in cloudless regions. Possibly the most comprehensive definition is just the one that associates CAT with high-altitude turbulence encountered outside of convective clouds [545]. Moreover, Chambers [158] defined CAT as high-altitude aircraft potholes in regions devoid of significant cloud cover and away from thunderstorm activity. In 1966, the National Committee for CAT under the U.S. Department of Commerce adopted the following CAT definition [752]:

> CAT comprises all turbulence in the free atmosphere of interest in aerospace operations that is not in, or adjacent to, visible convective activity. This includes turbulence found in cirrus clouds not in, or adjacent to, visible convective activity.

According to this definition CAT comprises all bumpy flight conditions away from convective clouds that primarily affect airplanes, aerospace vehicles, and in general vertical and/or short take-off and landing (VSTOL). A more modern definition considers CAT to be a form of turbulence found at altitudes of about 5.6 km where the atmospheric pressure is ≈ 500 hPa[1], or even higher, either in cloud-free conditions or within stratiform clouds [277]. This definition excludes low-level disturbances, which are often attributed to boundary layer and topographic effects, such as lee waves [934]. As was pointed out above, CAT is a class of turbulence that has a strong impact on aircraft in flight because it is not visible to pilots and no equipment allows its identification analysis to prevent pilots from entering a region with this class of turbulence. For this reason, it has been necessary to first identify the physical sources of CAT. One main source of CAT is the KH instability described in Chapter 1. In the free atmosphere, upper-level frontal zones and jet streams are the main synoptic environments conducive to KH instability [575]. On the other hand, KH instabilities are also associated with gravity waves and therefore they could manifest within clouds [791, 948]. In particular, KH instabilities are likely to play a role in the formation, maintenance, and dissipation of cirrus clouds affected by these instabilities [733].

The effects of CAT on civil aviation ranging from minor annoyances to passengers, who are forced to remain seated with their seat belts fastened, to serious structural damage to the airplanes, which may also cause serious injuries to passengers and crew, can be largely avoided by continuously monitoring the atmospheric conditions of air routes [182, 827]. For instance, the ever increasing air traffic has imposed the need to develop better security systems to make air routes more efficient, maintain flight schedules, provide comfort to passengers, and what is more important, avoid eventual catastrophic accidents [934]. As an atmospheric phenomenon, CAT must be studied in the context of meteorology. Since the atmosphere is an open and complex system, the phenomena that take place involve multiscale processes that are often the result of the interaction between the Earth's rotation and local thermodynamic processes. In general, such interactions are so complex that the prediction of atmospheric phenomena, including CAT and climate change, is difficult if not impossible. In a broad sense, the interest in understanding the dynamics of the atmosphere from a scientific point of view and in developing new technologies for prediction and forecasting has increased the number of research programs around the world, a large part of which are focused on CAT detection and prediction as well as on improving and implementing new technologies in airplanes to increase flight safety. Despite this effort and the investment of enormous economic and human resources, CAT detection remains an open problem because it involves the presence of turbulent flows with chaotic dynamics coupled with density, pressure, and temperature gradients and a wide range of spatial and temporal correlations. This makes the task of finding precursor signs, detection, measurement, and forecasting extremely difficult. CAT detection possibilities also rely on pilot reports (PIREPs), which

---

[1] 1 hPa (hectopascal) is equivalent to 100 Pa.

remain today one trusted CAT locator. However, the NASA in partnership with the *Federal Aviation Administration* (FAA), the *National Oceanic and Atmospheric Administration* (NOAA), the aviation industry, and the research community have developed new technologies for the cockpit presentation of graphical information on weather for turbulence prediction and warning, automated airborne *in situ* weather reporting, and data linking of weather information between airplanes in flight and users on the ground (see Ref. [876] and references therein).

Observations indicate that CAT spans distances between ~ 80 and 500 km in the wind direction and between ~ 20 and 100 km perpendicular to the wind. Typically, it has temporal scales of a few hours and falls within the range of mesoscale phenomena [752]. It is observed to occur more frequently near jet streams than away from them. Most CAT encounters have been reported in baroclinic zones with a stable stratified lapse rate. Encounters are also more frequent over mountains than over flat terrain and oceans [185, 545], thus suggesting that the terrain configuration accompanied by convective activity can feed with finite amounts of energy small atmospheric flow perturbations that can give rise to CAT. In particular, Clodman's [185] findings indicate that standing gravity waves are a primary cause of CAT over mountains, while over flat terrain and oceans traveling wave disturbances might be the main reason. Above 5 km of altitude from the ground, typical eddy sizes involved in CAT range from about 20 to 200 m. In the free atmosphere, the inertial sub-range of turbulence extends to slightly larger eddies (~ 100 to 200 m) than near the ground. Beyond this sub-range of turbulence, a variety of conditions may occur. For example, the turbulent energy in the isotropic sub-range is fed by the kinetic energy of the flow over rough terrains in a near-neutral stratification, which could be the case of low-level turbulence. In the free atmosphere, eddy sizes beyond the isotropic inertial sub-range may be influenced by positive and negative buoyant forces, while in the presence of vertical wind shears in a thermally stable layer CAT may be generated by eddies produced in the shearing layer. This seems to be the case for most of the CAT encountered in the vicinity of jet streams and in the lower stratosphere.

3.3 Dynamical meteorology

According to the Cambridge dictionary, meteorology is "the scientific study of the processes that cause particular weather conditions". In a broad sense, it is a sub-discipline of the atmospheric sciences that focuses on weather processes and forecasting as a result of the large-scale atmospheric circulation and the interaction of the oceans with the atmosphere [513]. From a scientific point of view, the study of the dynamic activity of the atmosphere is the focus of *dynamic meteorology*. It is a branch of fluid dynamics and forms the primary scientific basis for weather and climate prediction [410]. Its main areas of study include the development and evolution of transitory meteorological disturbances and the dynamic mechanisms conducive to intra-seasonal and inter-annual climatic variations associated with turbulence, acoustic waves, atmospheric tides, balanced flow, baroclinic instability, Coriolis force, critical layers, hydraulic flow, KH instability, Kelvin waves, potential vorticity, Rossby waves, orographic and thermally forced standing waves, inertial instability, and wave-mean flow interaction, among other phenomena.

The use of the principles of fluid mechanics for studying the dynamics of the atmosphere requires detailed knowledge of its thermodynamic properties and chemical composition. In particular, the atmosphere is composed of dry air which is mainly distributed along the first 25 km from the ground. Dry air is a mixture of gases, whose major constituents are nitrogen, oxygen, and argon. The combined proportions of these gases make up more than 99.9% of the dry air composition by mass and volume [787]. Table 3.1 lists the most abundant constituent molecules of dry air.

The presence of water (H_2O) in the atmosphere plays an important role since it can exist in its three phases (i.e., gas, liquid, and solid), although not necessarily coexisting with each other. This indicates that there are processes that can give rise to phase changes. Phase transitions are capable of causing fluctuations in the atmosphere since any phase change requires the absorption or emission of large amounts of heat. On the other hand, the atmosphere is also composed of pollutants, such as dust, smoke, sulfur dioxide, methane, and nitrogen oxides, that are highly harmful to human health. Many of these pollutants are produced by industrial activities and traffic flow in large cities. Pollutant particles can change up to heights of 25 km due to their interaction with the ultraviolet radiation from the Sun. The gases that are most affected by these interactions are N and O because these molecules break down allowing the formation of ozone (O_3) in the medium atmosphere (i.e., at heights from the ground of approximately 20 to 50 km) and of atoms and ions (charged particles) in the upper layers of the atmosphere (at heights greater than ~ 80 km).

Table 3.1: Composition of dry air. Taken from Saha [787].

Constituent gas	By mass (%)	By volume (%)	Molecular weight
Nitrogen (N_2)	75.51	78.09	28.02
Oxygen (O_2)	23.14	20.95	32.00
Argon (Ar)	1.30	0.93	39.94
Carbon dioxide (CO_2)	0.05	0.03	44.01
Neon (Ne)	1.20×10^{-3}	1.8×10^{-3}	20.18
Helium (He)	8.00×10^{-4}	5.2×10^{-4}	4.00
Krypton (Kr)	2.90×10^{-4}	1.0×10^{-4}	83.70
Hydrogen (H_2)	0.35×10^{-5}	5.0×10^{-5}	2.02
Xenon (X)	3.60×10^{-5}	0.8×10^{-5}	131.3
Ozone (O_3)	0.17×10^{-5}	0.1×10^{-5}	48.0
Radon (Rn)	—	6.0×10^{-18}	222.0

Excluding the exosphere, which is the outermost atmospheric layer (at heights of 700 to 10000 km from the ground), the atmosphere has four primary layers, which from lowest to highest altitudes from the ground are: the *troposphere* (from 0 to 12 km), the *stratosphere* (from 12 to 50 km), the *mesosphere* (from 50 to 80 km), and the *thermosphere* (from 80 to 700 km). The bound between the troposphere and the stratosphere is called the *tropopause*, that between the stratosphere and the mesosphere is called the *stratopause*, and that between the mesosphere and the thermosphere is called the *mesopause*. It is well known that in the troposphere the temperature drops with height at a rate of [951]

$$\frac{dT}{dz} \sim -6.5°C \ km^{-1}, \tag{3.1}$$

where T is the temperature in Celsius and z is height in kilometers, which is also known as the environmental lapse rate. Figure 3.1 displays a picture of the temperature variation with height in the atmosphere.

The troposphere contains about 80% of the total mass. It is relatively well mixed and continually cleansed from aerosols and other pollutants by cloud droplets and ice particles, some of which subsequently fall to the ground as rain or snow. Its negative temperature gradient leads to convective and turbulent mixing that provides the weather. The tropospheric layer closest to the Earth's surface is known as the atmospheric (or planetary) boundary layer and its height is determined by vertical or buoyant transport due to thermal convection. As the

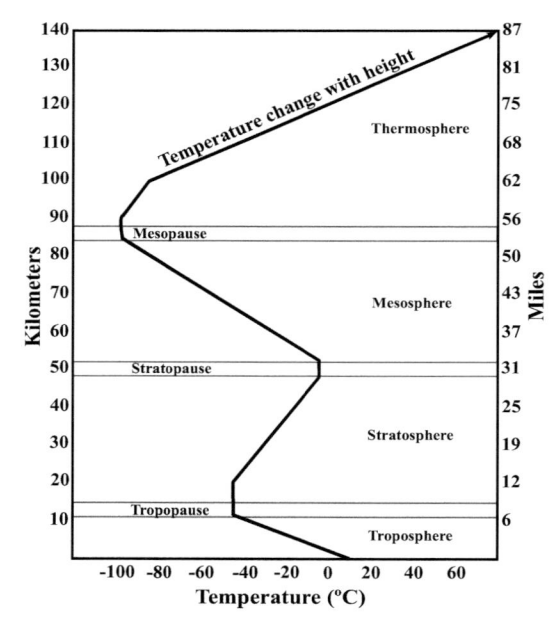

Figure 3.1: Variation of the temperature with height through the atmospheric layers.

Sun heats the surface, the air rises and the volume of the atmospheric boundary layer grows due to the lower pressure at higher altitudes. Its height varies from less than ~ 500 m in winter to about 2000 m in summer [172]. In contrast, the stratosphere is almost completely free of clouds or other forms of weather and is characterized by a positive temperature gradient (see Fig. 3.1). Such temperature profile creates very stable atmospheric conditions with very low water-vapor concentrations. Although the higher layers, i.e., the mesosphere and thermosphere are of meteorological interest, most of the phenomena related to CAT occur within the troposphere. Therefore, fluid dynamic research focused on CAT is mainly limited to this region.

The atmospheric dynamics are governed by the physical laws of mass, momentum, and energy conservation under the action of pressure gradients, gravitational forces, and the effects of viscosity. The interactions occur at different scales from a microscopic scale (i.e., between individual particles) to a macroscopic one where a large number of particles participate. Due to Earth's rotation, non-inertial frames also play an important role in atmospheric behavior. Such frames are the sources of fictitious forces, such as the Coriolis force, which yields a deflecting force on winds and oceans in a way that the resulting fluid patterns are clockwise in the northern hemisphere and counterclockwise in the southern hemisphere. In general, the collective interaction of clouds of molecules gives rise to long-range correlations which are the cause of many atmospheric phenomena, including large-scale circulation. Like all systems in nature, the atmosphere as a whole also tends to maintain its hydrostatic and thermal equilibrium at any moment. For example, fluctuations in the distribution of heat, such as those occurring with a heat source over a warm low-pressure area or a heat sink over a relatively cool high-pressure area, lead to processes, which try to evolve toward equilibrium, give rise to atmospheric circulation as the result of the interaction between different parts of the system. From a mechanical point of view, turbulence is caused by forces, such as the Coriolis force, and friction effects between adjacent fluid layers. The combination of these effects produces exchanges of energy between parts of the system that end up with complex and chaotic dynamics. Most of our understanding of atmospheric dynamics relies on theoretical and experimental models. The former is based on the design of mathematical models for the numerical solution of the governing fluid-dynamics equations (see Chapter 2, where the Navier-Stokes equations are discussed). Realistic models must account for molecular motions that give rise to rapid vibrations on horizontal scales of hundreds of kilometers and vertical scales of the order of the depth of the troposphere and on temporal scales of one or more days.

In response to the gravitational force, the atmospheric pressure, p, is defined as the weight, $w = mg$, of the overlying air column per unit area of the surface at a height h,

$$p(h) = \int_h^\infty \rho g \delta z, \tag{3.2}$$

where δz denotes variation along the vertical coordinate and ρ is the mean air density, which in general is a function of time, height, and temperature. The maximum pressure is achieved on the Earth's surface, where $p \sim 1000$ hPa at the sea level. The pressure decreases with height, reaching values of $\lesssim 100$ hPa in the tropopause, below 1 hPa in the stratopause, and values as low as 1.0×10^{-3} hPa just above the mesopause in the lower thermosphere, and $\lesssim 1.0 \times 10^{-4}$ hPa at heights above 100 km from the ground deep in the thermosphere [951]. The total variation of pressure, δp, on a parcel of air during a time interval, δt, can be written in Cartesian coordinates (x, y, z) as

$$\delta p = \left(\frac{\partial p}{\partial t}\right) \delta t + \left(\frac{\partial p}{\partial x}\right) \delta x + \left(\frac{\partial p}{\partial y}\right) \delta y + \left(\frac{\partial p}{\partial z}\right) \delta z. \tag{3.3}$$

In the limit $\delta t \to 0$, Eq. (3.3) becomes

$$\frac{dp}{dt} = \frac{\partial p}{\partial t} + v_x \frac{\partial p}{\partial x} + v_y \frac{\partial p}{\partial y} + v_z \frac{\partial p}{\partial z}, \tag{3.4}$$

where $\mathbf{v} = (v_x, v_y, v_z)$ is the fluid velocity, while on the right-hand side the first term denotes the barometric tendency at a fixed location, the second and third terms represent the horizontal pressure variations, and the last term accounts for the vertical pressure variation in the atmosphere. The pressure gradient force acts from regions where the pressure is high to regions where it is low. In the atmosphere, the driving force for wind is one of the causes of local pressure fluctuations at different heights. This force, which enters as a source term of the form $-\nabla p/\rho$ in the momentum conservation law, is balanced by the gravitational force, maintaining on average hydrostatic equilibrium. Under the assumption that the atmosphere can be regarded as an ideal gas, the pressure variation with height can be written as

$$\frac{\delta p}{p} = -\frac{g}{\mathcal{R}T} \delta z, \tag{3.5}$$

where $\mathcal{R} = 8.31446261815324$ J K^{-1} mol^{-1} is the universal gas constant. For example, in the troposphere the temperature decreases almost linearly with height so that $T = T_0 - \beta z$, where β is the lapse rate defined by Eq. (3.1) and T_0 is a reference value usually taken to be the ground temperature. Replacing this expression for T into Eq. (3.5) gives

$$\frac{\delta p}{p} = -\frac{g}{\mathcal{R}(T_0 - \beta z)} \delta z. \tag{3.6}$$

Integration of the above equation between $z = 0$ and z yields after some simple algebraic steps the expression

$$p(z) = p_0 \left(1 - \frac{\beta z}{T_0}\right)^{g/(\mathcal{R}\beta)}, \tag{3.7}$$

where p_0 is the pressure at $z = 0$, usually taken to be the ground pressure.

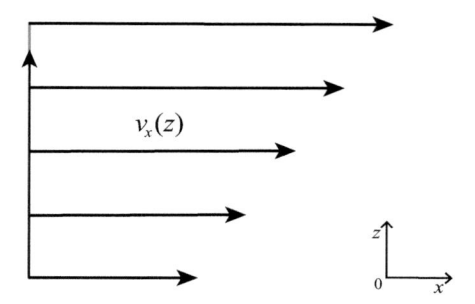

Figure 3.2: Schematic picture showing the relative steady-state motion of adjacent air layers in the troposphere.

Since air in the upper tropospheric layers moves faster than the lower ones, a net transfer of downward horizontal momentum is caused by molecular motion so that both layers experience the same shear stress. As a consequence, the net horizontal force acting between the layers is zero (see Fig. 3.2). Then, the non-conservative friction forces between the layers in relative motion are responsible for the KH instability and ultimately for turbulence. If according to Fig. 3.2, the relative motion takes place along the x-coordinate direction, then the magnitude of the horizontal frictional force is given by

$$F_x = \frac{1}{\rho} \frac{\partial \tau_{zx}}{\partial z} = \frac{1}{\rho} \frac{\partial}{\partial z} \left[\eta \left(\frac{\partial v_x}{\partial z} \right) \right]. \tag{3.8}$$

If the coefficient of shear viscosity, η, is constant, Eq. (3.8) is just the viscous force, $\nu \nabla^2 \mathbf{v}$, appearing on the right-hand side of the Navier-Stokes equation (2.8) for incompressible flows.

It is clear from the above exposition that dynamical meteorology provides the basis for the development of tools for the prediction of weather and climate. The atmosphere stratification coupled to the Earth's rotation are the primary ingredients for the atmospheric movements. Therefore, when the atmosphere is modeled in the framework of fluid theory, this makes a big difference between the fluid dynamics of the atmosphere and that of traditional engineering.

3.4 Aviation meteorology

Aviation meteorology focuses on the study of weather from the perspective of its impact on the aeronautical industry. It aims to contribute to the guarantee of safety standards, economy, and efficiency of flights. It covers many interrelated subjects and engineering topics of importance in the study of geophysical, fluid mechanical, and thermodynamic processes [366, 368]. Among the many topics, it includes ground-based operations that are related to adverse effects on aircraft performance, like frost, ice, snow, and visibility, as well as in-flight icing and turbulence. A detailed understanding of the dynamics of jets, bursts, eddies, high-amplitude gravity waves, wind shear, including low-level wind shear, as well as the physics and thermodynamics of clouds, fog, and precipitation will all contribute to the development of applications for aviation meteorology and forecasting, which will represent relevant information for aviation operations.

There are three main stages associated with aviation meteorology, namely the meteorological conditions that impact directly on aviation operations, the preparations operated in ground-based taking off and landing, and during flight. At the first stage, the main troubles that can occur during ground-based operations for taking off are those coming from the effects of frost, ice, snow, visibility, in-flight icing, and turbulence. As a consequence of these effects, the aircraft can suffer decreased thrust and lift, increased drag and stall speed as well as modified handling characteristics. In contrast, during the flight stage aircraft are subject to atmospheric fluctuations, like wind shear and CAT, which can be present any-where during a flight path [368]. In the landing stage, the aircraft can be subject to conditions of wet snow, freezing drizzle, and rain. Such conditions occur fre-quently at temperatures around $0°C$. However, temperatures between $8°C$ and $14°C$ could also be critical for aviation operations [896]. The types of aviation hazards encountered along with a brief description of their possible impacts in flight are summarized in Ref. [79]. Weather hazards have a significant negative impact on aircraft safety. Statistical records indicate that about 20–30% of world-wide crashes are due to adverse weather conditions and approximately 8% of these crashes are due to thunderstorms [367, 198].

Types of aviation hazards include (1) vertical wind shears, which may af-fect the performance capabilities of modern aircraft; (2) microburst wind shears, which have a direct impact on air terminals, demanding runway changes during take-off and landing; and (3) gravity/shear waves as well as terrain-induced dis-turbances and lee waves. The former may cause deviations from glide slopes, while the latter may be responsible for aircraft structural damage and deviations from the assigned flight altitudes. The same is true when flying in the vicinity of thunderstorms. Icing represents another aviation hazard, which may produce flight disruption accompanied by structural damage, increased drag, and reduced lift and stall angle. In addition, aircraft wake vortices are also highly dangerous because they induce roll moments that can disrupt the flight of the following air-craft. On the other hand, altimeter errors can be the cause of significant deviations from assigned or expected flight altitudes. The most common error is when the altimeter indicates a lower altitude than the aircraft's true altitude. However, it could also happen that when flying in air that is colder than standard temperature, the aircraft will be lower than the altimeter indicates. There are several factors that can impact the accuracy of altimeters. In general, weather changes affecting temperatures and air pressures may cause complications in understanding and using an altimeter.

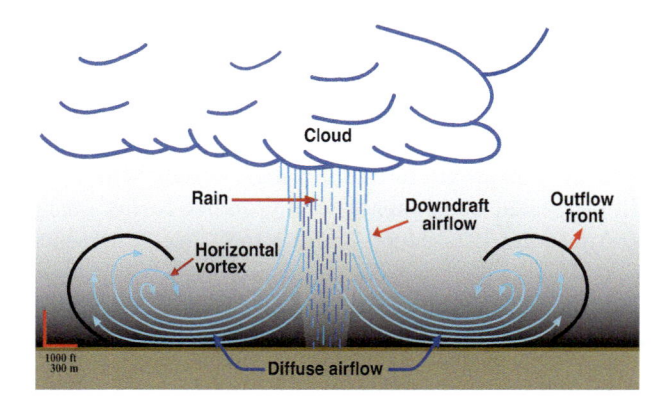

Figure 3.3: Schematic of an axisymmetric microburst airflow.

3.4.1 Atmospheric gravity currents

In fluid dynamics, a gravity (or density) current is the flow of a fluid within another one caused by relatively small differences in density between the two fluids [623]. An important type of atmospheric gravity current is the outflow from a distant thunderstorm downdraft. Since the downdraft region is usually quite large in diameter (typically 10 km, or more), the wind shears near the downdraft are relatively weak. However, the wind speed and the wind direction change that accompany the leading edge can cause significant air speed changes with consequences for aircraft safety. In the atmosphere, thunderstorm outflows and sea-breeze fronts are gravity currents of relatively cold dense air [836]. On the other hand, atmospheric-suspension gravity currents can include avalanches of airborne snow particles as well as fiery avalanches and base surges formed from gases and solids issuing from volcanic eruptions. Accidental or premeditated release and subsequent suspension of industrial pollutants in the atmosphere by human activity can also give rise to buoyancy-driven flows and gravity currents.

3.4.2 Microbursts

A microburst is a localized downdraft region within a thunderstorm of size less than 4 km and duration often less than 10 min. There are two types of microbursts: wet microbursts, which are accompanied by significant precipitation, and dry microbursts, which are produced by high based thunderstorms that generate little to no surface rainfall. In this latter case, the precipitation evaporates aloft within the downdraft. In the vicinity of a microburst, strong winds with velocities > 50 m s^{-1} and rapid wind direction changes of up to $180°$ can occur. After hitting the ground, the air spreads out in all directions producing damaging winds of over 240 km h^{-1}. Winds this high can produce major damage to homes and other ground-based structures. Forecasting for microbursts can be done from 6 to 12 h before convection is expected to develop. Doppler radar can detect microbursts once the radial outflow is established within the mid levels of the thunderstorm. However, they are so short-lived that their detection is often difficult and can sometimes occur between radar scans.

The performance of an aircraft can change rapidly when flying through a microburst. In particular, encountering a microburst during take-off or landing is dangerous if the pilot has reduced power and lowered the nose in response to the headwind. The danger comes from this leaving the aircraft in a low-power and low-nose configuration, which will make it difficult to recover when the wind switches to a tailwind. In response to this, many airports are now fitted with Low-Level Wind Shear Alert Systems (LLWAS) to detect microbursts, which measure wind speed and direction around the airport. When a warning is generated, it is transmitted to the air traffic control, which then relays the information to the pilots.

3.4.3   Vertical wind shear and gravity/shear waves

Vertical changes in wind speed and direction can also represent danger, especially for lower-level flight operations. A ground-based inversion is often accompanied by calm winds near the surface and strong winds just above the cooler and stable near-surface air. Aircrafts descending or ascending through such layers can encounter strong wind and turbulence as well as rapid fluctuations associated with gravity/shear waves. These waves range from tens to hundreds of meters, resulting in aircraft interaction times of seconds or less. In general, vertical wind shear conditions are particularly important for aviation operations if, for instance, the relative air speed is suddenly reduced on a low-level approach to the airport or during take-off. In those airports where wind shear above stable air is a frequent problem, boundary layer wind profilers or acoustic sounders can provide valuable real-time monitoring capabilities.

3.4.4   Icing

The accumulation of ice on aircrafts before take-off has long been recognized as one of the most significant safety hazards affecting aviation [368]. Aircraft icing is the accretion of supercooled water onto an airplane during flight. The formation of ice in flight begins at a microscale level, causing the growth of water droplets as well as their collision and adhesion to the aircraft structure. Icing can increase drag, decrease lift, and cause control problems. The added weight of accreted ice is generally a factor for light planes. For instance, larger droplets are more likely to strike airfoils because they do not easily follow the flow streamlines and pass around an obstacle as do smaller droplets. Icing hazards can be insidious because of the following two factors:

- Only a small amount of ice deposition can have deleterious effects upon lift and drag, thus reducing the aircraft performance.

- Icing and the degradation of performance can increase slowly and imperceptibly until an emergency exists.

The most common causes that are related to air accidents are: the accretion of ice during a hailstorm, which can affect the wing aerodynamics of a plane, and the encounter with big hailstones, which can cause severe physical damage to the airplane body.

3.4.5 Terrain-induced atmospheric turbulence

Terrain-induced turbulence is a class of low-level turbulence, which can be defined as temporal and spatial fluctuations in the atmosphere that are mechanically produced by terrain irregularities, such as mountains and mountain complexes [918]. The flow situations can range from lee waves, bora flows, which are a form of density currents, and rotors to mechanically induced turbulence. At times, organized instabilities can occur in the form of vertical or horizontal axial vortices. Such obstacle-involved situations can be exceptionally complex when the terrain-induced fluctuating flows interact with other meteorological factors, such as lee-side inversions. Numerical weather prediction models are powerful tools for forecasting low-level turbulence and providing timely turbulence warnings to pilots [159]. In many airports, this class of turbulence is monitored using remote-sensing instruments, including Doppler Light Detection and Ranging (LIDAR) systems and the usual wind profilers around the airport area.

3.4.6 Thunderstorms

Thunderstorms are very well-known meteorological phenomena. The Encyclopedia Britannica defines thunderstorms as "violent short-lived disturbances associated with lightning, thunders, dense clouds, heavy rains or hail, and strong gusty winds". They normally arise when layers of warm air meet as swift updrafts in the higher and cooler regions of the troposphere. For instance, thunderclouds induce zones of turbulence due to erratic updrafts and downdrafts, while in summer low-density air pockets can be created causing CAT, which is an unpredictable aviation hazard not connected with thunderstorms.

3.4.7 Aircraft wake vortices

When a body moves through a fluid, a wake is formed behind the body. For Reynolds numbers well above the critical value, the wake is a region of turbulent flow [521]. As any body moving in a fluid, an aircraft in flight will generate wake turbulence in the form of two counter-rotating vortices trailing behind the

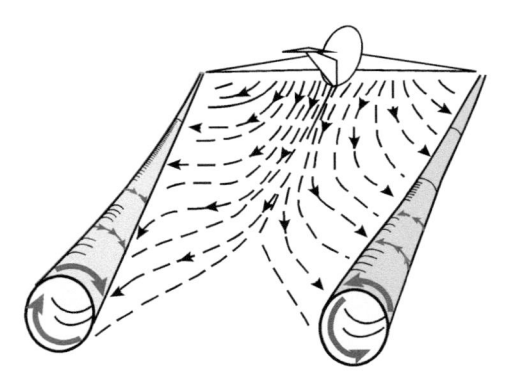

Figure 3.4: Schematic picture showing the wake vortex turbulence formed behind an airplane in flight. The left vortex is in clockwise rotation and the right one is in counterclockwise rotation.

aircraft as shown schematically in Fig. 3.4. An airfoil produces a lifting force acting perpendicularly to the airstream and a dragging force acting in a direction parallel to the airstream. The vortices are formed any time an airfoil is producing lift, which in turn is created by a pressure differential over the wing surfaces [331]. The strength of the vortices depends on the weight, speed, and shape of the wing of the generating aircraft.

Each vortex develops a circulatory motion around a core, whose size will depend on the type of aircraft and can vary from a few centimeters to a few meters in diameter. The core itself is surrounded by a differentially rotating external region of diameters as large as about 30 meters [673]. As a matter of fact, heavy aircrafts produces the strongest vortices when they fly slowly in a clean and stable configuration. Inside the cores of such vortices the air swirls with velocities of up to 100 m s^{-1}. For instance, the greatest hazard from a turbulent wake is the induced roll and yaw, which are particularly dangerous during take-off and landing when there is little altitude for recovery [808]. The intensity of the vortices beyond a distance equivalent to 10 to 15 wingspans from the plane depends on the atmospheric stability, the wind strength and direction, the ground effect when flying at low altitudes, and the mechanical turbulence. The effects of such vortices are considerably more hazardous for small planes flying beneath the path of a larger aircraft, resulting in a high probability of severe rolling motions with complete loss of control. In general, wake vortices decay rapidly within a few minutes. However, they could be more persistent in stable atmospheric conditions under light wind and near the ground. For a much deeper insight into the characteristics and effects of the turbulent wake produced by flying aircraft, the interested reader is referred to Ref. [673], which is a frequently revised booklet on wake turbulence prepared by the *Civil Aviation Authority of New Zealand*.

The characterization of wake vorticity requires knowledge of the following parameters [331]: (a) the time t_0 at which the vortex pair, shed by the aircraft, propagates downward over a distance of one initial vortex spacing; (b) the tangential velocity $v_\theta = (v_y^2 + v_z^2)^{1/2}$; (c) the axial vorticity

$$\omega_x = \frac{\partial v_z}{\partial y} - \frac{\partial v_y}{\partial z}; \qquad (3.9)$$

(d) the circulation

$$\Gamma(r) = \oint v_\theta ds, \qquad (3.10)$$

where r is the distance from the vortex center and ds is a differential length of a small element of a concentric circular path; (e) the core radius, r_c, which defines the distance from the vortex center where v_θ is maximum; (f) the dispersion radius, r_d, which provides a measure of the axial vorticity dispersion in the (y, z)-plane; (g) the vortex separation, b, which describes the local distance between the centers of both vortices; and (h) the vortex Reynolds number [57]

$$\text{Re}_\Gamma = \frac{\Gamma}{v}, \qquad (3.11)$$

where v is the kinematic viscosity. For the analysis of a general aviation take-off accident reported by Schwarz et al. [809], the initial vortex separation is estimated to be

$$b_0' = \frac{\pi}{4} b, \qquad (3.12)$$

where b is the wing span. For steady flight, the strength of the wake vortices is calculated according to

$$\Gamma_0 = \frac{4W \cos(\gamma)}{\pi \rho b v}, \qquad (3.13)$$

where W is the aircraft weight, γ is the flight path angle, ρ is the air density, and v is the air speed. The initial vortex descent speed, v_0, is given by

$$v_0 = \frac{\Gamma_0}{2\pi b_0'}, \qquad (3.14)$$

while the characteristic vortex time corresponding to the time the vortex pair takes to descend one initial vortex separation is defined as

$$t_0 = \frac{b_0'}{v_0}. \qquad (3.15)$$

These simple formulas represent the most important parameters for the characterization of aircraft wake vortices.

3.5   Synoptic meteorology

Synoptic meteorology is the branch of meteorology dealing with the analysis of collective data observations made simultaneously over a wide region for weather forecasting. Disturbances of interest in synoptic meteorology are frontal depressions, tropical cyclones, and anticyclones, which have sizes of many hundreds to a few thousand kilometers in the horizontal directions and lifetimes of days [757]. The word synoptic means *view together* or *view at a common point*. Therefore, in general, the focus of synoptic meteorology is to observe the weather at a specific region at different times [942].

Atmospheric processes involve a wide range of spatial and temporal scales. The shorter spatial scales correspond to the sizes of individual particles, which can display random and chaotic motion leading to tiny eddies which can then evolve to large turbulent motions. Such small eddies represent the so-called microscale atmospheric motion and are characterized by sizes of a few meters or less. In increasing order of scales, the mesoscale represents the middle scale, where typical circulations range from a few to a hundred kilometers in diameter. These include local winds, thunderstorms, tornadoes, and small tropical cyclones. The larger scales deal with circulations around high- and low-pressure areas leading to the formation of cyclones and anticyclones at middle latitudes and large tropical cyclones at lower latitudes. This last scale concerns the synoptic scale or weather-map scale. Typically such formations span areas of hundreds to even thousands of square kilometers and have lifetimes of days and sometimes weeks. Hurricanes and typhoons fall into this category as do the mid-latitude storm systems that bring rain, snow, and wind [19]. Table 3.2 lists the horizontal scales of motion in the troposphere [951]. However, there is a disagreement between the *American Meteorological Society* (AMS) and the *World Meteorological Organization* (WMO) regarding the sizes of these scales. For example, the synoptic scale is taken to be in the range 400–4000 km (for the AMS) and 1000–2500 km (for the WMO), while the mesoscale ranges from 3 to 400 km for the AMS and 3 to 50 km for the WMO and the size of the microscale is between 0 and 2 km for the AMS and 3 cm to 3 km for the WMO.

Table 3.2: Horizontal scales of tropospheric motion. Table adapted from Wallace and Hobb [951].

Size larger than	Scale designation	Scale name
20000 km	macro α	planetary scale
2000 km	macro β	synoptic scale
200 km	meso α	mesoscale
20 km	meso β	mesoscale
2 km	meso γ	mesoscale
200 m	micro α	boundary-layer turbulence
20 m	micro β	surface-layer turbulence
2 m	micro γ	inertial subrange turbulence
2 mm	micro δ	fine-scale turbulence
0.3 μm	viscous	dissipation range
0.003 μm	molecular	mean-free path between molecules
0 μm	molecular	molecule sizes

In general, the atmospheric processes are strongly dependent on the geographic location, the effects produced by the rotation and translation of the Earth, the Sun radiation, the topographic conditions of the Earth's surface, and the orographic characteristics of each region. In addition, the processes that occur over the oceans are very different from those that take place over the continental surfaces due to terrain irregularities and the presence of headlands and mountains. Valuable information for the analysis of meteorological conditions in wide areas is provided by the geographical weather (or synoptic) maps, which display observational meteorological data. In particular, such maps are constructed by drawing a set of isolines, which on surface charts are curves of constant atmospheric pressure (i.e., isobars) described by the equation $p = p(x, y)$, which connect points of equal barometric pressure normalized to the sea level[2], while plotted on upper air charts are curves connecting points of equal geopotential height (i.e., isohypses). Figure 3.5 displays an example of an upper level chart of the pressure surface p = const. [942]. A geopotential height can be thought of as the distance above sea level. For instance, isohypse lines or contours depend on the average air temperature and the average moisture[3] content of the air underneath the pressure level of interest. As shown in Fig. 3.5 closed isobars in a synoptic map will show zones of high pressure (marked with an "H"), zones of low pressure (marked with an "L"), and fronts, where the latter are the leading edges of current weather systems. A front[4] is a transition zone or boundary between air masses of different pressure and temperature. A cold front refers to cold air advancing into warm air, while a warm front is encountered when warm air is advancing into cold air. In general, high pressure systems indicate fair weather and little precipitation, while low pressure systems are indicative of low temperatures accompanied by a cloudy sky and precipitation. However, according to Schultz and Blumen [802], the above definition is not complete because fronts do not always form along the edge of air masses. In some cases, they form within air masses and therefore their boundaries could not be so clear. On the other hand, not all fronts are as large as the synoptic or horizontal scale. Indeed, some of them can have mesoscale dimensions in the along-front direction and perhaps of microscale dimensions (i.e., from ~ 2 to 20 km) in the across-front direction [478].

In 1936, Petterssen [701] introduced the term *frontogenesis* and defined it as the Lagrangian rate of change of the magnitude of the horizontal potential temperature gradient due to the horizontal wind velocity ($\mathbf{v}_h = v_x \mathbf{i} + v_y \mathbf{j}$, where $\mathbf{i}$ and $\mathbf{j}$ are the unit vectors along the x- and y-directions of a rectangular coordinate system)

$$F = \frac{d}{dt} |\nabla_h \theta|, \tag{3.16}$$

---

[2] Similarly, isobaric surfaces are surfaces of constant pressure and are described by $p = p(x, y, z)$.

[3] Indicators of tropospheric moisture are the relative humidity, the dewpoint, and the precipitable water.

[4] The definition of fronts was first introduced by Norwegian meteorologists during the 1910s and 1920s when studying the structure and evolution of extratropical cyclones [92, 93].

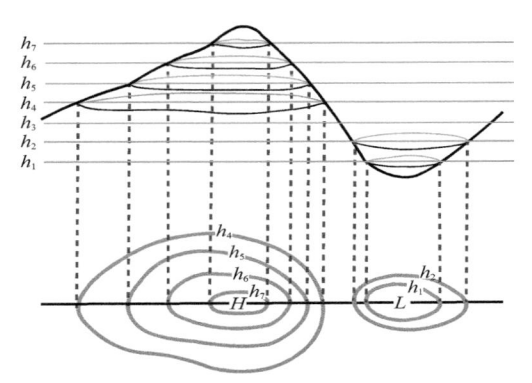

Figure 3.5: Schematic example of an upper level chart. Figure adapted from Vorobyev and Tarakanov [942].

where

$$\frac{d}{dt} = \frac{\partial}{\partial t} + v_x \frac{\partial}{\partial x} + v_y \frac{\partial}{\partial y}, \qquad (3.17)$$

and

$$\nabla_h = \mathbf{i}\frac{\partial}{\partial x} + \mathbf{j}\frac{\partial}{\partial y}. \qquad (3.18)$$

This mathematical definition of the front emphasizes that both the temperature and wind fields are responsible for producing frontogenesis. From this point of view, a front can be better defined as a region characterized by frontogenesis and containing both temperature gradient and vorticity maximum [802]. Therefore, frontogenesis by itself can be defined as the generation or intensification of a front. Such intensification occurs when warm air converges onto colder air and the horizontal temperature gradient amplifies by at least an order of magnitude. The intensity of a front is measured in terms of fluctuations in the temperature and wind across the transition zone. The strongest front intensities are obtained near the ground and around the tropopause. The buoyancy force[5], the pressure gradients, the Coriolis force, and frictional forces are all factors that determine the control of the physical processes that give rise to fronts. For example, Newton's second law for the vertical motion component reads

$$\frac{dv_z}{dt} = -\frac{1}{\rho}\frac{\partial p}{\partial z} - g + C_z + F_z, \qquad (3.19)$$

where C_z and F_z are the corresponding vertical contributions of the Coriolis and friction forces, respectively, and v_z is the vertical velocity component

$$v_z = \frac{dp}{dt}, \qquad (3.20)$$

---

[5]The buoyancy force is the result of a small imbalance between gravity and the upward vertical pressure gradient force. Hydrostatic equilibrium is restored when the vector sum of both forces equals zero.

given by the time rate of change of pressure experienced by air parcels as they move along their three-dimensional trajectories through the atmosphere. For a deeper and more rigorous scale analysis of fronts, the interested reader is referred to Holton's [409] book.

Within the synoptic scale, the fronts are classified as *cold fronts*, *warm fronts*, *stationary fronts*, and *occluded fronts*. A cold front can be identified with a strip of instability that occurs when a cold air mass approaches a warm air mass. Since the cold air is denser, it creates a wedge and gets underneath the warmer and less dense air. The passage of a cold front is always accompanied by a temperature drop as the cold air advances and replaces the warmer one [789, 801, 803]. In addition to a temperature drop, conceptual models typically consider traditional cold frontal passages to be also accompanied by a cyclonic wind change, a pressure minimum, and a decrease in dew point temperature, all coinciding with a line of deep convective clouds. In contrast, a warm front forms when warm air meets and replaces a cold air mass. In a modern perspective, the warm fronts indicate that the flow over the front cannot be characterized by a smooth ascent. Instead, the flow may consist of convective elements embedded within the ascent [657] or may be characterized by banded convective precipitation [669]. A stationary front arises when the boundary between two air masses, neither of which is strong enough to replace the other, is at rest, or is moving very slowly. This type of front may reflect either a change in the synoptic-scale circulation pattern that halts the translation of the frontal zone or the heterogeneity of the Earth's surface that provides the necessary conditions to fix a frontal transition zone to a preferred location [802]. On the other hand, an occluded front occurs when a slower moving hot front is followed by a faster moving cold front. The wedge-shaped cold front catches up with the hot one and pushes it forward so that both fronts continue to move one behind the other. The line between them is what forms the occluded front. Nevertheless, this definition is not completely adequate because it leaves out important characteristics. For example, much of the length of some occluded fronts are not shaped only by the catching up of one front with the other, but also by the deformation due to the differential rotation around the cyclone. Many cyclones continue to deepen after the formation of the occluded front and a few well-documented examples of cold-type occlusions have been reported in the literature; yet this is not explained by the traditional model [802]. A more modern interpretation that ascribes occluded fronts to the wrap up of the cold and warm fronts around an extratropical cyclone has been reported by Schultz and Vaughan [805]. Just to mention, three other categories of fronts can be found within the mesoscale, namely the sea-breeze fronts, the gust fronts, and the drainage fronts.

Air masses are boundary layers that form over surfaces of several million square kilometers and extend up through the troposphere. Temperature differ-

Table 3.3: Classification of air masses by geographical type.

Type	Arctic	Polar	Tropical
Continental	Extremely cold, dry, stable; ice- and snow- covered surface	Cold, dry, stable	Hot, dry, stable air aloft; unstable air surface
Maritime		Cool, moist, unstable	Warm, moist; usually unstable

ences between neighboring air masses cause baroclinicity[6] [597]. In particular, atmospheric zones of high baroclinicity are characterized by the frequent formation of extratropical cyclones [416]. Such zones correspond to expanses of air with distinctive properties that derive from their residence over a specific *source region*[7] and are still recognizable for some time after the air has moved into a different geographical setting. For example, air that has stayed over a continent at high altitudes during winter tends to be cold and dry. The main source regions are the high-pressure belts in the subtropics, which give rise to tropical air masses, and around the poles, where polar air masses originate. These two types of air masses can have a maritime or continental track after leaving their source regions. A summary of the classification of air masses by geographical type is displayed in Table 3.3.

3.6 Aircraft pollution

In addition to the risks faced by aviation due to meteorological factors, the aviation industry also contributes to pollution levels and climate change due to the rapid growth in demand for air transport. Since 1960 the growth rate has been almost 9% annually. This increase in air transport has generated an increase in fuel consumption of 13%. Such a constant growth rate has raised concerns about possible environmental problems with important impacts on future aircraft operations. Unlike land transport, more than 85% of the aircraft emissions are released above the atmospheric boundary layer and about 70% is released between the upper troposphere and the lower stratosphere. Consequently, the main polluting effects of airplanes in flight are expected to occur around the tropopause.

The fuel burned by jets is made up of fossil hydrocarbons. In fact, aircraft engines emit mainly carbon dioxide (CO_2) and water with minor contributions from nitrogen oxides (NO_x), sulfur oxides (SO_x), unburned hydrocarbons (HC),

---

[6]A baroclinic fluid is a stratified one for which the surfaces of constant pressure and density do not coincide. Therefore, baroclinicity can be defined by the condition $\nabla p \times \nabla \rho \neq 0$. In contrast, a fluid for which $\nabla p \times \nabla \rho = 0$ is called barotropic because in this case the isobaric surfaces are also isodensity surfaces [341].

[7]Those parts of the Earth's surface where the air stagnates and gradually attains the properties of the underlying surface are called source regions.

and soot. All these species of exhaust gases are pollutants. For example, CO_2 and H_2O are greenhouse gases that directly affect the Earth's climate, while NO_x and HC are reactive gases that affect the levels of atmospheric ozone and methane. In addition, SO_x, HC, and H_2O are also precursors to aerosols and clouds that affect ozone and climate. Although the gases emitted by aircraft are released in geographically narrow flight paths and corridors, their penetration into the zone comprised by the upper troposphere and the lower stratosphere indicates that their polluting effects may be felt on a regional and perhaps global scale. The dynamics of this region is different from that in a boundary layer since in the former case there is less vertical mixing and less diurnal variation in the wind direction. This implies that pollutants emitted around the tropopause can stay there for longer times and spread over considerable longitudinal and, in some cases, latitudinal distances. On the other hand, since the residence times are long, some pollutants, such as NO_x, remain circulating subject to catalytic chemical reactions that create or destroy the ozone with a consequent impact on weather and climate change.

3.7 Atmospheric gravity waves

Gravity waves are usually observed on the free surface of liquids in equilibrium in the Earth's gravitational field when some external action perturbs the surface. As a result of the perturbation, the surface leaves its equilibrium position and motion will occur in the liquid, which will then propagate over the whole surface in the form of waves [521]. These wavy motions of the surface are called *gravity waves* because they are due to the action of the gravitational field in an attempt to restore equilibrium[8]. The source of the perturbation can be instantaneous and located in a point of space, as occurs for an explosion, or it can be continuous in time over an extended space, as occurs for flow over a mountain range. In any case, the energy associated with the perturbation moves away from the source. In the stably stratified atmosphere, this energy is moved away by gravity waves.

The main two sources of gravity waves in the troposphere are ground topography and convective processes. For example, when a stable stratified airstream encounters a topographic barrier, a variety of perturbations can be produced, including gravity waves (when the barrier is a mountain), downslope winds, and

---

[8]The theory of gravity waves has been reviewed by Landau and Lifshitz [521]. In the limit when the amplitude of the oscillations is small compared with the wavelength, they obtained the dispersion relation $\omega^2 = kg$, relating the frequency and the wavenumber, $k = 2\pi/\lambda$, of a gravity wave, where λ is the wavelength, and derived the expressions for the paths of fluid particles in the wave, the motion diminishing exponentially with increasing depth in the liquid. In the opposite limit of long waves, wave motion is governed by the wave equation

$$\frac{\partial^2 \zeta}{\partial t^2} - gh_0 \left(\frac{\partial^2 \zeta}{\partial x^2} + \frac{\partial^2 \zeta}{\partial y^2} \right) = 0,$$

where $\zeta = \zeta(x, y, t)$ is the vertical displacement of the surface and h_0 is the equilibrium depth of the liquid.

low-level blocking [618]. Mountain waves exert a decelerating force on the large-scale flow that is parameterized in large-scale weather and climate models [486]. However, the precise impact of this forcing is not yet well understood. As mountain waves steepen and break, they generate CAT, posing serious hazards to aviation [53]. On the other hand, maximum gusts in downslope winds have been observed to exceed speeds of 56 m s^{-1} [118], posing a threat to communities in the immediate lee of steep mountain barriers. There is also evidence that low-level blocking on the upstream side of a mountain may exert a major influence on orographic precipitation and pollutant transport [422].

It is well known that atmospheric gravity waves may arise when the airflow is forced to move upward by a tall mountain range or other obstacles such as ridges, hills, and depressions in the terrain such as canyons, basins, and valleys. They can also originate from the airflow in the updraft from a powerful thunderstorm. The deflected air will then move from a denser atmospheric layer to a less dense one. The heavier masses of air that went up will then fall back down due to the Earth's gravitational force, resulting in periodic oscillations capable of transporting energy and momentum over large distances, i.e., from the troposphere to the stratosphere and mesosphere. Such waves have sizes that vary from a few hundred meters to a few hundred kilometers and therefore they are considered to be the smallest atmospheric waves. However, their cumulative effects can be hazardous and affect forecasting and predictions on many timescales. The amount of energy that these waves transport through the atmosphere represents an essential factor in the production of CAT.

Convective processes can also lead to gravity waves either through buoyancy driven by latent heat or by airflow over clouds in a sheared environment. The scales and frequencies of gravity waves that arise this way are determined by the geometry and timescale of the convective motions. For example, low-frequency gravity waves with small vertical scales are mainly excited by shallow and broad convection, whereas high-frequency gravity waves with larger vertical scales will require deep and narrow convection to be excited. The deeper and faster convective gravity waves have sufficiently large horizontal phase speeds to penetrate the mesosphere and thermosphere. On the other hand, gravity waves in the atmosphere can also be excited by jet streams that can appear at great altitudes through a process known as a spontaneous imbalance. In this process, a non-equilibrium flow can attain an equilibrium state by conserving energy through the emission of a gravity wave. These kinds of non-equilibrium flows are broad and relatively shallow, approximately of the order of 10 km deep and hundreds of kilometers across the jet stream, thereby producing gravity waves with large horizontal scales, small vertical group speeds, and near-inertial frequencies. Such waves are important because they comprise the most energetic part of the gravity wave frequency spectrum in the stratosphere and mesosphere and also because they can propagate over thousands of kilometers horizontally before reaching high altitudes, contributing to the global circulation.

It is worth mentioning that although weather researchers knew about gravity waves since the 19th century, they remained obscure until 1960 when Hines [403] introduced a widely accepted theory on how gravity waves work and how they can influence other events in the atmosphere. However, they were not taken into account in global weather forecasting models before 1980 [408]. However, early models were lacking sufficient spatial resolution to solve for the small-scale oscillations of tiny atmospheric gravity waves. This difficulty encouraged researchers to simulate the effects of gravity waves using parameterizations [684, 609], which are still being used today for modeling the effects of atmospheric gravity waves.

3.8 Wind shear

Wind shear is a change in wind speed and/or direction over a short distance. Such changes can occur either horizontally, vertically, or in both directions and are most often associated with strong temperature inversions or density gradients, which result in a tearing or shearing action. The effects of wind shear in a stratified medium have been amply discussed in Chapter 1 in connection to the KH instability as a precursor of CAT. Although wind shear in the atmosphere can occur at any altitude, it is particularly hazardous for aviation when it appears over a short period and below 2000 ft[9] from the ground. It can also be highly dangerous during take-off and landing [301].

In general, vertical wind shear consists of wind speed variations along the vertical axis of about 20 to 30 knots[10] per 1000 ft. In a flight, these changes can drastically alter the aircraft lift, the indicated air speed, and the thrust requirements when climbing or descending through wind shear layers [216]. A horizontal wind shear is defined as a change of horizontal wind direction and/or speed with the horizontal distance decreasing headwind and increasing tailwind, or a shift from a headwind to a tailwind of up to 100 knots per nautical mile [216]. These weather phenomena can occur at different atmospheric levels and can be associated with frontal sources, temperature inversions, jet streams (see next section), thunderstorms, microbursts, and mountain waves occurring sufficiently close to the ground, which is particularly hazardous for aircraft departing from or arriving at an airport.

The impact of wind shear on flight safety has long been a subject of enormous interest for the aeronautic industry [374]. Microbursts have been identified as a potential source of wind shear. Numerical simulations have recently started to appear to investigate the implications of wind shear following a microburst upon aircraft performance [479]. These simulations are a valuable source of information for aircraft designers, pilots, and aviation control authorities. How-

---

[9]One feet, abbreviated as ft, is equivalent to about 0.3048 m.

[10]A knot is a unit of speed equivalent to one nautical mile per hour, which is equal to 1.852 km h^{-1}.

ever, as was already commented in Chapter 2, DNS-based simulations of true wind shear are still beyond the present computational capabilities. On the other hand, many efforts have been addressed to the design and development of efficient aircraft instruments to detect weather effects. Also, recognition of the dynamical wind behavior has been aided by improving the wind shear forecasting, the training in wind shear recognition, and the widespread use of ground and airborne wind shear warning systems. Among the applied technologies are the *Terminal Doppler Weather Radar* (TDWR) and the *Low Level Wind Shear Alert System* (LLWAS). For example, a TDWR system detects and reports hazardous weather in and around the airport terminal. It identifies and warns air traffic controllers (ATCs) of low altitude wind shear hazards caused by microbursts and gust fronts, reports on precipitation intensities, and provides advanced warning of wind shifts. Moreover, LLWAS is employed to detect wind shear and associated weather phenomena, such as microbursts, close to an airport, especially along the runway corridors. The information is passed in real time to warn pilots and aerodrome services. In particular, an LLWAS consists of several anemometers strategically placed around and within an aerodrome. Onboard equipment, such as the *Ground Proximity Warning System* and the *Airborne Wind Shear Warning Systems* are also available in the aircraft. All these technologies and instruments are essential to provide timely warnings and help pilots respond appropriately.

3.9   Jet streams and streaks

Jet streams are intense and narrow quasi-horizontal currents of wind associated with strong vertical shear[11]. These currents are located near the tropopause and blow from west to east all across the globe following a wavelike path. They are also referred to as upper-level wind flows and are usually faster in winter when the temperature differences between tropical and polar air are greater. Jet streams are characterized by wind speeds in excess of 30 m s^{-1} and in some cases, they can achieve speeds as high as about 100 m s^{-1}. Their typical lengths are half to an order of magnitude greater than their widths. It is well known that strong upper-level flows are often associated with strong vertical wind shear and at mid-latitudes, they are accompanied by strong horizontal temperature gradients. Therefore, jet streams tend to be stronger around the tropopause where the horizontal temperature gradient reverses.

Jet streaks are known to be localized regions of wind speed maxima embedded within a jet stream. Both jet streams and streaks are important in synoptic and mesoscale meteorology because of their implications for cyclone and precipitation development. For example, a jet streak can turn a beautiful day into a severe storm the day after [371]. A strong jet streak has winds over 100 knots.

---

[11]The MWO defines a jet stream as a flat tubular, quasi-horizontal air current, whose axis is along a line of maximum speed and which is characterized not only by high speeds but also by strong transverse speed gradients.

The upper-level tropospheric divergence is proportional to both the wavelength of troughs and ridges[12] as well as to wind shear [371]. If a jet streak happens on the left of a trough and winds are stronger to its left, the trough will amplify in the course of time and will dig in a southerly direction. Conversely, if a jet streak happens on the right of a trough and winds are correspondingly stronger to its right, then the trough deamplifies with time and lifts out in a northeasterly direction. However, if the winds are roughly the same on each side of a trough, it will stay the same [371].

3.10   Lee waves and mountain waves

As stated above, when the wind blows over hills or mountains, invisible large waves form in the atmosphere, which then propagate away as gravity (or buoyancy) waves [812, 847]. Such gravity waves are more commonly called *lee waves* or simply *mountain waves*. They move in the wind direction and frequently their presence is revealed by clouds with very characteristic features, such as smooth lenticular clouds[13] or ragged rotor clouds[14]. Large-amplitude mountain waves can generate regions of CAT and produce very strong winds that blow down the lee slope of ridge-like topographic barriers [696, 189]. Extensive reviews and textbooks have been written on lee-wave theory [848, 258, 559].

When the air in a stratified medium, like the atmosphere, flows over a hill or mountain, it first descends and then enters a series of evenly spaced ripples or lee waves. Figure 3.6 shows the flow streamlines for airflow over a mountain [898]. As shown in this figure, the rotors, which are vortices forming past the mountain near the ground, are an important phenomenon associated with trapped lee waves [943, 900, 899]. The time required to complete a full wave oscillation is determined by the degree of stratification of the air. In highly stratified and stable air, whose density decreases with altitude, the time for a full oscillation can be about 20 s, while, if the stratification is light, a full oscillation can take up to 5 min [813]. The dynamics of air flow in the troposphere is a complex one because most natural obstacles are not regularly shaped. When the air flow is perturbed by a mountain barrier, it results in a pair of counter-rotating eddies immediately downstream, which are commonly referred to as *lee vortices*. These

---

[12]Troughs and ridges are curves in the meandering shape of a jet stream. Troughs are analogous to low pressure and ridges to high pressure. In the northern hemisphere, a trough is a southward dip in the jet stream, while a ridge is a northward hump in the wind current.

[13]Lenticular clouds are lens-shaped orographic wave clouds that form mostly in the troposphere when stable moist air flows over a mountain or a range of mountains. In general, lenticular clouds form when the temperature at the crest of the wave drops to the dew point, which is the temperature at which air has to be cooled to become saturated with water vapor. In the same way as they are formed they can also disappear as moist air moves down into the trough of the wave and evaporates.

[14]These clouds are associated with severe turbulence and rather strong vertical motion. They look very irregular. They are a form of lee eddy since the air within them rotates about an axis parallel to the mountain range.

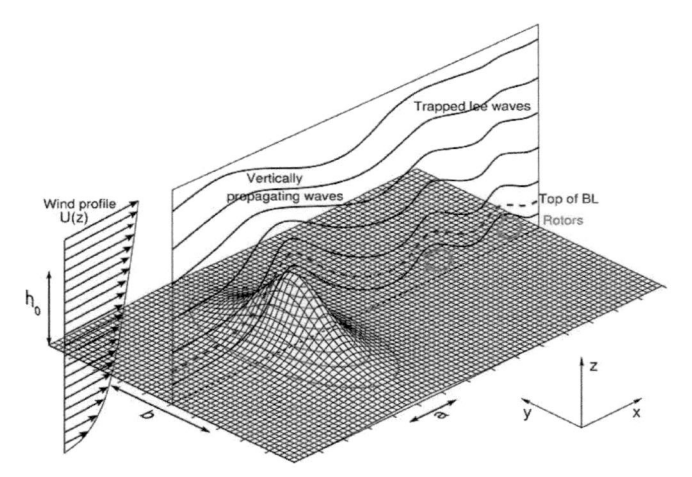

Figure 3.6: Development of mountain waves caused by flow over a mountain of height h_0 and half-widths a and b along the x- and y-directions, respectively. The flow is left to right along the positive x-axis. The flow streamlines (thick black lines), the boundary layer (dashed blue line), and the rotors (circular green arrows) are shown. Figure taken from Teixeira [898].

vortices evolve on time scales shorter than a day and have a typical length scales of the order of 10–100 km.

Many factors go into shaping the structure of lee waves, namely the size and shape of the mountain, the temperature gradient, the wind speed, and the humidity in the colliding stream. Early models of lee waves were calculated under the assumption that the vertical size of the mountain is small compared to the vertical wavelength so that linear analysis was justified [734, 812]. Linear theory has been continuously applied over the years to study mountain waves and is still a topic of interest (see, for example, Metz and Durran [619] and references therein). In most real-world mountain wave events, the cross-mountain wind speed has significant vertical variations. In linear theory, the vertical velocity, $v_z = v_z(x, z)$, is modeled by the two-dimensional, steady-state Boussinesq wave equation in the presence of vertical wind shear [259, 619]

$$\left(\frac{\partial^2}{\partial x^2} + \frac{\partial^2}{\partial z^2} \right) v_z + l^2 v_z = 0, \tag{3.21}$$

where x is the horizontal coordinate perpendicular to the ridge-line, z is the vertical coordinate, and l is the Scorer parameter given by

$$l^2 = \frac{N_{\mathrm{BV}}^2}{U^2} - \frac{1}{U}\frac{d^2 U}{dz^2}. \tag{3.22}$$

Here $N_{\mathrm{BV}} = N_{\mathrm{BV}}(z)$ is the Brunt-Väisälä frequency (or alternatively, the buoyancy frequency) and $U = U(z)$ is the cross-mountain wave speed. In particular, the Scorer parameter is an indicator of whether or not lee waves will develop.

The first term on the right-hand side of Eq. (3.22) is usually dominant, but occasionally the second term, which is the velocity-profile curvature term, can be of comparable magnitude. When l^2 is nearly constant with height, the conditions become favorable for the vertical propagation of mountain waves. If, on the contrary, l^2 decreases steeply with height, this will be an indicator that trapped lee waves are expected. This is especially true when the decrease with height suddenly occurs in the mid-troposphere, thereby dividing the troposphere into two regions: a lower layer of large l^2 (high stability) and an upper layer of small l^2 (low stability). The square root of l^2 has units of wavenumber. In fact, the wavenumber of the resonant lee wave lies between the l-value of the upper layer and the l-value of the lower layer, the equivalent wavelength generally lying between 5 and 15 km. Mountain ranges that are wide enough to force long wavelengths relative to the l-value in the upper layer produce vertically propagating waves with wavenumbers greater than the l-value in the upper layer. Small obstacles that force wavenumbers greater than the l-value in the lower layer produce waves that are evanescent with height.

Although there is fair evidence that linear theory provides a reasonably good approximation to the dynamics governing gravity wave propagation through much of the atmosphere [619], the linear approximation is not enough to model real systems, since it does not take into account the effects of the irregular shape of mountains, the temperature gradients, and the air density. Finite-amplitude effects remain small until wave breaking occurs. Therefore, below the wave-breaking threshold a fully nonlinear approach, yielding finite-amplitude solutions, would be needed to properly model complex real-world events [619]. However, the precise role of nonlinearity in setting the amplitude of mountain waves is not fully understood yet. Present day motivations resulting from errors of numerical weather prediction and climate models in the representation of subgrid-scale orography, have led to a growing effort to study the interaction between boundary layers and mountain waves [911, 525, 572, 573].

3.11 Temperature gradients

The radiation coming from the Sun has direct consequences on the temperature behavior across the atmosphere. The shape of the temperature distribution with height in the atmosphere is shown in Fig. 3.1. In the troposphere, the temperature decreases with height at a rate as given by Eq. (3.1). For most of the mesosphere, the temperature also drops with altitude, reaching values as low as $-80°C$ around the mesopause. In contrast, a thermal inversion occurs in the stratosphere and thermosphere, where the temperature is now seen to increase with altitude. Tropospheric thermal inversions may also occur at times when after a prolonged period of good stable weather the air stratifies according to its density, i.e., with the cooler and heavier air becoming warmer at higher altitudes from the ground. This phenomenon occurs frequently in winter, giving rise to persistent fog. Ther-

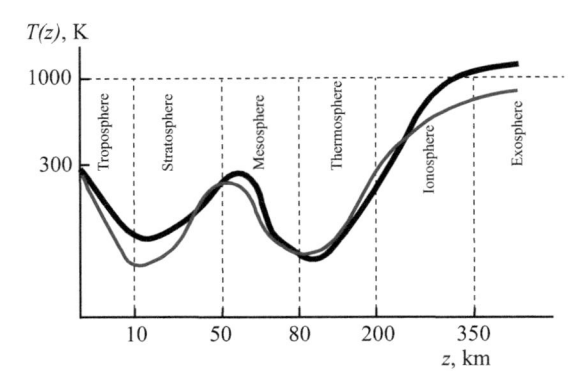

Figure 3.7: Comparison of the meteorological temperature distribution measurements (thin curve) and Gladkov's [348] theoretical dependence (thick curve). Figure taken from Gladkov [348].

mal inversion can also occur in some other situations as, for example, in cold front perturbations. In large cities, thermal inversion in winter can prevent the dispersion of pollutants, giving rise to smog[15].

It has long been recognized that the temperature gradient plays an important role in weather, climate change, and greenhouse gas among other atmospheric phenomena. More than a century ago, Humphreys [423] studied the vertical temperature gradient in a region of the upper inversion using balloons equipped with a suitable registering apparatus. It is known today that the air temperature decreases by about $0.6°C$ every 100 m up in the vertical direction. This value can be considered on average to be the normal thermal gradient in the lower atmospheric strata, apart from local fluctuations. The temperature gradient is responsible for differences in the air pressure, which in turn leads to the production of winds as the atmosphere tries to restore equilibrium. In general, larger temperature gradients lead to stronger winds.

The stationary atmospheric temperature distribution as a function of height from the ground has been investigated experimentally and theoretically. A thermodynamic model to study the stability of the atmosphere is described in Tsonis [912]. An analytical model based on the assumption of the interaction between the convective flow and solar radiation concerning coordinate dependence of the atmospheric density was proposed by Gladkov [348]. This model was able to explain the meteorological measurements of the stationary temperature distribution. Figure 3.7 shows how Gladkov's [348] theoretical solution compares with the meteorologically observed diagram.

Winds are usually stronger in winter, especially along fronts between highly contrasting air masses. There is a stronger temperature gradient between the

---

[15]The word "smog" is the result of combining the words "smoke" and "fog", because it is actually a mixture of water droplets and solid particles made up of dusts and residual combustion products.

poles and the equator during the winter months because the poles receive minimal sunlight during this period of the year, while the tropics receive the same amount of solar energy year-round. This causes the jet stream to dip farther south and the winds to blow stronger on either side of the jet stream in winter. The jet stream separates the very cold air mass at higher latitudes from the warmer air mass around the equator. Temperature gradients between water and land can also cause local atmospheric circulations which affect winds. During a typical sunny summer day when the land heats up more quickly than water, heat-related low pressure causes rising air over land which moves over the water and cools, then returns to land as a cooling "sea breeze". At night, the water is often warmer than the land and the reverse circulation, which includes a breeze from land to the sea called a "land breeze", takes place.

3.12 Vortex rolls

It is well known that in the limit of zero viscosity, a free shear layer evolves into a vortex sheet. A distinctive feature of this motion is *roll-up*, that is, the fluid spiraling in the shape of an evolving vortex sheet [507]. The importance of vortex sheet motion in fluid mechanics is reflected in the comment made by Krasny [507]: "vortex sheet motion belongs to the larger field of vortex dynamics, one of the main approaches to understanding fluid turbulence". The dynamics of vortex sheet roll-up has been recently revisited by DeVoria and Mohseni [235]. In particular, they found that in a finite vortex sheet with an elliptical circulation distribution, the self-induced velocity is most relevant in those regions where both the curvature and the sheet strength are large. They also predicted for the KH instability of an infinite vortex sheet the critical time at which the sheet forms a singularity in curvature. After the critical time, the formed finite-valued cusp rapidly increases, which is just the impetus that initiates the rolling-up.

The topic of vortex rolls has already been touched in Chapter 1 in connection with the KH billows and then later in Section 3.3.7 where aircraft wake vortices are discussed. In particular, atmospheric vortex rolls, also known as *horizontal convective rolls* or *cloud streets*, are horizontal long rolls or eddies of counter-rotating air that are oriented approximately parallel to the ground in the planetary boundary layer. A representation of horizontal convective rolls and the formation of cloud streets are drawn in Fig. 3.8. These atmospheric formations deserved attention due to their hazardous effects on airplanes in flight. When an aircraft encounters such vortices, it undergoes perturbations in the acceleration as a consequence of the changing wind direction generated in the rolls. A schematic of this situation is illustrated in Fig. 3.9, which shows a linear flight trajectory entering a vortex rolls formation.

Some important features can be noticed from Fig. 3.9. First, not all the vortices interact with the trajectory, but only some of them do so effectively. This interaction will provide acceleration to the airplane by propelling it in an os-

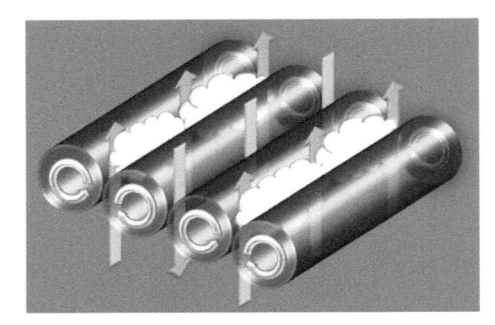

Figure 3.8: Drawing displaying atmospheric horizontal convective rolls and the production of cloud streets. Source: Daniel Tyndall from the Department of Meteorology of the University of Utah. URL: https://commons.wikimedia.org/wiki/File:Convrolls.PNG.

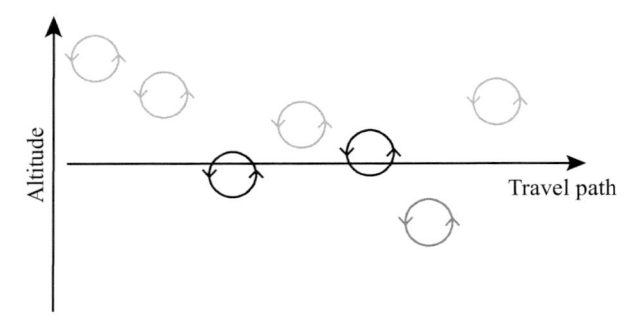

Figure 3.9: Schematic showing a flight line crossing an array of vortex rolls. Figure adapted from Parks et al. [688].

cillatory path as shown in Fig. 3.10. The severity of an encounter with CAT is quantified in terms of the aircraft's normal acceleration vector, which in an idealized situation is equal to the cross product between half the vorticity vector and the air velocity vector [578]. More realistic expressions for the airplane acceleration have been derived for aircraft interactions with wake vortices produced by other airplanes.

Satellite pictures have provided evidence of persistent cloud patterns indicating that the atmospheric planetary boundary layer is often organized into helical secondary circulations aligned parallel to the mean flow [124]. There is almost a formal agreement between theory and observations that these flow patterns originate from both convective motions in the presence of shear and dynamic inflection point instabilities of the Ekman layer. The scale of naturally occurring vortices in the atmosphere is greater than that characterizing aircraft wake vortices. Atmospheric vortex rolls may be aligned up to 30° to the left for stably stratified environments, 18° to the left for neutral environments, and nearly parallel to the mean wind for unstably stratified convective environments [122, 547, 123, 124].

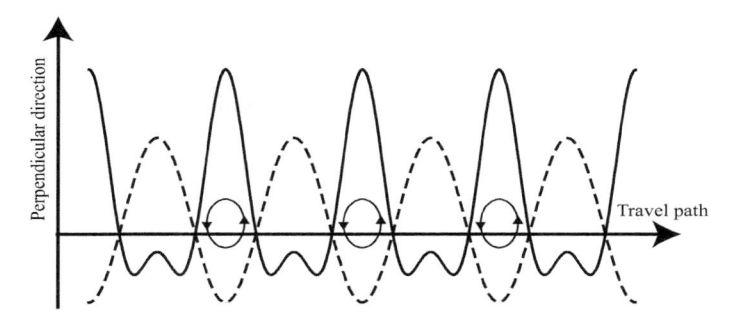

Figure 3.10: Drawing illustrating the effects of a sequence of vortex rolls on the trajectory of an airplane. The dashed line depicts the position of the plane and the solid line indicates the plane acceleration during the encounter with the vortices. The circles represent the vortices in counterclockwise rotation. Figure adapted from Lunnon [578].

In general, such vortices have a depth of 1 to 2 km and a lifetime ranging from hours to days. Given the fluctuating conditions in the atmosphere, there is not yet a definite theory that can explain exactly how horizontal rolls form. The Ekman model represents only an approximate attempt to explain the structure and dynamics of horizontal rolls [547].

3.13   Lighthill-Ford theory

In 1952, Lighthill [557] proposed a theory for the study of aerodynamically generated sound waves by taking into account the statistical properties of turbulent airflows, and later on, Ford [300] extended the problem to rotating shallow-water flow and inertia-gravity wave generation. In this theory, it is assumed that the waves are emitted spontaneously by the vortical flows. As emphasized by Knox et al. [491], this characteristic distinguishes the spontaneous imbalance problem from the initial value imbalance and gravity wave generation of geostrophic adjustment [776, 148] and from boundary condition perturbations leading to topographically forced gravity waves [847]. For example, gravity wave radiation is particularly important for the study and forecasting of CAT. Most of the bumpiness detected by an aircraft at high altitudes is in general caused by being flying in or near gravity waves [278]. The reader interested in the details and derivation of the theory is referred to Knox et al. [491].

In a note published in the Journal of the Atmospheric Sciences, Plougonven et al. [709] concluded that although the turbulence indicator introduced by Knox et al. [491] may be relevant for CAT and the possibility that gravity waves can play a role in triggering CAT, the interpretation of their CAT indicator, as the forcing coming from the nonlinear advection terms and their derivatives, is not founded, suggesting that their study does not bring any new element to the debates on the generation of gravity waves and their role in triggering CAT. Plougonven et al.

[709] also claimed that "a more rigorous interpretation will be needed to put the subject of CAT forecasting on a firmer theoretical footing". Today it is clear that the main mechanisms responsible for CAT are in order of importance: (i) the KH instability of shear layers, (ii) the mountain waves, (iii) the inertia-gravity waves from clouds and other sources, (iv) the spontaneous imbalance theory, and (v) the horizontal vortex tubes. An exhaust overview of these mechanisms as well as their implications for prediction is given by Venkatesh [934]. The lack of understanding of the underlying dynamical processes of aviation-scale turbulence has hampered in part accurate forecasting. More precise observations of turbulence encounters along with advanced theoretical research on the turbulence generation processes will certainly lay the foundations for improved CAT forecasting and avoidance. A review of the sources and dynamics of turbulence in the upper troposphere and lower stratosphere is given by Sharman et al. [829]. When applied to numerical model data, the Lighthill-Ford theory can help in the prediction of CAT areas. This idea is based on the hypothesis that spontaneously generated atmospheric gravity waves may trigger turbulence by locally modifying the stability and wind shear [606]. With this in mind, McCann et al. [606] improved Knox et al.'s [491] method for CAT forecasting, which they called ULTURB (i.e., Upper Level TURBulence). This method can forecast turbulence at all flight levels above the top of the atmospheric boundary layer, and since it is based on a self-consistent dynamical theory, it offers a better insight into the causes of CAT and enhances its predictability.

3.14   Fossil turbulence

Fossil turbulence is another phenomenon that over the years has drawn the attention of meteorologists. According to Woods et al. [998], fossil turbulence is defined as the three-dimensional microstructure of temperature, humidity, or any other scalar property that is advected by the mean flow. A further definition was provided by Gibson [336], who defined fossil turbulence as persisting perturbations of vorticity, temperature, density, etc. produced by a turbulent state at length scales that are no longer turbulent. In other words, it is a turbulent field that has since decayed. In 1999 Gibson's theory of fossil turbulence was revisited in the context of marine turbulence [337], where emphasis was focused on the existing confusion between fossilized microstructures and turbulent microstructures, which, in turn, may lead to underestimates of the average dissipation rates of kinetic energy and scalar variance along with the vertical fluxes in most ocean layers. Later on, this point was clarified by Gibson [338] himself by arguing that fossil turbulence must be any perturbation produced by turbulence and that persists after the fluid is no longer turbulent. Therefore, fossil turbulence waves preserve and propagate information about previous turbulent states. For example, fossil turbulence has a longer persistence than the turbulent motion from which it derives because it must mix the same velocity and scalar variances that existed in its progenitor turbulent motion, but with comparatively much smaller viscous and scalar dissipation rates.

Early work by Woods et al. [998] suggested that the existence of fossil turbulence in the atmosphere would produce sufficient radar backscattering as long as the three-dimensional refractive-index structure is contained by the fossil turbulent waves on a scale equal to half the radar wavelength, which is typically of 5 to 50 cm. They also concluded that fossil turbulence will affect the propagation of optical and acoustical waves as well as the line-of-sight radio propagation. However, after 50 years of Woods et al.'s [998] work most literature on the study of fossil turbulence focuses on the oceanic processes and little information exists on the atmospheric side. For example, high resolution detection of microstructures in the ocean reveals that fossil turbulence there contributes most to mixing [554]. Direct numerical simulations of turbulence and fossil turbulence waves formed by shear in a stratified medium were carried out by Pham et al. [702]. They succeeded to capture a fossil vorticity turbulence spectrum with slope $\approx -5/3$, which suggested a small-scale turbulent motion to the Ozmidov scale[16]. Further direct numerical simulations by Brucker and Sarkar [130] showed that although turbulence forms in both stratified and non-stratified turbulent wakes when the Obukhov energy scale exceeds the Kolmogorov scale, the non-stratified turbulent wake vanishes without a trace after cascading to length scales larger than the Obukhov scale. Conversely, fossilization of the stratified turbulent wake was observed at the Ozmidov scale l_O. These calculations have demonstrated that even when fossil turbulence waves radiate, their energy persists.

3.15 Is CAT increasing?

The phase of flight with the highest risk of injuries from CAT is the cruising phase above 10000 ft, mainly because during this phase passengers and crew are often unbuckled [826]. In general, the size of turbulent eddies that are most hazardous for aviation is typically in the range between 100 m and 1 km. Since CAT is invisible to pilots and cannot be foreseen by onboard radar, most of the operational turbulence forecasts rely on numerical simulation models. On the other hand, numerical calculation models of the impact of climate change on CAT have also started to appear.

The analysis of data from turbulence encounters with aircraft, turbulence diagnosed from reanalysis datasets, passenger injuries caused by turbulence, and

---

[16]The Ozmidov scale of turbulence is defined as

$$l_O = \left(\frac{\varepsilon}{N_{BV}^3} \right)^{1/2},$$

where ε is the dissipation rate of turbulent kinetic energy and N_{BV} is the Brunt-Väisälä frequency. Defined this way, the Ozmidov scale is the largest scale of turbulence that can overturn in a stratified flow and contribute to the energy cascade. The inverse of this scale defines the buoyancy wavenumber, $k_B = l_O^{-1}$.

turbulence diagnosed from climate models all point toward a CAT increase as a response to climate change. A search in the *National Transportation Safety Board's Aviation Accident Database* for the period from 1 January 1992 through 31 December 2001 for all records involving Part 121 air carrier turbulence-related accidents has revealed that the number of flight attendant turbulence-related injuries was on the rise during this period [915]. It was also reported that the most frequent injury was lower extremity fractures, especially the ankle. This trend provides clear evidence of the increase in CAT.

On the other hand, Williams and Joshi [987] have summarized most of our present understanding of the impact of climate change on CAT. For example, if the CO_2 emissions are increased, climate simulations predict an increase of CAT. The diagnostics show that the prevalence of light/moderate/severe CAT will increase 59%/94%/149% at cruising altitudes on transatlantic flights in winter. Similar results are also predicted on other flight routes and in other seasons [987]. Therefore, according to these predictions, it is not difficult to foresee that climate change will inevitably lead to bumpier flights later this century. As a consequence, flight routes may become more convoluted to avoid stronger and probably more frequent zones of turbulence. This will surely imply that journey times will lengthen and jet fuel consumption will increase accompanied by an increase in the travel ticket prices. Large increases in CAT by the period 2020–2050, especially in the mid-latitudes in both hemispheres were also reported by Storer et al. [875]. They found that the volume of severe CAT will approximately double over North America, the North Pacific, and Europe. The increase of CAT will have serious implications for aviation operations in the coming decades. While these studies have uncertainties in some sources as, for instance, the future emissions of greenhouse gases which depend on socioeconomic and political factors, future investigations should be addressed to quantify some of the uncertainties in order to increase the reliability of the predictions.

Chapter 4

Meteorological Evidence for CAT

4.1 Introduction

Most of the flight time of commercial, cargo, and military airplanes is spent at cruise altitudes around the tropopause, i.e., between the upper troposphere and the lower stratosphere. It is at these altitudes that CAT is frequently encountered, with serious consequences on the safety of air traffic. Since CAT is an unexpected and elusive phenomenon, it continues to threaten aviation today. This chapter begins with a description of the meteorological analyses that are often performed in the low stratosphere to predict and identify CAT events followed by a section that discusses how the probabilistic structure of CAT can be used to improve current models of turbulent motion in the atmosphere. Whereas CAT continues to be an elusive phenomenon to forecast, firm evidence exists that about two-thirds of CAT occurrences take place near jet streams and in regions of deep mesoscale convection and breaking gravity waves at mid-latitudes [278]. Therefore, a section on the relation of CAT with the mesoscale structure of the jet stream regions is also included. The chapter closes with a focus on the wind and temperature profiles around the tropopause and in the atmospheric surface layer[1]. In particular, the wind and temperature variations across the tropopause are breeding grounds for CAT. A section of turbulence at low altitudes was also mandatory,

---

[1]The atmospheric surface layer typically comprises the bottom 10% of the atmospheric boundary layer and therefore it is the tropospheric layer that is in contact with the Earth's surface. This layer is often turbulent and has a width of approximately 100 m. It is dominated by small-eddy turbulence within the wall shear flow and the wind speed varies nearly logarithmically with height [880].

first, because of the rather large amount of existing data about turbulent motions at heights below $\sim$ 1000 ft above the ground and, second, because for a flight over land one can expect to find turbulence with an over-all probability of at least 50% [668]. Due to its highly probable occurrence at low altitudes, this type of turbulence represents the most hazardous events in the lower troposphere with disastrous consequences during low-altitude flight, take-off, and landing operations [599].

4.2 Meteorological analysis of CAT around the tropopause

As was already mentioned, CAT is likely to occur around the tropopause, which is the thin layer separating the troposphere from the stratosphere. Over the Equator, the tropopause is at a height of approximately 65000 ft ($\approx$ 20 km) over the ground and descends step-wise to the poles, where it is at a height of about 20000 ft ($\approx$ 6 km) from the ground level [818]. In the period between 1967 and 2010, atmospheric turbulence was responsible for nearly 66% of the flight accidents, while about 56% of them were seen to occur during descent. Of these accidents 13% were attributed to CAT when cruising between the upper troposphere and the lower stratosphere, with only 7% of them occurring during descent [867]. At these altitudes, CAT can be associated with large-scale waves, jet streams, mountain waves, tropopause foldings, and upper-level fronts. Hazards may eventually occur just above the tropopause due to the presence of fluctuations and instabilities caused by abrupt changes in the temperature lapse rate across the tropopause. For example, mountain waves developing CAT may extend to up 5000 ft within the lower stratosphere and more than 100 miles downstream from the mountains. However, CAT events are more frequently detected in an upper trough on the polar side of the jet stream and also along the jet stream north and northeast of a rapidly deepening surface low [818].

In early 1966, the project HICAT (High Altitude Clear Ait Turbulence) became operational for the investigation of the statistical characteristics of high altitude CAT with the main objective of improving structural design criteria. For this project, digital instrumentation carried aboard an Air Force U-2 aircraft was employed to perform measurements of CAT in the wavelength range from 100 to 60000 ft and CAT velocity components were measured at altitudes of 45000 to 70000 ft in seven different geographic areas [210]. Important results were derived from HICAT. For example, it was deduced that mountain areas produce statistically strong turbulence, which can be explained in terms of stratospheric gravity waves [558]. From an analysis of stratospheric HICAT data over the west of the United States, it was found that areas with light-to-moderate CAT, as well as those without CAT, are associated with irregularities in the temperature profile, which included strong inversions and large lapse rates of temperature [811].

While these results suggested that probable CAT and non-CAT areas in the stratosphere can be determined from synoptic-scale data, it seems that it is simpler to distinguish between CAT and non-CAT events than between CAT areas of different intensity. The relation between CAT and the temperature inversions is due to an increase in the vertical wind shear, resulting from a sloping baroclinic layer, while positive buoyancy is responsible for CAT events when the lapse rate of temperature becomes large [261]. Further early analysis of HICAT data in the stratosphere, at heights between 13.7 and 21.4 km from the ground, has revealed that CAT can be associated with low Richardson (Ri) numbers as a result of either sharp drops of temperature with height or strong vertical wind shears [947]. The Ri number was further confirmed to be a much better CAT indicator than other parameters, such as the stability index, the vorticity, the vorticity advection, and the CAT index [195, 194, 811]. The Ri number was found to be a better parameter to outline areas of expected CAT than other parameters, such as the vertical or horizontal wind shear and the vertical or horizontal gradients of kinetic energy [510]. For example, values of $Ri < 1$ outlined about 40% of the reported occurrences of moderate and severe CAT.

In addition to HICAT data analyses, other studies were conducted by Kao and Gebhard [455], this time using data obtained over Colorado and New Mexico between November 30 and December 2, 1967. The data were collected with the aid of an aircraft equipped with an inertially stabilized turbulence measuring system and inertial equipment for the determination of the ambient wind velocity. The altitude range covered was from 15.7 to 18.2 km above the ground, while the aircraft average speed was 198 m s^{-1}, the mean air velocity was 11.5 m s^{-1}, and $Ri = 0.15$. Their results indicated that CAT is characterized by anisotropic turbulence with a rather intense lateral component associated with a strong thermally stable stratification. It was found that the power spectra of the temperature and wind speed were very similar and approximately proportional to $k^{-5/3}$, where k is the wavenumber, while the power spectra of the streamwise and lateral components of the velocity are $\propto k^{-2}$ and the spectrum of the vertical component of the velocity is $\propto k^{-1}$. On the other hand, measurements of wind shear, Ri number, and turbulent intensity were carried out successively by using sensitive balloon-borne instrumentation [58, 60, 59]. These studies revealed that, at these altitudes, CAT can appear within thin layers that are separated from adjacent layers by sharp boundaries, that the outer scale of the inertial subrange is an order of magnitude below the thickness of the thin layers, while the horizontal extent is several orders of magnitude above, and that CAT is tightly related to the shear instability. Even in the case of very light turbulence, significant mixing was observed to occur over depths that may be larger than the thickness of the CAT layer.

More recent analyses confirm that stratospheric turbulence occurs at altitudes from about 12 to 25 km from the ground and that it is characterized by patchy high-frequency fluctuations inside the stratospheric wind fields and long-lived energetic eddies with scales of a few hundred meters in the vertical direction [582]. The investigation of nine events of moderate-to-severe CAT observed on 2 April 2007 over South Korea in three different regions, namely the west coast, Jeju Island, and the eastern mountain areas, using the Weather Research and Forecasting (WRF) model, revealed that the CAT generation mechanisms were somewhat different in the three regions [483]. Localized turbulence was produced by the strengthening of an upper-level front over the west coast, by an intensification of both the inertial instability on the anticyclonic shear side of the enhanced jet stream and the ageostrophic flow in the lee side of the convective system in the Jeju Island, and by the vertical propagation of large-amplitude gravity waves produced by the mountain range and subsequent breaking down over the lee side of the topography over the eastern mountainous area [483]. Based on these studies, it is clear that gravity waves can be excited by different forcing mechanisms.

When flying in proximity of a tropopause fold[2], an aircraft can experience substantial CAT. PIREPs between June 2017 and December 2018 in the Romanian airspace have provided fresh new evidence of CAT events associated with tropopause folds. The analysis of the data was performed in conjunction with meteorological data, satellite imagery, and vertical profiles [867], revealing that out of 420 events announced by aircraft pilots, severe CAT was reported in 80 cases of which 13 were indeed associated with tropopause folding. In many cases, CAT can be detected horizontally over distances varying from 80 to 500 km in the wind direction and over distances between 20 and 100 km perpendicularly to the wind direction with lifetimes of 30 min to 24 h [682]. A numerical study of CAT encountered over the Yellow Sea between China and the Korean peninsula on 13 February 2013 had already concluded that a major cause for CAT production is the KH instability generated by strong vertical wind shear and tropopause folding induced by an upper-level frontal system [541].

Since the days of the HICAT project, it has been recognized that CAT is also associated with large horizontal temperature gradients. In this context, it has also been reported that CAT is correlated with irregularities in the temperature profile, including strong temperature inversions and large lapse rates of temperature [267, 631, 261]. Numerical predictions of horizontal and vertical wind shear along with the Ri tendency have been objectively applied in CAT forecasting techniques. From the definition of Ri in Eq. (1.35), the numerator is proportional to the temperature stratification and the denominator to the vertical wind shear so that it is clear that instability (Ri < 1) requires large vertical wind shear and a rather weak stratification. Based on the Ri tendency is the SCATR (Specific CAT Risk) index [469], which quantifies the energy dissipation rate resulting from turbulence as a function of the large-scale deformation forcing and the shear layer. For example, the magnitude of horizontal and vertical shear is correlated

---

[2]In general, a tropopause fold is an extrusion of stratospheric air into the troposphere. Such extrusions typically occur near an upper-level front beneath the polar and subtropical jet streams [821, 823].

to the probability of encountering CAT over flight segments of 100 km by the so-called Empirical Index (EI) [263]. During the first half of the 80s, analyses of satellite imagery and PIREPs released concrete evidence that strong CAT can be encountered along certain cloud boundaries, known as *deformation zones*[3] that can be found in regions of hyperbolic flow[4] [28, 273]. An objective forecast index based on horizontal deformation and vertical wind shear (VWS) was introduced by Ellrod [274, 275]. The deformation (DEF) parameter is defined in terms of the stretching deformation (DST)

$$DST = \frac{\partial v_x}{\partial x} - \frac{\partial v_y}{\partial y}, \tag{4.1}$$

where v_x and v_y are, respectively, the eastward and westward wind speed component forecasts produced by the *National Meteorological Center* (NMC) in Washington, D.C., and the shearing deformation (DSH)

$$DSH = \frac{\partial v_y}{\partial x} + \frac{\partial v_x}{\partial y}, \tag{4.2}$$

such that

$$DEF = \left(DST^2 + DSH^2\right)^{1/2}. \tag{4.3}$$

Since CAT is more often encountered in areas where both DEF and VWS are significantly large, the product DEF×VWS was used to define the Turbulence Index (TI) [587], which was subsequently validated against CAT PIREPs. However, rather than DEF × VWS the product

$$TI1 = VWS \times DEF, \tag{4.4}$$

resulted in the highest correlation with values between 0.43 and 0.48 [273, 274]. The TI1 index is the TI version of NMC, while the TI used by the *Air Force Global Weather Central* (AFGWC) at the Offutt Air Force Base in Nebraska is given by

$$TI2 = VWS \times (DEF + CVG), \tag{4.5}$$

where

$$CVG = -\frac{\partial v_x}{\partial x} - \frac{\partial v_y}{\partial y}, \tag{4.6}$$

is the convergence term[5] [273]. Although the TI1 and TI2 were developed in the 1990s to predict CAT, they are still in use today at many aviation forecasting

---

[3]Deformation is a kinematic property of any flow by which a circular patch of fluid transforms into an elongated shape. Deformation may contribute to promote or destroy horizontal temperature gradients and therefore upper-level frontal zones [682].

[4]Atmospheric regions of hyperbolic flow are associated with the exit region of the jet stream.

[5]This term becomes important when the fluid is compacted by the confluence of streamlines and/or the deceleration of air parcels. It can contribute to frontogenesis and, in some cases, can also generate gravity waves by perturbing the tropopause inversion.

facilities worldwide [276, 541, 874]. In particular, TI1 is one of the best forecast diagnostics for shear-induced turbulence and has been successful in forecasting about 75% of the atmospheric CAT events [874]. In part, the success of the TI1 index is due to the fact that most CAT events originate because of destabilization of the atmosphere by propagating gravity waves in zones of shear instability, leading to local Ri-values below $1/4$ and triggering KH instabilities that degrade into turbulence [261]. This explains why a deterministic TI1 forecast is routinely used by the *World Area Forecast Centres* (WAFCs). Case studies of CAT events over South Africa in the period 1993–1995 and then in 1998 have confirmed the reliability of the TI1 index as a CAT indicator in zones where strong vertical and horizontal wind shear are present [227]. Despite this, there is no complete confidence in the turbulence forecasting methods based exclusively on these indices. To turn this problem around, a collection of forecast runs is used today to predict aviation turbulence [344, 343, 131]. Together with this ensemble forecasting, the *Met Office Global* and the *Regional Ensemble Prediction System* (MOGREPS) was employed, which has been operational since 2008 [109]. Other turbulence predictors were included, like the Brown index [121] and the Dutton index [263], which are particularly useful to diagnose CAT associated with tropopause folding, the Lunnon index [764], the Richardson number, and two convective indices. Maximization of the forecast skill is usually accomplished by combining these predictors by using an iteration scheme [826, 344, 131]. Together with the Brown and Dutton indices, the TI2 and the horizontal temperature gradient (HT) [137] predictors are also useful to forecast tropopause-folding generated CAT. However, recent diagnostics have shown that the most useful indices for identifying tropopause-folding related CAT are the Brown and HT indices [867]. In passing, it is worth mentioning that the TI1 index may not be a good indicator of mountain-wave turbulence. A description of CAT forecast problems and early forecasting techniques of CAT associated with mountain waves can be found in Ref. [413]. In this respect, the use of the Graphical Turbulence Guidance (GTG) system has become a more reliable tool [826]. A new GTG system, dubbed GTG3, was introduced by Sharman and Pearson [828, 693] to improve diagnosis for mountain-wave turbulence associated with the mid-latitude westerlies and mountains perpendicular to the flow.

It has long been recognized that CAT is also relevant for cross-tropopause transport of chemical constituents [821, 469, 281, 434]. For instance, Esler and Polvani [281] performed numerical experiments on the linear and nonlinear dynamics of layers of anomalously high potential vorticity (PV), finding that the breakdown of PV layers due to a KH instability may represent an important mechanism for irreversible mixing of air masses between the troposphere and the stratosphere. The link between CAT and cross-tropopause mixing of air was further investigated by Jaeger and Sprenger [434] in a study of CAT indicators in the Northern Hemispheric tropopause region based on a reanalysis of 44-year climatology data set (ERA40), covering the period between September

1957 and August 2002, from the *European Centre for Medium-Range Weather Forecasts* (ECMWF). They used small Ri values, the square of the Brunt-Väisälä frequency, and the PV values as indices for KH hydrostatic and symmetric instability. The PV indicator is defined by [414, 804, 409]

$$\text{PV} = \frac{1}{\rho}\left(\nabla \times \mathbf{v} + 2\Omega \times \mathbf{j}\right) \cdot \nabla\theta \approx \frac{1}{\rho}\left(\zeta + f\right)\frac{\partial\theta}{\partial z}, \tag{4.7}$$

where ρ is the air density, $\mathbf{v}$ is the wind velocity, $\mathbf{j}$ is the unit vector in the horizontal y-direction in a Cartesian frame of reference, $\Omega = (0, \Omega\cos\phi, \Omega\sin\phi)$ is the Earth's angular velocity vector in the local Cartesian system, ϕ is the latitude, θ is the potential temperature, ζ is the vertical component of the relative vorticity, and $f = 2\Omega\sin\phi$ is the Coriolis frequency. In particular, this study has shown that in winter CAT appears with maximum frequencies over the North American east and west coasts. Local maxima were also identified over the Himalayas, central Europe. eastern China, and the western part of the North Atlantic and North Pacific. These frequencies are reduced in summer and shifted to the north. Over the 44-year period these trends were found to vary nonlinearly with increases of 40 to 90% over the North Atlantic, the United States, and Europe, while the interannual variability of CAT, as indicated by the empirical index TI [275] and the PV values were seen to be significant for the two phases of the North Atlantic Oscillation and the Pacific/North American flow pattern.

Early forecast verification essentially relied on subjective PIREPs. This verification test depended on the pilot's perception of the CAT intensity and also on the type of aircraft in the sense that small airplanes flying through the same turbulent air volume will experience stronger CAT than larger airplanes. To produce more reliable forecast verifications, the Derived Equivalent Vertical Gust (DEVG) and the Eddy Dissipation Rate (EDR) measures were introduced which are aircraft-independent [901, 388]. With the use of DEVG and EDR reports it is possible to combine different aircraft observations to produce a consistent observational database. Therefore, with this capability turbulence events encountered during commercial flights could be recorded, leading to improved forecast verifications.

4.3   The probabilistic structure of CAT

CAT differs from other complex atmospheric motions by its intrinsic random character. In fact, turbulence can be considered a sort of stochastic[6] (or random) process. The antonym of stochastic is "deterministic" or "certain". Thus a deterministic model predicts a single outcome from a given set of circumstances, while a stochastic model predicts a set of possible outcomes weighted

---

[6]The concept of stochastic processes represents a useful mathematical tool for the description of various physical and natural phenomena. These processes are nothing more than ways of quantifying the dynamic relationship of sequences of random events.

by their likelihoods (or probabilities). Therefore, the random character of turbulence makes deterministic forecasting a difficult task if not impossible. This explains why the probabilistic structure of CAT must be taken into account in any model aimed at allowing reliable predictions to be made and identifying hazards in flight [262]. Since randomness is present in any turbulent process, the aeronautical prediction systems rely on deterministic devices forced by random inputs.

Many models are based on the hypothesis that atmospheric turbulent motion can be described by a random Gaussian (or normal) process [262]. This hypothesis rests on the fact that the Kolmogorov scaling law assumes that turbulence is a Gaussian process at small scales [832]. On the other hand, the analytical properties of Gaussian processes are particularly desirable because they make life easier for statisticians and airplane designers. However, experiments at finite Reynolds numbers have demonstrated for some time the actual non-Gaussian behavior of turbulent motions [865, 944, 931, 717]. In particular, some of these experiments have demonstrated that the statistics of the Lagrangian acceleration is characterized by distribution profiles with long heavy tails, which are indicative of non-Gaussian processes [944, 717]. This implies that the energy spectrum of real-world turbulent motions does not exactly match the $-5/3$ slope predicted by the Kolmogorov scaling law [169]. Based on the observation that the Lévy β-distribution[7] represents a suitable approach to accommodate the heavy tail behavior observed in the probability density function of turbulence quantities [66, 888], Chen and Zhou [169] proposed to replace the Laplacian term, $\nabla^2 \mathbf{v}$, in the Navier-Stokes equations by the fractional Laplacian $-(-\nabla^2)^{\beta/2}\mathbf{v}$, with $0 < \beta \leq 2$, which serves as a stochastic driver and guarantees the positiveness of energy dissipation. From a dimensional analysis of the modified Navier-Stokes equations, they found that in the wavenumber space the energy spectrum obeys the law

$$E(k) = C\varepsilon^{2/3}k^{-3+2\beta/3}, \tag{4.8}$$

which for $0 < \beta \leq 2$, the exponent in the wave number ranges between -3, in the limit $\beta \to 0$ as observed in the experiments, and $-5/3$, when $\beta = 2$, corresponding to the Gaussian distribution and the standard Navier-Stokes equa-

---

[7]A Lévy distribution is a continuous probability distribution for a non-negative random variable, x. Over the domain $x \geq \mu$, the probability density function of the Lévy distribution is given by

$$f(x;\mu,c) = \left(\frac{c}{2\pi}\right)^{1/2} (x-\mu)^{-3/2} \exp\left[-\frac{c}{2(x-\mu)}\right],$$

where μ and c are, respectively, the location and scale parameters. In probability theory, a stable distribution is one such that if x and y are two random distributions coming from stable distributions, then their sum, $x+y$, will also have a stable distribution. As $x \to \infty$, the probability density function becomes

$$f(x;\mu,c) \sim \left(\frac{c}{2\pi}\right)^{1/2} x^{-3/2},$$

which exhibits heavy tail behavior falling off as a $x^{-3/2}$ power law.

tions. The result of Eq. (4.8) is known as the Lévy-Kolmogorov law. In order to represent anomalous dissipation, as observed experimentally and numerically [910, 155, 230], Chen and Zhou [169] replaced the fractional Laplacian dissipation term in the Navier-Stokes equations with the fractional time derivative of the Laplacian, namely $-\partial^{\alpha}(-\nabla^2)^{\beta/2}\mathbf{v}/\partial t^{\alpha}$, where $0 < \beta \leq 2$ when $\alpha = 0$ and $0 \leq \alpha < 1$ when $\beta = 2$. This way, the energy transfer splits up into three stages: sub-transport (fractional Brownian motion) before the Kolmogorov range, normal transport (Kolmogorov scaling), and super-transport (Lévy-Kolmogorov scaling). When $\beta = 2$, the energy spectrum that follows from dimensional analysis of the modified Navier-Stokes equations is given by

$$E(k) = C\varepsilon^{2/3}k^{-(5-3\alpha)/(3-\alpha)}, \tag{4.9}$$

where in the limit when $\alpha = 0$ the exponent of the wavenumber converges to the Kolmogorov law $-5/3$ and when $\alpha = 1$ the scaling converges to a -1 power law.

Returning to the statistical aspects of CAT, Lester [552] first noticed that intermittency is a fundamental characteristic resulting from the inhomogeneity of CAT. The fact that CAT is difficult to forecast may be due to its occurrence at small scales and its intermittent character. For example, the tendency for CAT intensity to change substantially along a flight path makes it hard to take measurements even with the use of current sensors. Evidence of intermittency in CAT records were first given by Dutton [260], which was characterized by non-Gaussian gust distributions. As said above, this finding was of great concern for the determination of aircraft response to CAT, which until then was supposed to be a Gaussian process. Further early graphical evidence of CAT intermittency was also collected during project HICAT [211]. These data revealed that intermittent CAT differs from the intermittent turbulence encountered in laboratory experiments and the atmospheric boundary layer because intermittency within a burst encountered during flight may also occur over small scales.

In laboratory experiments, intermittency is measured by the fraction of the flow field that is actually turbulent. This is done by means of the factor γ, which has been defined by the ratio [909]

$$\gamma = \frac{\alpha_4'}{\alpha_4}, \tag{4.10}$$

where α_4 is the kurtosis[8] of the first derivative of the velocity and α_4' is the kurtosis within the turbulent region, which is equal to 3 for a Gaussian distribution. Obviously, the γ factor calculated according to Eq. (4.10) provides in the best way a qualitative estimate of the fraction of time the flow is turbulent [66]. A second method that has been used in subsequent investigations of intermittency

---

[8]See Eq. (1.42) for the definition of kurtosis of the fluctuating part of the turbulent velocity field. The same expression is employed in Eq. (4.10) for the gradient of the velocity field.

in turbulent flows relates the γ factor to the time integral of an electronic signal, $I(t)$, which is generated during the turbulent periods [909], according to

$$\gamma = \frac{1}{T} \int_0^T I(t)dt, \tag{4.11}$$

where T is the total period of the records. When dealing with small-scale turbulence, the intermittency of the energy dissipation, ε, is an important aspect for the verification of theoretical predictions. Early studies predicted that ε should have a log-normal distribution [369], with a probability density function given by

$$P(\varepsilon) = \frac{1}{\sqrt{2\pi\sigma^2\varepsilon^2}} \exp\left[-\frac{(\ln\varepsilon - m)^2}{2\sigma^2}\right], \tag{4.12}$$

where m is the mean ensemble of $\ln\varepsilon$ and $\sigma^2 = \langle(\ln\varepsilon)^2\rangle - \langle\ln\varepsilon\rangle^2$. The statistical model proposed by Lester [552] consisted of a generalization of Dutton's [260] model, where useful indicators of CAT intermittency included the probability density functions, the γ factor, and the ratio of the mean deviation to the standard deviation. The frequency distributions and the statistical moments of real CAT data have allowed verifying the model predictions concerning the shape of the frequency distribution and the kurtosis (≥ 3). However, this model could not account for the significant skewness that has been observed in some cases. On the other hand, the intermittency indicators and the energy spectrum were seen to be sensitive to how the samples were selected. For example, the selection of larger or smaller samples, as well as slight changes of the sample, may lead either to increased or decreased intermittency. Data on the spectral distribution of turbulence energy in the free atmosphere near the jet-stream level have also been collected by other sources. In general, the power spectra obtained by different authors over the former USSR, Australia, and the United States follow the $-5/3$ law up to scales ~ 100 km, while in some parts of the obtained spectra there was significant deviations from this law. For instance, over the former USSR. under conditions of severe CAT the spectrum had a slope steeper than $-5/3$ at wavelengths > 600 m. The same were also observed under moderate CAT [704]. Such an increase of slope was attributed to the effects of turbulent energy dissipation by buoyant forces in a thermally stable atmosphere. Under conditions of stable stratification at scales > 600 m, the slope of the spectral curve was seen to approach -3 in agreement with the Lévy-Kolmogorov law given by Eq. (4.8) when the parameter $\beta \to 0$. An account of all these early observations is given by Pinus [704] and references therein.

The statistical properties of turbulence in the region between the upper troposphere and the lower stratosphere, where air traffic is most intense, are still not well understood today mainly because the available routine observations are insufficient to construct reliable CAT maps. Whereas the major benefits of these observations are to provide warnings for alerting of potentially hazardous conditions either in real time or for route planning, the capture of large-scale features conducive to CAT uses forecasting methods based on operational numerical prediction model outputs, which in turn are complemented by human input and real-time observations to calibrate the forecasts that provide indications on the expected levels of CAT intensity [828]. A particularly useful metric of turbulent intensity is the energy dissipation rate ε [824]. To improve on the lack of adequate routine observations, a project sponsored by the FAA has developed a software package that performs automatic estimates and reports CAT intensity levels as inferred from the cube root of the eddy dissipation rate ($\varepsilon^{1/3}$). This software has been tested on commercial flights over the United States providing an unprecedented amount of data from *in situ* reports [824].

4.4 Relation of CAT with the mesoscale structure of the jet stream region

In general, weather phenomena with a direct impact on human activities occur mostly on the mesoscale at altitudes between about 1 and 100 km in the vertical direction and over extensions of approximately 10 to 1000 km in the horizontal direction. Such phenomena include gap winds, thunderstorms, downslope windstorms, land-sea breezes, and squall lines. The duration of these phenomena ranges from minutes to days. Air currents within the jet stream with quasi-horizontal axes thousands of kilometers long, hundreds of kilometers wide, and several kilometers deep play an important role in triggering most of these weather phenomena. The winds in the core of the jet stream between the mid and upper troposphere can reach speeds higher than 30 m s^{-1}. On the other hand, the polar jets occur at altitudes of 9 to 12 km from the sea level, while the somewhat weaker subtropical jets, occur at relatively higher altitudes between 10 and 16 km from the ground [683]. It is well known that low-level jets play an important role in designing the dynamics of the atmospheric boundary layer, while the polar night jet is central in the dynamics of the middle atmosphere. In particular, the subtropical jet and the polar front jet at tropopause heights are among the most studied jet streams by meteorologists.

The study of the laws governing the dynamics of the jet streams is a rather difficult task because of the following reasons. First, the Coriolis force, which governs the geostrophic wind relationship depends on the Earth's latitude; second, the distribution of the heat sources and sinks around the planet is not uniform mainly because of differences in the heat capacities of continents and oceans and in the release of latent heat during precipitation; third, the atmospheric flow is strongly influenced by orographic effects as occurs, for example, when air masses are forced to flow over high topography[9]; and fourth, the heat sources and sinks

---

[9]When air masses are deflected by mountains and rise over them, the air cools and the water vapor condenses. This produces rainfalls that concentrate on the windward side of mountains and increase with altitude in the direction of storm tracks.

in the ozonosphere and ionosphere significantly affect the tropospheric dynamics. In addition to these complexities, seasonal and hemispheric differences may also play a role on the general circulation. As was pointed out by Reiter [751], the difficulty grows because the Earth's atmosphere does not necessarily behave like a laboratory where experiments can be repeated. Therefore, to understand the dynamics and effects of jet streams on weather, meteorologists have been compelled to resort to statistical treatments of weather events in the atmosphere and to numerical modeling with the aid of computers. An exhaust early overview on the dynamics of large-scale circulations and jet streams is provided by Reiter [751].

Aircraft observations of the mesoscale and the upper level jet stream-frontal zones systems have been in place for more than 40 years. Results of early research flights have been reported by Shapiro [820]. For example, observations of the chemical constituents in the atmosphere provide information about mixing processes and vorticity associated with mass exchange between the stratosphere and the troposphere in the vicinity of the upper level jet stream-frontal zones systems. In particular, commercial airline and synoptic aerological observations on April 1976 reported the presence of a jet streak velocity maximum over the Gulf of Alaska. By the end of March of the same year, a mesoscale structure characterized by a large-amplitude wave was detected over the central United States. In this case, the strongest jet stream was flowing out in a southwest direction ahead of the trough axis over Texas. These observations led to turbulent heat flux measurements in CAT regions above and below the layer of maximum wind, providing firm evidence of the importance of turbulent-scale processes in the generation and dissipation of potential vorticity [820]. Further evidence of stratospheric air intrusion into the upper troposphere has also been given by measurements of ozone concentration around the tropopause [821]. On the other hand, continuous MST/ST radar observations related to mid-latitude jet streams have also a long history dating back to early upper-level balloon observations [319, 320, 321, 527, 822]. These observations include KH instabilities, CAT, and gravity waves. In fact, the connection between the KH instability and the occurrence of CAT was first clarified using power UHF radar observations in the 1960s [42, 261]. The investigation of atmospheric KH instabilities via detailed wind measurements over a broad range of altitudes was possible thanks to the appearance of more powerful VHF and UHF Doppler radars [924, 489]. These observations have suggested a strong dependence of the wave activity on wind speed [653]. Therefore, the magnitude of CAT is enhanced under jet stream conditions as indicated by MST radar observations at higher altitudes [655, 849]. Ageostrophic winds at the entrance and exit regions of jet streaks have been described by Uccellini and Johnson [917] with the aid of numerical simulations. In particular, an ageostrophic component directed toward the cyclonic side of the jet streak is caused by confluent streamlines and downstream accelerations of geostrophic winds at the entrance region, while diffluent streamlines and downstream deceleration of geostrophic winds at the exit of the jet streak produce an ageostrophic component this time directed toward the anticyclonic side of the jet streak. These simulations predicted that severe convective storms can be forced by such mass and momentum adjustments during the propagation of an upper tropospheric jet streak.

The relation of CAT with the mesoscale structure of the jet stream region was also examined in an early work by Reed [746] for a case of widespread and persistent CAT detected by an aircraft of the *Air Force Cambridge Research Laboratories*. In this case, Ri < 0.25 for most of the turbulent region and a direct relationship could be established between CAT and mesospheric features, such as temperature, wind, and ozone concentration fields. While CAT was seen to occur just above the jet stream core, this study was in support of the hypothesis that all these features were induced by the turbulent heat flux. A more recent overview on the MST radar technique, including its extensions by the development of boundary-layer radars and the radio-acoustic sounding system (RASS) technique for atmospheric temperature measurements, has been reported by Van Zandt [923]. In this overview, applications for weather forecasting from data collected by networks of radars as well as to the study of CAT are thoroughly described. The more relevant observations on jet streams can be summarized as follows:

■ Jet streams are generally found in baroclinic zones associated with tropopause breaks. These include the following latitude bands: (a) the tropical tropopause from 16 to 17 km from the ground and extending to latitudes of 40° N, (b) the mid-latitude tropopause located at heights of 10 to 12 km from the ground in winter and ranging from latitudes of about 30° N to 60° N, and (c) the polar tropopause extending from 8 to 10 km in winter in the north of latitude 60° N.

■ The subtropical jet is found at an altitude of about 12 km near a latitude of 30° N, while the polar front jet is found at about 9 km from the ground level at latitudes between 40° N and 60° N.

■ The variability of the vertical wind induces a variability of the wave activity and location of the jet stream. In general, such variability is seen to occur from day to day [320].

■ A reliable index for the detection of CAT in the free atmosphere is given by the Ri value. The magnitude of CAT depends upon the background wave activity and wind intensity.

It is evident from the above observations that the jet stream is strongly dependent on the meteorological conditions associated with the hemispheral latitude as well as other thermodynamic parameters. Although radar techniques have contributed to our understanding of the relationship between CAT and mesoscale features,

CAT remains an open problem. The latest advances to improve its knowledge continue to come from *in-situ* observations and numerical simulation models. For example, dropsonde observations in the vicinity of an intense jet stream/upper-level frontal system on February 2001 and numerical simulations provided further evidence of the relationship between the KH instability and the generation of CAT by wave breaking [523].

Parallel to radar observations, photogrammetric measurements of wave formation at the cirrus-cloud level have also been conducted from the ground to explore the conditions that operate in the atmospheric meso- and microstructure for CAT generation [754]. A case study of severe CAT reports over northeastern Texas on April 1962 has revealed by photogrammetric-cloud measurements the merger of two jet-stream branches, which were indicative of a further source of CAT activity. The merger of two jet streams results in layers of sharp turning of winds with height, which under thermally stable conditions can lead to CAT via a gravity-wave type mechanism [754]. These measurements also revealed a strong correlation between the CAT frequency and the presence of mountainous obstacles. Three detected cases of severe CAT over the Allegheny Mountains confirmed the occurrence of a CAT mechanism different from the one observed when two jet streams merge. In these latter cases, the CAT was generated by a strong jet stream crossing the mountain ranges. The relation between CAT and mesoscale features has also been analyzed by measuring the power spectra of CAT in the vicinity of jet streams [704]. Such measurements near the jet-stream level were performed by research flights with especially instrumented aircraft over the former USSR [940] and Australia [753]. These measurements provided good evidence of the processes that lead to CAT.

The intermittent nature of CAT has made observational studies challenging. However, the modern high spatial and temporal resolution of MST radars represent an effective means for measuring the effects of convection on turbulence generation. MST radar observations from different sites around the globe have already shown that during convective events there is sustained CAT activity [108, 792, 698, 781]. More recent MST radar observations conducted by Hansen et al. [378] have found significant wave activity in summer in the upper troposphere at White Sands, New Mexico. They interpreted the wave activity as generated by deep convection associated with the summer monsoon in the southwestern United States. Using 50-MHz MST radar observations and surface weather reports, they also found that there can be enhanced wave activity even in the complete absence of convection and that the presence of a thunderstorm does not necessarily indicate enhanced wave activity [379]. Atmospheric turbulence parameters, such as eddy dissipation rate, eddy diffusivity, and horizontal scale length were obtained for the first time for the monsoon and winter seasons using MST radar observations at Gadanki, India [221]. However, Turbulent energy dissipation rates were previously observed during the MIDAS/MaCWAVE campaign in summer 2002 using the ALWIN VHF radar at Andøya/Northern

Norway [279]. The turbulent energy dissipation rates as determined from the energy spectra were seen to vary from 5 to 100 m W kg^{-1} in the altitude range between 80 and 92 km and to increase with altitude. To explore the structure, the background wind field, and the microphysics at cloud boundaries, Kumar et al. [512] focused on several passages of tropical cirrus using MST radar observations simultaneously with a co-located polarization LIDAR at the *National MST Radar Facility* at Gadanki. These observations revealed the presence of anisotropic turbulence at the boundaries of cirrus, which was associated with rather strong horizontal winds, enhanced vertical shear in the horizontal winds, and reduced vertical velocities. More recent MST radar observations have focused to examine turbulent altocumulus layers in the absence of strong wind shear or breaking gravity waves in undisturbed weather. In particular, Aberystwyth MST radar observations of two case studies from April 1996 to March 2000 have shown layers of turbulence where there is no wind shear or breaking gravity waves. Reduced stability by the presence of saturated air in these layers was the cause of induced convective motions and KH instability leading to CAT [999]. Since small-scale atmospheric turbulence is known to be spatially and temporally intermittent, the fraction of turbulent motion in the atmosphere is a key parameter not only for interpreting CAT radar measurements, but also for evaluating the transport properties of such small-scale motions. In fact, Wilson et al. [990] introduced a method, based on indirect estimates of the dissipation rates of the turbulent kinetic and potential energy, to provide unbiased estimations of the turbulent fraction from simultaneous radar measurements of reflectivity and Doppler broadening within a sampling volume.

4.5 Wind and temperature profiles

The description and modeling of wind and temperature profiles have been a long-standing problem in meteorology. The wind profiles and the temperature trends in the lower atmosphere are not independent factors. Even at low altitudes, the wind and temperature profiles, as well as the height of the boundary layer, cannot be measured on a routine basis. In general, information about these meteorological parameters is obtained using indirect methods, including the effects of heat, momentum, and moisture fluxes at the Earth surface.

CAT data collected from about 50 XB-70 aircraft flights in the stratosphere over the mountainous terrain of the western United States from March 1965 to November 1967 and analyzed by 27 rawinsonde[10] system stations have demonstrated that the mountain-wave amplitude depends on the curvature of the wind

---

[10]In contrast to a *radiosonde*, which is a balloon-borne instrument for the measurement and transmission of pressure, temperature, and humidity, a *rawinsonde* refers to a method for observation of upper air. It evaluates the wind speed and direction as well as the temperature, the pressure, and the relative humidity by means of a radiosonde tracked by a radar or radio direction finder [237]. Today, most rawinsondes are no longer tracked by radar but instead they use global positioning systems.

profile in the troposphere and that CAT, in turn, depends on the wave amplitude [719]. According to Corby and Wallington [200], the amplitude of a lee wave is proportional to the amplitude factor

$$a_n = \frac{\left(l_i^2 - l_s^2\right)\sin^2\phi}{(n\pi - \phi + \tan\phi)\sqrt{l_s^2\sin^2\phi + l_i^2\cos^2\phi}}, \tag{4.13}$$

for $n = 1, 2, 3, \ldots$, where l^2 is the Scorer parameter [812] defined by

$$l^2 = \frac{g}{v^2}\frac{1}{\theta}\frac{\partial\theta}{\partial z} - \frac{1}{v}\frac{\partial^2 v}{\partial z^2}, \tag{4.14}$$

the subscripts i and s mean that quantities in Eq. (4.13) refer to a lower and upper layer, respectively, v is the horizontal wind speed, θ, as before, is the potential temperature, and $\phi \in (0, 2\pi)$ as required to restrict the lee-wave number between l_i and l_s. Based on Eq. (4.13), large-amplitude lee waves occur in airstreams where $l_i^2 - l_s^2 \approx l_i^2$ and l_i^2 is sufficiently large, which can actually be satisfied for a variety of wind and temperature profiles. This aspect is important because it has long been pointed out that mountain waves of sufficiently large amplitudes can propagate upward and deep into the stratosphere [353, 270]. The analysis performed by Possiel and Scoggins [719] established the importance of the curvature of the wind profile at heights of 10000 to 25000 ft from the ground level to determine CAT and non-CAT regions at altitudes of 12–21 km and found that large-amplitude lee waves in the troposphere may produce favorable conditions for the generation of CAT in the stratosphere. In addition, it was concluded that about 89% of the turbulent regions detected in the stratosphere are related to mountain-wave areas and around 79% of them were found where the vertical gradient of curvature was positive. This conclusion was further confirmed by the fact that nearly 74% of the non-CAT regions were outside expected mountain-wave regions and that about 98% of these non-CAT areas happened where the vertical gradient of curvature was negative.

At the beginning of the last century, the discovery of the increase of the average temperature and sound speed with height in the upper stratosphere was made using the method of remote sounding, which uses infrasound[11] recordings from explosions at regional distances along with the effect of total internal reflection of sound waves from the nonhomogeneous atmosphere [972]. Studies of the structure and dynamics of the atmosphere can be found in Refs. [241, 537, 539, 38].

---

[11]Infrasound waves, also referred to as low-frequency sound waves, are acoustic waves characterized by frequencies of around 20 Hz, which are below the lower limit of human hearing. These acoustic waves are known to be produced by natural events, such as volcanic eruptions, earthquakes, and ocean swell. Human activities such as mining and surface explosions may also generate infrasound waves [538]. Infrasound waves have the ability to get around obstacles with very little dissipation and propagate through the atmosphere. They can be recorded by ground-based stations.

Continuous monitoring of the wind velocity variations in the stratosphere, mesosphere, and the lower thermosphere is accomplished by measurements of temporal variations of the azimuths and times of arrivals of infrasound waves propagating along refracting raypaths in the stratospheric and thermospheric acoustic waveguides [539, 38]. Vertical profiles of large-scale wind velocity variations are retrieved by the inversion method [38], which is based on parameterization of wind profiles with a set of orthogonal functions obtained from a self-empiric Ground-to-Space (G2S) model of the atmosphere [253, 251, 252]. However, since inverse methods are based on ray tracing they cannot capture fine-scale wind velocity variations. More recent studies of the wind velocity structure in the upper stratosphere, mesosphere, and lower thermosphere are based on the method of infrasound probing of the atmosphere [179, 180], which relies on estimating wind field fluctuations at such altitudes with high vertical resolution using volcanoes and surface explosions over the globe. The retrieval of wind structure in the upper stratosphere, mesosphere, and lower thermosphere from the infrasound signals recorded from the volcanoes Tungurahua (Ecuador), Karymsky (Kamchatka), Etna (Italy), and surface explosions in Russia have revealed that the mesoscale and lower thermosphere layers can exhibit vertical wind velocity gradients up to 10 m s^{-1} per 100 m [179].

Although much attention has recently been paid to the extratropical stratospheric dynamics and their coupling to the troposphere [711, 165, 630, 481, 94, 370, 694, 463], the amount of observational data flowing to numerical weather prediction centers is still insufficient. Therefore, much more data assimilation[12] studies are required to better represent middle atmospheric winds. Recent estimations of tropospheric and stratospheric winds using infrasound from 598 surface explosions from over 30 years in northern Finland were performed by Blixt et al. [95]. They showed, using infrasound waves on their paths between Finland and a ground-based station in Northern Norway, that direct estimations of atmospheric cross-winds can be made from these data using propagation time and back azimuth deviation observations. A framework for assimilation of tropospheric and stratospheric wind information based on infrasound data has been developed by Amezcua et al. [26], paving the way for further infrasound observations. An inversion methodology where infrasound observations are used to update atmospheric models has also been recently reported by Vera Rodriguez et al. [935].

In contrast to the upper atmosphere, extensive data on wind shear and temperature gradients exist for the boundary surface layer so that the theory for this substratum of the lower troposphere is sufficiently well developed (for instance, see Refs. [553, 952] and references therein). The Monin-Obukhov similarity (MOS)

---

[12]Data assimilation consists of combining different sources containing incomplete and imperfect information to produce a better estimate of an atmospheric variable of interest [36]. In most cases, it works with the uncertainty of the information sources.

theory is often applied in modern micrometeorology[13] for the modeling of the atmospheric surface layer [297]. In this theory, the dimensionless wind shear and potential temperature gradient in a horizontally homogeneous surface layer are given by

$$\left(\frac{kz}{v_\star}\right)\frac{\partial \bar{v}}{\partial z} = \Phi_m\left(\frac{z}{L}\right), \qquad (4.15)$$

$$\left(\frac{kz}{\theta_\star}\right)\frac{\partial \bar{\theta}}{\partial z} = \Phi_h\left(\frac{z}{L}\right), \qquad (4.16)$$

where $v_\star$ and $\bar{v}$ are the observed frictional velocity and the mean horizontal wind speed in units of m s^{-1}, while $\theta_\star$ and $\bar{\theta}$ are the temperature scale and the mean potential temperature at height z above the zero-plane displacement in kelvin (K), respectively, and k is the von Kármán constant. The other parameters are the stability parameter z/L, the Obukhov length $L = T_0 v_\star^2/(kg\theta_\star)$, where T_0 is a reference temperature in the surface layer and g is the gravitational acceleration, and the profile functions Φ_m and Φ_h correspond, respectively, to momentum and heat. Equation (4.16) was rewritten in terms of a turbulent Prandtl number, Pr_t, as [406]

$$\left(\frac{kz}{\theta_\star \mathrm{Pr}_t}\right)\frac{\partial \bar{\theta}}{\partial z} = \Phi_h\left(\frac{z}{L}\right), \qquad (4.17)$$

with $\mathrm{Pr}_t = K_m/K_h$, where K_m and K_h refer to the eddy diffusivities of momentum and heat, respectively. In the above relations, the profile functions must be derived from experiments and among the several forms that have been proposed, the most widely accepted ones are

$$\Phi_m\left(\frac{z}{L}\right) = \begin{cases} 1 + \beta_m\frac{z}{L}, & \text{if } \frac{z}{L} > 0, \\ \left(1 - \gamma_m\frac{z}{L}\right)^{-1/4}, & \text{otherwise,} \end{cases} \qquad (4.18)$$

and

$$\Phi_h\left(\frac{z}{L}\right) = \begin{cases} \mathrm{Pr}_t\left(1 + \beta_h\frac{z}{L}\right), & \text{if } \frac{z}{L} > 0, \\ \mathrm{Pr}_t\left(1 - \gamma_h\frac{z}{L}\right)^{-1/2}, & \text{otherwise,} \end{cases} \qquad (4.19)$$

where β_m, γ_m, β_h, and γ_h are numerical coefficients that may vary depending on the underlying surface [568]. In particular, from an analysis of turbulent flows and wind and temperature profiles at the Tazhong station over the hinterland of the Taklimakan Desert in China, the von Kármán constant in near-neutral stratification was found to be 0.4, in agreement with many other studies for different underlying surfaces, while the Prandtl number was ≈ 0.75 and the coefficients β_m, γ_m, β_h, and γ_h were found to be 5.4, 13, 6.1, and 22, respectively [568].

---

[13] According to the AMS, micrometeorology is "the part of meteorology that deals with observations and processes in the smallest temporal and spatial scales, i.e., local processes occurring over spatial scales less than about 1 km and lasting for less than a day." Therefore, the subject of micrometeorology is the bottom of the boundary surface layer, namely the surface layer.

4.6 Turbulence at low altitudes

Data on the turbulence spectra at low altitudes over land is of fundamental importance for commercial aviation operations. Encounters with severe low-altitude turbulence are particularly dangerous and one form of this kind of turbulence that is often associated with convective storms is the so-called buoyancy wave. Investigation of numerous documented aircraft incidents and some fatal aviation accidents at low altitudes are due to buoyancy wave-induced turbulent motions [90]. For example, Fig. 4.1 illustrates the result of an analysis of PIREPs data provided by the *National Center for Atmospheric Research* (NCAR) for turbulence encounters at different altitudes during the whole 2002 year over the Corridor Integrated Weather System (CIWS)[14] domain. A larger number of PIREPs of turbulence encounters occurred at altitudes between 1000 and 6000 ft, with a clear maximum at 2000 ft, implying that approximately 62% of moderate to severe turbulent encounters have occurred within these altitudes [90].

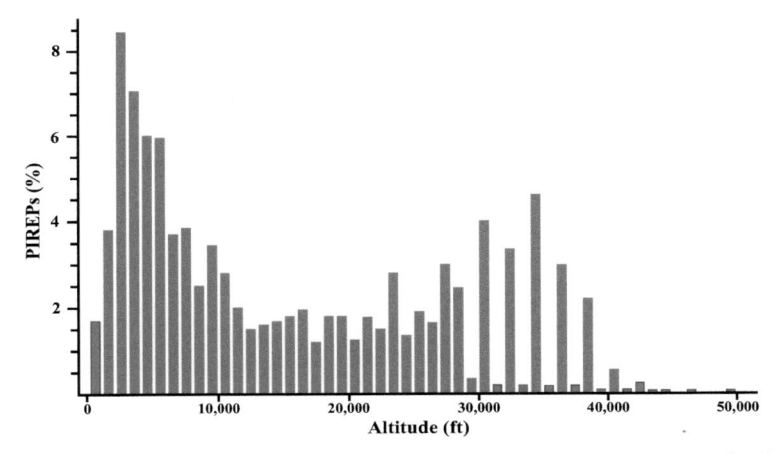

Figure 4.1: Distribution of PIREPs' data with altitude for turbulence encounters for the 2002 calendar year over the Corridor Integrated Weather System (CIWS). Figure of Bieringer et al. [90] reproduced from data provided by the National Center for Atmospheric Research (NCAR).

In the atmospheric boundary layer, turbulence evolves as a response to the local topography and the temperature fluctuations of the Earth's surface. As a result of surface friction, winds in the atmospheric boundary layer are usually weaker than in the upper troposphere and tend to blow toward areas of low pressure. For example, over desert areas, the altitude of the atmospheric boundary layer oscil-

---

[14]The CIWS is an automated system for weather analysis and forecasting, which was designed to improve convective weather decision support for congested en route airspace [488]. The system automatically generates graphical information of current severe weather situations and provides updated forecasts of future weather locations for forecast times from zero to two hours.

lates between 4000 and 5000 m (i.e., from 13100 to 16400 ft), while over the sea it is less than $\sim$ 1000 m (i.e., about 3300 ft) thick. At low altitudes, turbulence typically occurs within the first 1000 ft from the ground level and the effects of these turbulent motions have been of great concern for the aviation industry and operations for more than 60 years. It has long been recognized that turbulence just above the terrain is the result of complex interactions among a number of meteorological factors, the most important of which are the presence of winds blowing over the roughness of the terrain and the thermals, such as those rising from the surface cooling and heating [668]. The turbulent motion generated this way redistributes heat, moisture, the drag on the wind, and ultimately the pollutants produced by human activity. The strength of this mixing is extremely variable and modulates the temperature, humidity, wind strength, and air quality among other factors impacting the weather as we experience it daily.

When flying within a turbulent area, an aircraft may undergo Dutch roll motions, which are a combination of tail-wagging (yawing) and rocking from side to side (rolling) as well as pitching – the up and down motion of the nose of the airplane about its pitch axis – and heaving – the linear vertical up and down movement of the aircraft [642]. However, most modern jet-powered and propeller-driven aircraft are equipped with jaw damper systems to smooth out the tendency that an aircraft will have to experience oscillations in a repetitive rolling and jawing motion. This will not only reduce the pilot workload but will also provide a more comfortable flight to passengers and crew by preventing such undesirable sideslips and bumps. When a craft moves through the atmospheric boundary layer, the relative air velocity is the vector sum of the wind velocity and the craft velocity relative to the ground. Hence, when moving through a turbulent area at speeds much higher than the mean wind velocity, the turbulence perceived by the aircraft, i.e., the relative intensity of turbulence is reduced [960]. Such a reduction in the perceived turbulence intensity is accompanied by an increase of frequency. For flight along the x-direction of a Cartesian reference system, the relative intensity of turbulence is defined by [960]

$$J_x = \frac{v_{x,\text{rms}}}{v_{\text{rel}}} = \frac{\sqrt{\overline{v_x'^2}}}{\sqrt{(v_x + u)^2 + v_y^2 + v_z^2}}, \qquad (4.20)$$

where $\mathbf{v} = (v_x, v_y, v_z)$ is the mean wind velocity vector, v_x' is the fluctuating part of the wind velocity along the flight direction, and u is the aircraft speed measured with respect to the Earth. If u is much higher than the mean wind speed, the denominator in Eq. (4.20) increases in magnitude over the numerator, leading to a reduced longitudinal turbulence intensity. Since commercial airplanes spend most of their flight duration above the atmospheric boundary layer at relatively high speeds, less attention has been paid to this aspect for low-altitude flights. However, in the last decade or so the development of technologies was related to the design and construction of unmanned micro air vehicles[15] (MAVs) [516]. MAVs can be classified into three different configurations, namely fixed wing, rotary wing, and flapping wing. Fixed wing aircraft are highly efficient and can be used for long time flights, while rotary wing models have vertical take-off and landing capabilities and can move in any direction with shorter flight times compared to fixed wing models. Moreover, the flapping wing models are bio-inspired and are either bird- or insect-sized MAVs with high maneuvering capabilities [622, 699]. In general, the instruments used for collecting data for meteorological and wind engineering purposes, such as dynes or ultrasonic anemometers are sometimes larger and heavier than the typical dimensions of MAVs. However, the use of unmanned air systems and MAVs for meteorological studies in the atmospheric boundary layer has been growing in recent years thanks to the incorporation of onboard sensors capable to measure pressure, temperature, humidity, solar radiation as well as to detect low altitude clouds, fog, and turbulence. For example, preliminary results from flight campaigns of several unmanned air vehicles equipped with meteorological sensors for the study of fogs as part of the VOLTIGE[16] project have shown that important data on the daily evolution of the boundary layer height, the level of humidity, and the temperature gradients at low altitudes can be gathered, which among other meteorological predictions are also useful for fog forecasting [385].

The dynamic response of flexible aircraft in low-altitude atmospheric turbulence is a further aspect of great interest in the aviation industry. For example, today's large commercial airplanes are characterized by high flexibility. This results in a stronger interaction between the flight control system and the structural modes [132]. Much effort in this area is addressed to the development of sophisticated analysis tools capable of capturing the relevant interactions between flight dynamics, aerodynamics, structural dynamics, and control. This involves different disciplines such as flight mechanics, aeroelasticity, and flexible body dynamics among others. The balance between a very low take-off weight and a very large wing area have led to very flexible planes of greater endurance that show an enhanced coupling between their rigid-body and structural dynamics response. Long-endurance aircraft have also been constructed for high-altitude missions, including atmospheric science research, telecommunication relay service, and military applications, and are envisioned to serve as atmospheric satellites, with missions lasting from several months to years. In contrast to conventional aircraft, such long-endurance air vehicles typically have flexible wings with very high aspect ratios [163]. With the advent of fast computers and the use of fiber composite materials design engineers have been pushed to implement

---

[15]These are a class of unmanned air vehicles of small sizes and weights that can be used for multiple purposes, including civilian and military applications, such as ground surveillance, armed attacking, search and rescue operations, scientific research, image recording, and sometimes transportation. MAVs are remotely or autonomously controlled and have low noise and low production costs.

[16]Vecteur d'Observation de la Troposphère pour l'Investigation et la Gestion de l'Environnement.

ever more efficient optimization methodologies with the aid of complex mathematical models that simultaneously take into account aspects of flight mechanics, aeroelasticity, and control. For an exhaustive overview of the multidisciplinary design of flexible aircraft, the interested reader is referred to Ref. [56]. With this modern evolution, the role of atmospheric turbulence has become of primary importance in aircraft design. A survey on the state-of-the-art of the nonlinear aeroelasticity of very flexible aircraft with high aspect-ratio wings, as is indeed required by commercial aircraft, is provided by Alfonso et al. [22]. A common trend in commercial aircraft design has been to increase the wing aspect ratio to reduce the lift induced drag, improve fuel consumption, and reduce emissions. However, slender wing configurations have increased flexibility and are subject to higher deflections under the same turbulent conditions, which may result in increased instability. To improve the aeroelastic response and reduce unstable behavior, accurate numerical computer codes must be employed to ensure proper coupling between the aerodynamics and the structural aspects of the models in the presence of nonlinearities [22]. As was pointed out recently by Deskos et al. [234], design criteria based on studies of the interaction between the atmospheric turbulence and the elasticity of very flexible aircraft often rely on the von Kármán spectrum, which is derived from isotropic turbulence arguments. However, it is well known that atmospheric turbulence is not isotropic. In particular, as was already pointed out above, at low altitudes turbulence is strongly modulated by the presence of rough surfaces and heat from solar radiation. This makes turbulence in the lower part of the atmospheric boundary layer to be characterized by large-scale eddies and hence to depart from isotropicity. In contrast, the von Kármán view may be better represented at much higher altitudes by CAT, which is characterized by eddies of smaller scales [664, 234]. On the other hand, since high-aspect ratios enforce a strong coupling between rigid-body and elastic motions [692, 735, 794, 866], improved flexibility may induce dynamic instabilities of significant magnitude under the presence of large-scale turbulence structures [22]. Experimental studies have shown that the inherent complexity implicit in the interaction between wind gusts and aircraft aeroelasticity would demand additional theoretical and experimental analyses [568]. An investigation of low-altitude atmospheric turbulence models for aircraft aeroelasticity has been performed by Deskos et al. [234], where the industry-standard, one-dimensional von Kármán model, the stochastic, two-dimensional Kaimal model [448], and full three-dimensional LES models [718, 103] were assessed in simulations of turbulence near the ground. They concluded that LES was able to simulate realistic turbulence close to the ground much better than the other two models, with large scales extending in the streamwise wind direction and generating coherent structures and flow features typical of low-altitude atmospheric turbulence.

PART II

Chapter 5

Types of Forecasts for Aviation

5.1 Introduction

In general, a weather forecast is a statement providing information about expected meteorological conditions in a specific area or portion of airspace and for a specific time or period. Such reports are prepared by a meteorological office and include information on recent weather conditions, statistical data, analyses of the processes that may intervene in a change in weather conditions, climatic factors, and computer-based predictions. In particular, when dealing with aeronautical weather forecasts, the relevant information usually focuses on the following aspects: the level of visibility; detailed information on the wind speed, including its strength, its direction, and the occurrence of gusts as well as the presence of low-level wind shear; types and amount of clouds that can be found during flight as well as their vertical extent; the amount and type of precipitation; atmospheric parameters, such as pressure, temperature, and dew point; in-flight icing; CAT; extreme weather conditions such as those caused by thunderstorms, and intense rotating depressions over tropical ocean areas, such as typhoons, hurricanes, and cyclones, mountain waves, sandstorms, and volcanic ashes among other factors. Very often all this information is accompanied by a report on expected changes to forecast conditions.

A tool that provides invaluable meteorological information to commercial flight operators and crew is the so-called Significant Weather (SIGWX) charts. These are a vital flight planning tool for longer flights and when weather issues may be an important factor. For example, the production of SIGWX forecasts as specified by the *International Civil Aviation Organization* (ICAO) is the responsability of the WAFCs. In the world there are only two WAFCs: the WAFC London, based at the United Kingdom's Met Office headquarters in Exeter, and the WAFC Washington, based at the USA's Aviation Weather Service offices in Kansas City, Missouri. Depending on the height from the ground level there are three types of SIGWX forecasts, namely the high-, mid-, and low-level SIGWX forecasts. These fall in the category of the *area and route forecasts*. Other types of forecasts that are regularly used in aviation are the *aerodrome* and *special* forecasts. This chapter is devoted to a detailed description of all these types of forecasts.

5.2 High-level SIGWX forecasts

High-level SIGWX charts provide a forecast of significant en-route weather phenomena and encompasses airspace in a range of flight levels from FL250[1] to FL630 over the USA, Mexico, Central America, portions of South America, the western Atlantic, and the eastern Pacific. Such prognostic charts are derived for both domestic and international flights. The weather aspects that are usually depicted on these charts include active thunderstorms, tropical cyclones, severe squall lines, moderate and/or severe CAT as well as turbulence in clouds, widespread sandstorms, and moderate to severe icing. Also, they include information of clouds associated with active thunderstorms, squall lines, and cyclones in the range of altitudes between 10000 ft from the mean sea level and FL250 and in cumulonimbus clouds above FL250. The surface positions of convergence zones and frontal systems as well as their speed and direction of motion, the tropopause heights, the position of jet streams, and eventual volcanic activities are also indicated in high-level SIGWX charts. In this latter case, information is also given on the name of the volcano, its position in latitude and longitude as well as the date and time of the first eruption, including the airspace area occupied by the volcanic ashes.

Atmospheric events of interest for high-level SIGWX forecasts are depicted on prognostic charts using abbreviations and symbols. For example, thunderstorm activity and the presence of cumulonimbus clouds are identified by the abbreviation CB. This abbreviation does not refer to isolated or occasional (i.e., scattered) cumulonimbus clouds, but to either the occurrence or expected occurrence of an area characterized by widespread cumulonimbus clouds with no

---

[1]This notation is used in aviation and aviation meteorology to mean an aircraft altitude at standard atmospheric pressure. It is described by a number of two or three digits, which indicates the pressure altitude in units of 100 ft (30.48 m). For example, a pressure altitude of 25000 ft (7620 m) is referred to as "flight level" 250 or simply FL250. In addition, the pressure altitude is the height above a theoretical level where the weight of the atmosphere is 1013.25 hPa, as measured by a barometer. This is equivalent to the standard sea-level pressure in the International Standard Atmosphere (ISA). The pressure altitude is used in high-altitude flight above the transition altitude, which is the height from the sea level where an aircraft changes the use of barometer derived altitudes to the use of flight levels.

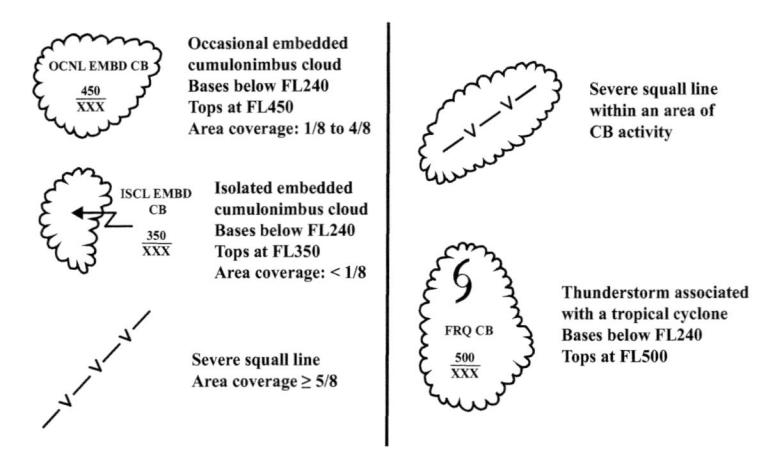

Figure 5.1: Symbols employed to indicate the presence of cumulonimbus clouds, squall lines, and tropical cyclones in high-level SIGWX charts along with a brief description of their meaning.

space between individual clouds or cumulonimbus clouds embedded in cloud layers[2] or concealed by haze[3] or dust. In general, the CB symbol warns the pilot about the presence of moderate to severe turbulence and icing. Figure 5.1 illustrates the symbology employed to indicate the presence of cumulonimbus clouds, squall lines, and tropical cyclones along with a description of the meaning [2, 1]. The scalloped lines in the chart define the area to which the forecast applies. The identifier ISOL EMBD CB means isolated embedded CB, OCNL EMBD CB indicates occasionally embedded CB, while ISOL CB in HZ and OCNLCB in HZ indicate isolated and occasionally CB in haze, respectively. For example, ISOL in coverage terms is less than 1/8, ONCL is from 1/8 to 4/8, and FRQ CB, which refers to frequent cumulonimbus with little or no separation, are more rare events with coverage exceeding 4/8, i.e., between 5/8 and 8/8. The figures above and below a short line indicate the top and base of clouds. For example, in the top symbol of Fig. 5.1, for ONCL EMBD CB, 450 refers that the cloud tops may reach FL450, while XXX means that the bases are below FL240. Tops above the upper limit of FL630 are indicated by ABV 630.

Turbulent areas, including moderate and severe CAT, are enclosed by bold dashed lines as shown in Fig. 5.2. In general, CAT includes all forms of turbulence not caused by atmospheric convective activity as, for instance, areas of turbulent motion induced by wind shear and mountain waves. The vertical extent of these areas is given in units of 100 ft from the standard sea level and within

---

[2]According to the NOAA's glossary, a cloud layer consists of an array of clouds whose bases are approximately at the same level. A cumulonimbus cloud is characterized by coverage, bases, and tops.

[3]Sometimes abbreviated by HZ, in meteorology haze refers to an aggregation of very fine particles, which may be solid or liquid, or both, in the atmosphere. Such fine particles give the atmospheric air an opaque appearance that represses colors.

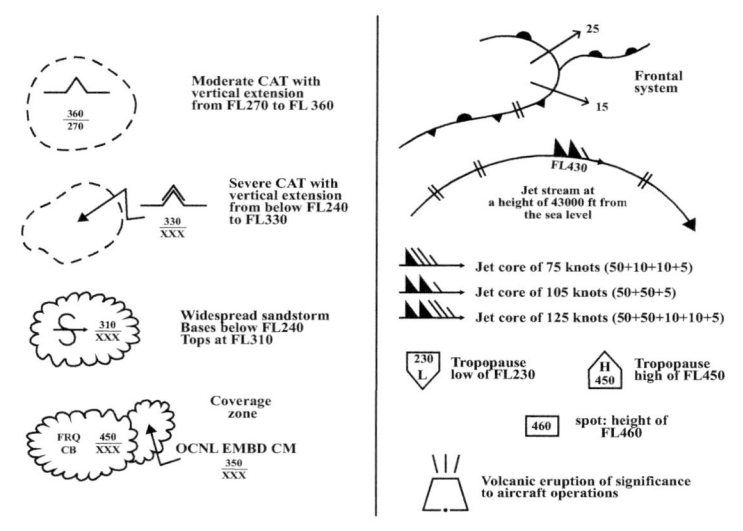

Figure 5.2: Symbols employed to indicate the presence of CAT, sand/duststorms, convergence zones, fronts, tropopause heights, jet streams, and volcanic activity in high-level SIGWX charts along with a brief description of their meaning.

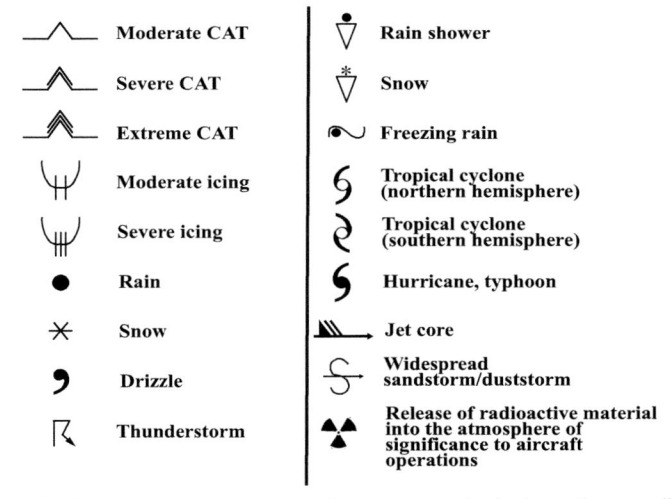

Figure 5.3: List of symbols employed to indicate meteorological weather conditions in use for forecasting purposes on high- and mid-level SIGWX prognostic charts.

the area enclosed by the dashed lines are the corresponding symbol indicating the turbulence intensity. Figure 5.3 shows the various symbols and their meaning for several atmospheric and weather events. Figure 5.2 also depicts the symbols for widespread sandstorms (BN) and duststorms (BD), whose areas are also enclosed by scalloped lines. The enclosed areas of convergence zones, which involve active thunderstorms are also represented by scalloped lines. As for the cumulonimbus clouds, the figures above the short line indicate the tops of the storms and convergence zones in flight levels and the XXX below the short line indicates that their bases are below FL240. In general, the base defines the lower bound of the forecast, while the top is above the upper bound of the forecast. The first row on the right in Fig. 5.2 shows the symbol for a frontal system, where the curved lines indicate the surface position, the arrows point in the direction of the front movement, and the numbers indicate the speed in knots. The tropopause heights are expressed in units of 100 ft from the standard sea level and are represented in a high-level SIGWX chart by a five-sided polygon. The example in Fig. 5.2 shows a low (L) tropopause height of 23000 ft over the mean sea level. Areas of very flat tropopause slope are represented by a small rectangle enclosing the height in hundreds of feet. On the other hand, jet streams whose core speeds are over 80 knots are displayed in high-level SIGWX charts. An example of the symbol employed to identify jet streams is now given in the second row on the right in Fig. 5.2. The height of the jet stream is expressed in flight levels. The core speed at the beginning of the curved arrowed line is always 80 knots, while the double, hatched lines to the left of the maximum speed (represented by an arrow with pennants and feathers) indicate a 20 knot increase of the core speed, while those to the right of the maximum speed indicates a decrease of 20 knots. The symbol in the last row on the right in Fig. 5.2 is employed to represent volcanic activity. The dot at the base of the trapezoid indicates the latitude/longitude of the volcano on the high-level SIGWX charts. Usually, any other information concerning the volcanic activity, such as the name of the volcano, the date and time of the first eruption, the airspace area covered by the ashes are also included in the legend of the chart. An example of a high-level SIGWX forecasting chart is shown in Fig. 5.4, which includes many of the atmospheric weather conditions described above. As was mentioned above, such charts are provided for en-route portions of international flights and are used by airline dispatchers for flight planning and weather briefings before departure as well as by crew members during flights. High-level SIGWX charts are valid at specific fixed times, namely 0000, 0600, 1200, and 1800 UTC. These charts are provided in accordance with the WMO and the *Weather Area Forecast System* (WAFS) of the ICAO. A suite of SIGWX forecasts for the WAFC Washington are provided by the *National Weather Service* (NWS) and the *Aviation Weather Center* (AWC) for different ICAO areas around the world, including the north and south Pacific, Europe, Africa, and the Americas. In addition, the WAFC London also issues high-level SIGWX charts for other geographical areas of the world.

5.3 Mid-level SIGWX forecasts

As in the case of high-level SIGWX charts, mid-level ones also depict several weather conditions that can be hazardous for aviation operations. However, these

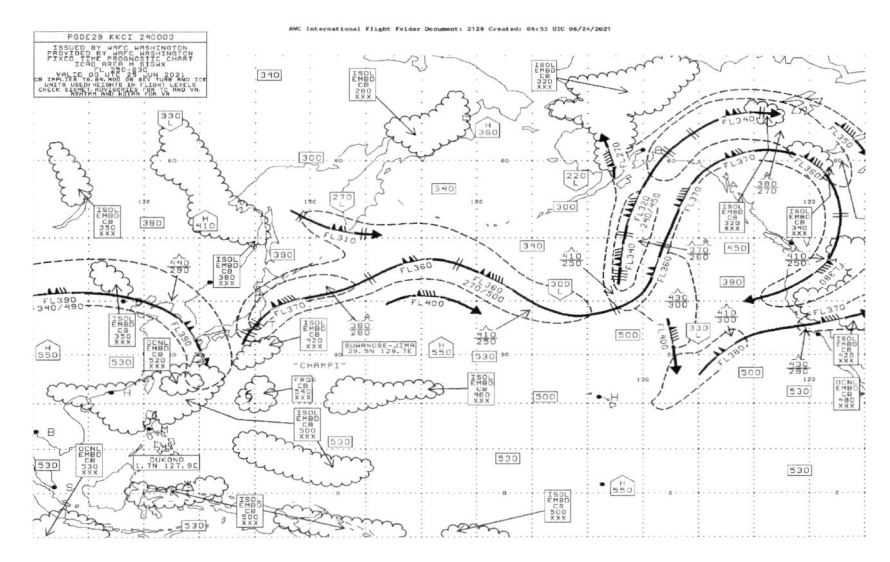

Figure 5.4: Example of a high-level SIGWX chart, which includes jet streams, CAT, CB clouds, thunderstorms, tropopause heights, squall lines, and tropical cyclones. Figure retrieved from the National Weather Service of the National Oceanic and Atmospheric Administration (NOAA/NWS) data and products at the URL: https://www.weather.gov/disclaimer.

charts are used for an overview of selected flight weather conditions between 10000 ft of altitude from the standard sea level and FL450. For example, the abbreviation CB is included only when an area of widespread cumulonimbus clouds, cumulonimbus embedded in cloud layers, cumulonimbus concealed with haze, or even cumulonimbus clouds with little or no space between individual clouds appear within this range of altitudes. The symbols for the representation of cumulonimbus clouds and area coverages are the same as shown in Fig. 5.1. The inclusion of CB or the thunderstorm symbol (see Fig. 5.3) should be understood to include all weather phenomena that are normally associated with these conditions, namely moderate or severe icing, moderate or severe CAT, and hail. The same symbols as those depicted in Figs. 5.1, 5.2, and 5.3 are also employed to represent jet streams, tropopause heights, tropical cyclones, moderate or severe icing, moderate or severe CAT, and volcanic eruptions.

Mid-level SIGWX charts include vertical depth forecasts for jet streams only when their maximum velocity exceeds 120 knots. In particular, the jet depth is given by the vertical depths up to the point in units of flight levels where the wind field above and below the axis of the jet stream drops to 80 knots. Codes for non-cumulonimbus cloud coverage are as follows: SKC for sky clear with 0/8 coverage, FEW for few clouds with area coverage from 1/8 to 2/8, SCT for scattered clouds with area coverage from 3/8 to 4/8, BKN for broken clouds with 5/8 to 7/8 coverage, and OVC for the overcast sky with 8/8 coverage. In addition, the following suffixes are used to identify other cloud types that can appear on a mid-level SIGWX chart: cirrus (CI), cirrocumulus (CC), cirrostratus (CS), altocumulus (AC), altostratus (AS), nimbostratus (NS), stratocumulus (SC), stratus (ST), and cumulus (CU). The presence or release of radioactive materials in the atmosphere is also depicted by the standard symbol along with information on the chart regarding the latitude/longitude as well as the date and time of the accident. As part of the WAFC Washington, the AWC issues a 24-hour mid-level SIGWX chart four times per day at 0000, 0600, 1200, and 1800 UTC for the North Atlantic Ocean Region (NAT). The *Bureau of Meteorology of the Australian Government* also issues high- and mid-level SIGWX prognostic charts at the same UTC times for the Australian region. The mid-level SIGWX forecast provided by the WAFC London between FL100 and FL450 covers the mid Atlantic, the Middle East, Africa, and the south of Asia.

5.4 Low-level SIGWX forecasts

A low-level SIGWX chart is a forecast for use in flight preparation and planning. It describes weather situations of significance to aviation operations for the lower airspace for altitudes between FL245 and the surface (SFC[4]). In the United States, the forecast domain covers the 48 contiguous states and the coastal waters, while in Europe it covers the central parts for a defined time. For example, the AWC issues low-level SIGWX charts four times a day. At each time, two prognostic charts are issued, one in advance of 12 and the other of 24 hours. These provide an overview of selected aviation weather hazards up to 24000 ft from the ground level and depict a snapshot of weather expected to occur at the specified valid time. For central Europe, low-level SIGWX charts are published every four hours, six times a day, and they are usually available three hours ahead of valid time, while outlooks are given seven hours in advance.

In general, low-level SIGWX charts highlight significant weather phenomena, displaying symbols for surface pressure centers, areas of maximum wind speeds, turbulent and icing areas, and surface fronts. In addition, they include relevant information concerning surface visibility, freezing levels, the presence of low-level clouds, and warnings about volcanic activity and the release of radioactive sources or materials in the atmosphere by disruption of a nuclear plant. The symbols employed to highlight these weather situations and warnings are the same displayed in Figs. 5.1, 5.2, and 5.3. Height information from the standard sea level is provided in units of hectofeet and when referred to flight levels it is explicitly marked by the suffix FL. The height of turbulent areas is depicted by two numbers separated by a slanted line. For example, a turbulent area on the chart indicated as 240/100 means that it can be expected from the top at FL240

---

[4]In the weather terminology, SFC means "surface", i.e., ground level. In weather reports SFC also stands for "surface wind speed".

to the base at 10000 ft over the standard sea level. On the other hand, turbulence reaching the ground level is forecast by omitting the base height. For example, a turbulent layer of height 7000 ft from the standard sea level will be identified by 070/, while turbulence associated with thunderstorms is never depicted on low-level SIGWX prognostic charts. Moreover, a freezing level is an altitude at which the air temperature drops below 0°C and a surface freezing level separates above-freezing from below-freezing temperatures at the Earth's surface. Above the surface, freezing levels are depicted in hundreds of feet from the standard sea level. In low-level SIGWX charts, they are represented by a two- or three-digit number. For example, 60 identifies a 6000-ft freezing level contour. For locations between contour lines, the freezing level is determined by simple interpolation. On the other hand, multiple freezing levels can be forecasted on prognostic charts where the temperature in Celsius is negative (i.e., below freezing) at the surface. On real charts, multiple freezing levels are possible below 4000 ft from the standard sea level, but the exact cut-off cannot be easily determined. Information is also given on the surface pressure centers. Figure 5.5 depicts some weather symbols that can appear only on low-level SIGWX charts. In particular, high pressure centers are represented by a bold H, while a low pressure center is depicted by a bold L. In Fig. 5.5, the figure 1020 on the bottom left of bold H identifies a stationary or moving (at < 5 knots) high pressure center with a mean sea level pressure of 1020 hPa. In the other two examples, the three- and four-digit numbers indicate the pressure in hPa, while the arrows indicate the direction of movement and the numbers near the arrowheads indicate the speed of movement in knots.

Within a weather area the prevailing conditions are marked in the predominant part of the area. In the case that variant weather conditions are present, which affect only part of the area, they are preceded by a location descriptor. The following acronyms are used: E (east), NE (northeast), NW (northwest), W

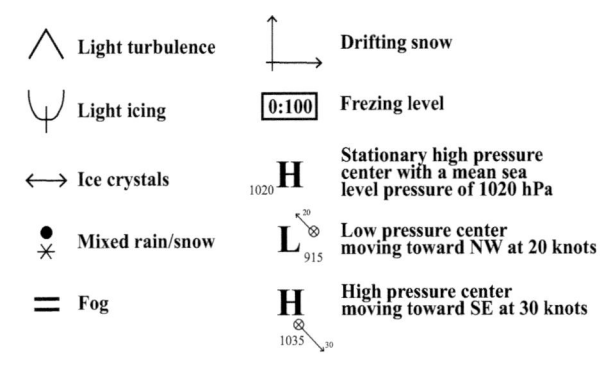

Figure 5.5: List of some symbols employed to indicate meteorological weather conditions in use for forecasting purposes on low-level SIGWX prognostic charts.

(west), S (south), SE (southeast), SW (southwest), COT (at the coast), MAR (at sea), MAR/COT (at sea and coast), LAN (inland), MT (mountains), MON (above mountains), VAL (in valleys), and LCA, which means locally, i.e., less than 50% of the area. Acronyms for significant weather phenomena are: SQL (squall line), TS (thunderstorm), FZ (freezing), GR (Hail), GS (small hail), SN (snow), RA (rain), DZ (drizzle), FG (fog), BR (mist), MTW (mountain waves), and LL-Jet (low-level jets). Combinations of these suffices may be formed as, for example, TSRA, which means a thunderstorm with rain, or SHRA, which mean shower rain. For example, when TSRA and SHRA occur, only the combined suffix TSRA is displayed on the chart. The quantitative indications for CB, TCU (towering cumulus), TS, and SH employ the same identifiers used in the mid- and high-level SIGWX charts. That is, ISOL, OCNL, FRQ, and EMBD means, respectively, isolated CB/TCU/TS/SH (area with coverage < 50%), frequent but clearly separated CB/TCU/TS/SH (areas with coverage in the interval between 50 and 75%), frequent CB/TCU/TS/SH with little or no separation (area with coverage ≥ 75%, and embedded CB/TCU. The visibility is another factor that is taken into account in low-level SIGWX charts and is expressed in meters.

Low-level prognostic charts are often created by aeronautical meteorological forecasters based on the results of numerical weather simulations and real-time observations and warnings. In Germany, a public institution responsible for providing such prognostic charts is the *Deutscher Wetterdienst* (DWD), which is Germany's National Meteorological Service. A sub-category of low-level SIGWX charts are the so-called Short-range surface prognostic (prog) charts, which provide a forecast of surface pressure systems, fronts, and precipitation every two days so that they can be used to overview the change in position, size, and intensity of weather conditions during the next two days. Since these charts are issued for forecast periods of 12, 24, 36, and 48 hours, the progression analysis is performed by comparing the observed conditions for these periods. In the United States, short-range surface prognostic charts are issued by the *Hydrometeorological Prediction Center* (HPC) in Camp Springs, Maryland. Prognostic charts for the North Atlantic and Europe are provided by the UK Met Office.

5.5 Aerodrome or terminal airfield forecasts (TAFs)

In aviation, a terminal aerodrome (or airfield) forecast (TAF) is a report on weather forecast information within an area of radius 5 statute miles[5] from the center of an airport's runway complex. For major civil airports, these prognostic reports are issued every 6 hours at 0000, 0600, 1200, and 1800 UTC and apply to periods ranging from 24 to 30 hours. In general, TAFs follow a slightly

---

[5]A statute mile is a unit of distance on land employed in English-speaking countries. It is equivalent to 1.609 km.

different format than METARs[6], which are the standard code formats for transmission of surface weather observations in the world. In fact, METAR reports are understood in most countries. However, TAFs are much easier to decode than METARs and NOTAMs[7] and are the standard code format for forecasts issued for airports and aerodromes.

In the United States, TAFs are produced by weather experts in 122 different weather forecast offices around the country. Usually, depending on how quickly the weather conditions change, there are three types of TAF reports: (a) the TAF normally produced, (b) the amended (AMD) TAF, which will always supersede the previously issued TAF, and (c) the corrected TAF (TAF COR). Only under particular conditions like fog, rain showers, or nearby thunderstorms, the TAF report may extend up to a radius of 10 statute miles from the airport's center. Any TAF will begin with the airport identifier or ICAO airport code. The airport code identifier is followed by the date and time origin or, failing that, by the time when the forecast was updated. The date/time group is given by the day of the month, the time starting with the hour, and the letter "Z" to remember that the TAF is not in local time but in "ZULU" time. For instance, the code "051730Z" indicates that the report was issued on the 5th at 1730 UTC/ZULU. The next piece of information is the so-called "valid time period", which is the amount of time covered by the TAF, starting with the day of the month and the first two numbers of the time. For example, the code "0518/0624" indicates that the report is valid from the 5th of the month at 1800 UTC until the 6th of the month at 2400 UTC. The heart of the TAF is given by the initial meteorological conditions, which provide information about the wind, the visibility, the weather, the sky conditions, and optionally about the wind shear. This part of the TAF is also contained in METAR reports, except that unlike the METAR, the TAF will only report cumulonimbus clouds (CB) and the lowest ceiling of the sky condition.

There are some specifics of a TAF that will never be seen in a METAR, such as Wind Shear (WS), Probability (PROB), Temporary (TEMPO), From (FM), and Becoming (BCMG). Regarding wind shear, the forecast includes the height of the wind shear followed by wind direction and speed at the indicated height. Wind shear is usually reported up to 2000 ft. For example, the code "WS010/18040KT" means that wind shear is reported at 1000 ft from the south at 40 knots. On the other hand, BCMG is used when the weather will change gradually over two hours or more. This is particularly useful for long-haul international flights. Unlike the FM group, the BCMG one will only indicate the change of one type of weather condition. The PROB group will appear when the chance of precipitation of thunderstorms is 30% or greater. However, only the

---

[6]METARs are formats for reporting information on weather conditions. They typically come from airports and/or permanent weather observation stations and are predominantly used by pilots and meteorologists, who assist weather forecasting by adding METAR information.

[7]A NOTAM (Notices to Airmen) is a notice containing information that is essential to personnel concerned with flight operations.

NWS uses PROB groups for a 30% chance and they do not include forecasts of high- and mid-level SIGWX in the vicinity or non-convective low-level prognostics. When a long list of FM and TEMPO indicators appear in a TAF, it implies that there will be changes to the main weather within the 24 hours covered by the report. For example, the TEMPO group is used when the weather prognostic has more than 50% probability to occur and covers less than half the period of the TAF. Otherwise, if the prevailing weather conditions are going to change rapidly, that is, in less than an hour, then the FM group is used. Anything after the FM will be now the new main forecast. In this case, the time format will start with the day and time when the new forecast will begin. Most of the discussion developed in this section relies on information about TAF decoding from Ref. [313].

5.6 Special forecasts

A special report or forecast (SPECI) is issued whenever significant changes have occurred or are expected to occur to the published report or forecast. These are published in the SPECI code format. The SPECI code is similar to the METAR/TAF code, except that it is unschedule and taken when any of the following features are observed [3]. The first criterion concerns the wind direction and speed. For example, wind shifts by more than $45°$ occur in less than 15 min and wind speed increases by 10 knots or more during the shift are reported immediately after the observation. Other criteria include the surface visibility when it decreases from or increases over the value previously reported by 1 to 3 miles and the runway visual range (RVR) when its value decreases from or increases over 2400 ft during the preceding 10 min. Important weather features include observations of tornadoes, funnel clouds or waterspouts; the begin and end of a thunderstorm; any kind of precipitation, including hail, freezing precipitation, and ice pellets, as well as any detected change of intensity of these phenomena; the eventual occurrence of squalls; and changes in the sky conditions due to cloud layers or obscurations below or above 1000 ft from the preceding METAR or SPECI report. Other meteorological situations caused by volcanic eruptions as well as situations designated by the responsible agency as critical will also be reported by SPECIs. However, the above SPECI criteria are not applicable to all stations, but only to those that are capable to evaluate the event in course. This is the case of visual tornadoes, which cannot be evaluated by automated stations working without staff.

Chapter 6

Effects of CAT on Aviation Operations and Aircraft

6.1 Introduction

Encounters with low and moderate air turbulence at cruising altitudes have become a normal event for most passengers and crews. Although such encounters are unsettling and occasionally they can be responsible for injuries, pilots are not worried about because modern aircraft are engineered to resist the effects of CAT that are normally encountered without showing obvious damage in most cases. In particular, this is true because commercial aircraft are designed for optimal performance and response at cruise speeds. On the other hand, experienced pilots are generally good at avoiding dangerous CAT on their flight paths aided by the use of SIGWX charts, real-time reports from other aircraft, radar returns, and periodic updates from the ground for long flights. However, severe and extreme wind shear at relatively low altitudes can cause serious aircraft damage and even crashes during approaches to airports, take-offs, and landings. Air turbulence is generally classified, according to its intensity, as light, moderate, severe, and extreme. While CAT is the type of turbulence that is generally encountered at high altitudes, other types of CAT include mechanical, wake, mountain wave, convective, and frontal turbulence.

In this chapter, the several categories and types of air turbulence are reviewed along with their impact on commercial aviation, including flight diversions, delays, and structural damages to airplanes. In general, atmospheric turbulence is short-lived and has a direct impact on aviation operations with consequences that range from low to very strong ones. Concerning the human aspect, reports by the FAA indicate that injuries and wounds suffered by passengers and staff on board have increased the payments by healthy and material affectations. The chapter ends with the key question of whether CAT can cause a plane crash along with updated statistical data referring to the trends of passenger and crew injuries as well as the operational costs faced by commercial airlines due to atmospheric air turbulence.

6.2 Categories and types of air turbulence

Atmospheric turbulence is often described as chaotic and random flow motions that occur between shearing air layers. Such motions are unpredictable and fragment into swirls and eddies that cascade into smaller and smaller sizes until they dissipate at the smallest scales. Aircraft wings are aerodynamically designed to provide a smooth and constant lift under steady wind conditions, enabling the airplane to fly. When the airflow becomes turbulent, the aircraft experience Dutch roll motions when it flies through the vortices, causing it and its occupants to shake. The violence of the shaking will depend upon the intensity of turbulence. Based on the reported perception of pilots[1] intensity has been graded from light to extreme, passing through moderate and severe.

Light air turbulence is quite common and present in any commercial flight. It causes slight bumpiness as the aircraft momentarily experiences very slight and erratic changes in altitude. As a consequence, passengers may feel a slight, and in some cases, imperceptible strain against their seat belts. Even though it may cause the displacement of loose and light objects, passengers can walk around the cabin with very little or no difficulty. Light turbulence is felt when the velocity of air fluctuates between 5 and 15 knots and the vertical gust velocity between 15 and 20 ft s^{-1}. Under moderate turbulence, the aircraft can go up and down distances of up to 6 m. During these more intense bumps, the pilots do not lose control of the plane and passengers find serious difficulties walking around and experience a much stronger strain against their seat belts. Drinks may spill and loose objects will move out from their places. Although it is less frequent than light turbulence, it may be encountered at least once in long, transoceanic flights when the airspeed fluctuates from about 15 to 25 knots and the vertical gust velocity from 20 to 35 ft s^{-1}. Severe turbulence, on the other hand, is encountered when abrupt changes in altitude of up to 30 m may occur accompanied by

---

[1]For example, according to the guidelines contained in the *Aeronautical Information Manual* (AIM) pilots flying ahead can pass on real-time information on turbulence to other flight crews by reporting the location, altitude, and aircraft type so that the latter can determine how their aircraft can be affected. When moderate to severe turbulence is encountered occasionally, the frequency of occurrence is reported by pilots as less than 1/3 of the flight time, while if turbulence is encountered intermittently or continuously it is reported with a frequency corresponding to 1/3 to 2/3 or more than 2/3 of the flight time, respectively.

airspeed fluctuations greater than 25 knots and vertical gust velocity variations between 35 and 50 ft s^{-1}. During these more violent up and down motions the airplane may momentarily be out of control and passengers are violently pressed against their seat belts. Under these circumstances, passengers who are unbuckled from their seat belts will be thrown out of their seats with a high risk of serious injuries and wounds, while unsecured objects will be tossed about. In general, extreme turbulence is confined to the strongest forms of wind shear and convective motions, which are usually characterized by rapid fluctuations in the airspeed (> 25 knots) and vertical gust velocity (> 50 ft s^{-1}). In commercial flights it is rarely encountered mainly because experienced pilots know how to avoid it. Under extreme CAT the aircraft is violently tossed about, becoming completely out of control and undergoing severe structural damage. However, the categorization of turbulence as severe/extreme is subjective since it would depend on the aircraft size and cruise speed. For example, moderate or severe CAT in a big commercial airplane can be perceived as extreme by a light plane. Similarly, the occupants of a heavy airplane flying at low speed may feel moderate turbulence as if it were severe compared to the same crew flying at a higher speed.

Based on its production mechanism, atmospheric turbulence can be classified into seven types, namely clean air turbulence, wake turbulence, mechanical turbulence, mountain wave turbulence, convective turbulence, frontal turbulence, and wind shear [350]. All these turbulent motions are indeed forms of CAT in the sense that they are not visible and therefore difficult to forecast and detect. In particular, clean air turbulence, which in most of the available literature is referred to CAT, occurs near the tropopause and is felt when a plane crosses the boundaries of a jet stream [862]. The wind shear associated with jet streams is generally a source of severe turbulence. Encounters with this class of turbulence are more frequent in winter when the jet stream lies at lower altitudes and latitudes [206]. Another invisible phenomenon that poses serious hazards to aviation is the pair of counter-rotating cylindrical vortices that form in the wakes of in-flight aircraft. Since these wake vortices grow in size with distance from the airplane, they could be highly dangerous for following airplanes if they are not kept at a safe distance. To prevent crashes, the FAA imposes a 6 miles distance in the airports between arriving and departing aircraft. At such distances, the wake vortices, produced by heavy aircraft when flying slowly under stable atmospheric conditions, will almost dissipate, thereby reducing the risk of accidents during take-off, landing, and approaching operations to the terminal area. A form of air turbulence that is important when flying at low altitudes is that caused by obstruction of air currents near the Earth's surface by the presence of tall buildings, terrain irregularities, such as hills and low mountains, and trees. In this case, air disruption occurs as the air flow suddenly changes direction at contact with the obstacle, forming large-amplitude vortices that can spread upward. This mechanical turbulence can be hazardous for light planes flying at relatively

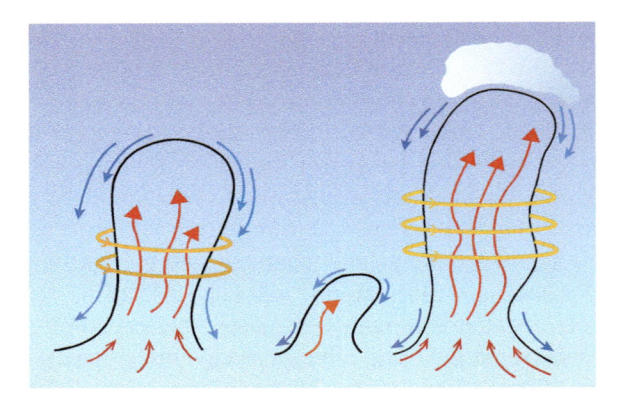

Figure 6.1: Schematic picture showing thermal bumps rising in the atmosphere.

low altitudes over cities and mountains. A further source of severe CAT is caused by mountain waves. For example, when air flows around and through the top of mountain ranges, gravity waves are produced which give rise to turbulent eddies downwind from mountain ridges. Such turbulent air masses can propagate to high altitudes and can be encountered at heights up to 20000 ft or more above a 12000-ft range [862]. A more detailed description of mountain waves has been given in Chapter 3, Section 3.10, where a picture showing the development and upward propagation of such waves is also depicted in Fig. 3.6.

Thermal (or convective) turbulence is the result of heat convection currents rising from the Earth's surface and mixing with cooler air flowing over the surface. This is similar to the natural convection arising from the Rayleigh-Bénard instability[2] [332, 579]. A similar mechanism occurs in the atmosphere, where solar heating of the Earth's surface causes bubbles of air in contact with it to heat up and rise. Since the sunlight heating of the surface is not the same everywhere, the vertical currents of warm air rise at different rates forming thermal bumps of different heights and acting as obstructions to the normal airflow as occurs through mountainous terrain, as shown schematically in Fig. 6.1. When wind flowing over the Earth's surface encounters such convective thermal bumps, it deviates around them resulting in turbulent eddies. In general, the turbulence intensity depends on the size of the convective thermals and the prevailing wind

---

[2]When a fluid with a free surface is heated from below, a regular pattern of convective cells or rolls forms. These cells are commonly referred to as Bénard cells. Because of gravity, warmer and less denser pockets of fluid rise, while cooler and denser ones sink. Convection occurs when the Rayleigh number

$$\mathrm{Ra} = \frac{g\alpha\beta}{\kappa v}H^4,$$

exceeds a critical value for which gravitational forces dominate over viscous damping forces, where g is the acceleration due to gravity, α is the coefficient of thermal expansion, κ is the thermal diffusivity, v is the kinematic viscosity, H is the height of the fluid, $\beta = (T_l - T_u)/H$ is the temperature gradient, T_l is the temperature of the lower fluid surface, and T_u is that of the upper fluid surface.

velocity. This is perhaps the most common form of atmospheric turbulence encountered at relatively low altitudes. On sunny days the strongest thermals are expected to form around midday and can reach altitudes as high as 13000 ft. When cumulus, towering cumulus, or cumulonimbus clouds are present, the thermal turbulent layer extends from the surface to the cloud tops. However, sometimes dry thermals are encountered when the thermals are not made visible by the presence of cumulonimbus clouds. When convective currents are strong enough, they produce thunderstorms which are often associated with severe turbulence. Since thermal turbulence can also be expected at low altitudes, especially when a cold air mass moves over a warm surface, it can be highly hazardous in the proximity of terminal areas during take-off and landing operations. Under these conditions, a landing aircraft may be displaced by the thermals from its normal glide path and forced to either overshoot or undershoot the runway.

Frontal turbulence is encountered when a mass of warm air is lifted by a fast-moving mass of cold air. If this is accompanied by abrupt wind shear between the air masses, then the frontal surface may become unstable and lead to the formation of severe turbulence. Figure 6.2 shows schematically the formation of turbulent flow across a moving cold front. In this case, the mixing of warm and cold air at the front surface together with the differences in the wind velocities and directions between both masses of air are all factors that contribute to increasing the intensity of turbulence. Other factors that can magnify the intensity of frontal turbulent motions are the width of the frontal zone, where mixing occurs, the temperature difference between the warm and the cold air, the moisture and degree of instability of the warm air, which will induce stronger vertical currents across the frontal surface, and the presence of associated thunderstorms. Since fronts are often observed to rotate around a low pressure center, in weather maps they are represented by closed isobars about the pressure center. A list of symbols used in SIGWX charts to represent different fronts is depicted in Fig. 6.3. For example, in cold fronts, a squall line of thunderstorms may sometimes form in the warmer side in advance of the front, which in most cases is triggered by wind shear and mixing in the front boundary. However, the squall lines

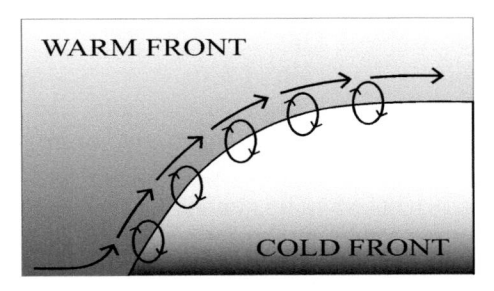

Figure 6.2: Schematic picture illustrating the formation of eddies just ahead of a fast-moving cold front.

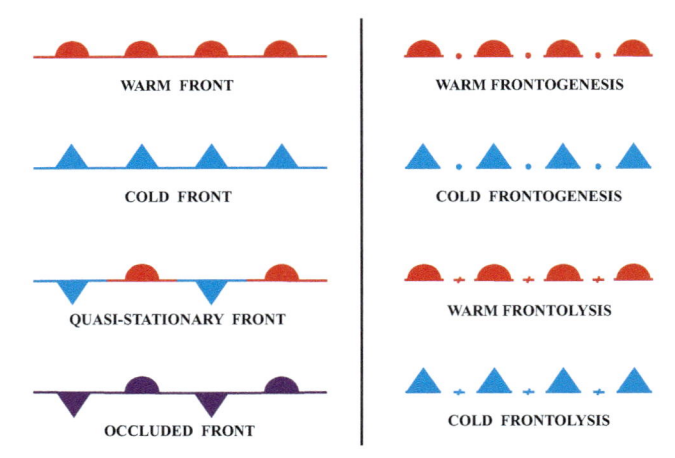

Figure 6.3: List of symbols employed to indicate the presence of frontal systems on SIGWX prognostic charts. Frontogenesis refers to the birth of a new front, while frontolysis refers to the death of an old front.

may also consist of thunderstorms produced on the colder side of the front. Along cold fronts, the winds tend to be stronger and gusty, while the pressure reaches a relative minimum. Ahead of the surface front, thunderstorm anvils often spread hundreds of kilometers and behind the front winds shift to a northerly direction so that colder air is being advected from the north. When sufficient moisture is present, either scattered cumulus or broken stratocumulus clouds are usually formed within the cold air. On the other hand, in warm fronts, a cloud shield can often form ahead of the front boundary, consisting of stratiform clouds extending hundreds of kilometers, while along the frontal zone there can be extended areas of low clouds and fog, which gives rise to hazardous travel conditions. In contrast to cold fronts, winds now shift to a southerly direction behind the warm front and advect into the colder side warmer, humid, and hazy air, leading to the formation of clouds and convective motions at relatively high altitudes.

As has been detailed in preceding chapters, wind shear is probably the most common mechanism leading to atmospheric turbulence at all levels. Wind shear is often defined as a change of wind velocity and/or direction and may be horizontal or vertical, or a combination of both. According to the ICAO, horizontal wind shear refers to changes in wind speed and/or direction as measured by two anemometers mounted at the same height, while vertical wind shear refers to changes of wind and/or direction as measured by two or more anemometers mounted at different heights from the ground. Wind shear may give rise to CAT at high cruise altitudes across jet streams and to low-level turbulence associated with moving frontal surfaces, to thunderstorms or convective clouds, to microbursts, or even to the lee of mountain waves. As explained in Chapter 1, the leading mechanism through which shear winds give rise to turbulence in any of these associated phenomena is the KH instability. Low-level turbulence may be particularly hazardous for take-off and landing operations. Typical scenarios could be either when a climbing aircraft departing from an airport encounters a microburst with strong downdrafts or when during take-off wind shear causes a sudden loss of airspeed and/or reduction in climb rate. In both cases, the lack of quick recognition and correct pilot response can lead to potentially calamitous consequences for the passengers and crew.

6.3 Impact of turbulence on commercial aviation

Atmospheric air turbulence has multiple impacts on the commercial aviation operations and industry. Considering that civil aviation has global transportation coverage of people around the world and that the network of connections becomes ever more dense due to the increasing demands of transoceanic, transcontinental, and domestic flights, the encounters with turbulence represent for the airline companies millions of dollars losses each year because of injury settlements, insurance premiums, and workers' compensations [350]. The problem could be magnified in the near future if, as predicted by recent research [434, 190, 986, 983, 875], the frequency and strength of the several forms of CAT will increase as a response to climate change [874].

Injuries to passengers and crew are of primary concern for commercial airlines. At cruising altitudes passengers are usually unbelted and occasional encounters with moderate to severe CAT may cause fatal injuries. As the air traffic increases, the injury statistics indicates that the number of fatal and non-fatal accidents has a clear tendency to increase with increasing number of turbulence encounters, which on the other hand, represents one of the most costly safety concerns [862]. Golding [350] reported that between 1985 and 2000 commercial airline encounters with CAT produced three deaths and about 850 injuries of which 70 were serious. This had a cost to airlines in compensation claims, accident investigations, and aircraft damage of about 100 million dollars each year. In addition to this, airlines are frequently involved in legal lawsuits and exposed to forced absences of crew members as they recover from injuries. Although CAT still represents a major threat to flight safety, crashes attributed to turbulence remain extremely rare. Instead, defective design, improper maintenance, careless construction, and retrofitted parts are among the many causes that can lead to structural failure and ultimately to fatal crashes. In fact, between 1980 and 2008, the FAA recorded 234 turbulence accidents, which resulted in a total of 298 injuries and only three fatal events underwent by passengers who were not wearing their seat belts [826]. The USA National Transportation Safety Board (NTSB) reports on average 58 injuries per year with most of these incidents occurring above 10000 ft from the ground level [826]. Most recent analyses indicate that the cost of turbulence to aviation is about 16 million dollars per year and that only approximately 5% of flights are forced to follow non-optimal routes [461, 826].

Encounters with severe turbulence may also induce structural damages to aircraft. For example, in the 1960s it was reported that after an encounter with severe turbulence a Boeing 707 had its vertical fin separated from the aircraft. Earlier planes had their wings fixed much more rigidly than modern aircraft and therefore they were subject under severe turbulence to possible detachment of their wings or even the breaking off of a wing by a jet engine. However, modern aircraft are designed to withstand extreme turbulence, making it highly unlikely that they will undergo significant structural damage under heavy turbulence.

As was mentioned above, changes in the climate system due to anthropogenic forcing will have significant impacts on turbulence and hence on the aviation industry. For example, predictions of an increase in latent heat release during convection will produce a warmer troposphere [190], while the lower stratosphere will cool down with the increase of greenhouse gases. As a consequence of the resulting larger temperature gradients between the upper troposphere and the lower stratosphere at cruising altitudes, the vertical wind shear will increase, thereby inducing more intense turbulence at mid-latitudes [231]. Since changes in the jet stream are predominant at such altitudes, this will increase the impact on the more congested flight routes in the world. In addition, the loss of atmospheric ozone (O_3) as a consequence of chlorofluorocarbons will also make the lower stratosphere cool down and the equator-to-pole temperature gradient increase. Climate model simulations also predict an increase of the moderate-to-severe CAT in the northern hemisphere [986, 983]. However, further numerical simulations have predicted that CAT will increase globally in all four seasons with the more pronounced increases at cruising altitudes around the jet stream [875]. Projections by 2050–2080 reveal that encounters with severe CAT in winter and moderate-to-severe CAT in summertime will be common. This certainly means that aircraft manufacturers and weather forecasters must be warned and prepared to prevent an increase in aircraft damage and injuries to passengers and crew, which would lead to increased economic losses for airlines.

6.4 Flight diversions and delays

When an airport is temporarily closed because of unplanned causes, the incoming flights are re-routed to other nearby air terminals. The air traffic in those nearby airports will increase and produce unexpected congestion with serious consequences to the efficiency, punctuality, and regularity of ground operations. Such flight diversions are generally unforeseen and determined by aircraft emergency, passenger emergency, mechanical failure, or poor weather conditions. In particular, flight diversions can cause tremendous annoyance and inconvenience to passengers who are prevented from completing their scheduled itinerary on time [817, 584]. When the alternate airports are not sufficiently close some other problems may arise. One of these problems occurs when the diverted aircraft does not have enough fuel to cover the distance to the nearest airport [783]. The choice of the nearest airport is another concern because it could have a lower capacity than anticipated or because it is not suitable for landing of the diverted flights. For example, some reasons may include the length of the runway, the absence of a base of handling operations, the lack of adequate apron facilities, or even the reduced capacity of spare parts. If, on the other hand, the alternate airports are suitable for receiving the affected flights, they should provide extra services as, for example, turnaround operations, which in this case involves the disembarkation of diverted passengers and boarding of scheduled passengers as well as refuelling of diverted aircraft. All these additional workloads will certainly have an impact on the regularity of operations at the diversion airport since, depending on the number of diverted flights, knock-on delays may arise due to affectation of the ability to process the scheduled traffic [585].

Flight delays and cancellations may well occur for other reasons beyond the control of the airlines, such as ground handlers' strikes. In the worst case, when all airport ground staff is on strike, flights are literally cancelled. However, if only part of the staff is on strike, the decrease in the number of operators may lead to departure delays that propagate in a cascade over the day, seriously compromising the punctuality and regularity of boarding operations [585]. Flight delays and cancellations may also represent additional costs for the airlines because of compensations and reimbursements to passengers. For example, the EU Air Passenger Rights Regulation (EUAPRR) regulates compensation payments for all flights departing from a European airport provided that the destination airport and the headquarters of the operating airline are within the EU. If the arrival to destination is delayed for at least three hours for unplanned airport closure, severe weather conditions, night flight ban, emergency medical service, or flight diversion, passengers are entitled to compensation of up to 600 euros per person. Also, in the case of flight cancellations, the amount of reimbursement depends on the flight distance.

The histograms in Fig. 6.4 depict statistical data from the U.S. *Bureau of Transportation Statistics*[3] (BTS) for on-time arrivals, arrival delays, and cancelled flights between 2012 and 2021, including all U.S. airports and carriers [8]. On-time arrivals were on average 80% of all flights during that period, while delayed flights oscillated between a maximum percent value of $\sim 22\%$ (in 2014) and a minimum of $\sim 11\%$ (in 2020). The worst year was 2014 for which only 74% of all flights arrived at the scheduled time with no delay. The percentage of cancelled flights were always less than 2.5% with peaks of 3.7% in 2014 and as high as 11.8% in 2020 due to the outbreak of the Covid-19 disease that forced many airlines to suspend or limit their flights for much of the year.

---

[3]This organization is part of the United States Department of Transportation. It is the source of statistics of commercial aviation. In addition to statistics, reports, and analysis it also provides direct access to the Federal government's most authoritative and comprehensive transportation data.

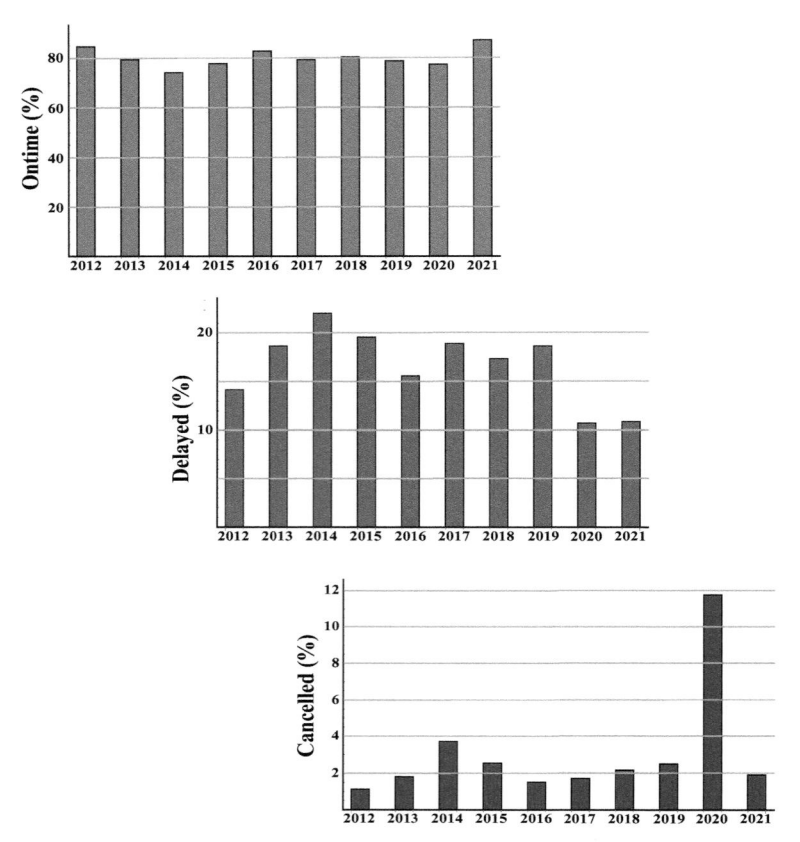

Figure 6.4: Historical records of on-time arrivals (upper panel), arrival delays (middle panel), and cancelled flights (lower panel) between 2012 and 2021, including all U.S. airports and airline carriers. Histograms reproduced from data of the Bureau of Transportation Statistics of the U.S. Department of Transportation [8].

6.5   Structural damage to planes

To guarantee the aircraft's structural integrity and its satisfactory performance, continuous maintenance must be required. Damage to metal aircraft structures must be repaired following the best available techniques to avoid potentially dangerous mishaps and accidents during flight. The lack of accurate maintenance and improper repair procedures may lead to potential danger under severe weather conditions, like severe and extreme CAT. For example, during routine maintenance and inspection operations, it is not uncommon to find damaged components and defects, which if not replaced or repaired could cause fatal accidents with loss of the aircraft and human lives. Although modern aircraft are designed to withstand the effects of atmospheric turbulence, under severe and extreme turbulence an aircraft is subjected to stress and fatigue, especially on the wings and on the horizontal and vertical stabilizers. For instance, failures associated with stress concentrations can occur because of design errors, defects in the microstructure of the material, and corrosion damage [291]. However, historical data on the frequency of failure modes have revealed that fatigue is usually the dominant mode. Under fatigue, a component may crack as a result of stress concentration due to bad design or material defects. The cracks eventually propagate and grow until the material can no longer support the applied loads and breaks suddenly. The failure mechanism by fatigue in aircraft components has a frequency of 55% against 16% by corrosion and 14% by overload [291].

A fixed-wing commercial aircraft consists of five major units: the fuselage, the wings, the stabilizers, the flight control surfaces, and the landing gear. A primary factor in the aviation industry is that all materials employed in the construction of an aircraft must be reliable so that unexpected failures are minimized. When an aircraft is on the ground or in flight, the various components of its structure are subject to pulling, pushing, or twisting forces, which may induce six different types of stress on the wings, fuselage, and landing gear, namely tension, compression, shear, bearing, bending, and torsion. In particular, when the airplane is static on the ground, the weight produced by the force of gravity is entirely supported by the landing gear, while during flight maneuvering of the airplane causes accelerations or decelerations that increase the forces and stresses on the wings and fuselage. Most modern aircraft use flexible wings, which under moderate to severe air turbulence are subjected to strong tension and bending. In particular, the four main forces acting on an aircraft in flight are the lift, the weight, the drag, and the thrust. The weight depends on the mass of all parts of the airplane plus the mass of fuel and any payload on board and is always directed toward the center of the Earth. In flight, the weight is overcome by the lift force, which is always directed perpendicularly to the flight path and opposed to the weight direction. The distribution of the lift force around the aircraft is very important to control it in roll, pitch, and yaw. On the other hand, the drag force acts on the aircraft along the flight path and is opposed to the flight direction, and like the lift it acts through the aircraft center of pressure[4]. The thrust is the force generated by the propulsion system of the aircraft and points toward the flight direction. It opposes and overcomes the drag force. When all these forces are in perfect balance, the aircraft cruises at constant velocity and if they become unbalanced, then the aircraft will accelerate in the direction of the dominant force. During actual flights, the magnitude of these forces varies. For example, depending on the wind speed there would be more or less drag, and as fuel is consumed the plane loses weight. Therefore, the engine thrust can be increased or decreased

---

[4]The center of pressure is defined by

$$\text{c.p.} = \frac{\int \mathbf{x} p(\mathbf{x}) d\mathbf{x}}{\int p(\mathbf{x}) d\mathbf{x}},$$

where $p(\mathbf{x})$ is the pressure distribution around the flying object.

as required. During take-off, the aircraft speeds up and gets off the runway. To do so the thrust and lift must dominate over the other two forces, while during landing the plane is lighter than it was during take-off and therefore the spoilers and airbrakes are working to produce more drag and slow down the aircraft.

It is well known in mechanics that tension is produced when two opposite forces act on a body trying to stretch it. In contrast, compression occurs when two forces acting on a body converge in the same straight line trying to squeeze it. Therefore, tension is like a pull and compression is like a push. For example, during take-off and flight, the weight and lift forces induce tension stresses along the vertical axis, while the thrust and drag forces do it along the longitudinal axis of the aircraft. When the aircraft is static on the ground, the landing gear struts are under constant compression stress induced by the aircraft weight. On the other hand, bending is a combination of tension and compression stresses. In flight, the wing spars of an aircraft are subject to bending stresses, which are transmitted to the fuselage. Also, when the aircraft is on the ground there is constant bending stress on the fuselage, which increases during a carrier landing. In particular, this bending produces compression on the top of the fuselage and tension on the lower part. Similarly, during flight the airflow impacts on the aircraft producing a bending action on the wings and empennage, which causes compression stress on the top of the wings and tension on the bottom. Shear is another type of stress that occurs in an aircraft when two parts, which are fastened, slide one over the other trying to separate. For example, the rivets and bolts that hold the aircraft members and parts together are susceptible to experiencing both tension and shear stresses. Moreover, torsional stresses arise when a body is subject to twisting forces. In turboprop aircraft, torsion is produced in the engine crankshafts and transmitted to the fuselage. The reason for this is that the torque created by a running engine tends to induce rotation of the aircraft fuselage in a direction opposite to the propellers' rotation. In modern jets, torsion of the fuselage is caused by the action of the ailerons under the aircraft maneuvering. Therefore, to minimize the effects of the various stresses and forces acting on an aircraft and reduce fatal accidents when encountering CAT or other severe weather conditions, the structural aspects and the materials employed in the aircraft construction must meet certain specified requirements, based on strength, weight, and reliability.

6.6 Can turbulence cause a plane crash?

It is a matter of fact that the risk of crashing during air turbulence encounters has decreased over time. For example, reports of the FAA between the years 1980 and 2008 indicate that U.S. air carriers underwent 234 turbulence accidents[5], which resulted in 298 serious injuries to passengers and crew, of which

---

[5]According to the NTSB an accident is defined as an occurrence in which a person suffers serious injury or death or in which the aircraft is substantially damaged.

184 involved flight attendants and 114 involved passengers, and only three fatalities, where the latter occurred to passengers who were unbuckled while the seat belt sign was illuminated. An interesting fact is that about two-thirds of the turbulence-induced mishaps and accidents occurred at cruise altitudes where CAT can pass from moderate to severe, and in rare circumstances to extreme. Moreover, according to the ICAO 2018 safety report there were 4.1 billion passengers traveling by air worldwide in 2017 with a total of only 50 fatalities for scheduled commercial departures compared to the 182 in 2016 and the 911 in 2014. However, the number of accidents remained approximately constant with a mean of ~ 88 accidents per year [4]. However, according to the ICAO 2020 report [7], the trend of the annual number of accidents between the years 2016 and 2019 has increased from 75 accidents in 2016 to 114 accidents in 2020, raising the mean of accidents per year to 93.4. Figure 6.5 shows the historical records of worldwide accidents from 2013 to 2019 (upper panel) and fatal accidents from 2008 to 2019 (lower panel) for scheduled commercial flights. The statistics show that although there was a sensible growth in the number of accidents in 2018 and 2019, fatal accidents showed a clear tendency to decrease over time. However, no crashes induced by turbulence encounters were reported in those years. Similar statistics are shown in Fig. 6.6 for the member states of the *European Union Aviation Safety Agency* (EASA) from 2015 to 2019 [5]. The histograms in the upper panel show the number of fatal accidents, non-fatal accidents, and serious incidents for all EU commercial flights and non-commercial flights for business operations from 2015 to 2019, while the lower panel shows the same for commercial flights operated by helicopters. During this period fatal accidents in commercial airline flights were almost non-existent, while non-fatal accidents were always below 30 per year. In contrast, a much larger number of fatal and non-fatal accidents were recorded for travels with helicopters. However, during the same period, there were more serious incidents when flying in commercial jets than in helicopters.

Despite of the above encouraging statistics, encounters with severe to extreme weather may still lead to catastrophic accidents and crashes under extreme circumstances. However, such catastrophic accidents are more likely to occur for light aircraft, such as single-engine planes, than for heavy commercial jets. In part, this is true because smaller airplanes are usually not equipped with weather radar, de-icing, and predictive warning systems. Moreover, commercial airlines and air traffic controllers are provided with technologies capable to monitor the occurrence of strong winds and storms. On the other hand, the risk of suffering turbulence-related mishaps or accidents is highly reduced because most modern commercial jets can spot severe threats and adjust their cruising speeds before entering a rough patch.

According to the ICAO, most aviation accidents actually happen on or near the runway during taxiing, taking off, and landing. Therefore, take-off and landing operations are statistically more dangerous than any other stage of a flight,

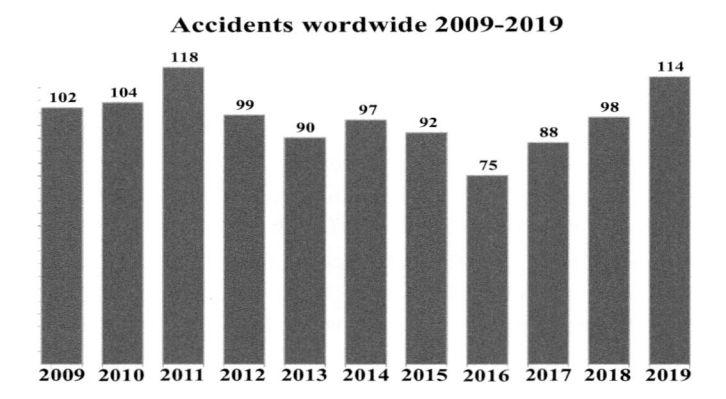

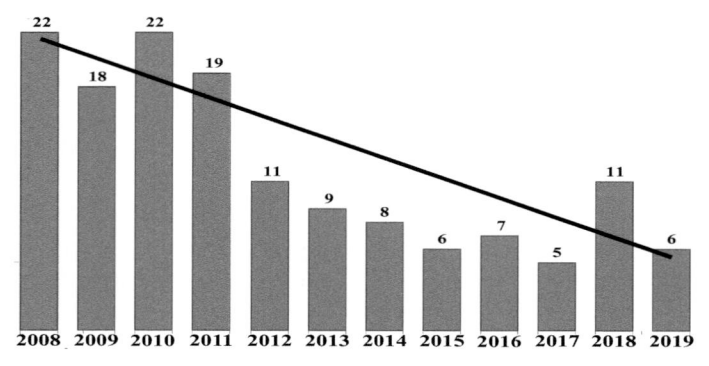

Figure 6.5: Historical records of worldwide accidents from 2013 to 2019 (upper panel) and fatal accidents from 2008 to 2019 (lower panel). Histograms reproduced from data of the ICAO 2018 [4] and 2020 [7] technical reports.

while probably there is less room for accidents on the runway due to increased control by ground crew. Other statistics from the Boeing research shows that nearly 49% of all fatal accidents occur during descent and landing, while only 14% occur during take-off and initial climbing [890]. One reason is that flying at very low altitudes in close proximity to the runway gives the pilot less time to maneuver and hence less chance to avoid a fatal accident. However, a common cause of runway accidents is pilot error. For example, a fatal accident during landing can be the consequence of an error of judgment due to fatigue or lack of experience of the pilot. However, traffic controllers may also be responsible after

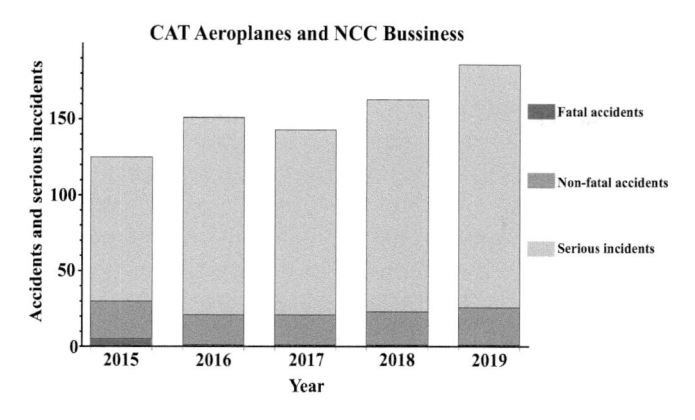

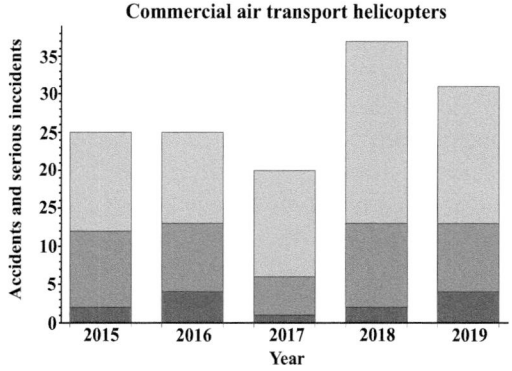

Figure 6.6: Historical records of accidents and serious incidents per year from 2015 to 2019 as reported by the European Union Aviation Safety Agency (EASA) from all member countries. The upper panel shows the statistics for commercial airlines, including non-commercially operated flights for business operations, and the lower panel shows the same statistics for commercial air transport operated by helicopters. Histrograms reproduced from data of the EASA 2020 technical report [5].

pilots for most runway excursions[6] and incursions[7]. For example, between 1995 and 2007 129 people lost their lives during five fatal runway incursions. The deadliest incursion in history occurred on March 27, 1977, at the Tenerife North Airport (former Los Rodeos Airport) in the Canary Islands when a KLM Boeing 747-200 collided with a Pan Am Boeing 747-100, killing 583 people. The recent deadliest excursion occurred on October 29, 2018 when a Lion Air Boeing

---

[6]A runway excursion is a type of runway accident that occurs when an aircraft veers off or overruns the runway surface. These accidents typically involve take-offs and landings and are due to several factors, including unstable approaches because of strong winds and bad weather as well as poor condition of the runway.

[7]A runway incursion refers to aircraft collisions with a vehicle, a person, or another plane on the runway. According to the *Flight Safety Foundation* (FSF), these are the second most common type of runway accidents.

737 Max 8, on a flight from Jakarta, Indonesia to Pangkal Pinang, crashed into the sea shortly after take-off, with the loss of all 189 lives on board the aircraft. Six months later, a second catastrophic accident involved an Ethiopian Airlines Boeing 737 Max 8 also crashed after take-off on March 10, 2019, killing all 157 people on board [5]. These two excursion accidents have led to one of the longest flight suspensions of a large commercial airplane type. In addition to runway excursions and incursions, runway confusion may also be another cause of runway accidents. They occur when an aircraft uses the wrong runway during take-off or landing. An example of this kind of accident was provided by the Singapore Airlines flight 006 on October 31, 2000, which took off in the middle of a typhoon using a closed runway and crashed shortly after, with the loss of 83 people of the 179 who were onboard. According to the FSF, nearly 96% of all runway accidents and 80% of deaths resulting from these accidents corresponded to runway excursions [540]. Other causes that may lead to runway accidents are a mechanical failure or defective design, maintenance error, and severe weather conditions in the proximity of air terminals. However, despite the danger carried by runway accidents, a study conducted by the NTSB revealed that passengers can actually have a 95% chance to survive and remain unharmed [890].

Even though the number of flights plummeted during 2020 due to the Covid-19 pandemic, on January 2, 2021, the BBC reported that more people died in commercial plane crashes. For example, the Dutch aviation consultancy To70 found that 299 people lost their lives in commercial crashes worldwide in 2020. The reported statistics includes fatal deaths in commercial plane accidents by unlawful interference, like the shooting down of a Ukraine International Airlines flight by Iranian armed forces on January 2020, where 176 people were killed. Also included are the deaths of 98 people when a Pakistan International Airlines flight crashed in the city of Karachi in May 2020. In this case, the accident was attributed to a pilot error. However, the number of fatal accidents remains small. For instance, in 2019 there were 0.18 fatal crashes of large commercial airplanes per million flights, while in 2020 this number rose to 0.27 fatal accidents per million flights. In 2020 there were 40 accidents involving large commercial passenger airplanes, of which only 5 were fatal, against 86 accidents in 2019, 8 of which were fatal. This decline in crashes coincided with a sharp decline in flights, which fell 42% due to the coronavirus pandemic. However, none of these crashes was attributed to encounters with CAT.

6.7 Turbulence injury trends

According to the FAA, air turbulence is the main cause of injuries on commercial flights. Every year numerous flight attendants and passengers are injured when flying through a turbulent region. During the sudden jerking of the aircraft, a seated passenger may suffer from neck strains and whiplashes as well as repeated blows to their heads against windows and seat backs. On the other hand, if the overhead compartments open suddenly during turbulence, some of the luggage could fall onto a passenger's head causing a serious head injury. Standing or walking passengers could lose their balance and fall violently to the ground with the enormous risk of broken bones and serious injuries, some of which could be fatal. Cabin crew is also affected, with hundreds of flight attendants having sustained serious or even fatal injuries.

According to the FAA regulations, the risk of injury during turbulence can be highly reduced if passengers and crew wear their seat belts. It is mandatory to have the belt fastened during take-off, landing, taxiing to the arrival gate, and in flight whenever the seat belt sign is turned on. However, injuries may also be due to defective individual plane components as, for instance, a defective latch on an overhead bin, which under sudden jerking at cruise altitudes or hard landing, may allow objects and luggage to fall. In order to avoid turbulence and prevent injuries, the *Commercial Aviation Safety Team* (CAST) has developed and integrated a data-driven strategy to reduce aviation fatality risks in the United States and is promoting government and industry safety initiatives throughout the world. Figure 6.7 depicts the number of seriously injured U.S. passengers and crew per year from 2009 to 2018 according to the FAA records, where it is understood that a serious injury requires hospitalization of individuals because of broken bones, hemorrhages, muscle or tendon damage, damage to any internal organ, and second- or third-degree burns, or any burn covering more than 5% of the body skin. However, a study on the epidemiology of turbulence-related injuries in airline cabin crew from 1992 to 2001 has revealed that the flight at-

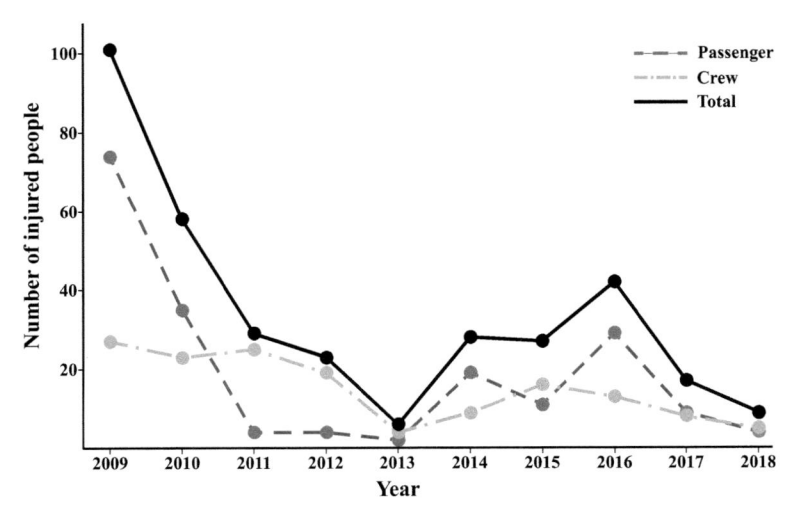

Figure 6.7: Number of injured people per year from 2009 to 2018 in U.S. commercial flights due to encounters with CAT. Only serious injuries are counted. Data for passengers are depicted with dashed lines, those for crew with dot-dashed lines, and the total number of injuries with solid lines. The data plotted are provided by the FAA and was retrieved from Ref. [6].

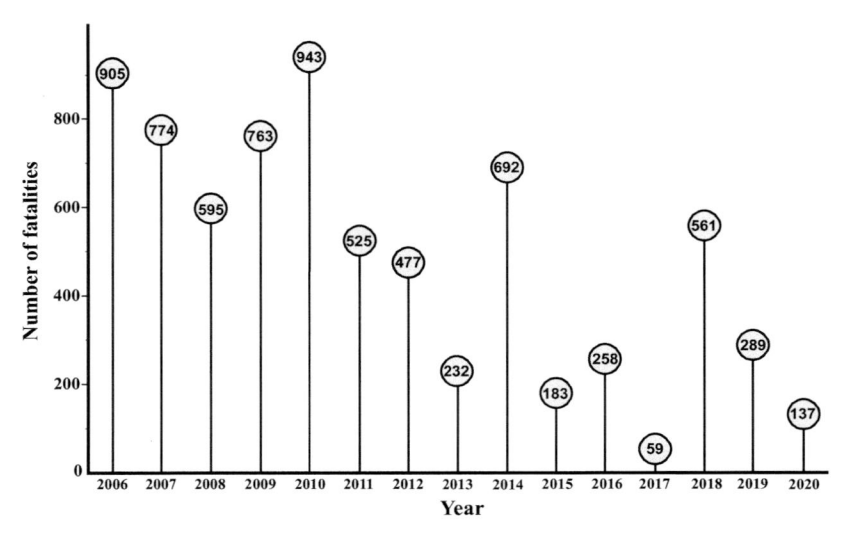

Figure 6.8: Number of worldwide fatalities per year occurred during commercial flights from 2006 to 2020. The data are provided by the *Statista Research Department* and were retrieved from the URL: https://www.statista.com/statistics/263443/worldwide-air-traffic-fatalities/.

tendants suffered from serious injuries in 45.8% of the identified cases, while minor injuries were involved in 54.2% of the cases from 92 reported accidents [915]. The most frequent type of injury among the flight attendants was ankle fractures. Figure 6.8 is a plot of the number of worldwide air travel fatalities[8] per year from 2006 to 2020. The number of reported deaths in 2020 excludes the 176 deceases from the shooting of the Ukraine International Airlines flight in Iranian airspace on January 2020 as well as those fatalities coming from corporate jets and military transport accidents. The broken line shows that despite the year-to-year differences in the number of deaths, the overall trend is for a reduction of the fatalities over time.

---

[8] According to the *Convention on International Civil Aviation* (CICA), the term "fatality" in civil air transportation is used to mean an incident where a person is fatally injured due to an occurrence associated with the aircraft operation. This definition covers any time from when a person boards the plane to when he/she disembarks.

PART III

Chapter 7

Earlier Methods of CAT Detection and Warning

7.1 Introduction

CAT encounters have always posed a threat to commercial and general aviation since their inception because most of these in-flight encounters happen inadvertently and without warning [380]. For example, at low altitudes, where buoyancy wave-induced air turbulence can be as severe as microbursts, such encounters cannot be detected in advance by simple wind shear detection systems. Therefore, extensive research directed to the implementation of new detection methods has continued for many years. In this chapter, earlier methods of CAT detection and measurement techniques will be reviewed. Most, if not all, of these earlier methods have inevitably relied on the measurement of some perturbed state variable of the atmosphere. During the 60s and 70s, sensing systems under consideration included passive and active optical techniques for detection of high-altitude CAT; optical stellar scintillation for measurements of the atmosphere's vertical turbulence profile; microwave scintillation of radio stars and satellite beacons; infrared and microwave radiometry; LIDAR (i.e., laser radar systems) backscatter for detecting and profiling atmospheric constituents, such as molecules, aerosols, and clouds; tropospheric scatter propagation of signals; and ultrasensitive radar detection. These systems included ground based, airborne, and satellite types of sensors [436]. In any CAT detection method, it would be desirable to obtain information on both the location and the intensity of turbulence. However, in some schemes, for instance, those based on scattering techniques, this information could not be extracted directly unless both the length of the CAT region along the line-of-sight of the detector and the full spectrum of eddy sizes were known.

To measure fluctuations of atmospheric variables due to turbulence, these sensors were all based on the use of electromagnetic waves in the visible, infrared, and microwave spectrum. In particular, passive infrared radiometers were employed for the remote detection of atmospheric thermal fluctuations [39, 441, 511]. Changes in the refractive index were monitored by active acoustic sensors to detect low-altitude turbulence [565, 603, 87]. On the other hand, the detection of the intensity of particulate matter in the atmosphere as a sign of the presence of turbulence has been measured using incoherent lasers [192, 535]. Monochromatic transmitters based on the Doppler principle and laser fringe anemometers were used to perform remote sensing of atmospheric turbulent motion as traced by particulate matter suspended in the air [420, 112]. Earlier detections of CAT echoes at altitudes up to 20 km were possible thanks to the microwave incoherent radar systems [40, 381, 556, 744]. However, as will be described in the next chapters some of these methods have been further evaluated and developed along with new technologies to achieve operational status in the detection of CAT and aircraft trailing vortices.

7.2 Passive and active acoustics

Passive acoustics refers to non-intrusive detection and differs from active systems in that a signal is not sent out to interact with the region of interest. Earlier passive optical techniques involved in the detection of CAT at high altitudes are older than active acoustic sensing for the measurement of low- and high-altitude turbulence. Therefore, passive acoustics will be discussed first. A special focus will be given on Rayleigh scattering, absorption-line shape, and infrared sensing.

7.2.1 Passive optical techniques

The evaluation of earlier passive optical techniques included the detection of CAT by the nonuniform behavior in scattered sunlight, the relative movement of stars, the atmospheric infrared emission, the infrared Schlieren imaging, and the optical stellar scintillation [637]. However, the method of optical stellar scintillation will be discussed separately in the next section.

7.2.1.1 Rayleigh scattering

Differences in density between a turbulent region and the undisturbed surrounding air or within a turbulent volume can be detected by measurements of the scattered sunlight. The detection of such density differences can be done using either a single detector or two detectors with crossed fields-of-view. The intensity of radiation, I, observed by the detector at one instant of time is given by the expression

$$I = \frac{\pi^6}{2} \frac{d^2}{\lambda^4} \left(\cos^2 \theta + 1 \right) \phi^2 |\alpha|^2 H_0 \int_0^\infty N(r) dr, \qquad (7.1)$$

where d is the collector diameter, θ is the scattering angle, λ is the wavelength of the incident radiation, ϕ is the angular field-of-view, α is the polarizability of air, H_0 is the flux density at the scattering volume, r is the distance from the detector, and $N(r)$ is the number density at distance r. In deriving the above expression, the Rayleigh scattering function for a single molecule

$$\beta(\lambda, \theta) = \frac{\pi^4}{\lambda} |\alpha|^2 \left(\cos^2 \theta + 1 \right), \qquad (7.2)$$

has been used. If turbulence is present at a later time, the number density and the intensity of radiation will change to $N'(r)$ and I', respectively, where the new number density can be decomposed into a non-turbulent and a turbulent component such that $N'(r) = N(r) \pm \Delta N(r)$. Therefore, turbulence is detected when non-vanishing values of $\Delta N(r)$ are measured. For example, spatial intensity fluctuations can be measured by an airborne detector scanning ahead of the aircraft as

$$S = K|I' - I|, \qquad (7.3)$$

at $\lambda = 0.5$ μm, where K is the radiant sensitivity of the detector photocathode. An important parameter for quantifying the eddy sizes is the signal-to-noise ratio, which can be typically evaluated to be [637]

$$\frac{S}{N} = 1.45 L_r, \qquad (7.4)$$

where L_r is the extent of the turbulent volume ahead of the aircraft along the radial distance r. Reversing the above expression gives

$$L_r \approx 0.69 \frac{S}{N} \qquad (7.5)$$

in kilometers. The signal received by the detector is the result of the integration along its line-of-sight and will include a large number of eddies. Each eddy in the detector field-of-view will contribute randomly so that the scanned fluctuating signal will be about an order of magnitude greater than that produced by a single eddy in the detector field-of-view. For example, according to Eqs. (7.4) and (7.5), the S/N-ratio for a single eddy is proportional to the eddy size, if the contribution of the largest eddies is considered, while the fluctuating signal increases as the square root of the number of eddies along the detector line-of-sight. However, since then it was known through PIREPs that at typical subsonic jet speeds eddies of size between $\approx$ 30 and 600 m will represent the greatest hazards [736]. The precise knowledge of the effects of eddy size on the structural fatigue of subsonic aircraft due to repeated CAT encounters depends on several factors and is still a subject of research.

At altitudes of ~ 10 km, the radiance[1] of a unit volume of the atmosphere is such that the angle-integrated volume scattering coefficients for molecular and particulate matter (i.e., aerosols) are roughly the same [941]. This led to the conclusion that doubling the contribution of the intensity of radiation from molecular scattering will suffice to take into account the effects of particulate scattering. Today, it is known that this formula is not as good as it had appeared at the time because of differences in the Rayleigh angular function of unpolarized scattered light by molecules and typical Mie angular functions for unpolarized scattered light by aerosols in the atmosphere.

In molecular scattering theory, the radiance of a scattered volume of length dl along the detector line-of-sight for incident unpolarized light is [637]

$$dB = \left(\frac{\sqrt{2}\pi}{\lambda}\right)^4 (1 + \cos^2 \theta) |\alpha|^2 I_0 N(l) dl, \tag{7.6}$$

where N is the number density in the scattered volume, α now refers to the molecular polarizability, and I_0 is the incident flux density. Integration of the above expression over the path l and solid angle ω_s extended from the collecting surface of the optical detector leads to the irradiance

$$I = \left(\frac{\sqrt{2}\pi}{\lambda}\right)^4 (1 + \cos^2 \theta) |\alpha|^2 I_0 \omega_s \int_{\text{path}} N(l) dl. \tag{7.7}$$

Under the assumption of spherical symmetry and the complete absence of CAT, the number density will be only a function of the path elevation angle. With respect to some reference altitude h_0, the number density as a function of altitude h takes the exponential dependence [919]

$$N(h) = N(h_0) \exp\left[\frac{-(h - h_0)}{H}\right], \tag{7.8}$$

where $N(h_0)$ is the number density at height h_0 and H is the scale height. In practice, the path integral in Eq. (7.7) is not easily assessable. However, since the relation between l and h involves the path elevation angle, a simplification in the presence of CAT is to perform the integration along a near-horizontal path, as shown geometrically in Fig. 7.1. From this figure it follows that $l^2 = (h - h_0)(2R_\oplus + h - h_0)$, where $R_\oplus$ is the mean radius of the Earth[2]. If the

---

[1]Radiance can be defined as the amount of light that a detector receives from an observed object. It has units of watts per steradian per square meter, W sterad $^{-1}$ m $^{-2}$.

[2]The Earth radius ranges from a maximum of nearly 6378 km (the equatorial radius) to a minimum of nearly 6357 km (the polar radius) so that the mean radius of the Earth is taken as the average between equatorial and polar radius, $R_\oplus \approx 6367.5$ km.

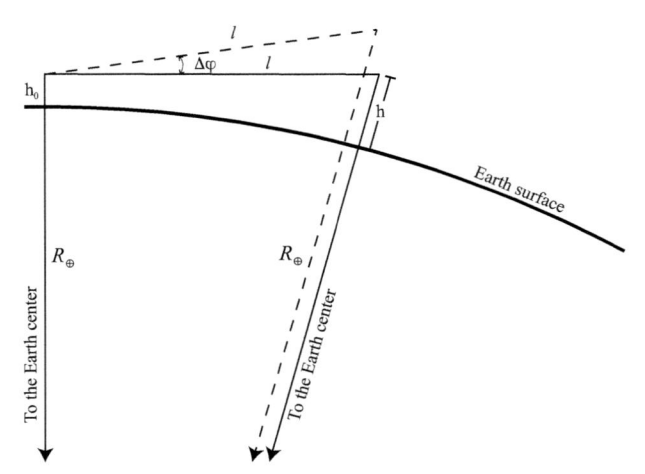

Figure 7.1: Schematic picture illustrating a near-horizontal travel path over the Earth's surface for CAT detection. Figure adapted from Montgomery and Weigandt [637].

altitude is measured from the standard sea level ($h_0 = 0$) and h is taken to be the tropopause altitude, then $2R \gg h + h_0$ and $h - h_0 \approx l^2/(2R_\oplus)$ from which it follows that

$$N(h) = N(h_0) \exp\left[\frac{-l^2}{2R_\oplus H}\right]. \tag{7.9}$$

Substitution of this expression into Eq. (7.7) yields after straightforward integration the expression for the irradiance

$$I = 2^{3/2} \pi^{9/2} \left(\frac{|\alpha|}{\lambda^2}\right)^2 (1 + \cos^2 \theta) \, \omega_s I_0 N(h_0) (R_\oplus H)^{1/2}. \tag{7.10}$$

On the other hand, aerosol scattering can be treated according to the above theory when the size of the particles is smaller than about 0.03 μm. For particles of larger sizes, the effects of diffraction must be accounted for. This will require the solution of Maxwell's equations for electromagnetic radiation, adding a high degree of complexity to the problem. An approximate solution was proposed by Mie [624] by considering a plane monochromatic wave incident on a sphere of radius r. In the far field, i.e., at sufficiently large distances from the particle the intensity of radiation is given by $I = I_0 f(r, \lambda, n, \theta)$, where n is the refraction index relative to the dispersion medium. In particular, the intensity of unpolarized light incident with a scattered angle, θ, at a distance, d, from the spherical particle is defined as

$$I = 2k^2 d^2 I_0 [i_1(\delta, n, \theta) + i_2(\delta, n, \theta)], \tag{7.11}$$

where $\delta = 2\pi r/\lambda$ is the size parameter, $k = 2\pi/\lambda$ is the wavenumber, and i_1 and i_2 are the intensity functions. For applications to atmospheric scattering, the angular-volume scattering coefficient is defined as the ratio of the flux scattered by a unit volume of homogeneous aerosol (with number density N per unit solid angle in the θ direction) to the incident flux on the unit cross-section of this volume, given by

$$\beta(\delta, n, \theta) = \frac{N}{2k^2} \left[i_1(\delta, n, \theta) + i_2(\delta, n, \theta) \right]. \tag{7.12}$$

Since particles suspended in the atmosphere have different sizes, the β coefficient must be expressed as an integral over the size distribution. From diffraction theory, the sums of the partial wave solutions define the intensity functions

$$i_1 = \left| \sum_{m=1}^{\infty} \frac{(2m+1)}{m(m+1)} \left[a_m P_m'(\cos\theta) + b_m P_m'(\cos\theta) \right] \right|^2,$$

$$i_2 = \left| \sum_{m=1}^{\infty} \frac{(2m+1)}{m(m+1)} \left[a_m P_m''(\cos\theta) + b_m P_m''(\cos\theta) \right] \right|^2,$$

where $P_m'(\cos\theta)$ and $P_m''(\cos\theta)$ are the first and second derivatives of the Legendre polynomials of degree m, respectively, and the coefficients a_m and b_m are explicit functions of the size parameter, δ, and the refractive index, n [138]. At aircraft cruise altitudes when atmospheric variability and the distribution of particles are highly inhomogeneous, the above method cannot provide an accurate description of particulate scattering since it is suitable for perfectly homogeneous, spherical particles of small size. Therefore, it was a common procedure to employ experimental results to describe the expected scattered intensities. In particular, the difficulties for large-particle scattering were partially alleviated using data obtained from airborne measurements, where sky brightness and polarization were both measured at altitudes between 6 and 21 km from the standard sea level [183]. These early measurements revealed that at heights of 30000 ft (≈ 9 km), aerosol scattering dominates over molecular scattering and exhibits pronounced departures from Rayleigh symmetry. On the other hand, changes in the photometer line-of-sight can lead to changes in the fluctuation of the scattered light. For example, any pitching of the aircraft will result in a change of the photometer line-of-sight, which in turn will produce changes in the detected signal at the frequency of the angular change. This situation is depicted in Fig. 7.1, where the line-of-sight, which is taken along the local horizontal, is shown after an angular change $\Delta\varphi$. For small values of $\Delta\varphi$, $\Delta h \approx l\Delta\varphi$. Taking $N_0 = N(h_0 = 0)$ as the number density at sea level ($h_0 = 0$) in Eq. (7.8) and differencing the result gives

$$\Delta N(l) = -N(l) \frac{l}{H} \Delta\varphi. \tag{7.13}$$

Therefore, the fractional change in signal is

$$\frac{\Delta S}{S} = \frac{\int_0^\infty \Delta N(l) dl}{\int_0^\infty N(l) dl} = \frac{-\frac{\Delta\varphi}{H} \int_0^\infty l \exp\left(-\frac{l^2}{2R_\oplus H}\right) dl}{\int_0^\infty \exp\left(-\frac{l^2}{2R_\oplus H}\right) dl} = -\sqrt{\frac{2R_\oplus}{\pi H}} \Delta\varphi. \tag{7.14}$$

With $R_{\oplus} = 6367.5$ km and $H = 8$ km, $\Delta S/S = -22.51\Delta\varphi$. For a one degree change, i.e., for $\Delta\varphi = \pi/180$ rad, the fractional fluctuation in the signal is $\approx$ -0.393. This result was found to be consistent with isovolume plots for the sky brightness at altitudes ≥ 30000 ft from the standard sea level [183].

The scattering technique was implemented using airborne systems with one detector or two detectors. For example, a single detector equipped with a two-dimensional scan can be used to provide information on the angular extent of the turbulent volume. The relation between the range, r, and the angle subtended by the turbulent area, φ, is $\varphi = s/r$, where s is the length of the subtended arc at distance r. The change in the angular extent at two different times will then be $\Delta\varphi/\varphi = -\Delta r/r$. Solving for r gives $r = -\varphi\Delta r/\Delta\varphi = v\Delta t(\varphi/\Delta\varphi)$, where $\Delta r = -v\Delta t$ is the distance traveled by the aircraft during the time interval Δt toward the turbulent area. However, relying on angular changes in cross section would only be appropriate for long-range turbulence detections because CAT generally extends from 10 to 60 km horizontally with small vertical extents.

Systems with two detectors were usually employed by placing one detector in each wing tip of the aircraft. Figure 7.2 shows a schematic picture of the two-detector correlation when the wing span is assumed to be 30 m and the fields-of-view of both detectors is 1.7 milliradians. The shaded region in the figure encloses the portion of atmosphere ahead of the aircraft that is intersected by both fields-of-view and has a horizontal extension of 40 m along the path travel. The two detector signals, namely $I_1(t)$ and $I_2(t)$, can be both decomposed into a mean intensity and a fluctuating component such that $I_1(t) = \langle I_1(t)\rangle + i_1(t)$ and $I_2(t) = \langle I_2(t)\rangle + i_2(t)$. The cross-correlation between the two signals will be defined as

$$R(I_1, I_2) = \frac{\text{Cov}(I_1, I_2)}{\sigma_{i_1}\sigma_{i_2}} = \frac{\langle i_1(t)i_2(t)\rangle}{\sqrt{\langle i_1^2(t)\rangle\langle i_2^2(t)\rangle}}. \tag{7.15}$$

If both signals are perfectly correlated then $R(I_1, I_2) = \pm 1$, while if they are un-correlated $R(I_1, I_2) = 0$ because they will have a zero covariance. A deficiency of the above system is that the intensity fluctuations would be of low frequency.

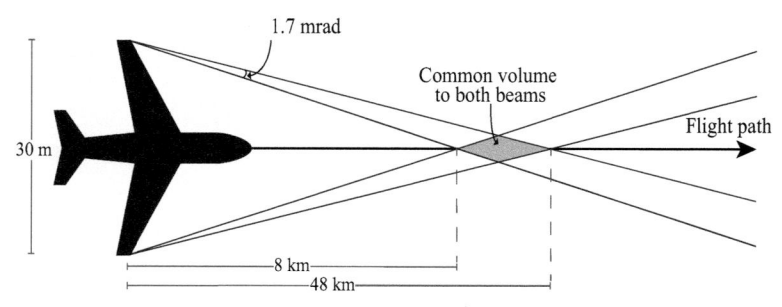

Figure 7.2: Schematic picture illustrating a two detector cross-correlation technique as applied to CAT detection. The shaded region is the common portion of atmosphere scanned by both detectors. Figure adapted from Montgomery and Weigandt [637].

This trouble has been in part overcome by allowing the detectors in each wing to oscillate such that they can scan together over the horizontal plane. However, the application of this technique would require that the angular flexing of the wings will not cause an uncrossing of the fields-of-view of the two detectors. Also, the frequencies associated with the scanning must be eliminated by electronic filtering. A drawback of the scattering technique is that it works only during the day, while it becomes useless during night flights because the S/N ratio decreases to less than one as the irradiance decreases to nighttime levels. Also, under extensive cloud cover along the line-of-sight, the method will also become useless because the amount of information that could be obtained with the correlation system is limited. However, above 30000 ft from the sea level, where clouds are encountered less than 3% of the time, the scattering technique can be applied with greater confidence. On the other hand, it must be emphasized that the correlation scheme assumes that only those signals coming from the common volume defined by the fields-of-view of the detectors are actually correlated, while all other signals, which are not originated in the common volume, are uncorrelated. For example, a large eddy external to the common volume can affect both detector beams. This will produce a false correlation of the signals coming from the common volume, thus indicating the presence of CAT at a distance where it does not exist.

7.2.1.2 Relative movements of stellar images

This technique, which is based on tracking the relative motion of at least two stars, resulted to be highly impractical and therefore it was quickly dismissed from consideration. The main reasons can be summarized as follows. First, the simultaneous track of at least two stars with an accuracy of a few arc-seconds must be acquired, which, on the other hand, would require the use of a sufficiently accurate and fast-responding inertial platform to take into account sudden aircraft motions, such as rolling, pitching, and yawing. Moreover, the continuous availability of stars would require the use of optical telescopes with apertures of at least 6 inches to allow imaging of stars with apparent visual magnitudes[3] less than seven to be observable. If, on the other hand, the observations are operated from a ground-based telescope, the resulting images could be lacking uniformity and will certainly change continuously with time. Also, because of star scintillation, the instantaneous positions of the stars cannot be determined with sufficient accuracy. Due to these difficulties, this method has been precluded from further study and development.

---

[3]In astronomy the magnitude usually refers to the apparent brightness of a celestial object as seen from the Earth. The apparent brightness is also called apparent visual magnitude. Higher positive magnitudes indicate fainter stars and higher negative magnitudes indicate brighter objects. For example, the dimmest star visible to the naked eye has a magnitude of +6.5, which is called the magnitude limit. The magnitude of a star will depend on its intrinsic brightness and its distance from Earth.

7.2.1.3 Atmospheric infrared emission

It is well known that atmospheric air turbulence can also be associated with horizontal temperature gradients and fluctuations. In particular, atmospheric air temperature variations can be scanned ahead of the aircraft by radiometric measurements of the infrared energy emitted in appropriate spectral bands. An early suggestion was to use the 6.3 μm water vapor band. However, this technique could not provide information on the range and magnitude of the temperature variation. To increase the amount of obtainable information it was suggested a technique based on spectral scanners of the short wavelength wing of the 15 μm carbon dioxide band [39]. Carbon dioxide has strong absorption bands, which are centered at 4.3 and 15 μm, and the emission approaches that of an ideal black body in the neighborhood of these bands. In addition, CO_2 is uniformly distributed in the atmosphere and its 15 μm band coincides with the thermal emission peak at atmospheric temperatures.

Based on the S/N-ratio, it was further demonstrated that the detection of large scale temperature discontinuities by this method was feasible under the assumption that the length scale of the CAT regions ranges from 80 to 160 km. For single detectors scanning ahead of the aircraft, the quality of the temperature measurements depended on the sensitivity of the detector and design of the radiometer and mounting platform. For example, if the aircraft is approaching a region of colder air, a radiometer pointing horizontally ahead of the aircraft will indicate that the signal relative to some reference value will decrease. However, more realiable measurements of the atmospheric temperature structure ahead of the aircraft were obtained using radiometers capable to perform angular scannings in both the horizontal and vertical planes [441]. In particular, the use of vertical scanners was incorporated successively to monitor fluctuations in the atmospheric lapse rate that could be associated with CAT.

The above problem can be put in analytical terms by dividing the field-of-view into elements along the line-of-sight such that the radiometer signal can be constructed by summing the product between the radiance of each element, $\mathcal{R}_i$, and the transmission of the atmosphere between the detector and each element, t_i, that is

$$\mathcal{R} = \sum_{i=1}^{n} \mathcal{R}_i t_i. \tag{7.16}$$

In the above relation, the radiance of each element is expressed in terms of Wien's displacement law approximation to Planck's law, which for a thin gas approximation, is given by

$$\mathcal{R}_i = \frac{A_1}{\lambda^5} \exp\left(-\frac{A_2}{\lambda T_i}\right) e_i N_i l_i, \tag{7.17}$$

where $A_1 = 3.74 \times 10^{-12}$ W cm^2, $A_2 = 1.443 \times 10^4$ μm deg, λ is the wavelength, T_i is the element temperature measured in kelvin, e_i is the element molecular extinction coefficient, and l_i is the element length. If the temperature of element i changes from T_i to, say T_i', the implied difference in radiance is

$$\Delta \mathcal{R}_i = \mathcal{R}_i - \frac{T_i}{T_i'} \exp\left(-\frac{A_2}{\lambda T_i'}\right), \tag{7.18}$$

where it has been assumed that the mixing ratio of CO_2 is constant. Since for the wavelength of the 15 μm CO_2 and for an altitude of 10 km in the upper troposphere, where the temperature is around 230 K, the spectral radiance difference is proportional to the temperature difference $T_i - T_i'$, Eq. (7.18) can be rewritten as

$$\Delta \mathcal{R}_i = C\,(T_i - T_i')\,e_i N_i l_i - \frac{1}{T_i} \exp\left(-\frac{A_2}{\lambda T_i}\right), \tag{7.19}$$

where C is the constant of proportionality that accompanies the temperature difference. Therefore, the difference in the detected signal when the aircraft passes from a region of almost uniform temperature to one where there is a temperature gradient is given by

$$\Delta \mathcal{R} = \sum_{i=1}^{n} t_i \Delta \mathcal{R}_i, \tag{7.20}$$

where as before t_i is the transmission between the ith atmospheric slab and the radiometer. The characteristics of the radiometer will ultimately determine the S/N-ratio with which the difference $\Delta \mathcal{R}$, given by Eq. (7.20), may be detected. Extensive flight test evaluations of passive infrared detectors in the 60s and 70s failed to provide an acceptable false alarm rate in CAT to justify their operational employment.

7.2.1.4 Infrared schlieren optics

Schlieren photography or imaging is a flow visualization technique that enables the detection of small changes and fluctuations in the refractive index of air and other transparent media. However, it is most commonly used to image air flows and therefore the method can be applied to observations of CAT over limited paths in the atmosphere. Although it has been around for over a century and a half, the method has experienced continuous developments over the years, becoming a widely-used imaging technique in engineering applications and the aeronautical sector.

Figure 7.3 is a schematic picture illustrating how a standard schlieren system works. The light from a source first passes through a slit. The slit is oriented in such a way that the light reaching the first concave mirror (or lens) is reflected, forming parallel rays that pass through the test section (i.e., the flow region of interest). As the parallel light beam passes through the test section, it is refracted and then collected by a second concave mirror (or lens), which then focuses it to a point at the knife edge. Most set-ups usually employ an ancillary (or third)

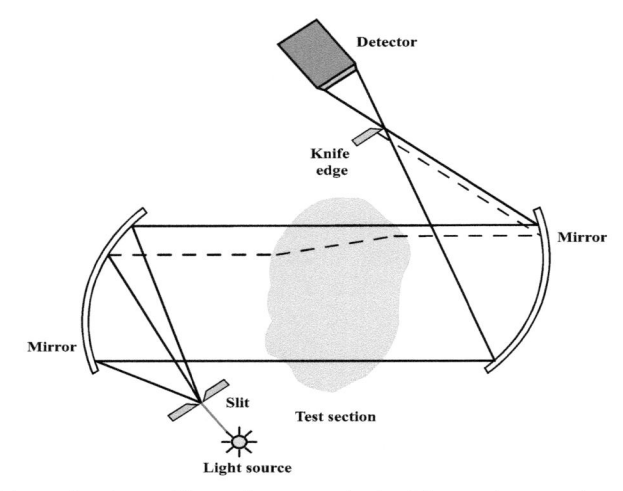

Figure 7.3: Schematic picture illustrating a standard schlieren photography system for flow visualization.

lens behind the knife edge to deviate the ray to a recording device like a video camera or photographic sensor. However, in the context of aviation, there would be some differences between the conventional schlieren system described above and the schlieren-type system used in the context of CAT detection in an in-flight aircraft. One difference is that the light source is replaced by the Earth's horizon. The radiance near the horizon will decrease rapidly with an increasing angle from the tangent line to the Earth's surface in spectral regions of small atmospheric absorption. When a sufficiently large radiance gradient is established, direct radiation from the Earth could be blocked with the use of a knife edge in the image plane of an airborne system so that the refracted radiation from the turbulent parcel could then be detected. Apart from being the most important greenhouse gas in the atmosphere, water vapor is found mostly in the troposphere. Therefore, the best region of the electromagnetic spectrum for this approach is near a wavelength of 11 μm where only weak continuum absorption by water vapor is present. To measure small angles (of $\sim$ 10 arcsec), a feasible approach is to image the radiance gradient and measure the redistribution of turbulent energy in the image.

7.2.1.5 Ozone detection

The existence of a correlation between the ozone concentration in the atmosphere and the presence of CAT was discussed by Rosenberg [774]. Extensive measurements of the ozone distribution in the vertical direction revealed that small differences in altitude correspond to large differences in ozone concentration [113]. Therefore, it was advanced that any redistribution that changes this behavior could be indicative of the presence of air turbulence. In particular, ozone has an absorption band of about 9.6 μm so that thermal detection is possible. For example, based on these early investigations on the vertical distribution of ozone, it was discovered that the ozone concentration can increase by factors of 3 or more when moving from one kilometer below to one kilometer above the tropopause. Therefore, at cruising altitudes, small vertical changes will encounter large changes in ozone concentration. For example, measurements of the ozone concentration in the cabin of a commercial jet aircraft at altitudes of 35000 ft have shown that the concentration in the cabin reflected the external ozone concentration with large changes in a matter of 10 min. Clearly, this has led to the conclusion that measurements of the ozone concentration may not be a feasible method for CAT detection. This is true even if there is a high correlation between the increased ozone concentrations and CAT.

7.2.2 Active acoustics

Early methods based on active acoustic sensing were mainly based on the use of microwave radar systems, laser Doppler radars for both ground and in-flight sensing, Doppler acoustic systems, and laser fringe anemometers for atmospheric measurement of aircraft vortices. In particular, active sources of acoustic power operating in both coherent wave (CW) and pulse modes for the measurement of low-altitude turbulence began to be used in the late 1960s and early 1970s [565, 603, 87].

7.2.2.1 Radar detection

Air turbulence in the atmosphere is accompanied by fluctuations in temperature, wind velocity, and humidity (water vapor). When an electromagnetic wave is propagated into the turbulent volume, such fluctuations produce variations in the refractive index that cause the scattering of microwaves and acoustic waves [969, 893]. This provides a means for remote detection of CAT. In particular, microwave radar systems have been employed extensively to detect CAT echoes at altitudes up to 20 km [40, 381, 256, 128, 126]. Microwave scattering in the atmosphere is mainly due to temperature and water vapor fluctuations. At low altitudes, the contribution to scattering of water vapor fluctuations dominates, while at higher altitudes, where the air is much drier, scattering depends primarily on temperature fluctuations and inhomogeneities. For example, direct images of atmospheric cross-sections by wave backscattering were obtained by radar sounding in a monostatic configuration. This allowed regions of turbulent activity and enhanced refractive-index variability to be scanned with acceptably good spatial resolution [680]. The description of the interaction between electromagnetic waves and the atmospheric structure has been documented by various investigators during the 60s and 70s [893, 634, 678, 679]. For example, Tatarski [893] derived an expression for the radar cross-section per unit volume (i.e., the radar reflectivity, η) in the backscatter direction as

$$\eta = \frac{\pi^2 k^4}{2} \Phi_n(k), \qquad (7.21)$$

where $k = 4\pi/\lambda$, λ is the radar wavelength, and $\Phi_n(k)$ is the three-dimensional spectrum of reflectivity. Further developments of this relation allowed the radar reflectivity to be expressed in terms of the normalized one-dimensional spectral density [793, 522, 679]. In particular, Ottersten [679] found that the deflection of electromagnetic waves across a CAT volume will allow radars to detect vortices in wind streams and that wave refraction can be accounted for using Bragg's theory of scattering so that the radar reflectivity will obey the relation

$$\eta \approx 0.38 c_n^2 \lambda^{-1/3}, \qquad (7.22)$$

where c_n is the refractive-index variability and λ is the radar wavelength. Results from the utilization of radars in the 60s and 70s for detection of CAT echoes produced by refractive-index variations have been reported in the literature by various authors [40, 381, 256], while further work focused primarily on improving the sensitivity of radars. A thorough review on the sensing of atmospheric waves and CAT with the use of sodar[4] and microwave radar detection systems along with a discussion of the experimental progress achieved during the 60s and 70s was provided by Ottersten et al. [681]. In general, radar and sodar methods are attractive because they allow sensing of the surrounding medium with acceptably good spatial and temporal resolution. Good spatial resolution is achieved by combining directive antennas and echo-delay ranging. The use of fixed antennas allows one-dimensional scanning of the surrounding atmosphere, while two- and three-dimensional scans can be obtained with the use of steerable antennas [681]. Later on in the early 1980s, detection systems based on Doppler technology started to be used and over time, the detection of wake vortices by radar systems gave rise to three methods, namely pulse Doppler, continuous wave Doppler, and radar acoustics [328]. Whereas Doppler frequencies in backscattered signals were used successfully in the sensing of storms, their interpretation and application to CAT detection barely began in the mid-1970s. However, Doppler sodar measurements rapidly became a powerful tool to study air turbulence in the lower atmosphere. For example, scanning of the horizontal and vertical air motions was performed since early times using monostatic and bistatic sodar, separately or in combination.

7.2.2.2 Doppler acoustic techniques

Doppler acoustic devices have been of great use to monitor variations in the refractive index caused by fluctuations and irregularities in the temperature and

---

[4]Sodar is the abbreviation used for *sound detection and ranging* in contrast to radar, which is the abbreviation for *radio detection and ranging*.

wind velocity. They have been classified into coaxial and bistatic systems. Coaxial systems are equipped with a transmitter and a receiver. These devices have been employed to measure the Doppler shift[5] of acoustic signals from variations of the refractive index caused by temperature fluctuations. Typical echo sounder parameters that have been used during early Doppler measurements of vertical velocities in the atmosphere are tabulated in Ref. [87]. On the other hand, Doppler shifts in atmospheric signals induced by wind velocity and thermal gradients have been measured with both monostatic and bistatic scatter systems. These methods are particularly convenient because they provide direct information on the velocity field associated with the CAT region, and therefore they appear to be more appropriate for the analysis of waves and turbulent features. In fact, Doppler radar measurements providing information on CAT were first documented by Browning et al. [129]. The investigation of small-scale fluctuations of importance in radio-wave scattering with the aid of fast-response multi-channel temperature, humidity, and refractive index sensors flown to 2000 m on a tethered balloon revealed that for frequencies above 1 Hz the power spectra data at different heights followed a slope close to $-5/3$, while at lower frequencies the slopes varied from -1.5 to -3.5 [347]. In addition, the method was successively applied to the remote sensing of the intensity of thermal plumes and wind shear as well as to the detection of aircraft trailing vortices [436, 85, 328]. Doppler sodar probing has also become an important tool for the detection of low-altitude motions in the lower boundary layer [87, 86, 75, 583, 447]. Acoustic sounder measurements of the abrupt transition from a laminar to a turbulent atmospheric boundary layer, which was then compared with meteorological measurements at altitudes of 10 and 137 m on an instrumented tower, were reported by Schubert [800]. A review of progress in the early 1970s on the understanding and interpretation of wind profiles and the CAT structure at altitudes between 30 and 150 m is given by Busch [144].

Following the discussion by Ottersten et al. [681] in their review paper on scatter geometry, the Doppler frequency shift between the scattered and incident waves is given by

$$\Delta f = f_s - f = \frac{1}{2\pi} \left(\mathbf{k}_s - \mathbf{k} \right) \cdot \mathbf{v}, \tag{7.23}$$

---

[5]The Doppler shift (or Doppler effect) refers to the apparent difference between the frequency of a sound (or light) wave leaving a source and the frequency at which the wave reaches an observer, resulting from the relative motion between the wave source and the observer. For a non-relativistic source moving with velocity v relative to an observer at rest, the frequency f' received by the observer is

$$f' = f \left(1 \pm \frac{v}{c_s} \right)^{-1},$$

where f is the frequency of the source at rest and v is the speed of sound. If the source is moving away from the observer (positive v), the received frequency is lower and the wavelength is greater (i.e., redshifted). Conversely, if the source travels toward the observer (negative v), then f' is higher and the received wavelength is shorter (i.e., blueshifted).

where $\mathbf{k}_s$ and $\mathbf{k}$ are the wave vectors of the scattered and incident signals, respectively. The difference between the received (scattered) signal, f_s, and the incident frequency, f, is proportional to the velocity, $\mathbf{v}$, of the scattering medium. The dot product on the right-hand side of Eq. (7.23) means that the mean component of the wind velocity along the direction of vector $\mathbf{k}_s - \mathbf{k}$ can be determined from the Doppler shift. Taking $k_s \approx k = 2\pi/\lambda$, Eq. (7.23) can be approximated by

$$\Delta f \approx \frac{2}{\lambda} \sin \left(\frac{\theta}{2} \right) v \cos \phi, \qquad (7.24)$$

where θ is the scattering angle and ϕ is the angle between the wind vector $\mathbf{v}$ and the direction of $\mathbf{k}_s - \mathbf{k}$. In most cases the error carried by the wind speeds as determined from Eq. (7.24) when refraction is neglected were always less than about 5% [184]. Under pure backscattering $\theta = \pi$ Eq. (7.24) simplifies to

$$\Delta f \approx \frac{2}{\lambda} v \cos \phi, \qquad (7.25)$$

where now ϕ is the angle between the wind velocity vector $\mathbf{v}$ and the radial direction of the beam. Monostatic Doppler radar and Doppler sodar are therefore capable of measuring only the radial component of the wind velocity. Thus, in the monostatic mode, the distribution of radial velocities within a resolution cell is represented by the Doppler spectrum so that, according to Eq. (7.25), every radial velocity $v \cos \phi$ corresponds to a Doppler frequency. For example, the mean and variance of the radar Doppler spectrum have proved to be useful quantities in the analysis of atmospheric motions. In particular, it has been demonstrated that the Doppler radar spectra of CAT can be employed to evaluate important CAT parameters and characteristics, such as the streamwise component of the mean wind velocity, the vertical shear of this component, and the turbulence intensity as estimated from the variance of the Doppler frequency spectrum [127].

7.2.2.3 Laser Doppler radars

Laser Doppler radars have been used for both ground and in-flight sensing of atmospheric processes, producing a wealth of information that was not previously available. Such applications have included wind shear, wind profiles, dust devils, and aircraft wake vortices [91]. Laser Doppler radars for ground sensing started to be developed during the early 70s. At that time a particularly promising detection technique was based on the use of a CW, CO_2 laser Doppler system for measuring atmospheric wind velocities. An advantage of this technique is that backscattering is provided by particles that are normally suspended in the atmosphere, which in turn serve as tracers of air motion.

The principle of laser Doppler radars for remote sensing is well established. It consists of illuminating the object of study with a laser beam so that information about the object is provided by the scattered signal. These specialized radars are coaxial sensors for which a new signal frequency is created by combining the frequency of the backscattered signal with the transmitter frequency in the photodetector[6]. By optical focusing most of the detected energy will be concentrated in the focal region, thus providing some degree of range resolution with a CW system. For example, when the laser signal is scattered from a rough surface of size structure comparable to the signal wavelength, then speckle effects will arise from the interference of the scattered field. In particular, if the scattering object is in motion, strong intensity fluctuations will be registered by the detectors. An interesting overview of remote sensing with coherent laser radar systems in Europe is provided by Vaughan et al. [930] and references therein. The detection of trailing vortices from low-altitude flying aircraft has been made with experimental detectors equipped with a 20 W stable single frequency CO_2 laser and 15 cm optics with promising results [436].

Extensive atmospheric and wind detection research was made at DLR[7] during the late 1970s and 1980s with the aid of CO_2 CW laser Doppler systems based on 30 cm optics and surface acoustic wave spectral analysis. In this case, pattern recognition from the digital signals was made using an on-line data processing system [23]. This work was included in the German Wake Vortex Program for field experiments during 1983–1985 and 1989–1990 in the extremely busy and congested Frankfurt airport, where the separation of only 520 m of the two parallel runways poses additional capacity limitations. During these years the ground effects of aircraft vortices were investigated for more than 1400 landings of heavy and large aircraft. A schematic picture of a ground laser Doppler system is shown in Fig. 7.4. For these experiments, the laser Doppler system was placed between both runways about 850 m from the threshold [501]. As shown in Fig. 7.4, a fast elevation scan at a fixed range setting ensures proper measurements of wake vortices during aircraft landing. After a vortex is sensed, the next one is chosen by changing the range setting. When the vortex reaches the sensor position, the measurement plane is turned around so that the vortex is tracked toward the parallel runway. The results of these experiments have revealed that a good number of the measured vortices showed a steep ascent toward the parallel runway because of bouncing from the ground. Since such vortices had a clear tendency to cross the parallel runway near the altitude of approaching aircraft, they may well pose serious hazards for ordinary landing operations at the Frankfurt airport [501, 930].

---

[6]The production of a signal frequency by combining or mixing two other frequencies using a signal processing technique is called *heterodyning* or simply *heterodyne or coherent detection*. The intensity of the resulting signal is proportional to the product of the intensities of the two mixed signals, while its frequency is proportional to the difference between the two signals. This is in contrast with the definition of *homodyne detection*, which in electrical engineering consists of extracting information encoded as modulation of the frequency of a signal by just comparing it with an identical standard oscillating signal carrying null information.

[7]Deutches Zentrum für Luft- und Raumfahrt.

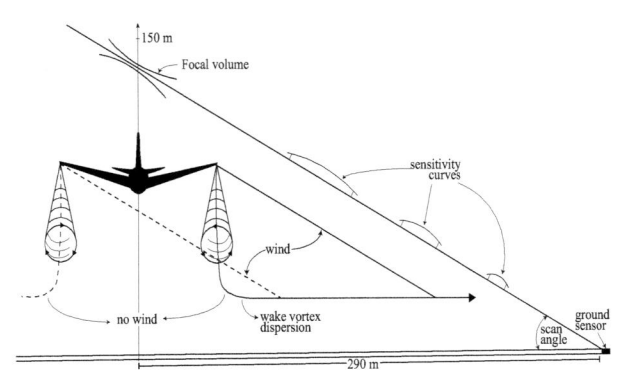

Figure 7.4: Schematic picture illustrating the ground sensing of aircraft wake vortices with a laser Doppler velocimeter placed between two closely separeted runways as operated in the Frankfurt airport. Figure adapted from Vaughan et al. [930].

On the other hand, airborne applications have been possible thanks to the design and construction of in-flight sensors. An early technique employed in airborne CAT detectors consisted of the development and use of a pulsed, laser Doppler radar for detecting the motion of particulate matter suspended in the atmosphere. These devices were long recognized to give the most meaningful information about the presence of atmospheric CAT and technological advances accomplished mostly during the 1970s made this sensing approach perfectly feasible. Such systems consisted of a highly stable CW CO_2 laser with a narrow spectral line employed as the reference oscillator and the coherent source for the transmitter output [436]. Modulation of part of the output is performed using a gallium arsenide crystal. This was necessary to produce 2, 4, or 8 μs duration pulses, which are then amplified in a pulsed CO_2 power amplifier to a peak power of 5 kW. For transmitting and receiving, a 12-in optics is employed and a photomultiplier detector is used to receive the backscattered signal coming from a polarized mirror, where it is optically heterodyned with the output signal of the CW oscillator. In accordance with the Doppler equation and depending on the aircraft's cruise velocity, an offset frequency between 15 and 60 MHz is provided by the aircraft's motion even if the atmospheric particles are at rest. Amplification of the output signal of the optical detector is performed using a pre-amplifier with a pass band of 15 to 60 MHz, which is then combined with an offset oscillator whose frequency can be adjusted as a function of the aircraft speed to produce a center frequency of 10 MHz. The amplified signal is passed through a filter bank for analysis and processed by means of a range-velocity display, which finally gives an indication of CAT and range ahead of the aircraft. Although this research model with 12-in optics was sensitive enough and able to provide feasible CAT determinations, it was not suitable for operational use in in-flight CAT detection because of its weight, size, and cost.

Further technical development of airborne systems in the late 1970s has led to the first coherent LIDAR[8] in Europe. This, which was called *Laser True Airspeed System* (LATAS), was designed and built by the United Kingdom's Royal Signals and Radar Establishment and flown in aircraft of the Royal Aircraft Establishment (RAE). LATAS was conceived to be a compact and robust system with the LIDAR head operating in remote mode in an unpressurized part of the aircraft. In brief, the LIDAR consisted of a 4 W CO_2 laser, comprising a 15 cm germanium output lens, polarizing optics with quarter and half wave plates, and a 100 MHz CMT detector cooled with high pressure air and Joule-Thomson minicooler. A full description of LATAS is detailed in Ref. [997]. After about 15 years of flying during far-off paths over Europe, the United States, and the North and South Atlantic the system proved to be highly reliable [930]. Another system based on a 3 W CW CO_2 waveguide laser, equipped with a CMT diode with over 200 MHz bandwidth, was constructed in the early 1960s in France by Crouzet S.A. [640]. The system consisted of polarizing optics with wave and Brewster plates. In order to provide a frequency shifted signal for helicopter operations, the local oscillator path included a double passage of a Bragg cell. Measurements to distances varying between 10 and 100 m ahead were allowed using an off-axis Dall-Kirkham telescope with an effective aperture of 7.5 cm and the total weight of the full system was 250 kg.

The LATAS system was employed in avionic applications and in atmospheric studies during the 80s. Airborne measurements of free stream airspeed and wind shear were first possible by installing the system in the nose of an HS125 jet. Useful warnings were provided with a focal range of approximately 300 m ahead [930]. In 1982, the LATAS system in the HS125 jet was included as part of the *Joint Airport Weather Studies* (JAWS) project in Colorado. Measurements from experiments and tests performed by flying through severe microbursts associated with thunderstorms have contributed to the development of a descending vortex ring model for microburst behavior, which has explained a number of features observed during the JAWS flights as, for example, the dust curtains rising to over 300 m around several microbursts. Further airborne sensing of air motion using a conically scanned optical Doppler technique has shown that the optical approach can be used to provide accurate measurements of the CAT spectrum at frequencies up to 10 Hz independently of the signal-to-noise ratio [465]. On the other hand, the CW Doppler LIDAR developed by Crouzet S.A. has also been employed for airspeed measurements on a variety of aircraft, including a Puma helicopter in 1984 and a Caravelle transport aircraft in 1985 and 1986 [640].

Among other uses, the LATAS airborne LIDAR has also been used as a tool for ground imaging [149] and for widespread investigation of atmospheric

---

[8]LIDAR is the abbreviation for *Light Detection and Ranging*. Sometimes the acronym is also employed to mean *Laser Imaging, Detection and Ranging*. In general, LIDAR is any method that can be used to determine variable distances (i.e., ranges) to a targeting test section (or object) with a laser and measure the time spent by the signal to reach the target and return to the receiver as a reflected signal [895].

air motions. In this latter aspect, much work has been focused on calibration [298, 929] and the development of algorithms for speeding up the processing of Doppler spectra and the calculation of backscatter coefficients [778, 777]. Direct comparisons of measurements were also carried out during the JAWS flights between the LATAS airborne LIDAR and the ground-based, pulsed, CO_2 LIDAR of NOAA's Wave Propagation Laboratory [105, 106]. Measurements of backscatter gradients as indicative of air turbulence were reported in 1987 for four different regions of the Northern Hemisphere [929], which supported the conclusions that airborne laser radars are reliable tools for wind shear detection and warning at low levels and for global wind fields at higher altitudes. From June 1981 to October 1983, airborne collected data for 40 vertical wind profiles over the United Kingdom were compared with similar data recorded by the NOAA-WPL ground-based LIDAR over Boulder, Colorado [779]. A compendium of airborne backscatter measurements at 10.6 μm from the *South Atlantic Backscatter Lidar Experiment* (SABLE) and the *Global Atmospheric Backscatter Lidar Experiment* (GABLE), where about 4×10^5 measurements from over 180 flight hours were analyzed statistically, is given in a lengthy paper by Vaughan et al. [928]. This compendium provided a climatology of atmospheric backscatter at 10.6 μm for the South Atlantic in summer and winter, the far-North Atlantic in spring, the mid-Atlantic between spring and summer, and the Northeast Atlantic in winter and summer during the period 1988–1990.

7.2.2.4 Laser Doppler velocimetry

Laser Doppler velocimetry (LDV), also called Laser Doppler anemometry, has been originally applied to liquid flow as well as to mixing and energy dissipation characterization in an extensive variety of flows. As a laser Doppler radar, a LDV system can perform fluid velocity measurements by detecting the frequency of light signals scattered by suspended particulate matter in the fluid as they cross a fringe or interference pattern. Early anemometers employed green light argon lasers operating at approximately 0.5 μm. Figure 7.5 shows schematically a typical LDV system operated in forward scatter mode. The signal emitted by the laser light source is collected by an optical arrangement, consisting of a beam splitter which divides the signal into two beams of equal intensity, and then transmitted

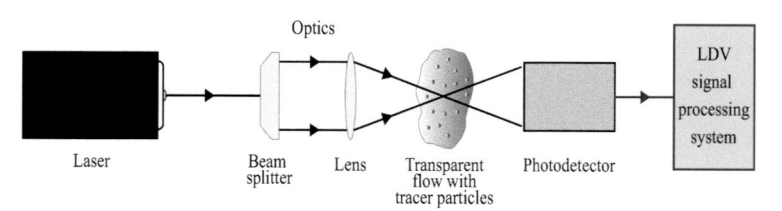

Figure 7.5: Schematic picture showing a standard laser Doppler velocimeter operated in forward scatter mode.

to a lens, where they are refracted so that they cross at a certain point, forming a local fringe pattern within a transparent flow system placed in front of the lens, where measurement of the velocity is required.

The interference fringe-pattern that forms at a single point in the fluid resembles that forming in the classical Young's two-slit experiment of quantum physics. Measurements of flow reversals within the fluid test section are possible if one of the incident beams is Doppler shifted. Finally, the incoherent light scattered from the volume of interest is collected by a photodetector, where the optical signal is automatically converted to an electronic signal. The scattered energy exhibits an amplitude modulation, which is caused by the optical fringes. As it is well known, the separation between two successive fringes[9] divided by the transverse particle velocity is equivalent to the time elapsed between two successive maxima in the signal[10]. This allows computation of the transverse particle velocity. The electronic Doppler signal is then finally passed to a LDV signal processing system. In the late 1960s and early 1970s, LDV systems were employed primarily to measure aircraft vortices in the atmosphere [112]. At that time, LDV systems for aircraft vortex detection were mostly operated with green light argon lasers because of their finer resolution scale measurements of the vortex velocity structure. Since the scattered light is shifted by an amount that is proportional to the flow velocity, both its mean and fluctuating part can be determined online from the record of the photodetector and the detected frequency shifts. In addition, local velocities can be resolved within fluid volumes of 10^{-6} to 10^{-4} mm^3. To obtain data at different positions across the fluid volume (i.e., the atmosphere), the LDV system must be turned around thereby allowing the split beams from the laser to intersect at a range of locations within the region of interest. In operational terms, this will require installing the system on a turntable, which may be handled manually or under computer control. For example, a single-channel LDV can measure only a single component of the fluid velocity, while the determination of the other two components would demand repeating the measurement by adjusting the orientation of the laser beams. On the other hand, devices equipped with a dual-channel LVD can measure simultaneously two velocity components, while the third component is measured separately at the same position by changing the orientation of the laser beams [242].

---

[9]The distance Δx between successive fringes is given by

$$\Delta x = \frac{\lambda}{2 \sin \phi},$$

where ϕ is the angle between the beams and the laser light wavelength, λ [560].

[10]That is,

$$\frac{\Delta x}{v} = f_D^{-1} \propto t_m,$$

where f_D is the Doppler frequency of the signal captured by the photodetector, v is the transverse particle velocity, and t_m is the time interval between two consecutive maxima in the signal.

In the late 1980s, the CW LIDAR built by Crouzet S.A. [640] was further developed by *Sextant Avionique* to become the ALEV3 Air Data Calibration equipment for airborne applications. The novelty of this system was that the output beam could be switched into three orthogonally directed, beam expanders with optical diameters equal to 75 mm [930]. This 3-axis CO_2 CW anemometer was subsequently certified by the *Aerospatiale Flight Test Division* and was made to operate since mid-1991 on various Airbus aircraft with successful results, accumulating more than 1200 hours of flight in two years. In particular, this LIDAR has been used as a precise and reliable reference for certification of other aircraft in flight test centers [930]. A consortium of European partners was subsequently created for studying the feasibility of an airborne system for the active onboard detection of wake vortices, which involved a project dubbed *Future Laser Atmospheric Measurement Equipment* (FLAME). This project ended towards the end of 1995, and a second project started on May 1, 1996, called *Multifunction Future Laser Atmospheric Measurement Equipment* (MFLAME), which concerned a multifunction version of the previous FLAME system and was completed in April 2000 [196].

7.3 Optical stellar scintillation

As was previously outlined, optical scintillation is a non-intrusive technique and belongs to the family of passive acoustic methods for CAT detection. Refraction of light coming from outer space occurs because the Earth's atmosphere has a refractive index that is different from that of a vacuum. Variations of the atmospheric refractive index with space and time give rise to a variety of refractive phenomena that require the use of refraction and diffraction theory to be understood and that can be separated into two main categories, namely regular (or normal) and random refraction. An early but no less comprehensive survey on this subject is provided by Meyer-Arendt and Emmanuel [621], where quantitative aspects of the regular and random refraction are thoroughly described. A light ray from a distant source and incident on a point within the atmosphere will not hit it with a perfectly straight path, but rather along a curved trajectory because the atmospheric density and refractive index decrease with altitude. This phenomenon, which is observed in a variety of situations, is called regular refraction. A well-known example is the apparent flattening of the Sun's disk during sunrise and sunset as the rays reaching the horizon from the top and bottom edges of the Sun are refracted by different degrees. In this sense, the Earth's atmosphere acts much like a lens, bending the light toward the ground. However, as was reported in 1963 by astronaut L. G. Cooper after his 22-orbit mission, the phenomenon looks much more enhanced when observed from an orbiting satellite. In addition to the bending of light bundles, the presence of air turbulence causes small scale, irregular changes in the refractive index, which are responsible for the apparent twinkling of distant stars, frequently referred to as stellar scintillation. In other words, when a distant star is observed with the aid of an optical telescope of a small aperture, the star rather than appearing fixed and steady, oscillates about a mean point and its intensity fluctuates. It is said then that the star twinkles or scintillates, and the phenomenon is the result of random atmospheric refraction. However, when the star is imaged with a large-aperture telescope, image motion is space-integrated and the point-like star is seen as a diffuse, tremor disk whose diameter provides a measure of the motion amplitude. Henceforth, the term scintillation will be used strictly for the apparent rapid fluctuations of the stellar intensity owing to the irregular variations of the refraction index caused by small-scale turbulence. The interested reader is referred to the survey by Meyer-Arendt and Emmanuel [621] and references therein for other definitions related to star motion and sparkling.

In particular, the term *astronomical seeing* has been often employed as synonymous with scintillation. However, it is generally agreed that seeing can be defined as the combined effect and degree of the continuous changes of position, shape, and size of the image of a pointlike celestial object. However, astronomical seeing and stellar scintillation are independent of each other. For example, in contrast to scintillation, it has been commonly accepted that seeing depends on local air turbulence close to the telescope. When the telescope is subject to intense heating, as occurs in solar observations, its focus can be shifted because of the expansions and contractions of parts of the telescope. Seeing effects can also occur when the telescope is near and pointed towards zones of temperature convection. On the other hand, the varied meteorological conditions occurring in the higher atmosphere may also influence the seeing. For example, moist air gives strong absorption in the infrared, which in turn produces a greenhouse effect that affects the seeing.

When looking at a scintillating star with the naked eye or through a small telescope, the white image that is being observed is composed of a succession of colors that change at such a high frequency that the eye is not able to distinguish them. This phenomenon, which is sometimes called the spectral drift of starlight, is commonly referred to as color scintillation. This phenomenon is most pronounced for stars observed near the horizon and is almost absent for stars near the zenith [627, 625]. At different wavelengths, the frequency of the intensity fluctuations has been inferred to be around 50 MHz, which is about the same as for ordinary white light scintillation. In general, color scintillation is due to the combined effects of random and regular atmospheric refraction. This is because the light of different wavelengths coming from the same star will hit the same point on the Earth's surface after having followed different paths through the atmosphere. This can be explained in terms of the refractive index, n, as a function of density, ρ, which obeys the relation

$$n = 1 + f(\lambda)\rho, \tag{7.26}$$

where f is a function of wavelength. For varying wavelengths between 4000 Å to 7600 Å, the function $f(\lambda)$ varies slowly from 2.308×10^{-7} to $2.247 \times$

10^{-7}, yielding a rather weak dependence of the refractive index on the starlight wavelength for fixed density.

On the other hand, when looking down at the lights of the city from a hill or from a high tower or building, sometimes incandescent street lamps are seen to scintillate. This phenomenon is called terrestrial scintillation and it can provide more complete data than stellar scintillation observations since it allows to monitor variations in the refractive index and wind velocities in an easier way. Such data were very useful because they were used to perform comparisons with theories of scintillation. Several experiments have been made in the past years to correlate measurements of terrestrial scintillation with air turbulence at low altitudes [621]. A typical setup employed for photographing air turbulence along paths at low altitudes from the ground is shown schematically in Fig. 7.6, where a light bundle incident on a concave mirror is reflected in the objective lens of a movie camera, which is, in turn, focused on the mirror. The exposure is controlled by a series of neutral density filters arranged in front of the camera. During a clear, moonless night, a pattern of shadow bands can be seen on the screen, which is the result of turbulent eddies in the atmosphere.

Based on early experiments and investigations, the effects of meteorological conditions on stellar scintillation can be summarized as follows. First, in zones where the air density is higher, scintillation is seen to increase. This explains why at lower altitudes scintillation is more pronounced. Second, on typical sunny days where high temperature gradients settle, scintillation is well-marked and tends to increase with increasing temperature gradients. Third, the degree of scintillation reduces during a cloudless day and night because under these conditions the magnitude of the vertical temperature gradients is greatly reduced. Fourth, increased wind velocity enhances the degree of scintillation. This occurs up to a certain point beyond which scintillation is seen to decrease, even on a sunny day. Similarly, a moderate wind can mix the lower air layers to a greater extent and height during the night, resulting in lower temperature gradients and

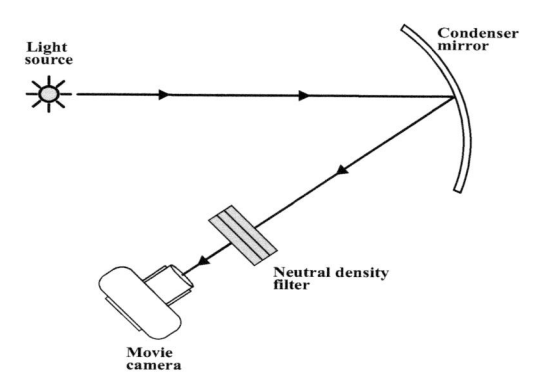

Figure 7.6: Schematic picture showing a standard setup used for photographing low-altitude air turbulence.

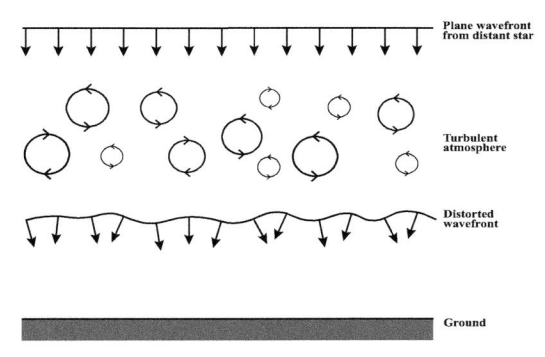

Figure 7.7: Distortion of a starlight plane wavefront passing through a turbulent atmospheric region.

as, a consequence, in reduced scintillation. Fifth, the presence of fog, rain, and snowfall are all conditions that result in a reduction of both stellar and terrestrial scintillation. Moreover, if between a light source and its recorded image there is a turbulent zone, the resulting image will degrade in several ways [421]. For point-like objects, such as a distant star, the image will be seen to fluctuate randomly in intensity (scintillation) and position (dancing), and therefore it will shift in and out of focus (i.e., the point will appear to be pulsating). If, on the other hand, the imaged object is an extended one, then its image will appear somewhat blurred as a whole. Figure 7.7 illustrates the distortion of a plane wave front coming from a star as it passes through a CAT region in the atmosphere.

The contrast reduction in the image occurs because the light coming from the object is deflected out of its normal path by scattering and attenuated by absorption. The balance between scattering and absorption determines the luminance of the image and therefore the reduction in the observed contrast. Numerous experiments carried out over the years have corroborated the relation existing between CAT and image distortion. In fact, image resolution is worst when the average vertical temperature gradient increases and light winds are present and is best in the absence of thermal stratification. For example, under a clear sky, image resolution typically deteriorates when the windspeed increases up to 8 km h^{-1} [784].

7.3.1 The refraction theory of scintillation

Light refraction in the atmosphere is at the base of the scintillation phenomenon. More specifically, the refractive index of the atmosphere is different from that of a vacuum and varies as a function of location. Therefore, when incident light coming from a distant star penetrates a zone where the refractive index is low, its velocity will be higher compared to the same light passing through a region where the refractive index is high. In the former case, the plane wavefront of the luminous signal will advance farther, while in the latter case, it will be retarded.

These inhomogeneities intrinsic to the atmospheric refractive index cause a deformation of the wavefront into a corrugated form. The corrugation is such that the convex portions of the wavefront point to the direction of propagation because they move faster, while the retarded portions are in contrast concave. This causes the starlight to change direction and, as shown schematically in Fig. 7.7, hit the ground with a non-uniform intensity, which explains the formation of a shadow band pattern, consisting of alternating bright and dark bands.

Refractive anomalies are important in the lower stratosphere (i.e., within about 3 km from the tropopause), where CAT encounters are more likely, and within the first 100 m from the ground. Early empirical field observations have indicated that after crossing the lower stratosphere under small-scale air inhomogeneities, the wavefronts remain essentially plane with their normals oscillating with respect to the axis of the telescope, producing a dancing effect [773]. This occurs because CAT does not cause much of a focus drift. However, with large-aperture telescopes, this oscillatory motion transforms into a tremor disk. On the other hand, as the wavefronts reach the objective of the telescope close to the ground, the wavefronts are more likely to be distorted. These observations were further confirmed by a series of experiments [78]. For example, air turbulence as it occurs naturally may entail cells of hot and cold air, which in front of the telescope objective would then act as weak positive and negative lenses, causing a divergent-lens effect. On the other hand, scintillation has been observed to depend on the vertical temperature profile along a horizontal path. The irregularities associated with such turbulent cells have been termed *aerial blobs* [1007], which are defined as air volumes exhibiting locally different density, temperature, and possibly water content. They may be characterized by globular, lenticular, or even cylindrical shapes with sizes ranging from millimeters to several meters. Observations at low altitudes (up to about 50 km from the ground) have revealed that most blobs can be very stable and preserve their shape for many hours [271]. Therefore, if these disturbances settle in front of the telescope objective, they will certainly defocus and displace the image of pointlike distant stars. Early model experiments have shown that the whole spectrum of phenomena that may occur due to low-altitude air turbulence may well be explained by light refraction in such blobs [272]. Thorough theoretical support for refractive inhomogeneities as the cause of scintillation has been provided by Tatarski [893] and Reiger [750].

Further experimental support to light refraction as a cause of scintillation has come from determinations of the radius of curvature of incident wavefronts. As was outlined before, small telescope apertures of ~ 3 to 4 cm are appropriate to determine the radius of wavefront corrugations. Early measurements have revealed that such radii can vary between 1.8 and 20 km, with average values of ~ 6 km [996]. An expression for the radius of curvature of corrugated wavefronts was previously derived by Pernter and Exner [697] as

$$R_c = \frac{2f^2}{a},$$ (7.27)

where f is the focal length of the objective and a is the distance of the focus drift. Moreover, under the assumption that turbulent blobs in the atmosphere have a cylindrical shape and same pressure and composition of surrounding air but a temperature difference ΔT, a light ray incident with an angle α would then deviate by an amount given by the relation [27]

$$\tau = \frac{0.2p}{p_0} \Delta T \tan \alpha, \tag{7.28}$$

where p_0 is the normal atmospheric pressure. For a temperature difference as low as $\Delta T = 1°C$, an angle of incidence $\alpha = 45°$, and $p \approx p_0$, Eq. (7.28) predicts a deviation of 0.2 arcsec, which is approximately of the right order of magnitude as has been found under good seeing conditions.

7.3.2 The diffraction theory of scintillation

In 1951, Little [564] proposed that stellar scintillation as caused by air turbulence in the atmosphere can be attributed to Fresnel diffraction on the basis that the refraction theory of scintillation would require high atmospheric density gradients compared to diffraction theory. The principle of this theory can be summarized as follows. As shown schematically in Fig. 7.7, after passing through a turbulent region, the elemental areas of the corrugated wavefronts will be out of phase in a random way. The wavefront amplitude distribution across a plane immediately below the diffracting screen will appear uniform because only the phase of the wave motion will be affected by the screen. As the wavefront spreads away from the screen, fluctuations in the signal amplitude start to appear, which are randomly distributed across a plane parallel to the original diffracting layer. The minimum distance from the screen at which amplitude irregularities arise is completely determined by the distortions of the emergent wavefront. The Fourier transform of the wave autocorrelation function below the diffracting plane gives the average power spectrum observed behind the diffracting screen. On the other hand, the width of the autocorrelation function is found to be of the same order as the wavefront corrugations, provided that they do not exceed about one radian. For example, in a deformed wavefront in which a typical distortion is $n(> 1)$ radians, the width of the wave autocorrelation function will be at least a factor of $1/n$ times the width of the individual corrugations [566]. This means that the width of the striations seen in the shadow band pattern will be of the same size as the irregularities that caused the striations. In addition, if due to wind, the atmospheric inhomogeneities are anisotropic then the generalized autocorrelation function of the diffraction pattern at the ground level will be a function of the wind direction.

In his 1951 paper, Little [564] stressed that the main reason for using diffraction theory to explain stellar scintillation is that much smaller density gradients are required to produce much the same effect as in the case of refraction. In fact, for diffraction theory, atmospheric density gradients of only $\sim 0.0006\%$ per cm are actually required compared to about 0.5% per cm for refraction. In terms of temperature gradients, this implies $\sim 0.002°C\ cm^{-1}$ for diffraction against $\sim 1.5°C\ cm^{-1}$ for refraction [564]. However, in the absence of refraction, there will not be diffraction and, if the refractive index is the same everywhere, then no diffraction will occur. When a light ray passes through the atmosphere, much energy is diffracted out of the original direction, producing a series of intensity maxima. The resulting pattern is broadened and the intensity of zeroth-order, corresponding to the undeviated ray, is much reduced. This reduction in intensity compared to the original normal ray can be used as a measure of the thicknesses of the layers through which the light ray has traveled. It was also noticed that there is no conflict between the diffraction and refraction theory of scintillation and the validity of the geometric-optical approach (i.e., refraction) lies entirely in its being an approximation to wave theory. In general, a ray treatment of light will suffice in the case of large aberrations, while a wave optical treatment will be necessary for small aberrations. For example, geometrical optics will be a sufficient approximation if the quantity $s^2/(\lambda r) \gg 1$ [100], where s is a measure of the size of the turbulent element (or blob), λ is the incident light wavelength, and r is the distance of turbulence from the observer.

On the other hand, suppose that an obstacle of dimension L is placed in the path of a plane wavefront and that at a distance l from the obstacle its projected image (i.e., its shadow) will have the same dimension L. The angle of the diffracted wave by the obstacle will be approximately $\theta \sim \lambda/L$ so that the size of the diffracted bundle can be estimated to be $l\theta \sim l\lambda/L$. Therefore, the geometric shadow of the obstacle will not change appreciably if the condition holds [893]

$$(l\lambda)^{1/2} \ll L. \tag{7.29}$$

If L_0 is a lower limit to the size of turbulence, then the condition

$$l \ll \frac{L_0^2}{\lambda}, \tag{7.30}$$

must be satisfied by the distance l in order for the geometrical optics to be a valid approach for describing scintillation.

7.3.3 Analysis of turbulent fluctuations

The existence of pressure, density, and temperature gradients may lead to a more or less continuous turbulent motion in certain atmospheric layers. Possibly, the gradients of other parameters may also contribute to trigger air turbulence. The variations of the refractive index, Δn, can be expressed in terms of variations of the density, $\Delta \rho$, of a given air blob and its surroundings according to the relation [872]

$$\Delta n = (n_0 - 1) \frac{\Delta \rho}{\rho_0}, \tag{7.31}$$

where n_0 and ρ_0 are, respectively, the refractive index and density of atmospheric air at normal pressure ($p_0 = 1$ atm) and temperature ($T_0 = 20°C$). If it is further assumed that the velocities involved in the turbulent motion are far less than the speed of sound in air, then the relative density differences will be the result of temperature variations. Under the assumption that the air behaves as an ideal gas, then $\Delta\rho/\rho = \Delta T/T$ and relation (7.31) takes the form

$$\Delta n = (n_0 - 1)\frac{\rho}{\rho_0}\frac{\Delta T}{T}. \tag{7.32}$$

According to Tatarski [893], turbulence is characterized by irregular and aperiodic distributions of the vorticity in space and time, by the transfer of kinetic energy from larger to smaller eddies until it is converted into random molecular motion at the smallest scale, and by the mean separation distance between neighboring fluid elements increasing with time. In this sense, except for strong inversion layers, the atmosphere can then be thought of as a turbulent medium everywhere. However, the intensity of atmospheric turbulence exhibits strong spatial and temporal variations. In the framework of Kolmogorov's theory, when Re$\rightarrow \infty$, the energy spectrum of turbulence becomes independent of the viscosity and is solely determined by the kinetic energy that flows per unit mass, ε, down the cascade. As shown in Fig. 1.8, this state corresponds to the inertial subrange or Taylor microscale. Equation (1.65) estimates the characteristic size of the viscous eddies, $\eta \sim (v^3/\varepsilon)^{1/4}$, where the kinetic energy is being dissipated into heat at the Kolmogorov scale or dissipation range (see Fig. 1.8). Estimates of the eddy sizes in the atmosphere indicate that the largest ones are associated with length scales of hundreds of meters, or even more, while the viscous eddies are of the order of about 1 cm at the most [883], implying that the inertial subrange in the atmosphere for which Kolmogorov's second hypothesis applies is indeed very large.

The influence played by aerial blobs on the refraction and diffraction of light coming from distant stars has been discussed before. The vertical acceleration of such blobs can be calculated under simplifying assumptions. In particular, by assuming that (a) the blobs move around with no countervailing motions of the surroundings, (b) they move without mixing with their environment, and (c) the environment is in hydrostatic equilibrium, the blob movement can be described by the differential equation for the blob acceleration in the upward direction [397]

$$\frac{d^2z}{dt^2} + \left(\frac{\Gamma_d - \gamma}{T_0}\right)gz = 0, \tag{7.33}$$

where T_0 is the blob temperature at height $z = 0$, g is the gravitational acceleration, and Γ_d and γ denote the environmental and adiabatic lapse rates, respectively. If $(\Gamma_d - \gamma)g/T_0 > 0$, Eq. (7.33) admits a sinusoidal function $z = z(t)$. This corresponds to a stable solution in which case the blob will oscillate about its equilibrium position with a period

$$P = 2\pi \left[\frac{g}{T_0} (\Gamma_d - \gamma) \right]^{-1/2}. \tag{7.34}$$

In contrast, if $(\Gamma_d - \gamma)g/T_0 < 0$, the solution is an exponential function of time, corresponding to an unstable case in which the blob will move without ever stopping. In the trivial case when $\Gamma_d = \gamma$, the blob acceleration is identically zero, which corresponds to a neutral case.

Since the actual distribution of density and refractive index in the atmosphere is not explicitly known and neither is the distribution of the light on the ground, the scintillation must be treated statistically regardless of whether geometrical or wave optics is employed. For light propagating in a vacuum, the optical path length traveled by the light during time t is $l = ct$, where c is the speed of light in a vacuum. In a medium of constant refractive index n, the optical path length will be $l = nr$, where r is the distance between the source and the receiver. This expression, however, is no longer valid for light propagating in an inhomogeneous (turbulent) medium. For the moment suppose two adjacent moving wavefronts of surfaces S =constant and $S + dS$=constant, which are separated by the element of distance $d\sigma = dS/n$ measured along the wavefront normals. This relation is equivalent to writing $dS/d\sigma = n$ for the directional derivative of S and is related to the eikonal equation as

$$\| \nabla S \|^2 = \nabla S \cdot \nabla S = n^2, \tag{7.35}$$

which is known as the infinitesimal formulation of Huygens' principle. This equation is a nonlinear, first-order, partial differential equation to be solved for S as a function of position, and it belongs to the Hamilton-Jacobi class of differential equations. If the refractive index in a given parcel of turbulent atmospheric air is given by $n = n_0 + \Delta n$, where, as before, n_0 is the refractive index of air at standard conditions and Δn is its variation due to turbulence. Working with a Cartesian coordinate system and putting the source of light at the origin $(x = 0, y = 0, z = 0)$, the solution of Eq. (7.35) can be expressed as the power series expansion

$$S = n_0 r \;\; + \;\; \varphi \int_0^r n' \left(r'\frac{x}{r}, r'\frac{y}{r}, r'\frac{z}{r} \right) dr'$$
$$- \;\; \frac{\varphi^2}{2n_0} \int_0^r \left[\frac{1}{rr'} \int_0^{r'} r'' \mathbf{r} \times (\nabla n')_{r'} dr'' \right]^2 dr' + \cdots, \tag{7.36}$$

where φ is the scattering angle, and $(\nabla n')_{r'}$ is the gradient of the refractive index evaluated at point $(r'x/r, r'y/r, r'z/r)$ lying at a distance r' from the source on the line that joins the source to point $\mathbf{r} = (x, y, z)$. From Eq. (7.36) it is possible to show that to first order the actual distance traversed by the light is equal to its mean optical path length across a randomly inhomogeneous medium [621].

Under the assumption of isotropic turbulence, the fundamental relations between the atmospheric turbulence and the intensity distribution in the seeing image are given by [468]

$$\rho(\xi,\eta) = \frac{1}{I_0} \int_{-\infty}^{\infty} \exp\left(2\pi i \frac{\xi\theta}{\lambda}\right) d\theta \int_{-\infty}^{\infty} I(\theta,\phi) \exp\left(2\pi i \frac{\eta\phi}{\lambda}\right) d\phi$$
$$= \exp\{-4\pi^2[S(0,0) - S(\xi,\eta)]\}, \tag{7.37}$$

where λ is the wavelength, $I(\theta,\phi)$ is the angular intensity of radiation, $\rho(\xi,\eta)$ is the autocorrelation function of the electric field strength of the incoming radiation, and

$$S(\xi,\eta) = \frac{2F}{\lambda^2} \int_0^{\infty} \sigma(\xi,\eta,\zeta) d\zeta, \tag{7.38}$$

where $\sigma(\xi,\eta,\zeta)$ is the autocorrelation function for fluctuations in the index of refraction, F is given by

$$F = \sum_m (\Delta z)_m \delta^2 f_m, \tag{7.39}$$

$(\Delta z)_m$ is the thickness of the mth turbulent layer, and $\delta^2 f_m$ is the corresponding mean square fluctuation in the index of refraction. Following Keller and Hardie [468] and assuming cylindrical symmetry, the autocorrelation function can be defined only in terms of the independent variable ξ so that

$$\rho(\xi) = \frac{1}{I_0} \int_0^{\infty} \exp\left(2\pi i \frac{\xi\theta}{\lambda}\right) T(\theta) d\theta, \tag{7.40}$$

where

$$T(\theta) = \int_{-\infty}^{\infty} I(\theta,\phi) d\phi. \tag{7.41}$$

The quantity $T(\theta)$ can in principle be measured by just measuring the amount of light passing through an infinitely long and narrow slit placed in the focal plane of a perfect telescope. Actually, there is a close statistical relationship between the structure of stellar shadow band patterns, as observed on the surface of a telescope objective, and the scintillation of the total starlight received. In particular, this implies that the light and dark bands in the shadow patterns are actually related, at least statistically, with the hot and cold air patches in the upper atmosphere [466]. The relative amplitudes of fluctuations of various dimensions in the shadow patterns can be expressed statistically as a power spectrum of power spectral density, which is represented by a function $B(k)$ corresponding to the Fourier cosine transform of the autocorrelation function [872], where k is the wavenumber of a Fourier component. In particular, for the isotropic case the function $B(k)$ has the form

$$B(k) = 8\pi\langle h^2\rangle(\Delta r)_0^2 \exp\left(-2\pi^2(\Delta r)_0^2 k^2\right), \tag{7.42}$$

where $\langle h^2 \rangle$ is the mean-square amplitude of the intensity fluctuations in the pattern. The sum of the squares of all Fourier coefficients of the intensity fluctuations with wavenumbers between k and Δk can be calculated as $2\pi k B(k)\Delta k$. From Eq. (7.42) it is easy to see that $B(k)$ decreases rapidly when $k > 1/(\Delta r)_0$. In addition, if very severe and extreme CAT are excluded, then in most cases of atmospheric air turbulence the power spectral density for turbulence, say $b(k)$, and that for the observed shadow band patterns, $B(k)$, can be related by the expression [467, 872]

$$B(k) = \left[\frac{4\pi h_0 \sin\left(\pi z \lambda k^2\right)}{\lambda} \right]^2 (\Delta z)b(k), \qquad (7.43)$$

where the various quantities have the same meaning as before. This relation encloses the manner in which atmospheric turbulence is related to the characteristics of the shadow patterns. For example, large temperature fluctuations are responsible for large refractive index variations, and therefore for large values of both $b(k)$ and $B(k)$, while a thick layer yields larger values of $B(k)$ compared to a thin layer. On the other hand, if a distant star is not at the zenith, Eq. (7.43) holds if z is taken to be the inclined distance along the light path from the telescope to the layer and Δz to be the distance through the layer. Thus, at large zenith distances, Δz becomes large and the sine term in Eq. (7.43) increases until it starts oscillating in the interval $[-1, 1]$. This means that scintillation is usually more pronounced for stars at large zenith distances [725]. For light coming from stars near the zenith the argument of the sine in Eq. (7.43) becomes much smaller than unity, therefore $\sin(\pi z \lambda k^2) \approx \pi z \lambda k^2$ and so Eq. (7.43) takes the simpler form

$$B(k) \approx \left(4\pi^2 z k^2\right)^2 (\Delta z)b(k). \qquad (7.44)$$

In this case, the dependence on the wavelength, λ, disappears and scintillation becomes color independent. Futhermore, as k decreases, $B(k)$ decreases rapidly, explaining the lack of large element sizes in observed shadow patterns. It was common practice to determine the statistical properties of stellar shadow band patterns photoelectrically by either of two techniques. The first technique consisted in finding the autocorrelation function of the pattern using a twin telescope of variable separation [466, 725], while the second one relied on finding the spatial power spectrum of the pattern using an optical Fourier analyzer [726, 727].

An alternative approach to infer the effective mean size of turbulent elements is based on the so-called *after-effect function* [318, 269], which is defined by

$$\Delta(\tau) = \overline{f(t + \tau) - f(t)}, \qquad (7.45)$$

where the average is taken over a sufficiently long time. The reduced after-effect function is defined as

$$\delta(\tau) = \frac{\Delta(\tau)}{\Delta(\tau \to \infty)}, \qquad (7.46)$$

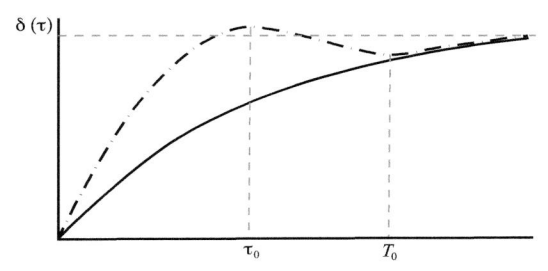

Figure 7.8: After-effect function $\delta(\tau)$, as defined by Eq. (7.46), corresponding to a purely random process (solid line curve) and to a process formed by an almost irregular sequence of pulses of duration t_0 each and average interval T_0 (dashed line curve). Figure adapted from Fürth [318].

which varies from zero when $\tau = 0$ to unity when $\tau \to \infty$. Contrary to what it might seem, the after-effect function can be calculated from observational data with little difficulty, and in many situations, it provides a direct indication of the type of elementary process from which a stochastic phenomenon is built. If $f(t)$ is strictly periodic, then its after-effect function will also be periodic. That is, if, for example, the elementary processes consist of a regular succession of pulses of duration t_0 each and separated from each other by equal time intervals T_0, then the function $\Delta(\tau)$ is also periodic with period T_0. However, the presence of any irregularity in the sequence of pulses will destroy the long range order of the diagram, and the after-effect function will take the form of the solid line curve for a purely random process, as shown in Fig. 7.8 [318]. The dashed line curve in the figure indicates an intermediate stage, where the position of the maximum marks the approximate pulse duration (t_0) and the minimum the average time interval (T_0) between consecutive pulses. It is worth noting that this method has been successfully applied not only to stellar scintillation, but also in the analysis of records of Brownian motion as well as of density and electrical fluctuations [269]. There can be little doubt that the size of atmospheric turbulence elements range from molecular dimensions up to kilometers and that the smallest element's sizes are still larger than the wavelength of light. It is well known that large elements will act as prisms as light crosses them, giving rise to dancing, while small elements will cause diffusion and boiling of the star image. However, these latter effects will greatly depend on the telescope aperture. For an element that is larger than the telescope aperture, the time length, t_p, of a pulse produced by such an element moving by is

$$t_p = \frac{D}{Lv}, \tag{7.47}$$

where D is the diameter of the aperture of the telescope objective, L is the horizontal size of the element, and v is its velocity. Estimates of the size of turbulent elements provided by various authors between 1949 and 1962 yielded values ranging from ~ 1 to 40 cm [621]. Using balloon-borne lights at different heights,

Gardiner et al. [324] confirmed that low frequency (1–10 Hz) scintillations in the image intensity of distant stars are primarily detected at heights as high as 30 km, whereas high frequencies (150 Hz) seem to be associated with lower altitudes (from 3 to 12 km), i.e., at or near the tropopause. In particular, they found that at heights of 7.5 km the balloon-borne lights started to twinkle and when the balloon was at heights as high as 15 km and about 30° distant from the horizon, that is, well above both the tropopause and the fast tropospheric winds, the twinkling of the lights was seen to approximately match that of adjacent stars. These authors have also explained why low frequency scintillations are associated with great heights and low frequencies with lower altitudes. By claiming that tropospheric westerly winds ordinarily prevail at all heights in middle latitudes and achieve maximum speeds near the tropopause and that the winds become easterly and light in the lower stratosphere (just above the tropopause), they realized that in an atmosphere at rest with the Earth, the density inhomogeneities will cross the light path from west to east because the light path between the star and the telescope is turning about 15° per hour from east to west. Although this effect is negligible at low altitudes, it becomes significant at greater heights.

On the other hand, Fürth [318] noted that diffraction theory must play a relevant part in the estimation of the sizes and height of the turbulent elements. He showed by diffraction theory that the size L of the turbulent elements, the size d of the shadow striations on the ground, the height z of the turbulent layer, and the wavelength λ of the incident radiation are related as

$$\frac{L}{z} \sim \frac{\lambda}{d}. \tag{7.48}$$

For example, L cannot be smaller than d because then diffraction effects would be irrelevant and, on the other hand, it cannot be larger than d because then the diffraction pattern on the ground would bear no relation to the atmospheric irregularities, and as a consequence, no correlation would exist between the patterns produced with different wavelengths. This implies that L and d must be of the same order of magnitude so that the ray theory equation $L = d$ must hold approximately. It then follows that the height z of the turbulent element of size L can be estimated from the relation

$$z = \frac{L^2}{\lambda}. \tag{7.49}$$

Further observations from balloon flights pointed to the conclusion that nearly all stellar scintillation is generated at altitudes below 24 km [626]. It was often referred to as a single-layer model by attributing scintillation to a thin and highly turbulent layer in the troposphere [161]. According to this model, estimates of the lifetime of elements are generally too small, i.e., of the order of a few milliseconds to tens of milliseconds [726]. However, the actual lifetime would be much longer if the whole atmosphere, and not just a single layer, is considered a contributing factor [146, 749]. In particular, it was realized that according to the experimental evidence thin, highly turbulent layers cannot be the principal cause of starlight scintillation. Needless to say, if such layers exist, they will only be responsible for a fraction of the total scintillation.

Measurements of the atmospheric optical turbulence profiling at high altitudes by stellar scintillometers continued over the years. From 15 March to 20 April 1985, two identical optical profiling instruments were operated at the Air Force Maui Optical Station (AMOS) on Mount Haleakala, Maui, to perform a series of measurements of the refractive index structure, C_n^2 [176]. In this case, the measurements were made in a collaborative instrument evaluation experiment between the *AVCO Everett Research Laboratory* (AERL) and the *Rome Air Development Center* (RADC), employing as the instruments two NOAA model II stellar scintillometers [672], which at the time were popularly known as *Star Sensors*. Each of these sensors is an optical detection instrument package that fits over the exit port on the back of a conventional Celestron-14 telescope. The results of these measurements were presented in terms of profiles, where each turbulence profiling corresponded to a set of seven different altitudes from 2 to 20 km. Daily turbulence averages for both instruments were seen to correlate with one another, even though the data acquisition as well as the processing systems were somewhat different, as were certain details of the electro-optical hardware. However, persistent and statistically significant differences were found in the absolute turbulence strengths as measured by each sensor, which were most pronounced at the lower altitudes, even when the sensors were operated simultaneously along common paths. Not least, when each instrument was employed to look at different stars, different turbulence strengths were measured at altitudes above 10 km. These observations revealed a non-uniform CAT distribution along the horizontal direction.

Stellar scintillation is generally quantified by calculating the variance of irradiance at a point of a telescope aperture given by [80]

$$\sigma_I^2 = \frac{\langle I^2 \rangle}{\langle I \rangle^2} - 1. \tag{7.50}$$

Scintillation is currently well understood in the regime of weak turbulence (i.e., when $\sigma_I^2 < 1$), where it can be described by Rytov's theory for the propagation of Helmholtz-Gauss (HzG) beams in turbulent atmosphere [665]. In this regime, the variance of irradiance depends on the total altitude-weighted turbulence along the light propagation path [893], i.e.,

$$\sigma_I^2 \cong \sigma_R^2 \equiv \exp\left[2.25 k^{7/6} \sec^{11/6}(\phi) \int_0^\infty C_n^2(h) h^{5/6} dh \right] - 1, \tag{7.51}$$

where C_n^2 is a measure of turbulence strength at height h, ϕ is the zenith angle, and k is the wavenumber. These relationship between σ_I^2 and the Rytov standard deviation, σ_R^2, breaks down in the regime of severe and extreme CAT, where the variance of irradiance saturates (i.e., $\sigma_I^2 > 1$). In this regime, the variance reaches a peak, which is commonly referred to as the strong-focusing regime, and then decreases slowly with increasing total turbulence [355]. In the decades of the 70's to the 90's, considerable progress was made in understanding the saturation of scintillation, particularly for horizontal propagation through homogeneous turbulence. For instance, the inner-scale effect of irradiance variance as measured for weak-to-strong atmospheric scintillation has been elucidated [197], which later was found to be more significant for spherical waves than for plane waves [295]. Also, the probability density function of irradiance has been evaluated numerically and its deviations from a log-normal distribution have been studied [294, 402]. In contrast, little was known about saturation because in general, the irradiance variance associated with starlight scintillation is small enough [249]. However, this trend changes at large zenith angles [136, 690, 689], where Rytov's theory is no longer applicable. Investigations of how the saturation of stellar scintillation differs from that associated with light propagation in a homogeneous atmosphere have been reported by Camparo [150]. To do so, the optical propagation through turbulence at the tropopause level and then through both the tropopause and low-altitude tropospheric turbulence was simulated numerically. Fits of the numerical data produced the empirical formula

$$\frac{\sigma_I^2}{\sigma_0^2} = 1 - \exp\left(-\frac{\sigma_R^2}{\sigma_0^2}\right), \tag{7.52}$$

where σ_0^2 is a parameter indicating the level of saturation in the regime of severe-to-extreme CAT and the value of σ_R^2 where saturation sets in. It was found that under strong CAT the vertical profile of turbulence at the tropopause level is responsible for some degree of scintillation. However, the combination of high-level CAT with low-altitude turbulence was seen to produce much more scintillation in the strong-focusing regime than high-altitude CAT alone. Camparo [150] interpreted these results in terms of optical wave decoherence, relative to the tropopause Fresnel distance[11], as the starlight wave propagates through the lower stratosphere and penetrates the tropopause.

---

[11]The Fresnel distance, or length, is defined as the minimum distance that is traveled by a ray light along the linear path before diffraction. For instance, if a light wave passes through an aperture of radius a and then travels a distance L to a screen, the Fresnel number is given by

$$N_F = \frac{a^2}{L\lambda},$$

where λ is the wavelength of the incident light. The Fresnel distance is then approximately given by $N_F L = a^2/\lambda$. Note that when $N_F < 1$, Fraunhofer diffraction occurs. In this case, the far-field diffraction pattern is shown in the screen. In contrast, when $N_F \gtrsim 1$, Fresnel diffraction occurs and the near-field diffraction pattern is seen on the screen.

7.4 Microwave scintillation of radio stars and satellite beacons

Radio astronomy began as a rapidly growing area of research in the years following World War II because of the advent of large antennas, the continuous improvement of receivers, and a reorientation of the astronomy community. Scintillations in the intensity image from radio stars were first observed by Hey et al. [399] in 1946, who reported that the Cygnus star fluctuated at 68 MHz. The scintillation of radio stars is in many aspects similar to that of visible stars and different in some other aspects. An entire bibliography exists for observations of radio star scintillations. In particular, references throughout the 1950s and early 1960s on such observations can be found in the technical notes by Nupen [670] and Meyer-Arendt and Emmanuel [621]. A review of radio-scintillation observations until the middle of the 90's is provided by Aarons [11].

As for visible stars, radio scintillation gives rise to shadow bands moving across the Earth's surface. They can be observed best by comparing the radio wave amplitudes from a single radio star at two different points on the surface about 1 km apart. For example, phase and amplitude variations at one point can be attributed to the steady drift of an irregular wave pattern over the ground [398]. Such irregularities have lateral extensions ranging from 2 to 10 km and are about 400 km from the ground, as was confirmed from satellite measurements, and move in a wind-like fashion with velocities of 100 to 300 m s^{-1}. Much larger irregularities were also observed later on by Lawrence et al. [533], having widths up to 200 km. These much wider irregularities were recognized to be responsible for the slow oscillatory motion or wandering effect in the light optical case, which arises from local, nearby turbulences. At frequencies $\lesssim 1000$ Hz, radio scintillation arises in the ionosphere, where gradients in the refractive index are caused by differences in the ionization density, while in the troposphere, where most optical scintillation originates, radio refraction is much less significant. In contrast, tropospheric radio refraction occurs primarily at higher frequencies. Moreover, at meter wavelengths, the intensity of radio waves carries fluctuations of the order of $\sim 10\%$ [533]. As for stellar optical scintillation, the intensity fluctuations increase rapidly with increasing zenith distances of the radio source [398, 949]. This trend was usually observed for zenith distances greater than $30°$ toward the horizon.

Early observations revealed that the mean amplitude of radio scintillation varies as the square of the observing wavelength up to the point where the lower frequency index saturates at 100% [173]. At the low-frequency end of the VHF band, the scintillation index approaches unity, while at the opposite end of the VHF band, i.e., at higher frequencies, the scintillation index is usually small and increases with decreasing frequency [99]. In the same way as for optical scintillation, in the framework of diffraction theory, the ionosphere behaves as a thin diffracting screen. Since for the frequencies that are typically employed for the

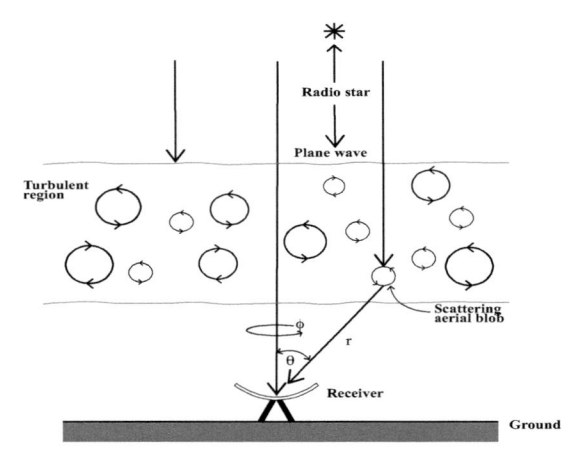

Figure 7.9: Schematic drawing of the scattering geometry for radio star scintillation. Figure adapted from Muchmore and Wheelon [648].

observation of radio stars, the absorption of radio signals in the ionosphere is negligible, there will be no variations in the amplitude of the emerging wavefronts and so only their phase will be allowed to change [949, 117]. Fluctuations in the amplitude will develop only after the wave has propagated beyond the screen, which will then appear as scintillation. At any point of observation, the wavefronts will be the sum of the unscattered waves and the scattered ones by the irregularities of the refractive index in the ionosphere [970, 648, 534].

A sketch of the scattering geometry for radio star scintillation is depicted in Fig. 7.9. The undisturbed plane wavefront induces a given voltage in the receiver, and this sets the reference phase. In addition, the sum of all scattered waves consists of many in- and out-of-phase components, while the actual signal received is the vector sum of these wave components with the primary, or undisturbed, wave [648]. Following Booker [99], two processes can be distinguished, namely "single scattering", which is due to a thin turbulent layer in the atmosphere, and "multiple scattering", which occurs when the mean-square phase deviation happens to be greater than one radian. In the former case, the projected atmospheric irregularity structure at ground level provides information about the scale of the irregularities, while in the latter case, the correlation distance on the ground is less than the scale of the atmospheric irregularities. Both processes are responsible for random fluctuations in the apparent position of a radio star viewed through the ionosphere. It is important to recall that the scale of the pattern on the ground is the same as the ionospheric irregularities provided that the mean phase changes of the incident radio wave are less than one radian [10, 534]. Ionospheric scintillations dominate at low frequencies and low angles of elevation, while at higher frequencies the shadow pattern on the ground is a projection of the inhomogeneous and irregular structure of the lower atmosphere.

7.4.1 Scintillation theory

A comprehensive review of the existing ionospheric radio scintillation models for high and equatorial latitudes is given by Priyadarshi [723]. It is a matter of fact that radio scintillation affects many radio communication and navigation systems operating through the auroral or equatorial ionosphere [207]. As was outlined above, the scintillation of radio signals is produced by random fluctuations of the refractive index associated with irregularities in the ionospheric electron density. Such fluctuations cause the original wavefront of the signal to distort, leaving a randomly phase-modulated wave [762].

Several ways have been proposed to aid in the interpretation of radio scintillation data, namely the weak-scatter theory, the Rytov approximation, the single, thin, or multiple phase screen, and the multiple-scatter theory. Here the discussion closely follows that provided by Priyadarshi [723], which focuses on the full structure of the complex signal. Starting from the assumptions that the wavelengths and the periods of the radio signals are much smaller than the characteristic scale sizes and the characteristic scale of the temporal variations of the refractive index fluctuations, respectively, the scalar wave equation holds for the electric field

$$\nabla^2 E + k^2 \left[1 + \varepsilon_1(\mathbf{r},t)\right] E = 0, \tag{7.53}$$

where $(\mathbf{r},t)$ denotes the spatial and temporal position of the electron density irregularities, $E = E(\mathbf{r},t)$, $\varepsilon_1(\mathbf{r},t)$ is the fluctuating part of the dielectric permittivity, and k is the wave number, with $k^2 = k_0^2 \langle \varepsilon \rangle$, where k_0 is the wave number in vacuum and $\langle \varepsilon \rangle$ is the average dielectric permittivity. This equation provides the basis of the scintillation theory. Under the further assumption that radio signal is a plane wave incident normally on a plane-parallel irregular slab, the solution of Eq. (7.53) for a radio wave incident along the z-axis of a Cartesian coordinate system can be written as

$$E = E(\mathbf{r},t) = \mathcal{E}(\mathbf{r},t) \exp\left(-ikz\right), \tag{7.54}$$

where $i = \sqrt{-1}$ and $\mathcal{E}(\mathbf{r},t)$ is the complex amplitude of the incident radio wave. Substitution of solution (7.54) into Eq. (7.53) yields the following differential equation for the wave amplitude

$$\nabla_\perp^2 \mathcal{E} - 2ik\frac{\partial \mathcal{E}}{\partial z} = -k^2 \varepsilon_1(\mathbf{r},t)\mathcal{E}, \tag{7.55}$$

where $\nabla_\perp^2$ is the transverse Laplacian for which the term $\partial^2 \mathcal{E}/\partial z^2$ is neglected under the assumption that the radio wavelength $\lambda \ll l_0$, where l_0 is the characteristic length scale of the irregularities. In Eq. (7.55) the Laplacian term represents the diffraction of the wave, while the term on the right-hand side represents the random phase shifts caused by the fluctuations in the refractive index. The parabolic equation (7.55) can be further simplified if $r_0^2 \gg L/k$, where r_0 is the outer scale of the irregularities and L is the thickness of the slab. In this case, the

Laplacian can be neglected as compared to the gradient term[12] and Eq. (7.55) becomes

$$\frac{\partial \mathcal{E}}{\partial z} = -i\frac{k}{2}\varepsilon_1(\mathbf{r},t)\mathcal{E}. \tag{7.56}$$

Therefore, after integration the complex amplitude is given by

$$\mathcal{E}(\mathbf{r},t) = A_0 \exp[-i\phi(\mathbf{r},t)], \tag{7.57}$$

where now A_0 is a real amplitude and $\phi(\mathbf{r},t) = \phi(\mathbf{p},z,t)$ is the wave phase given by the integral

$$\phi(\mathbf{p},z,t) = \frac{k}{2}\int_0^z \varepsilon_1(\mathbf{p},z',t)dz' = -\lambda r_e \delta N_T(\mathbf{p},t), \tag{7.58}$$

where $\mathbf{p}$ is a position vector in a plane perpendicular to the wave propagation direction, r_e is the classical electron radius, and δN_T is the fluctuation of the ionospheric electron content between the transmitter and the receiver on the ground. Equations (7.57) and (7.58) correspond to the phase screen model [100, 740, 763]. Below the screen, the radio wave propagates in free space and undergoes phase mixing. Mathematically, this is expressed by setting the right-hand side of Eq. (7.55) equal to zero. Using the solution (7.57) as an initial condition, this latter equation admits a solution given by the Fresnel diffraction formula

$$\mathcal{E}(\mathbf{p},z,t) = \frac{ikA_0}{2\pi z}\int_{-\infty}^{\infty}\int_{-\infty}^{\infty}\exp\left\{-i\left[\phi(\mathbf{p}',z,t) + \frac{k}{2z}|\mathbf{p}-\mathbf{p}'|^2\right]\right\}d\mathbf{p}'. \tag{7.59}$$

Since scintillation is a stochastic phenomenon, validation of any theory aimed at describing it will require comparing the observational data with the derived theoretical statistics. This is most easily done by calculating the so-called scintillation index, S_4, defined as the normalized standard deviation of the signal power, and the phase standard deviation, σ_ϕ, which requires calculating the moments of the complex amplitude $\mathcal{E}(\mathbf{r},t)$. This can be accomplished by assuming that the random electron density fluctuations obey a Gaussian probability distribution with zero mean. Under this assumption, the first moment of $\mathcal{E}(\mathbf{r},t)$, namely $\langle\mathcal{E}\rangle$, is given by [1002]

$$\langle\mathcal{E}\rangle = \exp\left(-\frac{\phi_0^2}{2}\right), \tag{7.60}$$

where ϕ_0 is the variance of the phase introduced by the irregular medium. Defining $\mathcal{E}^*$ as the complex conjugate of $\mathcal{E}$, the average signal intensity (or power) is given by

$$\langle\mathcal{E}\mathcal{E}^*\rangle = A_0^2, \tag{7.61}$$

---

[12]The condition $r_0^2 \gg L/k$ is equivalent to the requirement that the size of the first Fresnel zone, $\sqrt{\lambda L}$, is much less than r_0 so that the effects of diffraction can be ignored in a good approximation.

whose value remains the same as the wave propagates through the irregular medium, implying conservation of wave energy. For the derivation of the higher moments of $\mathcal{E}(\mathbf{r},t)$, the interested reader is referred to the review by Yeh and Liu [1002] and references therein. Within the framework of the phase screen approach, described by Eqs. (7.57) and (7.59), at higher frequencies the statistics S_4 and σ_ϕ can be approximated by assuming that on a screen $\phi_0^2 \ll 1$, which is known as the shallow screen approximation. Expressing the complex amplitude as

$$\mathcal{E}(\mathbf{p},z,t) = A_0 \exp\left[\chi(\mathbf{p},z,t) - iS_1(\mathbf{p},z,t)\right] = A_0 \exp\left[\psi(\mathbf{p},z,t)\right], \qquad (7.62)$$

where χ is the logarithmic amplitude and S_1 is the departure phase of the radio signal, the power spectra of χ and S_1 for a shallow screen are given by [1002]

$$
\begin{aligned}
F_\chi(\mathbf{k}_\perp) &= \sin^2\left(\frac{k_\perp^2}{2k}\right) F_\phi(\mathbf{k}_\perp) = 2\pi\lambda^2 r_e^2 \sin^2\left(\frac{k_\perp^2}{2k}\right) F_N(\mathbf{k}_\perp,0), \\
F_{S_1}(\mathbf{k}_\perp) &= \cos^2\left(\frac{k_\perp^2}{2k}\right) F_\phi(\mathbf{k}_\perp) = 2\pi\lambda^2 r_e^2 \cos^2\left(\frac{k_\perp^2}{2k}\right) F_N(\mathbf{k}_\perp,0), \quad (7.63)
\end{aligned}
$$

where $\mathbf{k}_\perp$ is the transverse spectral wave vector, while F_ϕ and F_N are, respectively, the phase and density power spectra.

Among the several existing theories, the weak-scatter theory was formulated in 1950 [100] within the framework of a weak phase-changing screen and was improved later on by a number of authors [740, 107, 117, 133, 134]. For example, at the edge of the phase-changing screen, z_0, the complex amplitude is written as

$$\mathcal{E}(\mathbf{p},z_0) = \exp\left[i\phi(\mathbf{p};z_0)\right], \qquad (7.64)$$

which resembles the first Born approximation solution to the vector wave equation. In expression (7.64), $\phi(\mathbf{p};z_0)$ is obtained by integration along a straight line [763]. The variance of $\phi(\mathbf{p};z_0)$, namely ϕ_0^2, is equal to the product $\sigma^2 L$, where σ^2 is a linear scattering coefficient and L is the layer thickness. In particular, for ionized media, this product can be written as

$$\phi_0^2 = \sigma^2 L = \frac{\lambda^2 r_e^2 L}{4\pi^2}\sec^2\theta \int\int \Phi_{\Delta N_e}(\mathbf{k}, -\tan\theta \mathbf{a}_{\mathbf{k}_T}\cdot \mathbf{k})d\mathbf{k}, \qquad (7.65)$$

where $\mathbf{k} = (2\pi/\lambda)\mathbf{a}_\mathbf{k}$ is the wave vector, θ is the incidence angle, $\mathbf{a}_{\mathbf{k}_T}$ is the wave vector projected on the (x,y)-plane, and $\Phi_{\Delta N_e}$ is the three-dimensional spectral density function of the electron density inhomogeneity. The intensity of the radio signals is defined as $I \approx \mathcal{E}\mathcal{E}^\star = |\mathcal{E}|^2$, and only its fluctuations are indeed measurable with

$$R_I(\Delta\mathbf{p};z) \triangleq \frac{\langle II'\rangle - \langle I\rangle\langle I'\rangle}{\langle I\rangle\langle I'\rangle}, \qquad (7.66)$$

being the main observable, where $I = |\mathcal{E}(\mathbf{p},z)|^2$ and $I' = |\mathcal{E}(\mathbf{p}',z)|^2$. The scintillation index S_4 can be expressed as the square root of $R_I(0;z)$. The weak-scatter theory is a scalar wave field of the form $\mathcal{E}(\mathbf{p}, z) \approx 1 + \Psi$ (with $\Psi \ll 1$), where Ψ is the perturbation function. In contrast, the Rytov approximation takes the form $\mathcal{E}(\mathbf{p}, z) \approx \exp(\Psi)$ [763]. On the other hand, in the Gaussian phase screen model, $\phi(\mathbf{p}, z_0)$ is assumed to be a Gaussian random process. In contrast to the previous approximations, all observables can be calculated with this model [763]. For example, if $\phi(\mathbf{p}, z_0)$ is assumed to be a Gaussian field, the wave field can be described by second-order statistics using the mutual coherence and complementary functions, while the first moment of the complex amplitude is given by Eq. (7.60). A multiple-scatter theory was also formulated by deriving a system of differential equations for the moments of $\mathcal{E}(\mathbf{r}, t)$ from the parabolic approximation of the scalar wave equation (7.53). A brief account of all these models can be found in the review by Priyadarshi [723] and a much deeper insight in the references therein.

7.4.2 Coherence

The effects of radio refraction can be studied by comparing the phase ϕ_r of radio waves received over a path with their phase ϕ_t at the transmitting antenna [667]. If S_E is the electrical length of the path and v is the radio frequency, the phase difference is given by

$$\Delta\phi = \phi_r - \phi_t = \frac{2\pi}{c} v S_E, \tag{7.67}$$

where c is the speed of light in vacuum. Accurate measurements of the variations in the electrical length of the propagation path were first obtained through measurements of $\Delta\phi$ by transmitting controlled CW frequencies v over line-of-sight paths [395] so that

$$\Delta S_E = \frac{c}{v}\left(\frac{\Delta\phi}{2\pi}\right). \tag{7.68}$$

When a light bundle passes through a slit, an interference phenomenon is produced. Constructive interference occurs when the phase difference between the waves is an even multiple of π, while destructive interference is obtained when the phase difference is an odd multiple of π. As shown schematically in Fig. 7.10, for demonstrating interference, it may be desirable to know within what size of the cone of light coherent light is present. The coherence condition can be expressed as [710]

$$a \sin\theta \ll \frac{\lambda}{2}, \tag{7.69}$$

where a is the diameter of the pinhole, θ is half the vertex angle of the cone of light subtended from the pinhole, and λ is the light wavelength. For $\theta \ll 1$, $\sin\theta \approx \theta$, and from Fig. 7.10 it follows that $\theta \approx D/(2f)$, and the coherence condition (7.69) becomes

$$\frac{aD}{f} \ll \lambda. \tag{7.70}$$

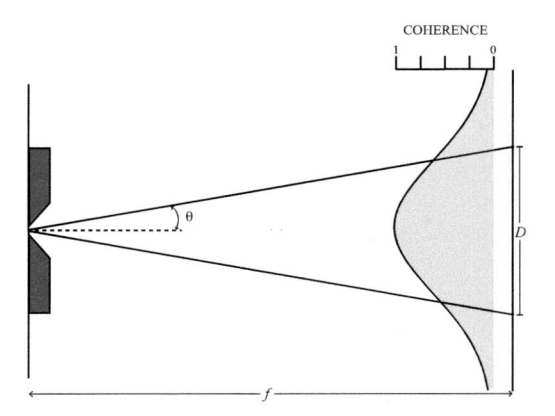

Figure 7.10: Schematic drawing showing the relation between a light bundle limitation and the degree of light coherence. Figure adapted from Meyer-Arendt and Emmanuel [621].

A wave train entering an interferometer is divided into two wave trains, which will interfere with each other at the point of observation only when $\Gamma < \Delta s$, where Γ is the optical path difference and Δs is the length of the wave trains. If this condition is not reached, one wave train will pass the point of observation before the other wave train component arrives. The coherence time is defined according to $\Delta t = \Delta s/c$. In a conventional light source, this time is of the order of 10^{-8} s, which according to the above relation corresponds to coherence lengths of ≈ 3 m. However, the coherence time for light emitted from a gas laser can be up to about four orders of magnitude larger ($\sim 10^{-4}$ s), yielding a coherence length ten times longer. On the other hand, the amplitude of light emitted from a conventional source will remain constant only during a time interval $\delta t < 1/\Delta v$, where Δv is the effective spectral width. The maximum value of δt is therefore the coherence time $\delta t_{\max} = \Delta t = \Delta s/c$. However, it must be remembered that a definition of coherent light should entail a measure of the coherence length in space and time. As explained above, the coherence length is given by

$$\Delta s = c\Delta t \sim \frac{c}{\Delta v}, \tag{7.71}$$

and interference effects will come out only when the path difference between the two arms of the interferometer satisfy the condition $\Gamma < \Delta s$.

7.4.3 The advent of satellite beacons

Radio-star satellite observations commenced with the launch of the low-altitude Soviet satellites Sputnik 1, in October 1957, and Sputnik 3, in May 1958. The observations were made using a simple antenna, a preamplifier, and a communication receiver. The low frequencies of 20 and 40 MHz allowed to perform surveys and long-term measurements of radio scintillation and intensity over the globe. Results on the morphology of E- and F-layer[13] irregularities were also emerging. For example, the Sputnik 3 satellite was transmitting at a 20 MHz frequency from within the F-layer of the ionosphere.

After the Sputnik 1 and 3 missions, a plethora of radio wave observations followed with the aid of the BEB and BEC satellites, with their 40 and 41 MHz transmissions at a nominal altitude of 1000 km from the Earth's surface. These satellites were employed over long periods to obtain information about the ionospheric irregularities that cause radio scintillation [687, 64, 769, 739] and the total electron content [504]. The 40 MHz signals of the BEB and BEC satellites were also used to measure the width of the equatorial scintillation belt in several longitudinal regions across Thumba in India [160], across Huancayo in Peru [166], and across several African longitudes [837]. However, one problem with using the 40 MHz data is that it is contaminated with strongly scattered signals.

The Applications Technology Satellites (ATS) were a series of experimental satellites launched by NASA between December 1966 and May 1974. These satellites were designed to act as communication satellites and carry equipment related to meteorology and navigation, and were placed into geosynchronous orbit at various longitudes. Their transmissions at 136 MHz allowed for Faraday-rotation investigations and studies of the total ionospheric and photospheric electron content. Out of the several satellites that were used for radio propagation studies, the ATS-6 [223], the Defense Nuclear Agency (DNA) WIDEBAND [308, 569], the HiLat [307], and the Polar BEAR satellites have all been used extensively to carry radio beacons. The WIDEBAND satellite was largely concerned with ionospheric equatorial irregularities as an example of the radio wave propagation difficulties to be expected after atmospheric nuclear detonations. On the other hand, HiLat and Polar BEAR were mostly used to study the polar regions. In particular, they carried UV images to identify regions that cause scintillation. The NAVSAT or NNSS (for Navy Navigation Satellite System) series of Transit satellites with their 150 and 400 MHz transmissions have been used to track phase scintillations and the distribution of irregularities in the Northern polar ionosphere [477, 476].

Most of these early satellite beacons contributed to clarify the morphology of ionospheric irregularities. The fading of the 40 MHz signals emitted by the Sputnik 1 satellite was produced by the passage of the signals through irregularities present in the F-layer of the ionosphere [472]. This investigation has also led

---

[13]The ionosphere consists of three main layers, namely the D-, the E-, and the F-layer. The D-layer is the lowest layer of the ionosphere and usually forms in the upper mesosphere. The E-layer is the middle layer of the ionosphere and extends from 90 to 150 km above the Earth's surface. This layer can only reflect radio waves at frequencies lower than 10 MHz. Only during sporadic E events (E_s-layer), this layer can support the reflection of radio waves up to 50 MHz. The F-layer, also known as the Appleton-Barnett layer, extends in the upper ionosphere from about 150 to more than 500 km above the Earth's surface. This layer has the highest electron density and consists of one layer (F_2) at night and an additional layer (F_1) appearing during the day. The day and night F_2-layer is responsible for the propagation of star radio waves and long distance high frequency (HF) radio communications.

to an estimation of the size and distribution of these irregularities, which were found to be ~ 1 km in size and to occur at altitudes above 250 km from the Earth's surface. They are more concentrated toward the north and usually appear at latitudes greater than 50°N. The fading of radio waves is more rapid for satellites than for radio stars because in the former case it depends on the satellite velocity rather than the ionospheric drift velocity [601]. Using both radio stars and satellite beacons, fading continued to be observed mainly in the equatorial region [505, 837, 842]. Later on, it was found that the most intense fading was observed to occur most frequently around the magnetic equator during years of high solar flux [13]. In parallel, magnetic activity gradually emerged as responsible for the distribution of ionospheric irregularities and the research community started to believe that they were responsible for radio-stellar scintillation and spread-F echoes. Moreover, further observations of satellite beacon signals at 40 and 54 MHz during the period between 1961 and 1966 have indicated that the lower boundary latitude of the irregularity region decreases on average about 1.6° per unit change in local K index[14] [12]. By examining the data available at that time from high-inclination and synchronous satellites, Aarons and Allen [12] realized that the change in latitude with the K index is a function of time. Scintillation observations during the high-solar-flux period of December 1957 to November 1958, at geomagnetic latitude 55° 33′ N in Canada, which is a location particularly convenient for radio star observations looking both north into the auroral zone and south away from it, revealed that well to the south scintillation has lower amplitudes and rate indices compared to the north and that the transition between the relatively quiet lower latitudes and the highly perturbed auroral scintillation occurs within a few degrees of latitude [782]. In particular, this observation was very important because it demonstrated that radio-star studies also led to the same conclusion. Further investigations of the ionospheric mid-latitude trough using satellite *in situ* probes and topside sounders have shown a movement of the trough toward lower latitudes as dawn progresses [475]. The recording and comparison of radio signals in the VHF range with ionograms over a wide range of southern latitudes during a few equinoctial months have shown that the equatorward edge of the auroral scintillation oval extends into mid-latitudes for high values of the magnetic K-index, suggesting that ionospheric inhomogeneities are associated with the equatorward propagation of traveling disturbances generated in the auroral zone [375].

7.5 Infrared and microwave radiometry

The relationship between the atmospheric temperature change and CAT at jet altitudes above 25000 ft was reported several years ago in a technical paper by

---

[14]In meteorology, the K index is a measure of thunderstorm potential based on the vertical temperature lapse rate as well as on the amount and vertical extent of low-level moisture in the atmosphere.

Kadlec [446]. For this study, the data were collected on 146 flights during a year and the aircraft instrumentation consisted of a portable test instrument to detect temperature variations. In particular, a rate of temperature change of 1°C per minute was found to be a reliable criterion for indicating flight conditions in the majority of cases. Based on the inversion of a series of radiometric measurements at narrow wavelength intervals on the wing of the 15-μm CO_2 absorption band, a method for the determination of atmospheric temperature profiles from satellites was also simultaneously developed at that time [456, 956]. Shortly after, the method was used for the remote sensing of CAT by just taking advantage of the thermal radiation properties of gases. According to the Stefan-Boltzmann law, the radiant emittance[15] of a grey body is directly proportional to the emissivity, ε, and the fourth power of the body's thermodynamic temperature, T, i.e.,

$$J = \varepsilon \sigma T^4, \tag{7.72}$$

where $\sigma = 2\pi^5 k_B^4/(15c^2 h^3) \approx 5.670374 \times 10^{-8}$ W m^{-2} K^{-4}, k_B is the Boltzmann constant, h is the Planck constant, and c is the speed of light in free space. In contrast to a perfect black body for which $\varepsilon = 1$, a grey body is one that does not absorb all incident radiation and therefore emits less energy than a black body so that its emissivity is $\varepsilon < 1$. For real materials and gases, the emissivity is generally a function of wavelength, $\varepsilon = \varepsilon(\lambda)$. However, compared to other materials the emissivity of gases is generally very low except in absorption bands, where its absorptivity approximately equals its emissivity. In these spectral regions, the gas behaves much like a black body. This way, the gas temperature can be inferred by radiometric measurements of the infrared energy emitted in the appropriate absorption bands. For example, carbon dioxide (CO_2) exhibits strong absorption bands centered at 4.3 and 15 μm. In the neighborhood of these bands, the emission approaches that of a black body. Since CO_2 is uniformly distributed through the atmosphere, its temperature will appear the same in all spectral regions of the band, and since its 15-μm absorption band coincides with the thermal emission peak at atmospheric temperatures, it is particularly suitable to compute temperature profiles along the line of sight over quite long distances.

The applicability of infrared radiometry to the remote sensing of CAT was first tested using an airborne Fabry-Perot type interferometer with a narrow field of view along the flight path, which was built in 1965 [39]. With this device, spectral scans from 13 to 14.5 μm were performed. The detection of distant temperature gradients was obtained by comparing the signal received in a low-absorption region with that in a high absorption region, where the latter measurement indicated the local or near air temperature. This device was thereafter employed in a flight evaluation program conducted by the National Research Council of Canada in the winter of 1966–1967. The results of this evaluation have shown

---

[15]The radiant emittance of a material body is defined as the total energy radiated per unit surface of the body across all wavelengths per unit time.

that temperature variations could be detected remotely at distances of at least 16 km. Shortly after in 1967 three narrow field infrared radiometers with their field of view servo controlled to remain horizontal were designated by the registered trademark IRCAT. These instruments were installed on three commercial airliners in a cooperative program between Pan American Airlines, Eastern Airlines, TWA, and Barnes Engineering Company with the primary objective of recording CAT signatures and several other relevant flight parameters, including air speed, altitude, and static air temperature. The measured data were accumulated during the winters of 1968–69 and 1969–70. A thorough description of the IRCAT instrument can be found in Ref. [39]. The radiometric sensitivity of IRCAT was inferred in terms of the noise equivalent radiance of the radiometer to be 0.6×10^{-7} W cm^{-2} sr^{-1}.

Although the radiometric technique was proposed as a valuable method for remote temperature measurement, its capability to detect CAT depends on its close correlation with temperature. However, even though there is no doubt about the existence of such correlation, a rigorous quantitative assessment of it has not been fully established yet. In fact, this was the main purpose of these early flight evaluation tests. Other factors that could limit the ability of IRCAT were the altitude, the airspeed changes, and the presence of clouds. Since clouds are essentially opaque to 14-μm radiation, they blind the system and do not produce output signals. In 600 hours of flight recording, only two cases of moderate CAT were encountered. No cases of severe or extreme CAT were found mainly because by that time CAT prediction techniques had already been sufficiently improved and therefore flight plans by then were designed to warn pilots of commercial flights about regions of severe and extreme CAT. However, many periods of light turbulence were recorded. On the other hand, the results of CAT detection with the IRCAT instruments were presented only for the cruise portions of the flight and not during climbing to cruising altitudes or descent for landing, mainly because during these operations the aircraft maneuvering accompanied by large air speed and temperature gradients made the recorded data too confused. Based on the flight experience accumulated during these programs it was found that temperature disturbances of 3°C and above can be detected at distances of 32.6 to 48.9 km from the static air temperature change. Approximately 90% of the light turbulence encounters were found to be associated with detectable temperature perturbations and much variable structure was observed in the vertical scan records, suggesting that considerable information is actually contained therein [39].

Radiometry at microwave frequencies uses the same measurement principles as infrared radiometry [798]. In particular, passive microwave satellite sensors became available since 1998. At present, they provide complementary information to infrared measurements. Applications of both technologies have been developed not only for estimates of the atmospheric temperature variations, but also for measurements of the sea surface temperature [663]. The spatial resolution of microwave sensors (25 km) is poorer than that provided by infrared detectors (between 250 m and 4 km). However, in contrast to infrared sensors, microwave satellite sensors penetrate non-precipitating clouds, thereby providing an all-weather, high-resolution temperature variation product. While microwave wavelengths are not affected by clouds, the sea surface temperature microwave retrieval is generally prevented in regions with sunlight, rain, high wind speeds, and in regions within 50 to 100 km from the ground. This way only a small amount of data is actually missed [712]. Microwave and millimeter-wave radiometric remote sensing of the troposphere is a more recent issue and therefore it will be discussed in Chapter 9, where an overview of more modern methods and technologies for CAT detection from the year 2000 onwards is presented.

7.6 LIDAR

As introduced in Section 7.2.2.3, LIDAR is an optical radar technique using laser energy. It is a generic technique that can be applied in different forms to a broad range of research and operational problems. After the advent, in 1960, of the laser as a source of energy and the giant pulse, or Q-switched, laser in 1962, scientists began to recognize the applicability of this device to atmospheric studies [354]. Among the research applications, LIDAR was used to investigate dust in the high atmosphere as well as air movement and turbulence as revealed by cirrus and other clouds [225]. In addition, it has applications to boundary layer phenomena as well as to turbulence and diffusion processes using suitable indicators and, what is more important, to CAT detection and investigations of the effects of cirrus and other particulate layers on measurements of radiation in and through the Earth's atmosphere. On the other hand, ceilometry, transmissometry, and the monitoring and tracking of atmospheric pollutants filled the list of operational applications. A first atmospheric study involved the use of ruby laser radar to detect echoes at atmospheric altitudes up to 140 km in 1963 [293, 192]. At the same time, a program was initiated at the Standford Research Institute to probe the lower atmosphere and study meteorological phenomena with a similar pulsed ruby laser radar system [193, 192]. Shortly after, the same technique was applied to map and track concentrations of atmospheric particles and to study density profiles by reference to gaseous backscattering. Meanwhile, multiple wavelength LIDARs were employed to determine by reference to differential absorption of the atmospheric gas composition. In parallel to this, Doppler techniques were used to determine atmospheric motions and temperatures [191].

7.6.1 Atmospheric sounding

A generic LIDAR hardware consists of three subsystems, namely the transmitter, the receiver, and the detector or recording system as depicted in the block diagram of Fig. 7.11. The transmitter, which is composed of a laser and a beam

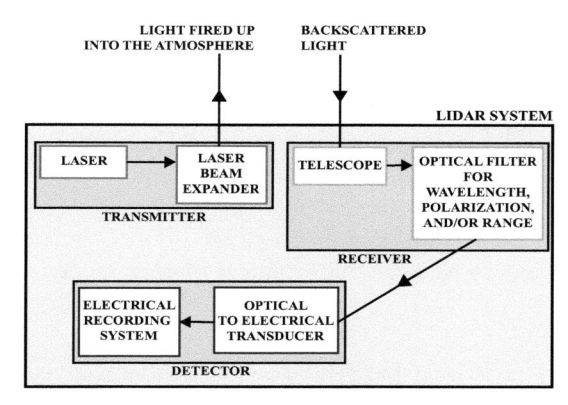

Figure 7.11: Block diagram showing the subsystems of a generic LIDAR. Figure taken from Argall and Sica [32].

expander is used to generate light pulses and direct them into the atmosphere. The scattered laser light is collected by the receiver, where it is processed before being conveyed to the detector. Depending on the purpose for which the LIDAR was designed, processing of the received light can be based on optical filtering of the wavelength, polarization, and/or range. For example, processing of the backscattered light based on wavelength consists of reducing the background using a narrow-band interference filter, while signal separation based on polarization is often employed in measurements of atmospheric aerosols. On the other hand, processing based on the range is often performed to protect the detector from the intense near-field returns of high-power LIDAR systems [32]. The light coming from the receiver is finally sent to a detector where a permanent record of the measured intensity is produced as a function of altitude. In the first LIDAR systems, the detector consisted of a camera and a photographic film, while in most modern devices light is converted into an electrical signal, which is processed and recorded by a microcomputer. Photomultiplier tubes are the most common devices used as detectors for incoherent LIDAR systems working in the visible and ultraviolet. For example, wind speeds can be determined by measuring the Doppler shift of backscattered light. Incoherent detection consists of measuring the wavelength of the transmitted and received luminous signals independently, using a spectrometer. In this case, the Doppler shift is determined from these two measurements. In contrast, coherent detection uses a local oscillator and a narrow-band continuous-wave laser to set the frequency of the transmitted pulses.

For atmospheric targets the essential features of LIDAR detection can be summarized by the following relation between the received, P_{rec}, and transmitted, P_{trans}, power [191]

$$P_{\text{rec}} = \frac{c\tau A\beta_{180}}{8\pi r^2} P_{\text{trans}} \exp\left[-2\int_0^r \sigma(r')dr'\right], \qquad (7.73)$$

where c is the speed of light in free space, τ is the pulse duration, A is the effective aperture of the receiver, r is the range, β_{180} is the atmospheric volume backscattering coefficient at range r with dimensions of area per unit volume, and $\sigma(r)$ is the extinction coefficient at range r. Therefore, a typical LIDAR observation consists of evaluating the received radiation power, P_{rec}, in terms of range and direction. The number of photons detected as pulses at the photomultiplier output can be derived from the so-called LIDAR equation by including all forms of scattering. In particular, the number of photons per laser pulse is given by [32]

$$\mathcal{N}_{ph} = \mathcal{A} \int_{\Delta\lambda} \mathcal{P}(\lambda)\tau_t(\lambda)\tau_r(\lambda)Q(\lambda) \int_{r_1}^{r_2} \frac{\xi(r)}{r^2}\tau_a(r,\lambda)^2 \sum_i \frac{d\sigma_i}{d\Omega}(\lambda)N_i(r)drd\lambda,$$

(7.74)

where $\mathcal{A}$ is the area of the telescope, $\mathcal{P}(\lambda)$ is the convolution of the number of photons emitted in a single laser pulse and the function $S(\lambda)$, which accounts for any wavelength shift during scattering, including Doppler and Raman shifts, $\Delta\lambda$ denotes the wavelength interval for which $\mathcal{P}(\lambda) \neq 0$, τ_t and τ_r are, respectively, the optical transmission coefficients of the transmitter and receiver optics, $Q(\lambda)$ is the quantum efficiency of the photomultiplier, as before r is the range, while r_1 and r_2 are, respectively, the minimum and maximum ranges for a range bin, $\xi(r)$ is the overlap factor, τ_a is the atmospheric optical transmission along the laser path, $d\sigma_i/d\Omega(\lambda)$ is the backscatter cross-section for scattering of type i, and $N_i(r)$ is the number density of scattering centers, which cause scattering of type i.

7.6.1.1 Rayleigh LIDAR

Rayleigh LIDARs have been primarily used to measure the intensity of light backscattered by atmospheric molecules at altitudes from ~ 30 to 100 km from the ground. Relative density and absolute temperature profiles are derived from the measured intensity profiles. As described in Section 7.2.1.1, the scattering of light by atmospheric molecules whose sizes are smaller than the wavelength of the incident radiation is referred to as Rayleigh scattering. This has to be distinguished from Mie scattering, which, in contrast, describes the scattering of light by aerosols that are larger than the wavelength of the incident radiation. In particular, the color, the intensity, and the polarization of the blue sky can be explained in terms of the Rayleigh scattering by atmospheric molecules. The atmospheric Rayleigh backscatter cross-section below 90 km obeys the relation

$$\frac{d\sigma_R}{d\Omega}(\theta = \pi) = \frac{C}{\lambda^4} \text{ m}^2 \text{ sr}^{-1},$$

(7.75)

where C varies from $\sim 4.75 \times 10^{-57}$ to about 5.0×10^{-57}, depending on the adopted value of the air refractive index. At heights higher than 90 km, Eq. (7.75) becomes less accurate as the concentration of atomic oxygen increases and the refractive index of air changes. Equations (7.74) and (7.75) can be used to determine the intensity of the backscattered light for a particular Rayleigh LIDAR. The Rayleigh technique cannot be applied in the stratospheric aerosol layer at altitudes between 25 and 30 km from the ground. However, it can be applied at higher atmospheric altitudes because the aerosol concentration drops. In actual operations, the expected signal strength can be calculated from the number of collected backscattered photons by applying Eq. (7.74) to a Rayleigh LIDAR system. For practical applications, Eq. (7.75) can be written in the much simpler form [32]

$$\text{SIGNAL} = \frac{C}{r^2} N_a(r) \delta r, \tag{7.76}$$

where as usual r is the range, δr is the width of a range bin, $N_a(r)$ is the number density of air, and the constant C includes all constant terms coming from Eq. (7.74). From this equation, it is clear that after correcting for range, the strength of the Rayleigh LIDAR's signal is proportional to the number density profile. However, since the determination of the constant C in Eq. (7.76) is subject to uncertainties, this equation cannot be used to determine an absolute density profile. Once a relative density profile is measured, the corresponding relative pressure profile can be calculated by direct integration of the hydrostatic equilibrium condition. An absolute temperature profile is finally calculated using an ideal gas law from the relative density and pressure profiles.

7.6.1.2 *Aerosol LIDAR*

Mie theory provides a useful approximation for describing the scattering of light from atmospheric aerosols. The presence of aerosols and clouds affects the energy budget of the atmosphere in a very complex manner because they scatter and absorb radiation coming from both the Sun and the Earth's surface. Aerosols and clouds in the troposphere and lower stratosphere have been studied since the early 1960s using several LIDAR systems installed at various stations around the globe.

In September 1994, the NASA sent a space shuttle mission (STS-64), which was equipped with the *LIDAR In-Space Technology Experiment* (LITE) instrument. LITE was the first space-based LIDAR designed to measure both tropospheric and lower-stratospheric clouds and aerosols as well as the surface reflectance on a global scale. In particular, wind velocities can be measured by light backscattered from aerosols. This light experiences the same Doppler shift due to wind as light backscattered by atmospheric molecules. However, the latter exhibits much broader spectral lines than the light backscattered from aerosols because of differences between the masses of molecules and aerosol particles. The narrower spectral broadening corresponding to scattering from aerosols in the lower atmosphere allows the use of coherent detection for the determination of wind velocities.

7.6.1.3 *Differential-absorption LIDAR (DIAL)*

The differential-absorption LIDAR (DIAL) technique has been successfully used to measure with good spatial and temporal resolutions the concentration of trace species in the troposphere, such as NO, H_2O, O_3, SO_2, and CH_4, as well as the concentration of ozone in the stratosphere. The method is a laser remote-sensing technique for range-resolved measurements of atmospheric species concentrations [120]. DIAL was first used in 1966 for remote sensing of water vapor and since then it has been applied to measurements of other atmospheric gases as well as many pollutant gases, such as nitrogen dioxide (NO_2), ammonia (NH_3), mercury (Hg), carbon monoxide (CO), and hydrocarbons. Later on, in the 1970s and early 1980s, the main thrust of the DIAL applications was pollution monitoring. In fact, the first airborne applications with DIAL focused on studying the concentration of tropospheric O_3 over the east coast of the United States in 1980.

The DIAL relies on determining the average gas concentration over some range by monitoring the backscattered light intensity for wavelengths tuned on and off an absorption peak of interest [889]. The concentration of a species of interest is determined from the ratio of the signals at the on/off wavelengths. Airborne DIALs can operate in nadir and zenith modes with a vertical resolution of 50-200 nm through the entire troposphere. In particular, they have been applied to observations of the antarctic ozone hole as well as other areas. Measurements of the ozone concentration were also performed using an infrared DIAL system using transverse excited CO_2 lasers.

7.6.1.4 *Raman LIDAR*

The Raman LIDAR (RL) is an active, ground-based laser remote detector that provides height- and time-resolved measurements of vertical profiles of water-vapor mixing ratio as well as several cloud- and aerosol-related quantities. When monochromatic light is scattered by gaseous or liquid molecules, the spectrum of the scattered light contains lines corresponding to different wavelengths compared to the incident radiation. This is known as the Raman effect and is due to the interaction of the incident radiation with the quantized rotational and vibrational energy levels of the scattering molecules. This process is an inelastic one because it involves an energy transfer between the radiation and the molecule. The method is particularly well suited for sensing molecular species since the measurement of the scattered Raman light intensity allows an estimate of the abundance of a particular molecular species, and is primarily used for the sensing of atmospheric water vapor and temperature. For example, molecular nitrogen profiles determined by Raman LIDAR can be used to determine temperature profiles with the aid of the Rayleigh method, even in aerosol regions. The elastically scattered light from aerosols can be separated from the Raman nitrogen backscatter by spectral filtering, which is therefore proportional in an approximate sense to the number density profile.

The RL operating mode is similar to the other types of LIDAR. In particular, the RL operates by firing up short pulses of ultraviolet (UV) laser light into the atmosphere. As the transmitted radiation propagates, a small fraction of its energy is backscattered to the LIDAR receiver, where it is converted into a time-resolved signal. The distance to the scattering region can be inferred by the instrument from the delay between the transmitted outgoing pulse and the backscattered signal.

7.6.1.5   *Resonance-fluorescence LIDAR*

A variety of geophysical phenomena that take place in the middle and upper atmosphere are recognized to play a role as sensitive indicators of global climate change. For example, the ablation of meteors in the upper atmosphere produces layers of alkali metals, such as Na, K, Ca, Li, and Fe, at altitudes of about 90 km from the ground. In addition, to these meteoric layers, the polar mesospheric clouds and noctilucent clouds, the airflow layers of OH, O, and O_2 as well as the planetary, tidal, and gravity wave activities influence the global atmospheric circulation. On the other hand, the warming of the lower atmosphere due to the absorption of infrared radiation by greenhouse gases, such as CO_2 and CH_4, results in cooling of the middle and upper atmosphere [766].

Meteoric metals are not abundant. However, they have very high resonant scattering cross-sections. Resonant scattering is an elastic process in which an atom absorbs a photon and instantaneously emits another photon having the same energy and frequency. The probability of occurrence of resonant scattering is much higher than that for Rayleigh scattering. For example, the cross-section of resonant scattering for Na at a wavelength of 589 nm is around 10^{15} times larger than that for Rayleigh scattering by air at the same wavelength. Sodium is more abundant in the mesosphere at altitudes from 80 to 110 km, with a peak near the mesopause. In particular, this layer is a good tracer of wavy motions and Na LIDAR has made important contributions to the study of gravity waves near it [325]. Sodium LIDAR systems have been used to estimate Na abundance profiles at heights between 85 and 105 km, with an altitude resolution of a few hundred meters and a time resolution of tens of seconds [32]. At these altitudes, the presence of wave motions induce density perturbations as have been revealed by the sodium density profiles. In addition, the spectral resolution of the resonance-fluorescence scattering of Na allows the determination of temperature and wind profiles across the Na-abundant layers around and near the mesopause. On the other hand, resonance-fluorescence LIDAR can be used to measure the concentration of individual alkali metals in the upper atmosphere. Apart from its applications to sodium, resonance-fluorescence LIDAR has also been applied to calcium (i.e., Ca and the ion Ca^+), potassium (K), lithium (Li), and iron (Fe).

7.6.2 Backscatter

The several aspects and definitions regarding LIDAR backscatter are very well described in the review paper by Platt [707]. Therefore, part of the description given here was guided by the discussion in that paper. Most LIDAR systems are monostatic in the sense that they comprise a telescope receiver, which is close to, or coaxial with, a laser pulse transmitter (see Fig. 7.11). When a pulse of light is directed to the atmosphere, it is scattered in all directions by molecules, aerosols, and clouds, while only a small amount is backscattered and returns to the receiver. The scattered light in any direction, θ, with respect to the forward direction of the transmitted pulse forms a pattern that can be described by the scattering phase function $P(\theta)$. Figure 7.12 shows a schematic of a backscattered light ray by a particle. The scattering efficiency, Q_{scat}, of an atmospheric particle[16] is related to the phase function at a scattering angle $\theta = \pi$ and determines the amount of radiation that is scattered in all direction by a particle. The scattering efficiency is defined by the simple relation

$$Q_{scat}(r, \lambda) = \frac{I_{scat}}{I}, \tag{7.77}$$

where r is the size of the scattering particle, λ is the wavelength of the incident radiation, I_{scat} is the scattered light, and $I = I_{scat} + I_{abs} + I_0$ is the transmitted radiation, which is the sum of the scattered light, I_{scat}, the absorbed light, I_{abs}, and the portion of the transmitted light that passes straight through, I_0. For non-absorbing particles of size larger than the wavelength $Q_{scat}(r, \lambda) \to 2$.

In contrast, the backscattering efficiency, denoted by $Q_{\pi}(r, \lambda)$, is defined as the part of the scattered radiation I_{scat} that is backscattered in a cone of unit solid angle. Under the assumption of a perfectly spherical particle and isotropic scattering, the LIDAR backscatter efficiency will be $(4\pi)^{-1}$, or ≈ 0.08. In addition, the radar backscattering efficiency is defined as 4π times the LIDAR backscattering efficiency, thus yielding a value of unity for isotropic scattering. Another important quantity pertaining to LIDAR backscatter is the volume backscatter coefficient given by

$$\beta(\lambda) = \pi \int_0^{\infty} Q_{\pi}(r, \lambda) n(r) r^2 dr, \tag{7.78}$$

where $n(r)$ is the number density of particles having size r and again the assumption is made that the particles are perfectly spherical. However, in the case of ice crystals, their size will depend on the type and habit of the crystals. For example, for hexagonal shapes with length l and diameter D, the above definition still holds with $n(r) r^2 dr$ replaced by $lDdl$ in the integrand if the ice crystals are large enough. In actual situations where the ice crystals are tumbling, an effective dimension must be determined and the calculation of the volume backscatter coefficient will be a bit more complicated than described above.

---

[16] A particle may be a molecule, a water drop, an aerosol, or an ice crystal.

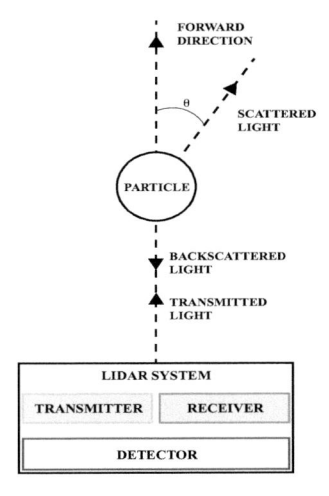

Figure 7.12: Schematic drawing of backscattering by a particle in the atmosphere. Figure taken from Platt [707].

The intensity of the radiation scattered by a particle at an angle θ from the transmitted direction is given by the phase function. The total scattered radiation obeys the normalization condition

$$\frac{1}{4\pi} \int_0^\pi P(\theta) \cos \theta \sin \theta d\theta = 1. \tag{7.79}$$

Thus, the backscattering efficiency will be determined by the backscattering phase function and the scattering efficiency. The interested reader is referred to the review paper by Platt [707], where some phase functions for different atmospheric constituents and their corresponding backscatter phase function and efficiency are considered. Another quantity of relevance to LIDAR backscatter is the backscatter-to-extinction ratio defined as

$$k = \frac{\beta}{\sigma} = \frac{Q_\pi}{Q_{\text{scat}}} = P(\theta = \pi), \tag{7.80}$$

where σ is the volume extinction coefficient. This quantity is crucial to LIDAR backscatter measurements and is needed to solve the backscatter LIDAR equation for volume attenuation of the LIDAR beam. The reciprocal of k defines the LIDAR ratio, S, which is measured in steradians.

LIDAR uses pulsed lasers that are linearly polarized to a high degree and depending on the type of atmospheric particle depolarizing effects in the backscatter may or may not occur. For example, molecules exhibit no polarization effects in the backscatter, while water drops have only very weak depolarization and ice crystals depolarize the backscattered light by various degrees. As explained by Platt [707], a scattering matrix can be used to describe the light scattering, which for nonspherical particles that are randomly oriented in space and which contain a plane of symmetry, as is the case of typical ice crystals, can be expressed by the diagonal matrix [790]

$$\mathbb{F}(\pi) = \begin{pmatrix} F_{11}(\pi) & 0 & 0 & 0 \\ 0 & F_{22}(\pi) & 0 & 0 \\ 0 & 0 & F_{33}(\pi) & 0 \\ 0 & 0 & 0 & F_{44}(\pi) \end{pmatrix}, \qquad (7.81)$$

where the components can be measured by optical instruments. For macroscopically anisotropic media, non-vanishing off-diagonal elements will appear in the scattering matrix. For macroscopically isotropic and symmetric media, the above matrix holds and the depolarization ratio can be defined as

$$\Delta = \frac{F_{11}(\pi) - F_{22}(\pi)}{F_{11}(\pi) + F_{22}(\pi)}. \qquad (7.82)$$

For perfectly spherical particles $\Delta = 0$ because all diagonal elements are the same. The extinction-to-backscatter ratio $S = k^{-1}$ is important for the retrieval of the backscatter coefficient of a given layer. To see this, let us start from the LIDAR equation (7.73) written in terms of the power $P(r)$ measured as a function of range r:

$$P(r) = \frac{EA\beta c}{r^2} \exp\left[-2\int_{r_1}^{r_2} \sigma(r)dr\right], \qquad (7.83)$$

where E is the energy of the transmitted pulse, A is the telescope area, c is the speed of light in free space, $\sigma(r)$ is the layer volume extinction coefficient between the ranges r_1 and r_2, and the backscatter coefficient β is given in terms of the sum of the $\beta_{pa} + \beta_{pe}$ of the two polarization components. Defining $\beta'(r) = T^2\beta(r)$ with

$$T^2 = \exp\left[-2S\int_{r_1}^{r_2} \beta(r)dr\right], \qquad (7.84)$$

where the relation $S = k^{-1} = \sigma(r)/\beta(r)$ has been used, it can be demonstrated that

$$\beta(r) = \beta'(r)\left[1 - 2S\int_{r_1}^{r_2} \beta'(r)dr\right]^{-1}, \qquad (7.85)$$

which clarifies the importance of the extinction-to-backscatter coefficient in the LIDAR solution.

Early LIDAR applications were primarily addressed to observations of stratospheric aerosols. Some of these LIDAR observations responded to large volcanic eruptions [473]. In particular, the cloud of aerosol mass (ashes and soot) around the Earth that is produced by volcanic activity is easily observed with LIDAR. The total backscatter consists of a mixture of aerosols and molecules, with the amplitude of the aerosol backscatter increasing and decreasing over the years as volcanic eruptions appear and dissipate. In addition to the stratosphere, the atmospheric boundary layer also exhibits large amounts of aerosols due to smog from large urban areas, dust from desertic regions, and ashes and soot from forest fires around the world. Ground-based pulsed CO_2 LIDARs have also been used to probe the troposphere [462]. In particular, coherent pulsed CO_2 Doppler LIDAR measurements of backscatter coefficients taken for almost a year through the lower troposphere have yielded typical summer and winter backscatter values in the ranges of $10^{-7} - 10^{-8}$ and $10^{-8} - 10^{-9}$ m^{-1} sr^{-1}, respectively, while measurements beyond the lower troposphere yielded occasional values of 4×10^{-11} m^{-1} sr^{-1} during the winter months when the absorption is at a minimum level [780]. Later on, an analysis of 10.6 μm backscatter data was reported in terms of probability distributions of the backscatter mixing ratio as a function of altitude [779]. For stations in the United Kingdom (RSRE, Malvern) and Boulder, Colorado (NOAA/WPL), a background mode in the mixing ratio of approximately 10^{-10} m^2 kg^{-1} sr^{-1} was identified. This value for the 10.6 μm backscatter ratio was further compared with the 1 μm extinction data from the SAGE II satellite experiment, whose results were used to produce climatological information on the variation of free tropospheric aerosol characteristics on a global scale [474]. The comparison yielded an extinction-to-backscatter ratio of 7.4×10^{-5} sr^{-1}. A direct experimental comparison of *in situ* 10.6 μm backscatter and 1 μm satellite occultation measurements was previously reported by Vaughan et al. [927]. They found an approximate linear relationship between the aerosol extinction and the backscatter-to-extinction ratio $\beta_{10.6}/\sigma_{1.0}$.

The detection of CAT by LIDAR as a function of backscattering by dielectric inhomogeneities has also attracted much attention. However, it was soon demonstrated that for temperature and molecular number density values typical of the upper troposphere and for a large temperature structure coefficient, representative of turbulence in the atmospheric boundary layer, the backscattering due to air turbulence at ruby wavelengths is about seven orders of magnitude less than the resulting one from molecular backscattering [650]. In contrast, variations in the turbidity of CAT may readily be detected by LIDAR. Vertical LIDAR observations have shown that the turbid air close to the ground is separated from the overlying CAT at the level of the temperature inversion base at approximately 300 m [191].

A combined Raman elastic-backscatter LIDAR using a XeCl excimer laser as the radiation source was developed in the early 1990s and employed for nighttime measurements of the vertical humidity distribution up to the tropopause, particle extinction, backscatter, and extinction-to-backscatter ratio height profiles [30]. For example, this system was able to produce time series of humidity profiles up to 8 km in the upper troposphere at nighttime with a time resolution of ~ 10 to 15 min and a spatial resolution of ~ 100 to 300 m. The simultaneous observations of cirrus water-vapor above 8 km were also possible. In general, LIDAR pulses can often penetrate cirrus clouds (because of their semitransparent nature) with sufficient backscattered photons to be detectable [708]. Height profiles of the particle extinction and backscatter coefficients were measured independently in clouds as well as in the atmospheric boundary layer and the stratospheric aerosol layer. An important result derived from this study indicates that the LIDAR ratio varies with height, especially in ice clouds. In contrast to elastic-backscatter signals, only inelastic Raman LIDAR can be used for reliable profiling of the aerosol extinction coefficient. On the other hand, the analysis of LIDAR signals from optically thin clouds was also addressed by Young [1003]. It was demonstrated that the solution of the LIDAR equation for profiles of backscatter and extinction in such clouds is constrained by the cloud transmittance values determined from the elastically scattered signals above and below the cloud. Optical properties of mid-level, mixed-phase clouds were reported for the first time in 2000 [1004]. These properties included cloud infrared emittance and absorption coefficient, effective backscatter-to-extinction ratio, and LIDAR depolarization ratio. It was found that at temperatures between $-10°$C and $-20°$C, ice crystals are often hexagonal plates and float horizontally in the air, implying large cross-sectional areas that do not depolarize in the back direction.

7.7 Tropospheric radio scatter

The properties of light scattering by irregularities in the troposphere depend upon its refractive-index structure. The complex phenomenologies associated to troposcatter propagation have been studied intensively for the past 60 years, or more, both theoretically and experimentally [222]. The models that have been proposed can be grouped into three main categories, i.e., models based on turbulence theory – the mixing-in-gradient theory of Villars and Weisskopt [938] and the mixing theory of Obukhov [893] –, models derived from the fitting of the experimental data – those that account for anisotropy and nonstationarity by specifying correlation coefficients with different mathematical functions[17]–, and models that consider scatter mechanisms as, for example, the specular and diffuse reflection from thin layers [311, 254, 629]. However, an examination of the various tropospheric models are beyond the scope of the present section.

It has long been recognized that sharp discontinuities in the refractive index structure give rise to intense tropospheric turbulent motions. Such discontinuities were envisaged by a plurality of scatter mechanisms ascribed to the complex structure of the troposphere, which is the result of several space- and time-dependent turbulent phenomena, such as turbulence in narrow layers, turbulence due to the breaking of gravity waves, turbulence due to strong variations between stable layers, and undulations caused by gravitational interactions among others. For example, communications networks and early warning radar strongly depend

---

[17]An example of this category is given by Norton [666], who proposed an anisotropic model based on a modified Bessel function of the second kind. This model was able to reproduce approximately the law predicted by the mixing-in-gradient theory. Another example was provided by Gjessing [346], who assumed a power-law function for the wavenumber spectrum by leaving the exponent to fit the experimental data.

on the tropospheric scatter mode of radio propagation. It is well known that transhorizon VHF and UHF fields are subject to deep fading or large enhancements of several hours of duration. This is a consequence of the propagation mechanism that may alternate between partial reflection and scattering by turbulent atmospheric dielectric fluctuations that occur when strong refractive-index layers develop below ~ 1 km from the ground [632]. Under the presence of refractive gradients, the effects of the refractive layers drop exponentially with altitude in the microwave region, implying that reflections are unimportant for layers above ~ 1 km. On the other hand, there was no evidence of the existence of layering under normal transhorizon radio transmissions. However, partial reflection and scattering by dielectric fluctuations may coexist at higher altitudes under well-established layering, while turbulent scattering may dominate in the absence of layering at such higher altitudes [761].

In a 1950 paper, Booker and Gordon [100] postulated that the received power, P_{rec}, obeys the relation

$$P_{\text{rec}} \propto \frac{F(k)}{\sin^n(\theta/2)}, \tag{7.86}$$

where $F(k)$ is a function of the intensity of dielectric fluctuation at wavenumber $k = 4\pi \sin(\theta/2)/\lambda$, θ is the scattering angle, λ is the radio wavelength, and the value of the exponent n depends on the form of the function $F(k)$. It was found that for an isotropic radiator $n = 11/3$ [65], while for propagation with narrow beam antennas $n = 14/3$ [632]. However, the above expression fails for grazing propagation and ducting gradients. In this case, propagation is along the line-of-sight and power is maximized by disregarding focusing effects. The scattering angle was found to be strongly dependent on the signal level and a method for its determination in a multilayered atmosphere from the refractive index profiles was outlined by Moler [632]. On the other hand, it was also found that the spectral intensity of the refractive index at high wavenumbers in the inertial subrange of turbulence depends only weakly on atmospheric stability and refractive gradients [352]. From all this early work it has been concluded that the scattering angle and the intensity of dielectric fluctuations at high wavenumbers both depend on the refractive layering and the thermal stability of air masses, while the vertical velocity in the atmosphere is a key factor to determine the refractivity and stability. Based on refractometer measurements made during convective activity in the atmosphere by the Navy Electronics Laboratory (NEL) from January to mid-April 1958, it was further inferred that the direction and magnitude of the vertical velocity component is correlated with the upper-tropospheric wind velocity divergence [632].

A full description of the theory of partial reflection from finite layers as well as the comparison between the results of tropospheric scattering and experimental data is condensed in the work by Fehlhaber [286]. The form of the Doppler spectrum of scattered light from a vertically thin, horizontally homogeneous, refractively turbulent layer via a bistatic radio link having beam-swinging capability was analyzed by Atlas et al. [44, 45]. Although these assumptions implied a major simplification of the atmosphere, they provided analytical expressions for the relationship between the Doppler shift, the beam offset, and the crosswind velocity and, in addition, established the basis for the interpretation of meteorological significant observations. Further analysis of the efficiency of microwave scatter systems in probing the refractive-index structure of the troposphere was presented by Lammers [520]. In particular, in this analysis, three different tropospheric scatter systems were compared for sensitivity and spatial resolution. The systems were characterized by antenna-beam angular discrimination, or a combination of both antenna-beam angular distribution and signal-delay temporal discrimination. In addition, two of the radars employed were bistatic, while the third one was a monostatic radar. The scatter volume in one bistatic radar was defined by the intersection of the transmitter and receiver antenna beams, while the common scatter volume for the second bistatic radar was defined either by the transmitter or the receiver antenna beam and a spatial region of constant path delay. For the monostatic radar, the scatter volume was defined by the common transmitter-receiver antenna beam and the spatial region of a constant path delay. The details of the scatter geometry and scatter integral are discussed in Ref. [520] and will not be repeated here. The results of the comparison showed that the beam-delay system yielded good sensitivity and coverage, while the beam-beam system resulted in coarser resolution and no degradation in the midpath plane. However, depending on the antenna gains chosen, the beam-beam technique resulted to be much more sensitive than the beam-delay system in all positions. The capability of both techniques to monitor the troposphere at larger distances was depending on the transmitter power and receiver sensitivity. On the other hand, the bistatic radar was found to be more sensitive and coarser than the monostatic radar. In particular, the former system had more coverage at high altitudes, being suitable for high-level wind monitoring and CAT detection at the tropopause on a semisynoptic scale. In contrast, the monostatic radar was better suited for low-level turbulence detection and close-range coverage.

An evaluation of a two-dimensional matched filter technique for processing the signal from a pulsed coherent bistatic troposcatter link was further presented by Bures [141]. This technique was particularly useful to obtain high crosspath resolution maps, which showed the relative positions of individual radio scatterers. It has been shown that strong turbulence will degrade the image resolution in a rather severe manner. However, actual measurements of CAT show that the velocity fluctuations are no more than about 10% of the mean velocity and that the fluctuations are generally isotropic. This implies that the ratio between the standard deviation of the velocity over its mean will be smaller for the case when the horizontal velocity has no along-path component compared to the case when an along-path component is present and, as a consequence, the image degradation will be less. However, image degradation is useful to provide information about atmospheric turbulence. For instance, as argued by Bures himself, if turbulence is sufficiently strong, then the image spreading will approach the mean scatter spacing, and as a result, the image will be completely washed out.

7.8 Ultrasensitive radars

Ultrasensitive radars started to appear in the early 1960s as a need to optimize CAT detection. For example, if fluctuations in the refractive index in a given atmospheric region are large enough at a scale corresponding to half the wavelength of the radar, then the radar may detect the perturbed region. An analysis of the turbulent refractive spectrum in and beyond the limiting microscale reported by Atlas et al. [41] in 1966 has shown that the best radars available at that time could hardly detect the most reflective CAT at 18.5 km, corresponding to a coefficient $C_n^2 = 10^{-14}$ cm$^{-2/3}$ [41]. This was true for most conventional radars of limited antenna size even if they were equipped with maser receivers and the most powerful transmitters. Under this prospect, the detection of moderate CAT ($C_n^2 = 10^{-15}$ cm$^{-2/3}$) by ground-based radars would require an increase in sensitivity. From the dependence of the CAT reflectivity for various microscales, l_m, (or dissipation rates, ε) and C_n^2 values expected for CAT, as was derived by this analysis, the optimum wavelength for detection of weak CAT was found to be between 5 and 6 cm, where a 20-decibel improvement would be required to detect most CAT ($C_n^2 \gtrsim 10^{-16}$). The best radars at wavelengths of 5 to 6 cm will also detect typical cirrostratus cloud echoes, which are important tracers of CAT detection. The strength of CAT in the upper troposphere and lower stratosphere was found to depend on the magnitude of the wind shear (i.e., the vertical wind velocity gradient) just before breakdown [576]. Early efforts on CAT detection and warning has mainly focused on the development of airborne light compact devices working with various techniques, such as infrared and microwave radiometry, stellar scintillations, laser radar, electric field measurements, air temperature gradients, and VHF radar with rather little useful outcome. While some of these techniques have been useful to measure other atmospheric properties which are not necessarily correlated with CAT, ground-based ultrasensitive radar has been the only successful technique to remotely detect tropospheric CAT [400, 135, 128, 959].

As was mentioned in previous sections, a radio wave traveling through a medium with a nonuniform refractive index, such as the Earth's atmosphere, will lose some of its energy along its direction of propagation. For radio waves, the refractive index of air is given by [74]

$$n = 1 + N \times 10^{-6} = 1 + \frac{77.6}{T}\left(p + 4810\frac{e}{T}\right) \times 10^{-6}, \tag{7.87}$$

where N is the refractivity, T is the air temperature in kelvins, and p and e are, respectively, the air and water-vapor pressure in millibars. Since at the Earth's surface $N \sim 350$ and at the upper troposphere $N \sim 90$, the use of Eq. (7.87) shows that the refractive index is close to unity in both cases. Therefore, over vertical distances of 100 m, the changes in the refractive index are correspondingly much smaller. While the contribution of air pressure differences to the refractive index change is negligible at all altitudes, water-vapor pressure differences dominate in the lowest parts of the atmosphere and air temperature differences become important over all other contributions at tropospheric altitudes and beyond because the moisture content at such heights is very low. Partial reflection of radio waves is often produced by discontinuities in the air refractive index, which are sharper than the operating radar wavelength. Such irregularities in the refractive index cause incident radio waves to be scattered in all directions, and if the scattering is strong enough it will be detectable by a powerful radar system. The scattering of electromagnetic waves by fluctuating refractive index has been revisited in Section 7.2 and has been treated extensively by Tatarski [893] and references therein.

The reflectivity, η, defined as the scattering cross section per unit volume, is a measure of the strength of scattering of an air volume. According to Megaw [615], the reflectivity of a turbulent medium is given by

$$\eta = \frac{\pi}{8}\overline{(\Delta n)^2}k^2 F_n(k)\frac{\sin^2 \beta}{\sin^4(\theta/2)} \tag{7.88}$$

where $\overline{(\Delta n)^2}$ is the mean square fluctuation of the refractive index, k is the wavenumber, β is the angle between the direction of the receiver and the direction of the electric field of the transmitted wave, $F_n(k)$ is the one-dimensional spectral density of $\overline{(\Delta n)^2}$ as a function of k, and θ is the scattering angle. For backscatter with monostatic radars, for which the same antenna is used for transmitting and receiving, $\beta = \pi/2$, $\theta = \pi$, and the sinusoidal fraction in Eq. (7.88) tends to unity. Only those fluctuations with a wavenumber $k = (4\pi/\lambda)\sin(\theta/2)$ contribute to the spectrum $F_n(k)$, which corresponds to a scale $l = \lambda/[2\sin(\theta/2)]$, where λ is the radar wavelength. In particular, for backscatter $k = 4\pi/\lambda$ and $l = \lambda/2$. Under the assumption that l lies within the inertial subrange, the form of the fluctuation spectrum is given by [496, 893]

$$F_n(k) = \frac{2}{3}k_0^{2/3}k^{-5/3}, \tag{7.89}$$

where $k_0 = 2\pi/L_0$, with L_0 being the outer scale of turbulence. The smallest scale for which this form holds is the limiting microscale l_m. For weak and severe CAT, $l_m = 10$ mm and 6 mm, respectively [41]. Substitution of Eq. (7.89) into Eq. (7.88) yields

$$\eta = \frac{\pi}{12}\overline{(\Delta n)^2}k^2 k_0^{2/3}k^{1/3}\frac{\sin^2 \beta}{\sin^4(\theta/2)}, \tag{7.90}$$

where $\overline{(\Delta n)^2}$ is related to L_0 and the structure constant, C_n, by the relation

$$\overline{(\Delta n)^2} = 0.19L_0^{2/3}C_n^2. \tag{7.91}$$

Substitution of the above relation into Eq. (7.90) gives the more convenient form

$$\eta = 0.38 C_n^2 \lambda^{-1/3} \frac{\sin^2 \beta}{\sin^{11/3}(\theta/2)}. \tag{7.92}$$

Note that in the case of backscatter this form reduces to Eq. (7.22). The structure constant, referred to as the root-mean-square difference of the instantaneous refractive index at two different points separated by a unit distance, can be defined by [678, 679]

$$C_n^2 = A^2 \varepsilon^{2/3} \left(\overline{\frac{dn}{dz}}\right)^2 \left(\frac{d\bar{v}}{dz}\right)^{-2}, \tag{7.93}$$

where ε is the energy dissipation rate, $\overline{dn/dz}$ is the mean vertical derivative of the refractive index, $d\bar{v}/dz$ is the vertical derivative of the mean horizontal wind velocity, A^2 is a dimensionless constant factor, and z denotes the vertical coordinate. For example, the presence of water vapor in the lower troposphere gives rise to large gradients of the refractive index, and hence large values of C_n^2 and η will be possible together with relatively light turbulence. However, in the upper troposphere and above, the refractive index gradients are shaped by temperature gradients so that

$$\left(\overline{\frac{dn}{dz}}\right)^2 = f(p,T) \left(\overline{\frac{d\theta}{dz}}\right)^2, \tag{7.94}$$

where $f(p,T)$ is a smooth function of the mean absolute temperature T and pressure p and θ is the potential temperature [868]. According to the above relation, the larger the value of $\overline{d\theta/dz}$, the more intense is the CAT that can develop. In particular, in terms of ε, $\overline{d\theta/dz}$, and $d\bar{v}/dz$, the following relation can be derived for the structure constant [678, 679]

$$C_n^2 = \frac{A^2 f(p,T) T^2 \text{Ri}_c^2}{g^2 K_\text{h} (1 - \text{Ri}_c)^2} \varepsilon^{5/3}, \tag{7.95}$$

where Ri_c is the critical Richardson number (see Eq. 1.35) and K_h is the exchange coefficient for the turbulent diffusion heat. Since ε is a measure of the turbulence intensity within the inertial subrange, the above relation between C_n^2 and ε provides a close relationship between CAT and the detected radar echoes in clear air at high altitudes. In addition, Eq. (7.92) shows that the radar reflectivity increases for small values of the scattering angle. This suggested that for widely separated transmitting and receiving antennas, as in a bistatic system, the reflectivity should be much greater than for backscatter with a monostatic arrangement.

As was already described in Section 7.2.1.1, a measure of the sensitivity of any radar is given by the signal-to-noise ratio, S/N, where S is the strength of the echo detected by the radar from a target and N is the noise power of the receiver, which is generated by internal noise and thermal radiation from the ground and atmosphere. The ratio S/N must always be greater than unity for a target to be detectable. However, practical values of S/N can be affected by factors like unwanted echoes from the ground (i.e., ground clutter). As was outlined above, radars can operate in two different modes, the usual monostatic mode, and the bistatic mode. In what follows, the equation relating the signal echo strength to the radar parameters will be depicted for both modes of radar operation.

7.8.1 The monostatic radar equation

A monostatic radar operates by sending from a high gain antenna a short duration pulse of high power microwaves into the atmosphere and then by receiving with the same antenna the backscattered signals from the target region. Following Watkins and Browning [959], the power of the backscattered signal captured by the radar, dS, from a target at a range r in an angular direction (θ,ϕ) obeys the relation

$$dS = P\frac{G(\theta,\phi)A(\theta,\phi)}{16\pi^2 r^4}\sigma L, \tag{7.96}$$

where P is the transmitted power, $G(\theta,\phi)$ and $A(\theta,\phi)$ are, respectively, the gain and effective collecting area of the antenna in the direction of the target, σ is the radar cross section, and $L \ll 1$ is a factor accounting for energy losses and attenuation along the propagation path. Note that $L = 1$ corresponds to the ideal case when there are no losses. If the target is a point mass located on the axis of the radar beam, where the antenna has its maximum gain $G = 4\pi A/\lambda^2$, Eq. (7.96) reduces to the well-known radar equation [959]

$$S = \frac{PG^2\lambda^2}{64\pi^3 r^4}\sigma L. \tag{7.97}$$

If, on the other hand, the target is an extended thick scattering layer as shown schematically in Fig. 7.13, the radar will receive backscattered signals at any instant from all parts of the volume illuminated by the radar pulse. The shaded region within the thick layer defines the pulse volume. The effective radar cross section of an illuminated volume element is given by $\sigma = c\eta\tau r^2 d\Omega/2$, where τ denotes the duration of the pulse, c is the speed of light in vacuum, η is the reflectivity, $d\Omega$ is a small element of solid angle in the angular direction (θ,ϕ), and $r \gg c\tau/2$. Substitution of this form of σ into Eq. (7.96) and integration over the solid angle of the radar beam, Ω, the signal-to-noise ratio S/N for a single pulse can be written as [958]

$$\frac{S}{N} = f\frac{PAc\tau L\eta}{8\pi r^2 k_B TB}, \tag{7.98}$$

where f is a factor related to the shape of the radar beam, $N = k_B TB$, k_B is the Boltzmann constant, T is the system noise temperature, and B is the receiver bandwidth. For a paraboloid aerial $f = 0.28$ [360] and setting $B = \tau/2$ for near optimum performance of the radar, relation (7.98) for the signal-to-noise ratio

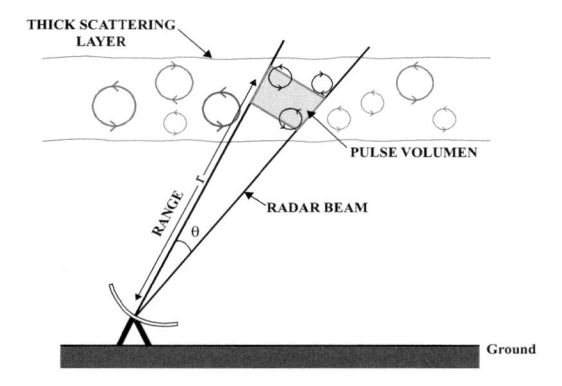

Figure 7.13: Schematic drawing showing a monostatic radar system operating within a thick turbulent layer.

takes the form

$$\frac{S}{N} = 1.2 \times 10^{29} \frac{PA\tau^2 L}{T} \frac{\eta}{r^2}. \tag{7.99}$$

An expression of the signal-to-noise ratio for CAT can be derived from the above relation by direct substitution of Eq. (7.35) for the CAT reflectivity

$$\frac{S}{N} = 4.5 \times 10^{28} F \frac{C_n^2}{r^2}, \tag{7.100}$$

where F is a term that contains all the radar parameters, i.e.,

$$F = \frac{PA\tau^2 L}{T\lambda^{1/3}}, \tag{7.101}$$

and is considered to be a figure of merit for the radar. It is important to note that expression (7.100) is strictly valid only when the pulse volume is completely filled by the scatterers. Otherwise, the signal strength is reduced. From relation (7.101), it also follows that the signal-to-noise ratio has a rather weak dependence on the wavelength provided that $l_m \ll \lambda/2 \ll L_0$, thereby allowing large antennae, high power transmitters, and sensitive receivers to be important factors to determine the operating frequency.

For a persistent target the signal-to-noise ratio (i.e., the radar sensitivity), as described above for single transmitted pulses, can be improved by either coherent or noncoherent integration of successive pulses, depending on the characteristics of the target. For coherent integration the phase of the transmitted pulse is first stored as a reference to allow the phase and amplitude of the backscattered signal to be measured. For example, if the Doppler frequency spread of the received signals is less than the bandwidth, b, of each channel of the spectrum analyzer, the improvement factor after coherent integration during a time $1/b$ is just $I_1 = r/b$, which is equal to the number of transmitted pulses. Therefore, from Eq.

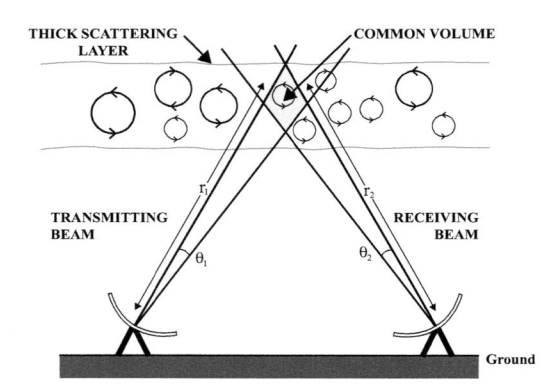

Figure 7.14: Schematic drawing showing a bistatic radar system operating within a thick turbulent layer.

(7.101) it follows for coherent integration that

$$FI_1 = \frac{\overline{P}A\tau L}{b}\frac{\lambda^{1/3}}{T},\tag{7.102}$$

where $\overline{P} = rtP$ is the mean power transmitted. Coherent processing provides information about the target motion through the Doppler frequency shift $f_D = 2v/\lambda$, where v is the target velocity. The motion of the scatterers within the pulse volume can be measured by the spread in frequency Δf_D. On the other hand, noncoherent integration applies if the successive echoes are uncorrelated either with the radar characteristics or with the received signals. For a large number of uncorrelated echoes the improvement factor is given by $I_2 \sim \sqrt{rt\delta t B}$, where t is the integration time and δt is the time interval at a given range over which the echoes are integrated. If $\delta t = \tau$ and since $\tau \sim 1/B$, then $I_2 \sim \sqrt{rt}$ so that the improvement factor is proportional to the square root of the number of pulses integrated. Therefore, from Eq. (7.101) the figure of merit becomes

$$FI_2 = PA\tau^2 L\frac{\lambda^{1/3}}{T}(rt)^{1/2}.\tag{7.103}$$

In the early 1970s, a much closer control of I_2 was achieved by using digital computers to record and store radar data. The interested reader looking for a more in-depth description of radar improvemnet in sensitivity by integration is referred to the classical textbook by Skolnik [843]. A review on early work on sensitive Doppler radar probing of the clear atmosphere has been presented by Gage and Balsley [319].

7.8.2 *The bistatic radar equation*

A bistatic radar system operates using two widely separated antennae, one for transmitting and the other for receiving, as shown in Fig. 7.14. In this case, the radar equation is given by [443]

$$\frac{S}{N} = \frac{PL\lambda^2\eta}{64\pi^3 r_1^2 r_2^2 K_{\mathrm{B}} TB} \int_{V_s} G_1 G_2 dV, \qquad (7.104)$$

where r_1 and r_2 are the ranges of the scattering volume V_s from the antennae of gain G_1 and G_2, respectively, and the other quantities are as before. It must be noted that the gain of both antennae is a factor that varies within the common volume and therefore the integral on the right-hand side of Eq. (7.104) does not admit an exact evaluation. Several numerical and approximate methods were proposed to evaluate this integral [43, 770]. As for the monostatic case, the overall sensitivity of a bistatic radar can be greatly improved through both coherent and noncoherent processing so that appropriate forms of the improvement factors I_1 and I_2 can be derived.

7.8.3 Limitations of radar sensitivity

Ground clutter and the occurrence of CAT in thin layers could affect signal integration as was described in previous sections. Ultrasensitive radars succeeding in detecting CAT are also sensitive to ground targets. Although his ground clutter lies outside the main beam, it can overwhelm and swamp the echoes associated with CAT. For some radar systems, the ground clutter is particularly limiting at short ranges and can even affect CAT measurements at ranges below 40 km. However, ground clutter can be reduced by using a coherent radar, which employs a Doppler frequency shift to separate echoes of interest from those due to stationary ground targets. A coherent processing called *Moving Target Indication* (MTI) has also been used for some time [843]. With this technique the strength of ground echoes is strongly reduced, while those from moving targets remain almost unaffected. Another troublesome aspect that could affect echoes from moving targets is when their Doppler frequency shift equals the pulse repetition frequency. On the other hand, CAT echoes can be contaminated by heavy rain occurring at lower altitudes at the same time as CAT. In this latter case, the MTI technique does not alleviate the problem.

Equation (7.100) can be used to perform successful predictions of the signal-to-noise ratio only when the pulse volume is completely filled with CAT. The same is true for the bistatic case, given by Eq. (7.104). For example, if CAT is occurring in thin layers, i.e., thinner than the vertical length of the pulse, this will also affect the sensitivity of the radar, reducing the signal-to-noise ratio by a filling factor $C = d/h$, where d is the thickness of the CAT layer and h is the vertical extent of the volume illuminated by the pulse. In particular, the value of h depends on the pulse duration τ, the radar-beam width θ, and the range and height of the CAT layer.

Chapter 8

New Technologies for Reducing Aviation Weather-Related Accidents

8.1 Introduction

Already since the second half of the 20th century, airline companies have been strongly interested in the development of onboard instruments capable of identifying hazards caused by atmospheric phenomena, such as CAT, in flight. The ever-increasing need to improve aircraft safety and keep passengers safe has driven the development of new communication technologies as well as the design of devices for installation in the cockpit, which can be operated by the crew to be aware of adverse weather and make decisions regarding safe routes. To handle the complex technologies on board the most modern aircraft, flight crews rely on computers to fly them and to monitor their systems.

The first autopilots began to be used in the mid-1930s and towards the end of the 1950s computers began to be small enough to be installed on board airplanes. Over the years, the development of aviation technology has been so sustained that today's digital computers can now fly aircraft in virtually any situation while ensuring that all systems are functioning properly. Today, the aircraft is required to carry onboard high-power equipment, electronic warfare, radar, and other advanced components. As a response, digital technology has enhanced safety and efficiency, while reducing the flight crew's workload.

8.2 Cockpit weather information systems

Atmospheric instabilities leading to severe weather conditions and CAT are responsible for many aviation accidents. Therefore, obtaining real-time information in-flight of weather changes has been a major concern for pilots. For example, pilots are sometimes called upon to make sudden decisions, such as planning route changes to avoid atmospheric turbulence and close encounters with adverse weather. Today, aviation technology has developed to the point of providing more sensitive and reliable instruments capable of detecting atmospheric effects. Such devices can be installed in the cockpit of any aircraft and have been designed and developed primarily for turbulence detection. An example of such systems is the *Cockpit Weather Information System* (CWIN), which is an integrated communication and navigation aeronautical system that delivers real-time graphical weather information with global coverage to the flight deck of an aircraft. Since its inception, the CWIN concept consists of two basic elements: generic system hardware and a functionally specific software [913]. The hardware provides a display control, data processing, memory, and aircraft interfaces. Weather data are received, stored, and processed by the CWIN system. All the processed data are mapped into a graphical interface so that they can be observed graphically on screens for analysis and interpretation. For example, the moving weather can be displayed in color and the user interface is designed for presentation to the flight crew. In particular, a coded color is defined for ground radar summaries to distinguish between different degrees of weather severity, lightning strikes intensity, and different degrees of category for surface observations and terminal forecasts.

In general, the information is presented in a graphical format on a 10.4-inch color LCD with touch panel interface and data selection. The CWIN graphical weather information is supplied to the pilot by the *Pilot Access Terminal*. The CWIN system can provide real-time ground and onboard radar data, wind shear warning images, surface observations, and terminal forecasts for all domestic and international airports. It is also able to operate as a *Traffic Alert and Collision Avoidance System* (TCAS) and display terrain and weather overlay, three-dimensional graphics, airport runway taxi maps as well as graphical and textual information of other flight operations. An overview diagram showing how the CWIN platform works is depicted in Fig. 8.1. The drawing is a modification of the figure presented in Ref. [913]. The CWIN system serves as an interface between aircraft-ground communication, the satellite communication system (SATCOM), and the navigation and satellite communication system (GPS/NAV). In particular, the SATCOM data link provides the ground data communication system to the CWIN aircraft and receives the meteorological data linked to the aircraft in graphical format. The meteorological information consists of radar-detection summaries, lightning data surface observations, and terminal forecasts. Updates to the weather information are made periodically and are automatically stored, thereby allowing instantaneous access to the latest in-

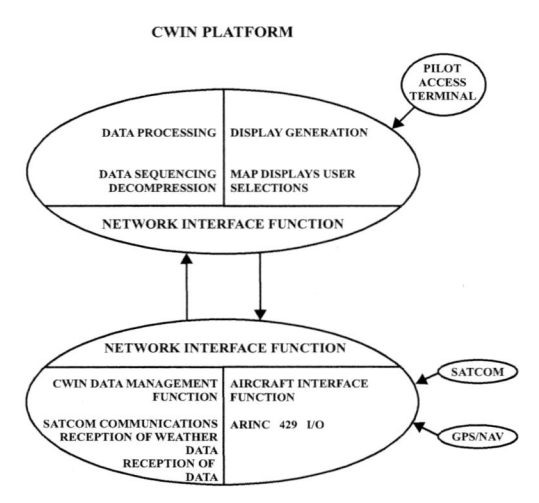

Figure 8.1: Schematic diagram showing how the CWIN platform operates. Figure adapted from Tu [913].

formation to the flight crew. The CWIN meteorological information is received from the SATCOM Satellite Data Unit (SDU) through a *Ground Earth Station Data Broadcast* (GSDB) digital data interface and through the touch panel interface. The screen generation function allows the user data display mode to be selected. The arrival patterns of each of the incoming CWIN data types include a national ground radar, which consists of an 8-pixel national radar mosaic that is transmitted to the aircraft every 15 minutes in the form of a compressed file. For instance, information on air beams is also transmitted to the aircraft every 15 minutes and consists of a 5-minute file with the latitude and longitude of each impact. The surface observations and forecast information of the terminals as well as the data management capacity take into account the reception of the transmission information from both the SATCOM SDU unit and the GPS/NAV unit, and the packaging of the data in a preset form for application-specific processing. In general, surface observation refers to the ceiling and visibility data for each of the airports listed in the CWIN database. It includes data on local precipitation, flight danger, and excess wind for each airport. On the other hand, terminal forecasts consist of similar data to surface observations, except that they consist of predicted values that are displayed in a color-coded text format for a selected airport.

Further technological advances have led to the improvement of cabin meteorological systems that are normally presented in a *Multi-Function Display* (MFD), while some recently installed systems allow for the superposition of these data in the *Primary Flight Display* (PFD). In fact, they provide meteorological data, such as those that are available on the ground as, for example, satellite weather and radar images, aviation routine weather reports (METAR), terminal

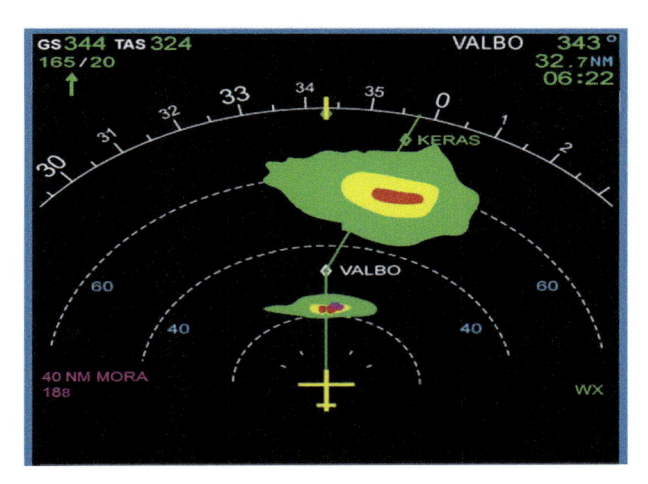

Figure 8.2: Example of radar data, aircraft position, and programmed route displayed on a MFD. Figure taken from Ref. [590]. URL: https://safetyfirst.airbus.com/optimum-use-of-weather-radar/.

weather forecast (TAF), significant weather information (SIGMET), and aviation weather information (AIRMET), just to mention a few of many other products that are now easily accessible at any time during flight [562]. Some MFD devices can register meteorological events, such as, for instance, thunderstorms and areas of precipitation. However, most MFDs are also capable of registering electrical activity as lightning, where the corresponding data come from two sources, namely the onboard and the broadcast weather systems [563]. Modern aircraft have all integrated devices to get real-time meteorological information and the MFDs have been designed to simultaneously present the position of the aircraft, the programmed route, and the radar data, as shown in the example of Fig. 8.2. The onboard weather radar system uses a radar antenna that can detect weather phenomena near the aircraft in real time and its coverage is similar to that of a flashlight beam, as illustrated schematically by Fig. 8.3. However, this class of radar cannot be used to detect CAT, lightning, or fog since it only shows areas with rain, sleet, snow, or hail and the covered area is limited in both direction and range. For example, in Fig. 8.3 the radar system fails to detect the cell that lies below and the two cells that lie beyond the shaded region representing the radar beam. Despite the technological progress and sophistication of most integrated systems, a common problem associated with the devices that receive and transmit information is the delay time and the loss of information especially when it comes in the form of images. In particular, this situation is of major concern when the pilot of an aircraft in flight is forced to make quick decisions. Among the factors that may influence this delay are those of electromagnetic origin. Other limitations occur during the early stages of thunderstorm formation since they are currently free of precipitations and it is very likely that the

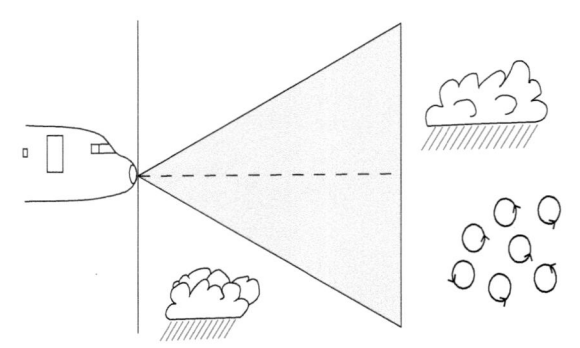

Figure 8.3: Schematic drawing shown the limited coverage of an onboard radar system. Although the tilt of the radar antenna can be adjusted upward and downward, the actual coverage of the radar is limited in both direction and range. Figure adapted from Ref. [562].

radar fails to detect forming storms as well as zones of convective wind shear, severe CAT, and ice formation, which are all phenomena that are characteristic of storms during the cumulus stage.

8.3 First-generation systems

In the late 1990s, the U.S. government set as a goal to reduce the fatal air accident rate by 80% over the next ten years. In response to this goal, NASA developed the *Aviation Security Program* (AvSP) to work up new technologies to increase aviation safety. This program was implemented to assist the FAA in the framework of the AvSP. Since about 30% of all air accidents are caused by climatic factors, the project was focused on the climate as a primary factor. For example, in the . period between 1982 and 1993, a total number of 5894 weather-related aviation accidents were reported, with 1750 fatalities [299]. During nearly the same period from 1983 to 1995, the NTSB identified the various weather instabilities as the principal cause of 112 commercial aviation accidents and incidents, leading to 13 fatalities, 58 serious injuries to passengers, loss of 8 aircraft, and substantial damage in other 34 aircraft. In view of these records, NASA promoted cooperative research efforts with industry-led teams. As a result of this cooperation, first-generation systems were produced from prototypes whose evaluations were led by the Boeing and the Honeywell companies for worldwide transportation operations and by the ARNAV Systems and the Honeywell-Bendix/King companies for U.S. domestic aviation operations. These systems were designed to use weather products that already existed, which were conditioned and reformatted in such a way to establish a data link and display in the cockpit. Moreover, the Honeywell International multinational conglomerate corporation, in a joint effort with NASA, developed a *Weather Information Network* (WINN) capable of providing graphical weather information to the cockpit of commercial and business aircraft flying worldwide [876]. This network included aerial displays, aerial and ground servers, multiple providers of meteorological products, and data link services. In its first stage of development, an open architecture was adopted to adapt the system to any type of data link technology and several tests were carried out on the satellite and terrestrial telephone links in the VHF/UHF range [876].

The *Aviation Weather Information* (AWIN) project[1] was created as a government-industry-academia cooperation team to develop prototype systems for the detection of atmospheric hazards and the enhancement of the flux of information to pilots about weather phenomena to help them make better reroute decisions. This was achieved by assembling different technologies for weather sensing, forecasting, and graphical displays. This original and innovative system used both ground-based and satellite infrastructures to deliver graphical weather information to the pilots, dispatchers, and controllers in an accurate, timely, and strategic manner for collaborative decision making. It also boasts of several technical achievements that were developed in a record two-year period. The prototype systems that were developed for both commercial and general aviation airplanes have been extensively flight tested and the user feedback is presently being incorporated into the second generation of prototype cockpit displays [299]. On the other hand, a *Flight Information Services Data Link System* (FISDL) was developed in 2002 by the FAA to provide link meteorological data in the U.S.

The first-generation systems for data link meteorological information were mainly employed for general aviation in the United States. During the winter of 2001, United Air Lines conducted more than 40 in-service (ISE) assessment flights with the WINN system incorporated into a prototype electronic flight bag. The meteorological devices included airport observations (METAR), terminal area forecasts (TAF), ground-based weather radar reflectivity (*Next Generation Radar*, NEXRAD), turbulence warnings, significant weather warnings (SIGMET graphics), and satellite cloud images. The use of this kind of instrumentation resulted in a saving of time and cost due to a better knowledge of the meteorological situation. In support of this, the FAA implemented a national network of radio transceivers, such as the U.S. Universal Access (UAT) to provide operational flight and traffic information, including weather information, linked to data in the cabin of the equipped aircraft. Moreover, a reduction in *Aircraft Communications Addressing and Notification System* (ACARS) message traffic (and therefore cost) was estimated to be around 40–50% [876].

8.4   Weather information presentation

The technology geared toward pilots has made significant strides. For example, advanced displays are available today where pilots can observe weather patterns in the cockpit through a data link. The use of these instruments has required

---

[1] https://flight.nasa.gov/events/chicago/awin.htm.

training pilots to mitigate possible errors and increase air safety. Conventional round dial instruments accompanied by aeronautical charts, approach charts, and flight service station summaries are some of the many separate data that must be accessed by pilots for a safe flight. Therefore, part of the pilot training is to be able to integrate all the available information into an accurate mental model of the exterior world. In this sense, new algorithms have been implemented and used in the devices to reduce the pilot's mental workload by just converting much of the integrated information as user-friendly and easy to interpret. To this end, flat-panel displays with terrain, traffic, routes, and weather overlaid on a single screen have been developed, thus fostering a more intuitive mental model of the "big picture" for the pilot. With such a reduced workload involved in the mental integration of multiple elements, a pilot can turn his/her attention elsewhere, such as high-level decision-making as well as judgment and situation assessment tasks [143].

Meteorological information is recorded with the aid of different devices, such as weather radars on the ground and on board or lightning detection systems. It can even be obtained by other means, for example, from another aircraft or directly from the ground. All these data must be transmitted to the crew in a legible, clear, and secure manner. For this purpose, NASA has developed and implemented formats and algorithms to send and receive data using the instrumentation in the cockpit. This information not only allows the pilot to avoid dangerous weather conditions but also to identify trends and changes in the meteorological patterns that could affect flight security and allow the pilots to make timely and efficient decisions [876]. The influence of the number of frames per loop, the total loop time, the pilot viewing time (i.e., the exposure time), and the inclusion of animated aircraft track history (i.e., the aircraft looping) on the optimal design of an animated NEXRAD mosaic was studied by Lemos and Chamberlain [548]. The main goal of this study was to determine the NEXRAD animation characteristics that optimize the pilot weather situation awareness under different workloads and to compare this trending technique with other in-flight trending presentations. The formats for the presentation of the meteorological information in the cockpit has also been examined by NASA from both the textual and graphical point of view and their effects on the pilot's navigation decisions have been thoroughly studied.

The position of the aircraft relative to the graphical representation of the weather pattern is one of the most important parameters for a pilot because it provides an indicator of the range (i.e., distance from the pattern) and the temporal information on the meteorological conditions. For example, the use of NEXRAD provides highly-resolved graphical representations of linked data that show how the navigation decisions of pilots would be affected by adverse weather conditions. In particular, the NEXRAD project consists of a network of 160 high-resolution Doppler S-band weather radars that are operated by the NWS. Its technical name is WSR-88D (meteorological surveillance radar, 1988, Doppler)

[212]. With such radar technology, the image resolution has increased to the point that the pilots could continue their flights with the expectation of much greater efficiency. Also, when different sources of information convey together, such as views from the window, radio voice communications, and data link displays, the convective meteorological situation during flight becomes clearer.

The trending information presented through the NEXRAD image loop and the *National Convective Weather Forecast* (NCWF) product visualization was found to provide a significant increase in situational awareness for the pilot concerning the location, proximity, and direction of the convective weather movement. However, over-reliance on the information presented by the data link system at the expense of accessing more conventional information sources, such as flight service stations (FSS), was found to outweigh the increased situational awareness to the extent that the decision to make was indeed not different with or without a weather display in the cockpit. On the other hand, the flight crew's confidence in the cabin weather information and the team's reaction to screens of impending adverse weather conditions have also been studied. A result of this study indicated that in general crews trust the onboard weather radar more than the data-linked information. However, when both systems agree, crews' confidence in the weather display linked to the data increases. When the onboard and NEXRAD displays did not match, crews become more confident with the onboard radar information, but still use the NEXRAD display to increase their overall situational awareness. As a matter of fact, the pilots were more likely to make correct detour decisions when the NEXRAD system showed impending adverse weather. Concerning this point, Burgess and Thomas [143] proposed an experiment, which they called the "cockpit weather experiment", aimed at determining the effects of using the NEXRAD looping and the NCWF product on the pilot workload and decision-making. The experiment was designed to investigate the potential misuse of these two techniques for display of hazardous weather information, so as to support the recommendations suggested by the FAA and the display manufacturers for the design and use of such displays. In this experiment, 48 pilots were separated into three equal groups: a control group and two treatment groups. All three groups were exposed to a simulated flight scenario involving convective activity, with the control group having access to conventional sources of pre-flight and in-flight weather products and the other two groups having access to a weather display that presented NEXRAD mosaic images as well as a graphic depiction of METARs and text METARs. In addition, one treatment group of pilots used a NEXRAD image looping feature, while the second group used the NCWF product overlaid on the NEXRAD display. The results of the experiment showed that both the treatment displays provided a significant increase in situational awareness and incomplete information as required to deal with dangerous convective weather conditions, implying that substantial more pilot training would be required to allow a safer and more effective use of such displays.

8.5 Next-generation systems

In 2003, the U.S. Congress created the *Joint Planning and Development Office* (JPDO) to face and solve aeronautical problems dealing with air traffic and delays, which afflicted most airlines in those years. Shortly after, in 2004, the JPDO created the *Integrated National Plan for the Next Generation Air Transportation System*, whose task was to transform the air transportation system in the United States [886]. In this context, the role of the U.S. government was to allow the participation of several companies to provide profitable solutions within a set of performance-based safety, protection, and environmental rules as well as promote and improve incentives to give rise to more efficient air traffic and airport services. On the basis to keep the world leadership in the aviation industry, several strategies were considered and published that literally consisted in the following items [886]:

■ Developing the airport infrastructure to meet future demand by empowering local communities and regions to create alternative concepts of how airports will be used and managed in the future.

■ Establishing an effective security system without limiting mobility and civil liberties by embedding security measures throughout the air transportation system – from curb to curb. Creating a transparent set of security layers will deliver security without creating undue delays, limiting access, or adding excessive costs and time.

■ Creating a responsive air traffic system by devising alternative concepts of airspace and airport operations to serve present and future aircraft. As new vehicle classes and business models emerge, such as remotely operated vehicles and spaceports, the safe and efficient operation of all vehicles in the National Airspace Sistem will be critical to creating new markets in aviation and beyond.

■ Providing each traveler and operator in the system with the specific situational awareness they need to reach decisions through the creation of a combined information network. All users of the system will have access to the air transportation system data they require for their operations.

■ Managing safety through a comprehensive and proactive approach that can integrate major changes, such as new technologies or procedures. This will be done in a timely manner and without compromising aviation's current superior safety record.

■ Introducing new policies, operational procedures, and technologies to minimize the impact of noise and emissions on the environment and eliminate ground contaminants at airports. This effort includes the exploration of alternative fuels, engine, and aircraft designs. These actions will result in reduced environmental impact and sustained aviation growth.

- Reducing the impact of weather on air travel through a system-wide capability for enhanced weather observations and forecasts, integrating them with the tools used by air system operators. This capability will substantially improve airspace capacity and efficiency while enhancing safety.

- Harmonizing equipage and operations globally by developing and employing uniform standards, procedures, and air and space transportation policies worldwide, enhancing safety and efficiency on a global scale.

According to the FAA, the Next Generation Air Transportation System, or abbreviated "NextGen", is the FAA-led modernization of the U.S. air transportation system to make flying even safer, more efficient, and predictable [662]. NextGen is not a technology, product, or goal. Its portfolio simply encompasses the planning and implementation of innovative new technologies and airspace procedures after thorough safety testing. To further establish the FAA's global leadership in aviation, the NextGen concept sets standards around the world through research, innovation, and collaboration. Today, the term NextGen is used to mean the continuing transformation of the U.S. National Airspace System (NAS), which was planned in stages between 2012 and 2025. In 2007, the FAA started working on NextGen improvements [661, 885], converting it into one of the most ambitious infrastructure projects in U.S. history. It is aimed at implementing new technologies and capabilities to offer a more efficient service to more than 2.7 million passengers and 44 thousand flights per day. Due to the ever-increasing number of passengers and flights, NextGen represents the evolution from a ground-based system of air traffic control to a satellite-based of air traffic management through the development of aviation-specific applications for existing and widely used technologies, such as the GPS, the technological innovation in weather forecasting, the data networking, and the digital communications [662, 844].

The expected results from the NextGen project include improvements in safety, reduction of delays, fuel savings, and a significant reduction in aircraft exhaust emissions. In order to reach further improvements, NextGen has included a shift to smarter digital and satellite technologies, which rely on the new interconnected systems created by the FAA. These have changed and improved communications, navigation, and surveillance in the NAS [662]. To what concerns communications, aircraft will have as a result a much greater capacity to receive instructions from ground systems in real time, which will allow pilots to identify their current positions and forecast their destinations in time and place. On the other hand, quick communication between air traffic controllers and pilots will also be enormously facilitated and possible communication losses due to saturation of the corresponding lines will be avoided. These conditions point towards an improvement of the air traffic dynamics. To facilitate navigation, the FAA has switched to satellite-enabled systems as they are by far more accurate. This makes it easy to create optimal flight routes anywhere on the NAS for all departure operations, altitude cruises, arrivals, and landing. These modifications all operate in the way of reducing fuel consumption, the emission of polluting gases, and the flight times. The continuous surveillance enables air traffic controllers to follow the precise position of any aircraft and have a clear view of the weather conditions surrounding the aircraft.

8.6 The route towards enhanced turbulence radars

The radar as a detection system was already introduced in Chapter 7. In this section, we shall review some basic principles before introducing ourselves to the concept of enhanced turbulence radar. The working principle of a conventional radar has been succinctly described by Brown [125] as:

> A conventional weather radar transmits a narrow pulse of electromagnetic radiation in the cm wavelength and then waits to see if any of the transmitted energy is backscattered from distant targets before the next pulse is transmitted [760]. All hydrometeors, such as raindrops, hailstones, and ice crystals, are targets of meteorological interest. The time delay between the transmitted and returned pulse determines the distance to the hydrometeors, while the energy received (i.e., the radar reflectivity) is proportional to the size and scattering characteristics of the hydrometeors within the pulse volume. Due to random fluctuations of hydrometeors within the pulse volume, tens of consecutive pulses are averaged together to obtain a representative measurement.

The basic radar structure has four main components as shown schematically in Fig. 8.4: a transmitter, which produces the energy pulse, a transmit/receive

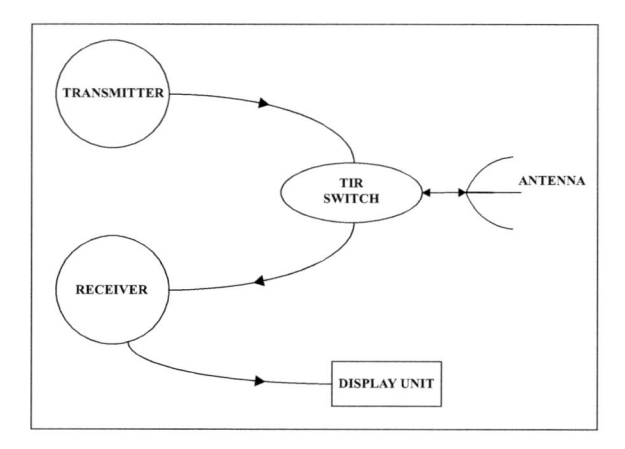

Figure 8.4: Schematic drawing showing the components of a basic radar structure.

switch, which tells the antenna when to transmit and when to receive the pulses, an antenna, which sends the pulses out into the atmosphere and receives the reflected pulses back, and a receiver, which detects, amplifies, and transforms the received signals before displaying them. As a meteorological device, Doppler radar techniques have long been applied to map the structure and speed of severe storms, showing in real time their evolution [142, 240, 244]. Radars have also been developed to detect and determine the position of aircraft using radio techniques. Over time new advances in communications technology have found other applications, for example, in the mapping of both the Earth's surface and atmosphere, and also for astronomical purposes [247]. In early 1941, the first storm detection using a microwave radar was carried out in England [247]. As was more extensively described in Chapter 7, this type of radar emits a microwave signal that can penetrate thunderstorms and clouds, thereby allowing to reveal their dynamical structure. For example, an important feature that distinguishes microwave radars from other radars is their ability to observe the rains and thunderstorms day and night. Since microwave signals are only slightly attenuated by rain and clouds for wavelengths > 0.05 cm, the intensity of the rain is directly related to that of the dispersed signal and the phase change ratio in time (i.e., the Doppler displacement) is a measure of the radial velocity of the raindrops. A field of radial speeds can be mapped by a single Doppler radar, but when two separate radars are used, it is possible to determine the wind velocity components by assuming that the terminal fall speed of precipitation particles is known. This assumption is no longer necessary if the data from three noncollinear Doppler radars are available [33, 555]. However, Doppler radars are not limited to the scanning of precipitation-loaded air. For example, they have been useful to map the kinematic structure of the planetary boundary even when the particulate material does not offer significant reflectivity [245]. Although they have represented a valuable tool for meteorological research applications, unfortunately, it has not been possible to employ them in routine operational applications. Table 8.1 lists the bands that are commonly used for meteorological radars according to the IEEE Standard 521–2002 [649].

From early experimental attempts to measure rainfall by using a Doppler radar at wavelengths of 10 cm [595], the following power-law relation between the radar reflectivity, Z, and the rainfall rate, r, was found

$$Z = ar^b, \tag{8.1}$$

where estimations of the coefficient a and exponent b are given in Ref. [596]. Nevertheless, it has been shown that these parameters can vary widely and are likely to be associated with synoptic conditions [649]. More general expressions

Table 8.1: Band designations, frequency ranges, and significant absorption lines. Table adapted from Munchak [649].

Designation	Frequency range (GHz)	Significant gas absorption lines
S	2–4	
C	4–8	
X	8–12	
K_u	12–18	
K	18–27	H_2O (22.235 GHz)
K_a	27–40	
V	40–75	O_2 (several lines 49–70 GHz)
W	75–110	
G	110–300	O_2 (118.75 GHz), H_2O (183.31 GHz)

for the radar reflectivity and the rainfall rate are given by [649]

$$Z = \int_{D_{\min}}^{D_{\max}} N(D)D^6 dD, \tag{8.2}$$

$$r = \frac{\pi}{6} \int_{D_{\min}}^{D_{\max}} N(D)v(D)D^3 dD, \tag{8.3}$$

where $N(D)$ is the drop size distribution and $v(D)$ is the drop fall speed. By assuming an analytical form for $N(D)$, Eqs. (8.2) and (8.3) can be combined to produce the power-law relation (8.1) between Z and r. The integral in Eq. (8.2) does not depend on the radar wavelength, meaning that this equation is valid only when the particle size is much smaller than the radar wavelength. In general, when larger particle sizes are considered, the equivalent reflectivity factor, Z_e, given by [649]

$$Z_e = \frac{\lambda^4}{\pi^5 |K|^2} \int_{D_{\min}}^{D_{\max}} N(D)\sigma_b(D,\lambda)dD, \tag{8.4}$$

is used, where K is a function of the complex index of refraction, λ is the radar wavelength measured in millimeters (mm), σ_b is the backscattering cross section (in mm^2), $N(D)$ is the number concentration of raindrops (in m^{-3}) per size interval, and Z_e is given in units of mm^6 m^{-3}. The backscatter cross section σ_b of a water drop with diameter D_i is small compared to λ. In the Rayleigh approximation for particles much smaller than the wavelength, where the individual dipoles that comprise the particle can be treated as coherent scatterers, the backscattering cross section is given by [649]

$$\sigma_b = \frac{\pi^5}{\lambda^4} |K|^2 D_i^6, \tag{8.5}$$

which describes the amount of electromagnetic radiation that is scattered towards the source of incident radiation. As the size parameter

$$\frac{\pi D}{\lambda} \to 1, \tag{8.6}$$

the Rayleigh approximation breaks down and Mie theory should be used because it provides the exact results for spherical drops.

Strategic and tactical planning for safe flights has required the implementation of airborne weather radar systems. With the use of these devices, pilots can evaluate in real time the intensity of convective motion in front of the aircraft. The weather radar technology has evolved significantly in recent years producing a range of improved products. The proper use of airborne radar systems provide pilots with the necessary information to substantially reduce encounters with adverse weather, increase flight safety, and minimize their workload [590]. Note that current airborne weather radars work only with reflectivity and cannot show turbulence. However, turbulent regions can be observed based on the spectral width detected by the radar. In these situations, radars are designed to color the region with magenta when the spectral width exceeds a certain threshold, while below the threshold no region is displayed. However, the spectral width threshold for a given level of turbulence depends on the aircraft's speed, altitude, weight, and characteristic response to turbulence. This represents a problem because different aircraft react differently to the same atmospheric turbulence. In addition, false alarms may also occur where turbulence is signaled as not significant and vice versa [724].

The correct interpretation of the weather radar display requires having a good knowledge of the radar principle. For example, Fig. 8.5 illustrates the basic working principle of radar reflectivity, while Fig. 8.6 shows the weather radar principle by including the color code of water meteorological states associated with the reflectivity intensity. Weather detection is based on the reflectivity of water droplets and the weather echo appears on the Navigation Display (ND) with a color scale that goes from red (high reflectivity) to green (low reflectivity). The weather radar echo returns the intensity as a function of the droplet size, composition, and quantity. For example, a water particle is five times more reflective than an ice particle of the same size [590]. In fact, weather radars can detect rainfall, wet hail, wet turbulence, and wind shear, while they fail to detect ice crystals, dry hail and snow, CAT, sandstorms (solid particles are almost transpar-

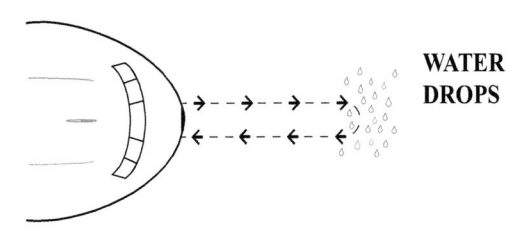

GOOD REFLECTIVITY

Figure 8.5: Schematic drawing showing the working principle of radar reflectivity when liquid water is found ahead of an in-flight aircraft.

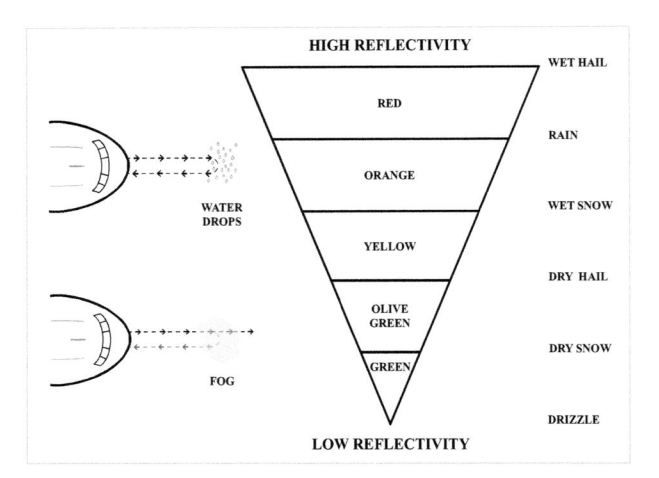

Figure 8.6: Schematic drawing showing the working principle of radar reflectivity in terms of the reflectivity intensity. Figure taken from Ref. [590]. URL: https://safetyfirst.airbus.com/optimum-use-of-weather-radar/.

ent to the radar beam), and lightning. However, recent weather radars offer hail and lightning prediction functions. Hence, the weather radar detection capability increases in the presence of liquid water. Since the amount of liquid water in the atmosphere decreases with altitude, thunderstorms beyond the radar top will have low reflectivity and therefore may not be detected by the weather radar. The drawing of Fig. 8.7 illustrates schematically the failure of the weather radar to detect turbulent areas above the radar top, where the atmosphere is drier. It is possible that convective clouds as well as their intrinsic threats can extend significantly above the upper detection limit of the radar top. This means that reflectivity is not directly proportional to the level of risk that may be encountered. Hence a convective cloud may be dangerous, even if the radar echo is weak.

The deficiencies and shortcomings in the ability of conventional weather radars to detect dry turbulence and CAT led the AeroTech section of the NASA Aviation Safety Program to develop the Enhanced Turbulence (E-Turb) Hazard Prediction algorithm for airborne Doppler weather radars to take and process the radar measurements with real-time aircraft information (i.e., weight, speed, altitude, etc.) and calculate predicted loads on the aircraft [724]. AeroTech's E-Turb algorithm presents the turbulent area to the pilot at the aircraft's current flight conditions and far enough ahead to allow the pilot to decide in time. Such a forewarning enhances pilots' ability to route aircraft around regions of CAT with no reduction in safety and can lead to potential savings in fuel and reductions in flight delays. It has also implications on aircraft maintenance since there will be less overall stress on the aircraft structure and an increase in the airframe life due to the ability to avoid turbulence. As a consequence, there will also be a reduction in the costs because of less unnecessary maintenance and fewer severe loads

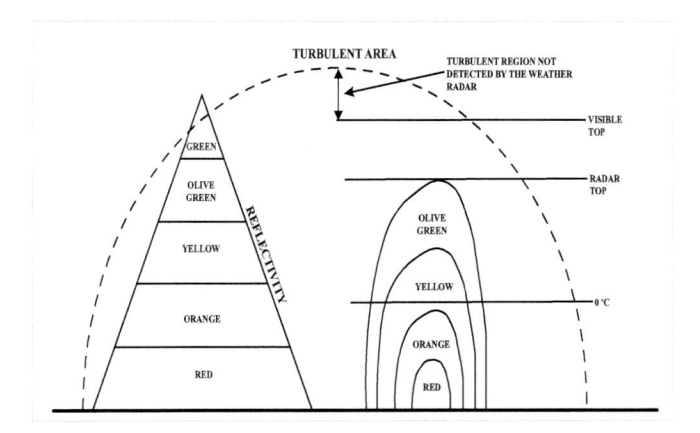

Figure 8.7: Schematic drawing illustrating the failure of the weather radar to detect turbulence at altitudes where the amount of liquid water in the atmosphere is low. Figure taken from Ref. [590]. URL: https://safetyfirst.airbus.com/optimum-use-of-weather-radar/.

inspections. The AeroTech's algorithms were first implemented on a Rockwell Collins WXR-2100 radar and installed on a Delta Air Lines B-737-800 aircraft. The Delta aircraft implementation displayed turbulence as a two-level magenta severity display, where speckled magenta indicates light to moderate turbulence and therefore the warning to passengers to fasten their seat belts, while solid magenta was indicative of moderate to severe turbulence and hence the warning to avoid if possible. This represented a new and innovative way to display turbulence in the cockpit allowing pilots, for the first time, to distinguish between moderate and severe turbulence hazards. A comparison between a current Wx radar and the AeroTech's E-Turb process is depicted in Fig. 8.8.

The wake vortices generated by aircraft during take-off and landing are another important aspect that has important implications on aviation safety and arrival and departure rates. Based on specific FAA rules, for flight safety aircraft are separated in distance when approaching a terminal airport. If the separation distance is too large, flight safety is increased at the expense of adversely affecting the capacity of the airport and hence the overall national airspace system, If, on the other hand, the separation distance is too small, then safety may be reduced and the airport arrival and departure rates are increased. Wake vortices are therefore a severe constraint to aircraft separation and airport capacity, which may be at or near limits at many major airports in the world. The AeroTech research has been awarded a NASA contract to develop an *in situ Wake Vortex Encounter Detection and Reporting System* (VEDARS) to optimize spacing aircraft and increase the airport capacity while maintaining a high level of flight safety during take-off and landing [724]. In particular, VEDARS will detect and transmit real-time reports of wake encounters from an aircraft to other aircraft and air traffic controllers, thereby increasing safety of operations around terminal

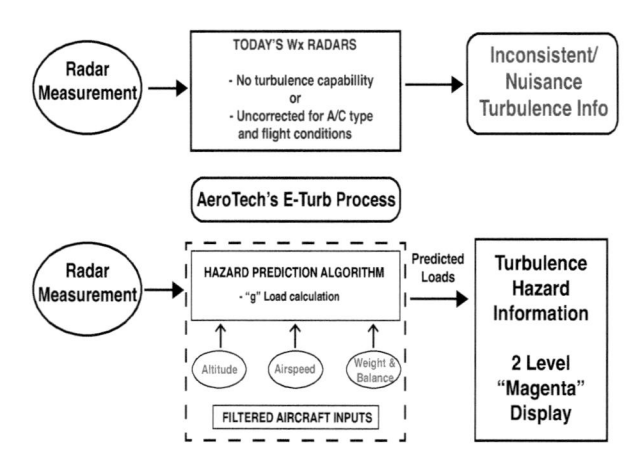

Figure 8.8: Comparison between a current weather radar and the AeroTech's E-Turb process for enhanced detection of turbulence. Figure adapted from the webpage of AeroTech Research (USA), Inc. http://www.atr-usa.com/index.html.

airports. AeroTech research is also working to improve the accuracy and reliability of reported wind speed and direction and, therefore, crosswind estimation by just improving and validating an estimator for sideslip angle. Such crosswind estimate is a key component in the transport of wake prediction.

8.7 Automated turbulence reporting

Before the advent of the new communication technologies, the encounters with turbulence were always reported by pilots to the controllers and then informants and dispatchers transmitted the information through voice communications to other pilots. This kind of communication was known as PIREP and represented the only means to transmit the information. However, PIREPs were often not consistent, accurate, and timely regarding the intensity and position of the turbulence. Therefore, they were not aware of the severity of turbulence and the damage it could cause to the aircraft. However, from 1998 to 2003 things started to change as AeroTech Research in partnership collaboration with NASA developed the concepts and first algorithms of various safety-related technologies under the NASA *Turbulence Prevention and Warning System* (TPAWS) as part of the *Weather Prevention (WxAP) Program* within the *Aviation Safety and Security Program* (AvSSP) [722]. As was described in the previous section, one of the first systems designed to identify and notify potential turbulence encounters was the Enhanced Turbulence (E-Turb) radar [876], which was developed as an automated system to drive information with aircraft. The development of an automated turbulence encounter reporting system was based on aircraft turbulence response algorithms that were developed to measure the E-Turb radar performance. The enhanced turbulence algorithm developed by AeroTech Research takes the radar estimate of the second moment[2] and calculates real-time estimates of predicted g-loads [246]. This tool can provide reliable turbulence information to the cockpit. For example, a small second moment indicates smooth air so that most of the particulates are moving with the same speed and direction, while a large second moment indicates large fluctuations of the particles' velocities indicating turbulent motion. However, as was already outlined above, a drawback of this technique is that it cannot differentiate between aircraft types. A smaller aircraft would experience a smaller second moment under severe CAT encounters than a larger aircraft. The indirect and often incorrect assessment of turbulence has led many pilots to believe that the systems were rather unreliable for warnings of rough skies ahead [209].

A second product from AeroTech Research was the *Turbulence Auto-PIREP System* (TAPS), which performs automatic reports on turbulence encounters from aircraft, downlinks them to the ground, and disseminates the information to aircraft operators and ATC. The software-only system uses sensors and computers on aircraft as well as existing data-link networks. The TAPS reports are processed and disseminated by a server network in real-time to participating airlines seconds after the report was generated [724]. TAPS is an autonomous system that was created as a combination of non-flight critical software applications located in the aircraft computer and in the ground station information systems. Besides being able to report all significant turbulence encounters, it also allows the interpretation of the reports for dissimilar aircraft. The information is displayed for use in the cockpit and by ground-based personnel. The TAPS algorithms have three components. The first one is the hazard metric, $\sigma_{\Delta n}$, which is quantified by a 5-second windowed root mean square of the vertical acceleration and

■ $0.1g \leq \sigma_{\Delta n} < 0.2g$ (Light turbulence),

■ $0.2g \leq \sigma_{\Delta n} < 0.3g$ (Moderate turbulence),

■ $0.3g \leq \sigma_{\Delta n}$ (Severe turbulence).

This metric is continuously calculated during flight. The second component is referred to as reporting logic that determines when a TAPS report should be generated. The third component is report generation, which is processed once the value of the hazard metric exceeds a particular threshold. A TAPS report contains information on the time, the aircraft location and flight conditions, and the maximum accelerations experienced by the aircraft during the CAT encounter. The report also contains a parameter that may be used to scale reports to other aircraft and determine the potential hazard to that aircraft. An additional benefit of TAPS is that the system reports any encounters when the aircraft exceeds the vertical load acceleration limits that require an airplane structural examination,

---

[2]This is a measure of the second spectral moment estimate from the radar signal processor, which provides a reflectivity-weighted variance of the radial velocities and a measure of the shear or turbulence within the resolution volume.

as defined by the aircraft maintenance manual [722]. It has been demonstrated that the use of TAPS has introduced significant cost savings, which include about 98% reduction of airframe inspections due to CAT encounters, a significant reduction of unexpected CAT encounters and related injuries because of the improved awareness of turbulence for controller and dispatcher operations, reduction in aircraft cruise altitude changes due to uncertainties of the level of turbulence along the flight path, and fuel savings over many flight routes, which in turn translates into cost savings and lower emissions. These benefits have also led to a drastic reduction in flight delays as well as improved collaboration between pilots and dispatchers/controllers and assistance to meteorologists in validating and/or enhancing their weather forecasts. In addition, the technicians in charge of inspecting the airplanes use the TAPS reports to determine whether to inspect aircraft after CAT encounters and the NTSB is currently using TAPS data to perform accident investigations [724]. In 2005, an estimation of the eddy dissipation rate was added to the TAPS data stream in recognition of the importance to the forecast community and starting from 2010, the suite of cutting-edge flight planning and flight following tools of the *Weather Service International* (WSI) have integrated TAPS report data.

8.8 Automated airborne weather reporting

Meteorological observations were collected from in-flight aircraft since World War I in 1919. Since 2003, 130000 meteorological observations from aircraft per day were reported, including information on temperature, winds, and in some cases humidity, vertical wind gust, and/or eddy dissipation rates. Up to now, automated airborne weather reports have been an important data source for numerical weather prediction models. Such reports have been taken into account for several decades and in recent times forecasting models benefit from the use of onboard electronic devices [636]. In early times and previously to these new devices, the observations were reported on cylindrical charts that were retrieved only after the aircraft landing. The temperature, pressure, and relative humidity data were then read from the charts and disseminated [384]. For example, it is well known that humidity in the atmosphere is located below 25000 ft from the ground and is one of the key risk factors in the development of dangerous weather for airplanes. In general, observing systems often provide little data at relatively low altitudes. Therefore, sensors with a more sensitive detection range of meteorological parameters at low altitudes have been developed as will be described in the next sections.

Accurate and timely meteorological information is a crucial aspect for air operations to be safer and more efficient because knowledge of the location of possible dangerous weather, as well as its evolution over time (forecasting), allows for making correct rerouting decisions. In addition, improved forecasting and broadcasting of hazardous weather locations enable aircraft pilots to avoid atmospheric hazards, such as icing, CAT, and thunderstorms. The *Meteorological Data Collection and Reporting System* (MDCRS) collects position, temperature, and wind data and transmits them to the ground from participating jet transport aircraft, and sends the information to the U.S. NWS for input to forecast models. Nevertheless, the network of atmospheric soundings is limited to the U.S. space. On the other hand, aircraft operating at low altitudes, rather than those flying at cruise altitudes, are then able to have the potential to significantly contribute to the improvement of weather products through the collection and dissemination of in-flight weather observations. Moreover, the implementation of an automated and *in-situ* airborne weather reporting system will require the use of viable sensors and an extensive data-link communication network.

8.9   The quasi-common-path method

In general, radars can warn aircraft pilots of potential weather hazards during flight. However, they cannot detect all possible atmospheric hazards, such as, for example, CAT because it does not contain rain droplets and radars cannot sense dry phenomena. Conversely, aerosols and gases in the air can be detected from their emitted infrared spectral signatures. To this end, passive infrared radiometers have been designed based on high-resolution Fourier transform spectrometry technologies for satellite remote sensing. Nevertheless, these instruments have never been installed in aircraft for in-flight operations [703]. The feasibility of using such forward looking interferometers (FLI) to detect CAT and other invisible hazards during take-off, cruise, and landing has been evaluated since 2008 by research scientists in the *Georgia Tech Research Institute* (GTRI). As for aerosols and gases in the air, these interferometers can detect the presence of invisible hazards by identifying the different infrared spectral signatures emitted by each of them. These include CAT, wake vortices, volcanic ash, low visibility, dry wind shear, and icing. The GTRI scientists have also developed and evaluated algorithms for hazard prediction and severity estimation. The results obtained were in good agreement with the model predictions and the tests were successful in detecting for the first time mountain waves with interferometric radiometry. Further tests were conducted to investigate the ability of the interferometer to detect tornado-like turbulence patterns and wake vortices. In particular, it was realized that wake vortices associated with take-offs and landings have a distinctive signature in the high-resolution infrared radiance spectrum that will allow their interferometric detection. In fact, the GTRI team succeeded to resolve a wake vortex about 80 seconds after a DC-9 aircraft passed overhead. Another important aviation hazard is represented by the presence of volcanic ashes. For example, in 1982 a British Airways 747 lost function in all four engines because of volcanic ashes and more recently, 44 Alaska Airlines flights were canceled because volcanic ashes drifted into the flight paths. Other field tests conducted to evaluate the interferometric detection of dry wind shear by examining temperature and humidity differences and ice on runways all produced successful results. Therefore, the combination of high spectral and temperature resolutions in the FLI will then enable sophisticated algorithms with high detection rates and low false alarm rates. This instrument can also work as an infrared imager and can provide real-time video displays during the night and in obscured conditions [968]. In addition, FLI not only provides highly resolved spectral measurements in the infrared but also the visible and ultraviolet bands of the electromagnetic spectrum [933]. It enables efficient light collection from extended light sources with weak radiation [433] and can perform multiplexed measurements – the property known as the Fellgett advantage [288] – so that the signal-to-noise ratio of the retrieved spectrum is improved.

The NASA Langley Research Center supported the FLI project to assess whether an FLI sensor can be used to detect and measure invisible hazards from in-flight aircraft with sufficient capability to improve flight safety [850]. For this purpose, the FLI concept was based on an aircraft-mounted Michelson Fourier transform spectrometer (FTS) capable of measuring atmospheric conditions ahead of the aircraft, performing hazard detection, and/or collecting any other information related to flight deck requirements [345]. Figure 8.9 shows a sketch of airborne remote CAT detection using the FLI concept. Since its introduction as a detection tool, it was long expected that the FLI project will finally provide a technologically advanced means to obtain accurate measurements of CAT on board that would alert pilots of potential weather hazards during all phases of flight [967]. The feasibility of detecting several types of aviation hazards in flight with an FLI was investigated more recently by Daniels et al. [219]. Their goal was to develop measurement methods and algorithms capable of detecting wake vortices in real time from either an aircraft or ground-mounted hyperspectral FTS instrument. They found that an imaging infrared instrument sensitive to five narrow bands in the spectral region between 670 and 3150 cm^{-1} will be sufficient for the detection of wake vortices. To explore the CAT sensitivity, further studies included the development of a high spectral resolution spherical shell radiative transfer model [968]. This model was based on a line-by-line

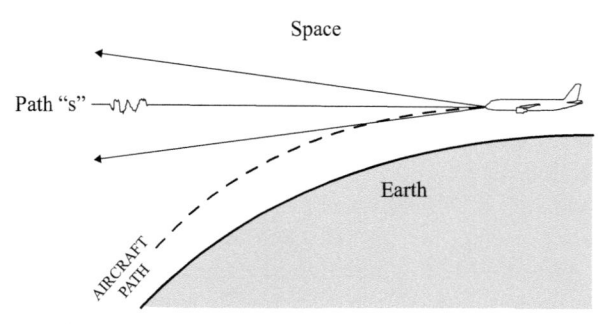

Figure 8.9: Schematic drawing showing the airborne application of weather hazard detection using the FLI concept.

radiative transfer model for simulating radiances from hyperspectral sounding instruments [187, 186]. These tests were performed at some distance ahead of the aircraft, as depicted in Fig. 8.9, using the model developed by West et al. [968]. In this model, the radiance R is calculated according to

$$R(v,A,h_0) = \int_0^{ToA} B[v,T(s)] \left(-\frac{\partial \tau(v,s,A)}{\partial s} \right) ds, \tag{8.7}$$

where ToA is the top of atmosphere, v is the frequency, A is the atmospheric state consisting of pressure p, temperature T, absorbing constituents (as, for example, the molecular constituents H_2O, CO_2, O_3, N_2O, CO, CH_4, O_2, NO, SO_2, NO_2, NH_3, HNO_3, clouds, and aerosols), h_0 is the height of the aircraft, and B is the frequency-dependent Planck function. The transmittance of the atmosphere between the instantaneous position of the aircraft and any position along the viewed path, s, is given by

$$\tau(v,s,A) = \int_0^s \exp\left[-\sigma(v,s',A)\right] ds', \tag{8.8}$$

where σ is the optical depth. The coordinate along the viewed path, s, is calculated as a function of height using the expression

$$s(h) = \left[(R_\oplus + h)^2 - (R_\oplus + h_0)^2 \right]^{1/2}, \tag{8.9}$$

where $R_\oplus$ is the Earth's radius and h denotes height from the ground.

The detection of atmospheric gases is based on the principle embodied by the Stefan-Boltzmann law, which states that the radiant heat energy emitted from a unit area in one second (i.e., the power per unit area), $P = \sigma_{SB}T^4$, where T is the absolute temperature (in kelvins) and $\sigma_{SB} = 5.670374419 \times 10^{-8}$ W m^{-2} K^{-4} is the Stefan-Boltzmann constant. While the Stefan-Boltzmann law assumes that the emissivity, $\varepsilon = 1$, real materials have emissivities $\varepsilon < 1$, which may also be spectrally dependent. The emissivity of gases is usually very low, except in certain spectral regions, known as absorption bands, where it is high as a consequence of the thermodynamic principle that the absorptivity is equal to the emissivity [39]. Within these spectral absorption bands, a thick slab of gas will emit exactly like a blackbody at the same temperature as the gas and obey the Stefan-Boltzmann law. This way, the gas temperature can be determined by radiometric measurements of the infrared energy emitted in appropriate spectral bands. For example, considering that the spectral emission of CO_2 in a spectral radiant plot appears centered at wavelengths 4.3 μm and 15 μm, it is then possible to build up a temperature profile along the line of sight over quite long distances ahead of the aircraft [39]. As was reported by West et al. [968], the investigations on the sensitivity for CAT detection included sensing temperature fluctuations in the wavenumbers of 6.5-7.8 m^{-1} in the CO_2 band and sensing water vapor fluctuations in the wavenumbers between 13 and 16 m^{-1}. Their simulations revealed detectable signatures in both the interferogram and spectral domains for the CO_2 temperature fluctuations at cruise altitudes. However, the maximum range appeared to be about 5 km, which was not as large as was claimed by early researchers. In addition, wavelengths between 5 and 8 μm in the spectral region were also used to detect water vapor disturbances [850]. For example, humidity anomalies in the case where data exist were detectable from 30 km, or with a warning time of about 2 min, assuming a lower instrument sensitivity of 1 K. On the other hand, wake vortices may well be detectable because of the entrained water vapor from the aircraft engine exhaust.

It is well known that interferometry is an optical methodology based on the principle of wave superposition. For instance, when two (electromagnetic) waves with the same frequency are combined, the resulting intensity pattern is determined by the phase difference between the two waves. The conditions under which wave superposition produces constructive (in phase) or destructive (antiphase) interference are very well described by the wave theory. However, not always the interference is totally constructive or destructive. In most general cases, the phase difference is such that intermediate interference patterns arise and interferometry uses this kind of pattern to evaluate the phase difference between the waves [382]. There are many kinds of interferometers that have been designed for specific applications. In the most general type of interferometers, both beams (the reference and test) follow widely separated paths. However, they present some problems associated with possible mechanical shocks and temperature fluctuations, thereby producing unstable fringe patterns in the observation plane, which makes it very difficult if not impossible to perform proper measurements [586]. However, the interferometers based on the common path method avoid this difficulty. With these interferometers, the reference and test beams traverse the same general path, and as an additional advantage, they do not require perfect optical components (the master) with the same dimensions as those of the system under test for producing the reference beam. The path difference between the two beams in the center of the field of view is, in general, zero, making the use of white light perfectly possible. Also, in some devices, the reference beam is directed through a small area of the optical system under test, which causes the system to be unaffected by aberrations. For example, Mallick and Malacara [586] explain that since the reference beam interferes with the test one, which has traversed the full aperture of the optical system, information about the system defects can be obtained. Also, the width of the beam can be divided if it is partially scattered with a refractive crystal or a semi-reflective surface. Because of its advantages, the quasi-common-path method has been implemented in interferometers. For example, Bai et al. [55] reported a quasi-common-path point diffraction interferometer (PDI) that allows fringe contrast and fringe spatial frequency to be adjusted conveniently. This device is a self-referencing interferometer wavefront sensor with the high spatial resolution because each pixel corresponds to a sub-aperture in the interferogram.

Due to the high demand for precision, a technical problem remains when targeting CAT, since noise associated with it limits the accuracy when the measuring range is wide [431, 903]. Unfortunately, mathematical algorithms cannot be used to either reduce or eliminate the influence of turbulence. However, there exist some other ways to reduce the error caused by CAT, such as, for example, the common path compensation method, which allows a reduction of the disturbance noise in adaptive optics. In particular, a quasi-common-path method to improve the stability of laser measurement instruments was reported by He et al. [389]. This method releases the strictness of the common-path structure and was found to compensate for the errors caused by CAT. On the other hand, integrated circuits of radio frequency as well as modular microwave components have opened the possibility for many different kinds of compact microwave systems. For example, McBeth and Jones [604] presented the analysis and design of a compact interferometric radar at 10 GHz (X-band) for sensing CAT, which was used in a *Naval Innovative Science and Engineering* (NISE) research project.

8.10 Sensor systems

Since the late 1930s, altitude data have been recorded using balloon-launched radiosondes. Although they are still in use today and provide useful information, they do not guarantee large spatial and temporal coverage. In addition, they are limited to only two launches per day, at 00:00 and 12:00 hours, in 92 stations around the United States, including Alaska, Hawaii, and Puerto Rico. However, this method leaves behind a large amount of uncovered data. To obtain more complete meteorological information, the U.S. *National Aviation Meteorology Program Council* (NAMPC) has recommended expanding and institutionalizing the generation, dissemination, and use of automated PIREPs for the entire spectrum of the aviation community, including general aviation. Because of this recommendation, NASA began cooperative research with the industry to develop an electronic pilot notification capability for small aircraft [217]. The research was focused on developing a small low-cost sensor capable of collecting useful meteorological data, downlinking the data in near real-time, and using the data to improve weather forecast by making observations below altitudes of 25000 ft, to provide information to air traffic controllers, flight service stations, airline weather centers, and pilots. Although there were some devices like the *Meteorological Data Collection and Reporting System* (MDCRS), which was already in use on commercial transport aircraft, many of the collected information was only for use by forecast modelers. This last instrumentation was further improved by developing the *Tropospheric Airborne Meteorological Data Reporting* (TAMDAR) [217], which was designed as a robust, compact, lightweight, and integrator sensor system. In particular, this sensor was able to perform automated measurements and reports of humidity, pressure, temperature, wind, magnetic heading, turbulence and averaged turbulent energy dissipation, and location of aircraft in flight [876]. It can also compute pressure altitude as well as indicate air speed, peak turbulence, ice accretion rate, winds, and turbulence scale. The data from these aircraft-based observations are commonly referred to as *Aircraft Meteorological Data Relay* (AMDAR) reports.

TAMDAR is a network of aircraft-based observations that was introduced by AirDat, LLC [218]. Although the TAMDAR sensor is intended for commercial purposes, it is yet affordable for smaller aircraft as well. Compared to AMDAR, it has several potential advantages. For example, AMDAR observations are likely to be present below 20000 ft above the ground level only around major airports [635], whereas TAMDAR observations are generally performed on smaller commercial aircraft. These can use smaller airports so that the potential of TAMDAR observations is to provide a much wider coverage below 20000 ft than other aircraft observations. A further advantage of TAMDAR over AMDAR is that it was designed to measure water vapor. In general, the TAMDAR sensor can perform direct measurements of pressure, temperature, humidity, heading, and ice, while the pressure altitude in two ranges (0–25000 ft) and (25000–50000 ft), the indicated airspeed (70–270 knots), the true airspeed (70–450 knots), the eddy dissipation rate ($\varepsilon^{1/3}$), and the winds aloft (peak and median) can be calculated as derived parameters. The probe body has the shape of a symmetric airfoil with a span of 4.05 inches and a cord of 2.6 inches. It is equipped with an internal electronic module of volume equal to 36 cubic inches. The sensor is designed as a probe that is installed in the external part of the aircraft with an attached signal processing unit. Both the probe and the electronic module are directly connected. These two components are mounted together at a single location either on the wing or fuselage and have an approximate weight of 1.5 lbs and aerodynamic drag of 0.4 lb at 200 knots [217]. The dynamic pressure is sensed via a port protruding from the leading edge, while the static pressure is measured via a port located on the trailing edge of the sensor body. An ambient air temperature sensor (compensated for Mach heating) is used to calculate true and indicated airspeed as well as pressure altitude.

The sensing of turbulence has been a longstanding problem for the aviation industry. In particular, Stickland [871] proposed that ideally turbulence should be measured in an objective, aircraft independent fashion. However, it has not been easy to establish an accurate scale for turbulence like, for instance, the Richter scale, which quantifies the energy in an earthquake. A turbulence metric should describe the turbulence energy state with a standardized, precise value [63]. For example, MacCready [581] proposed a universal turbulence standardization technique based quantitatively on the atmospheric turbulence rather than on the effects it produces on an aircraft. This method provides a single turbulence intensity number that can be measured continuously in flight in a variety of ways. In addition, taking into account the aircraft type and speed, it can be linearly related to the rms value of the aircraft's vertical acceleration. The development of the TAMDAR sensor includes the possibility of comparing both the acceleration-based method and the eddy dissipation rate method. The input from an external Heading Accelerometer/Attitude Module could also be used to quantify the turbulence-induced acceleration experienced by a specific aircraft. An additional algorithm is employed to compute the eddy dissipation rate as an aircraft independent measure of turbulence. Relative humidity is measured by allowing air to flow through a tube directly into a sensing cavity and using a thin-film (polymer) capacitor. The airflow from the sensor cavity is then discharged through holes near the base of the sensor. A leading edge notch incorporates infrared transmitters and detectors for ice detection, while a built-in GPS provides time and location for each observation as well as ground tracking, which is used with externally provided heading information to calculate winds aloft. All observation intervals are based on static pressure (altitude) with a timed default. The signal-processing unit computes the derived parameters from basic measurement data, which then serves as output to a data-link transceiver. The power consumption with the de-icing heaters powered off is 10 W and with the heaters on is 280 W. The sensor mounted under a wing is connected to a Heading and Accelerometer/Attitude Module, an optional cockpit annunciator for icing, a GPS receiver, and a data-link transceiver.

The performance of the TAMDAR sensor was verified by conducting ground-based and in-flight tests. The ground-based testing was performed using wind tunnels and environmental chambers at the *LeClerc Icing Research Laboratory* in New York City in December 2003. In this case, the ice detection capabilities of the sensor were verified by taking into account five parameters, namely water concentration, ambient static temperature, true airspeed, water droplet MVD, and response time, under three different conditions [220]. On the other hand, flight testing was accomplished using different atmospheric research aircraft. Computational fluid dynamics simulations were complementarily conducted to determine optimal locations for pressure ports and nozzle shapes. The results obtained were satisfactory enough to guarantee the installation of the sensor in aircraft. For a detailed description of these results, the interested reader is referred to the article by Daniels et al. [220]. However, Gao et al. [323] raised some important considerations about the TAMDAR operation limits. They estimated that TAMDAR errors are comparable to those of rawinsondes' rainfall for temperature. The errors appeared to be a little worse for wind speeds less than 15 m s^{-1} and a little better for speeds greater than 15 m s^{-1}, and perhaps a little better for humidity [747].

Beyond its use in commercial aviation, the TAMDAR sensor has also been of significant interest to the Army because its use guarantees a source of meteorological information in environments where available observations are very limited. The U.S. Army Research Laboratory and the Physical Sciences Laboratory's Center for Technical Analysis and Applications of the New Mexico State University was engaged to continue research work on the sensor. As a result, the AirDat LLC developed a modified TAMDAR sensor, which was called *TAMDAR-Unmanned Aerial System* (TAMDAR-U) for use in unmanned aerial systems (UAS) [691]. The data produced during the testing phase of this new device was assimilated by the advanced research version of the *Weather Research and Forecasting* model (WRF-ARW) [841], using its observational thrust capability and nested domains with a horizontal grid spacing of 3 and 1 km. This project was focused on applications of the WRF-ARW model to domains centered over southern California in early 2012, with the main purpose of investigating how to better assimilate single-level, above-surface observations (e.g., aircraft-based observations) as well as the relative value of various observation types [747]. The results indicated that the inclusion of the observations showed substantial day-to-day variability in the forecast improvements. It was also found that the inclusion of the observational data improves the temperature and water-vapor predictions, but degrades short-term (1 to 6 h) wind forecasts. In contrast, the use of local rawinsonde data improves short-term temperature and wind forecasts, but degrades the predictions on water vapor. On the other hand, the introduction of denser surface observations improves the surface forecasts of temperature, moisture, and wind, but above surface the results may be degraded.

8.11 Weather information communication

Weather conditions can seriously affect aircraft operations and the levels of service available to system users. Therefore, how the weather is observed, forecasted, disseminated, and ultimately used to make air traffic management (ATM) decisions are of critical importance to the operations in the U.S. National Airspace System (NAS) and international airspaces, especially in the oceanic domains.

As the United States moves towards significantly increasing the capacity of the NAS through the implementation of the NextGen, the integration of the weather information (and associated uncertainty) into the ATM decision-making process is of ever-increasing importance. Moreover, harmonization around the globe with partners, such as the *Single European Sky ATM Research* (SESAR) and the *Collaborative Action for Renovation of Air Traffic Systems* (CARATS) in Japan, plays an important role as well. In particular, NCAR/RAL contributes in various ways to these efforts by developing aviation weather hazard guidance products and proper means for their dissemination as well as the collaboration with users of these products, including the assistance with integration into decision support tools. In addition, RAL participates in many outreach and education activities.

The efficient communication of weather information allows for timely sharing of data and information between the ground and air domains and the information transfer between aircraft. Current investigation of the communication requirements and associated data-link architectures for the optimal delivery of graphical weather products to general aviation and commercial-air-transport cockpits are in continuous progress to satisfy mid- and long-term communication requirements.

Chapter 9

Modern Methods and Techniques for CAT Detection

9.1 Introduction

Over the last ten years or so, scientists and engineers have been working in the design and development of new technologies for more efficient and accurate detection and warning of wind shear, CAT, and volcanic ashes. For example, Boeing has recently been testing a new technology to help pilots detect turbulence in time and adjust the cruise altitude to avoid it, and halve aircraft accidents due to CAT encounters. The Japan Aerospace Exploration Agency (JAXA) has developed a technology on board the aircraft capable of delivering a warning up to a minute before impending CAT. This device, based on a liquid-cooled infrared laser called LIDAR, has been designed to be integrated into a forward display and warn the pilot about the intensity of turbulence and exactly when it will hit by just detecting tiny particles in the air (aerosols) several miles ahead of the aircraft. On the other hand, some American Airlines planes are now flying with sensors connected to a tracking system called *Total Turbulence*. This system can measure intensity levels of turbulence around an airplane and convey the observations in real-time to a data center, which then transmits warnings to aircraft flying over the same route. A new weather radar system, called *MultiScan ThreatTrack*, designed to warn pilots on moving thunderstorm cells into their aircraft routes and detect two different levels of CAT, is also planned for use in American Airlines. Moreover, in conjunction with the NWS, the Southwest Airlines are exploring the performance of new water vapor sensors on its 737 planes, which provide moisture readings at different altitudes during the day. The data from these sensors seem to be good candidates for replacing those coming twice a day from the balloons sent by the NWS into the sky at 100 locations around the United States.

In this chapter, a thorough overview is provided on recent methods and technologies that have been developed and tested over the last ten years for CAT detection as well as other invisible phenomena, such as wind shear and volcanic ashes. Some of these advances comprise methods and techniques that have been either extended and improved over the years or developed as new. They include airborne devices for turbulence measurements, ground-based Doppler radars for remote turbulence detection as well as the JAXA project, the *Airborne Coherent LIDAR for Advanced In-flight Measurement* (ACLAIM) project at NASA's Dryden Flight Research Center, and the European joint projects *Demonstration of LIDAR-Based CAT Detection* (DELICAT) and *Aircraft Wing with Advanced Technology Operation* (AWIATOR). For completeness in the exposition, the KMW method and the ULTURB algorithm for CAT forecasting along with the NCAR/NEXRAD turbulence detection algorithm will also be reviewed. The chapter ends with an overview of modern optical airflow meters and small unmanned aircraft systems.

9.2 Airborne measurements of turbulence

Since the mid-90s, efforts have been addressed to provide adequate real-time, comprehensive, and direct atmospheric turbulence measurements using fully automated algorithms that have been primarily designed to run on commercial airplanes. Such algorithms were operated using available data sources and computational capabilities. With this technology, it was possible to obtain estimates of the eddy dissipation rates from vertical aircraft accelerations and maps of detected CAT, replacing the intermittent and subjective PIREPs by providing increasingly more accurate warnings to pilots and air traffic controllers. One of the first methods for computing a CAT intensity metric for *in-situ* aircraft measurements was reported by Cornman et al. [205], which was based on a power-spectral, linear-system frequency response analysis of aircraft response to turbulence. Airborne instruments for *in-situ* detection of atmospheric aerosol particles, clouds, and radiation and related emerging technologies that have been in use on research aircraft around the world are overviewed by Baumgardner et al. [69]. Moreover, airborne laser sensors for civil and military applications include LIDAR, Laser Range Finders (LRF), and Laser Weapon Systems (LWS) [786]. More recently, the Multi-Purpose Airborne Sensor Carrier (MASC-3) for wind and CAT measurements in the atmospheric boundary layer has been validated by comparing its performance with meteorological towers and sodar systems [743]. Very recent *in-situ* measurements of wind and turbulence by a motor glider in the Andes provide a good example of how collecting data from soaring flights might be used for CAT detection in the upper troposphere and lower mesosphere [975].

9.2.1 In-situ measurements

Airborne *in-situ* measurements of turbulence refer to calculations of the turbulent kinetic energy or the energy dissipation rate (EDR) in terms of measurable flow variables, such as flow velocities, pressure, and temperature. In this sense, these turbulence intensity metrics are not directly measurable flow quantities. According to Cornman [202], *in-situ* measurements differ from forward-looking measurements, such as those from Doppler radars or LIDARs, in that the latter can be used for real-time warnings of impending CAT encounters, while the former can be employed for warning following aircraft, used in CAT forecasting and nowcasting algorithms [824]. It has long been recognized that the EDR, which refers to $\varepsilon^{1/3}$, can be used as a measure of turbulence intensity, while the absolute value of the peak vertical acceleration, namely $|\ddot{z}_{\text{peak}}|$, and its root mean square (RMS) value, $\sigma_{\ddot{z}}$, represent useful measures of aircraft response to turbulence [202]. The division of RMS by 1-g is often denoted by RMS-g, while the so-called derived equivalent gust velocity, U_{d}, is an estimate of the amplitude of the 1-cosine gust profile from the measured value of $|\ddot{z}_{\text{peak}}|$. In general, measurements of the atmospheric turbulence intensity can be obtained independently of the aircraft response to turbulence, but the converse is not true since the aircraft response is a function of the forcing mechanism. In mathematical form, this translates into a linear system relation, where the atmospheric turbulence is the input and the aircraft vertical acceleration response is the output. This concept can be put in more concrete form using the example provided by Cornman [202]. That is, when an aircraft, say A, performs a measurement of EDR and transmits the information to another aircraft, say B, the latter can either use the information as it is received, or it can translate it into a response metric, such as $\sigma_{\ddot{z}}$, from knowledge of its response function. Conversely, aircraft A can measure $(\sigma_{\ddot{z}})_A$ and transmit it to B. In this case, if aircraft B wants to convert it into EDR or $(\sigma_{\ddot{z}})_B$, it would also require to have information of the response function of A.

To simplify the modeling of turbulence, it has been a common practice to assume that turbulence is homogeneous and isotropic. The former assumption implies that the statistical properties of turbulence are invariant under translations of the coordinate system and the associated basis vectors, while isotropic turbulence means invariance under rotations of the coordinate axes. In contrast, inhomogeneous fields are referred to as discrete turbulence, as opposed to coherent structures, such as vortices or small-scale shears. However, the above simplifications can be improved by considering a limited view of inhomogeneous turbulence, which consists of locally homogeneous fields, whose intensity is allowed to vary smoothly, but sufficiently rapidly, with position [592]. On the other hand, the turbulent motion is said to be stationary when its statistical properties at a point remain the same over time. Since typically an aircraft goes through a turbulent area much faster than the field is allowed to evolve, Taylor's frozen hypothesis[1] [404] can be invoked so that a time series measured at a single point can be interpreted as purely spatial variations. This hypothesis works only when the relative turbulence intensity, calculated as the ratio of the eddy velocity, u, over the mean velocity, U, is small (i.e., $u/U \ll 1$). Then the substitution $t = x/U$ is a good approximation. In applications to operational *in-situ* measurements of turbulence $x = Vt$, where V is the aircraft airspeed. This allows to study the spatial properties of the velocity field from the time series of airborne measurements.

In the following, the discussion will focus on the modulated homogeneous turbulence model along with an empirical approach to EDR measurements from inhomogeneous turbulence described by Cornman [202]. For a homogeneous wind field, the time variation of the turbulent kinetic energy per unit mass, K_{turb}, is given by Eq. (1.59), which in Lagrangian coordinates can be rewritten as

$$\frac{dK_{\text{turb}}}{dt} = -\left(K_{\text{turb}} + \frac{p}{\rho}\right)\nabla \cdot \mathbf{v} - \mathbf{v} \cdot \nabla \left(\frac{p}{\rho}\right) + \mathbf{v} \cdot \mathbf{F} - v|\nabla \mathbf{v}|^2, \quad (9.1)$$

where the first term on the right-hand side is the variation of the kinetic energy due to advection in the mean flow, while the second, third, and fourth terms are, respectively, the variations of the turbulent kinetic energy due to pressure and density gradients in the fluid, to small-scale motions in the flow induced by the presence of a force acting on the fluid, and to losses due to viscous heating, which according to Eq. (1.60) quantifies the EDR. For homogeneous and isotropic turbulent velocity fields, Eq. (1.60) can be written as

$$\varepsilon = \frac{15}{2}v\left\langle \frac{\partial v_i}{\partial x_j} + \frac{\partial v_j}{\partial x_i} \right\rangle, \quad (9.2)$$

where the $i, j = 1, 2, 3$ subscripts denote velocity and coordinate components. However, in most airborne applications the EDR for a homogeneous and isotropic field is calculated as

$$\varepsilon = v \int_0^\infty k^2 E(k) dk, \quad (9.3)$$

where k is the wavenumber in units of rad m^{-1} and $E(k)$ is the three-dimensional energy spectrum for isotropic turbulence [404]. For sufficiently large values of k, the von Kármán energy spectrum matches asymptotically the Kolmogorov energy spectrum given by Eq. (1.70). When inserting this dependence into Eq. (9.3), the integral diverges as $k \to \infty$ because, as shown in Fig. 1.8, the Kolmogorov spectrum is valid only in the inertial subrange where $E(k) \propto k^{-5/3}$. For

---

[1]This hypothesis refers to the assumption that the contribution of turbulent circulations to fluid advection is small. Therefore, the advection of a turbulent field past a fixed point can be assumed as due entirely to the mean flow. In other words, the statistical properties of turbulence are not changed during the advection process.

large k, i.e., within the dissipation range, the wavenumber dependence of the energy spectrum strongly diverges from the $k^{-5/3}$ variation. However, the problem can be solved by simply demanding that the von Kármán spectrum matches the Kolmogorov one for large k. In particular, the von Kármán energy spectrum is given by [404]

$$E_{\mathrm{vK}}(k) = \frac{55}{9\pi} \frac{\sigma^2 L (k/k_0)^4}{[1 + (k/k_0)^2]^{17/6}}, \tag{9.4}$$

where

$$\sigma^2 = \frac{2}{3} \int_0^\infty E(k)dk, \tag{9.5}$$

is the velocity variance for a single component,

$$L = \frac{\pi}{2\sigma^2} \int_0^\infty k^{-1} E(k)dk, \tag{9.6}$$

is the integral length scale, and

$$k_0 = \frac{\sqrt{\pi}}{L} \frac{\Gamma(5/6)}{\Gamma(1/3)}, \tag{9.7}$$

where Γ is the gamma function. The power series expansion of Eq. (9.4) produces a $k^{-5/3}$-dependence for the von Kármán spectrum and equating this with Eq. (1.70) gives for the EDR the expression

$$\varepsilon^{1/3} = \frac{\sigma}{L^{1/3}} \left(\frac{55}{9C_2 \pi^{1/6}} \right)^{1/2} \left[\frac{\Gamma(5/6)}{\Gamma(1/3)} \right]^{5/6}, \tag{9.8}$$

where C_2 is the same proportionality constant in Eq. (1.70). Equation (9.8) shows that the EDR is directly proportional to σ, i.e., to the RMS of the turbulent velocities, and since σ is in turn directly proportional to the RMS of the aircraft vertical accelerations, these proportionalities are good reasons for using the EDR as a turbulence measurement metric for aviation.

Cornman [202] also describes a method for calculating EDR from a simplified class of inhomogeneous turbulence, called "modulated homogeneous turbulence". First, assume the x-direction as the flight path and consider the transverse velocity component along this same direction. The one-dimensional velocity power spectrum is then given by the Fourier transform of the associated velocity autocorrelation function, which for a homogeneous and isotropic field, is defined by the ensemble average

$$R_v(D) = \langle v(x)v(x+D) \rangle, \tag{9.9}$$

where D is the length of the displacement vector, $D = \mathbf{D} \cdot \mathbf{x}$, where $\mathbf{x}$ is the unit vector in the x-direction. The Fourier transform of the $R_v(D)$ defines the spectrum of v, namely

$$\Phi_v(k_x) = \frac{1}{2\pi} \int_{-\infty}^{\infty} R_v(D) \exp(ik_x D) dD, \tag{9.10}$$

where k_x is the x-component of the wave vector $\mathbf{k}$. It can be demonstrated that for large values of k_x (within the inertial subrange), the one-dimensional von Kármán spectrum of the homogeneous and isotropic wind field obeys the asymptotic form

$$\Phi_v(k_x) \to \frac{12}{55} C_2 \varepsilon^{2/3} k_x^{-5/3}, \tag{9.11}$$

which is proportional to the Kolmogorov spectrum. A window function must necessarily be employed in practical applications where only a finite sequence of data are available [686, 592]. Denoting X as the length of the truncation window, the truncated modulated and windowed data can be defined as $u(x) = w(x)m(x)v(x)$ if $|x| \le X$ and 0 otherwise, where $w(x)$ is the window function, $m(x)$ is the modulation function, and $v(x)$ is the homogeneous turbulence data. In this case, the autocorrelation for the modulated and windowed data is given by

$$R_u(D, X) = R_v(D) R_{w,m}(D, X), \tag{9.12}$$

where $R_{w,m}(D, X)$ is the correlation of the combined modulation and window functions. Under the assumption that these functions are deterministic and from the convolution theorem, it can be demonstrated that the Fourier transform of a product is equal to the convolution of their Fourier transforms. Therefore, the Fourier transform of Eq. (9.12) reduces to

$$\Phi_u(k_x, X) = \int_{-\infty}^{\infty} \Phi_{w,m}(\alpha, X) \Phi_v(k_x - \alpha) d\alpha, \tag{9.13}$$

where away from $\alpha = 0$, $\Phi_{w,m}(\alpha, X) \to 0$ and this occurs more quickly for larger X. For $k_x \gg \alpha$,

$$\Phi_v(k_x - \alpha) \approx \Phi_v(k_x) + O\left[\left(\frac{\alpha}{k_x}\right)^n\right], \tag{9.14}$$

where $n \ge 1$ is an integer and hence from Eq. (9.13) it follows that

$$\Phi_u(k_x, X) \approx \Phi_v(k_x) \int_{-\infty}^{\infty} \Phi_{w,m}(\alpha, X) d\alpha, \tag{9.15}$$

for large k_x. Moreover, since $\Phi_{w,m}(\alpha, X)$ is the Fourier transform of $R_{w,m}(D, X)$, the integral in Eq. (9.15) can be shown to be equal to $R_{w,m}(0, X)$. Therefore, from Eq. (9.11) it follows that

$$\Phi_u(k_x, X) \approx \frac{12}{55} C_2 R_{w,m}(0, X) \varepsilon^{2/3} k_x^{-5/3}, \tag{9.16}$$

for large k_x. This shows that the spectrum of the modulated and windowed homogeneous field is a scaled version of the homogeneous spectrum given by Eq. (9.11). Hence, the EDR for modulated turbulence can be defined as

$$\varepsilon_u^{1/3}(X) = \sqrt{R_{w,m}(0,X)}\,\varepsilon_v^{1/3}, \tag{9.17}$$

for large k_x, implying that the von Kármán homogeneous spectrum and the modulated turbulence spectrum have the same functional dependence. Since the EDR calculation for inhomogeneous turbulence occurs over a window of finite extent, the resulting EDR value will be a function of the window length X, via the correlation function $R_{w,m}(0,X)$.

The above analysis was performed over a continuous spatial domain. However, in actual airborne applications, the analysis must be performed on discrete samplings and finite window lengths in the temporal domain. If ΔT is the discrete sampling time interval, the Taylor hypothesis consists of setting $\Delta x = V\Delta T$, where V is the true air velocity averaged over the sampling window. Following the description of Cornman [202], in the time domain, the modulation function is defined as

$$m(q) = 1 + A_1 \exp\left[-\pi\left(\frac{q\Delta T - t_0}{T_\sigma}\right)^2\right], \tag{9.18}$$

where t_0 is the center of the temporal window, A_1 is a dimensionless amplitude, $T_\sigma = L_\sigma/V$, and L_σ is the spatial width for the modulation. The power spectrum of the modulated and windowed homogeneous velocities is given by

$$S_u(p) = 2\Delta T N_w \sum_{q=-M+1}^{M-1} R_{w,m}(q)R_v(q)\exp\left(\frac{2\pi i q p}{M}\right), \tag{9.19}$$

for $p = 0, 1, \ldots, M-1$. In this expression, N_w denotes the normalization factor of the window function defined according to

$$N_w = \left[\frac{1}{M}\sum_{q=0}^{M-1} w^2(q)\right]^{-1}. \tag{9.20}$$

The velocity variance is given by

$$\sigma_u^2 = \Delta f \sum_{p=0}^{M-1} S_u(p), \tag{9.21}$$

where $f = p\Delta f$ is the linear frequency and $\Delta f = 1/(M\Delta T)$. A graphical example illustrating the spectral scaling method is given in Fig. 5.1 of Cornman [202]. For large sample sizes or uncorrelated data, the EDR can be approximately calculated from the measured power spectra using the maximum likelihood method [824] so that

$$
\varepsilon^{1/3} = \left[\frac{1}{(p_2 - p_1 + 1)} \sum_{p=p_1}^{p_2} \frac{\hat{S}_u(p)}{S_u(p)} \right]^{1/2}, \tag{9.22}
$$

where $\hat{S}_u(p)$ is the estimated power spectrum, $S_u(p)$ is the expected periodogram given by Eq. (9.19) with $\varepsilon = 1$, and p_1 and p_2 are the lower and the upper cutoff frequency indices, respectively. It must be noticed that if the modulation function is known and the expected periodogram is included, Eq. (9.22) provides an estimate of $\varepsilon_v^{1/3}$. However, for operational purposes the interest is to estimate $\varepsilon_u^{1/3}$ rather than $\varepsilon_v^{1/3}$, so that the expected periodogram should correspond to the windowed homogeneous data and not that with the modulated function.

9.2.1.1 Aircraft response to turbulence

In actual aircraft operations, the aircraft acceleration response to turbulence is along its vertical body-axis, which is mainly due to the vertical wind velocity component. It is commonly assumed that to a good approximation, the aircraft/autopilot system behaves as a linear one, thereby allowing the use of the linear system theory. For example, the output spectrum in the frequency domain, $\Phi_{out}(f)$, is linearly related to the input spectrum, $\Phi_{in}(f)$, through the modulus square of the response function, $H_i^o(f)$ [82], which is related to the Fourier transform of the correlation function of the vertical accelerations. These are given by the convolution of the aircraft response function, namely $h(t)$, such that

$$
\ddot{z}(t) = \int_0^\infty h(t - \tau) v(\tau) d\tau. \tag{9.23}
$$

From the vertical wind as input, the power spectrum of the vertical acceleration as a function of linear frequency is given by

$$
\Phi_{\ddot{z}}(f) = |H_{wind}^{\ddot{z}}(f)|^2 \Phi_{wind}(f), \tag{9.24}
$$

where the wind spectrum expressed in terms of the wavenumber is assumed to be

$$
\Phi_{wind}(k) = \varepsilon^{2/3} \Psi_{wind}(k), \tag{9.25}
$$

which has to be converted to linear frequency using the relation $k = 2\pi f/V$ so that $\Phi_i(f) = (2\pi/V)\Phi_i(k)$. Using Eqs. (9.24) and (9.25), it can be easily demonstrated that the variance of the vertical accelerations is

$$
\hat{\sigma}_{\ddot{z}}^2 = \int_0^\infty \Phi_{\ddot{z}}(f) df = \varepsilon^{2/3} I, \tag{9.26}
$$

where the integral

$$
I = \int_0^\infty |H_{wind}^{\ddot{z}}(f)|^2 \Psi_{wind}(f) df, \tag{9.27}
$$

encapsulates all the information regarding the aircraft response and the wind spectrum. The square root of Eq. (9.26) provides an estimate of the EDR in terms of measured RMSs of the aircraft's vertical accelerations

$$\hat{\varepsilon}^{1/3} = \frac{\hat{\sigma}_{\ddot{z}}}{I^{1/2}}, \tag{9.28}$$

where effectively I contains information on the aircraft and operating conditions, such as, for example, airspeed, aircraft altitude and weight, autopilot mode, etc. The relation between the EDR and the RMS g-values is obtained by dividing the left-hand side of Eq. (9.28) by the acceleration due to gravity, g. For example, if two aircraft, say A and B, are flying through a turbulent area with the same EDR, then from Eq. (9.28) it follows that

$$\frac{(\sigma_{\ddot{z}})_A}{(\sigma_{\ddot{z}})_B} = \frac{(\sqrt{I})_A}{(\sqrt{I})_B}. \tag{9.29}$$

This relation shows that the relative aircraft response is equal to the ratio of the square root of the integral factors of both aircraft.

9.2.1.2 *Aircraft response function*

An accurate calculation of the aircraft vertical acceleration response to continuous turbulence is obtained using a short-period, two degree-of-freedom (2-DOF) model [415, 205]. In this model, an aircraft is assumed to be a rigid body with two degrees of freedom (i.e., vertical motion, z, and pitch angle, θ), the lift forces for the wings and horizontal tail surfaces are taken to be acting at their $1/4$-chord points and are given by the downwash velocities at their $3/4$-chord points, and first-order corrections are implemented for the Mach number, the aspect ratio, the sweepback, the unsteady aerodynamic gust penetration (Küssner) effects, and flexibility effects via effective lift curve slope values. In addition, the model includes quasi-steady wing downwash effects on the tail.

In terms of aircraft vertical velocity, w, and pitch angle, θ, in the stability axes, the Laplace-transformed equations of motion are given in matrix form by Cornman [202] to be

$$\mathbb{A} \cdot \begin{pmatrix} w \\ \theta \end{pmatrix} = \mathbf{B} w_{\text{gust}} + \mathbf{C} \delta_{\text{e}}, \tag{9.30}$$

where w_{gust} is the gust velocity and δ_{e} is the elevator deflection angle. For short-period motions, including only pitching and vertical displacements, $\mathbb{A}$ is a 2×2-matrix given by [614]

$$\mathbb{A} = \begin{pmatrix} s - Z_w & -Vs \\ -sM_{\dot{w}} - M_w & s^2 - sM_q \end{pmatrix}, \tag{9.31}$$

while **B** and **C** are column vectors given by

$$\mathbf{B} = - \begin{pmatrix} Z_w \\ sM_{\dot{w}} - \frac{sM_q}{V} + M_w \end{pmatrix}, \qquad \mathbf{C} = \begin{pmatrix} Z_\delta \\ M_\delta \end{pmatrix}, \tag{9.32}$$

where $s = 2\pi i f$ is the Laplace transform variable, Z is the aerodynamic force along the vertical z-axis in a frame of reference fixed with the aircraft, M is the pitching moment, $Z_w = (1/m)(\partial Z/\partial w)$ is the change of force Z due to w (with units of s^{-1}), $M_w = (1/I_y)(\partial M/\partial w)$ is the change of pitching moment due to w (with units of ft^{-1} s^{-1}), $M_{\dot{w}} = (1/I_y)(\partial M/\partial \dot{w})$ is the change of pitching moment due time variations of w (with units of ft^{-1}), $M_q = (1/I_y)(\partial M/\partial q)$ is the change of pitching moment due to incremental lift (with units of s^{-1}), where $q = \dot{\theta}$, $Z_\delta = (1/m)(\partial Z/\partial \delta)$ is the change of force Z due to surface deflections (with units of ft rad^{-1} s^{-2}), and $M_\delta = (1/I_y)(\partial M/\partial \delta)$ is the change of M due to surface deflections, where δ accounts for control surface deflections, which includes elevators, stabilizers, flaps, slats, dive brakes, etc. However, the elevator is the most important longitudinal control surface because through applied pitching moments it controls the angle of attack of the airframe in equilibrium and maneuvering flight. Other quantities involved in the above definitions are the aircraft mass, m, and the moment of inertia of the entire aircraft mass about the y-axis (perpendicular to the aircraft elongation) in the body-fixed frame, $I_y = \int (x^2 + z^2)dm$. A full account of the aircraft dynamics and automatic control, including the most general case of longitudinal and lateral dynamics in three-space dimensions for finite period motions, are described in the book by McRuer et al. [614].

The linear system (9.30) can be solved for w and θ and the vertical acceleration along the body axis of the aircraft can be approximated by

$$\ddot{z}(s) \approx s(w - V\theta), \tag{9.33}$$

and the frequency response function is given by

$$H^{\ddot{z}}_{w_{\text{gust}}}(s) = \frac{\ddot{z}(s)}{w_{\text{gust}}(s)}. \tag{9.34}$$

With no autopilot, the response function takes the form

$$H^{\ddot{z}}_{w_{\text{gust}}}(s) = -\frac{2\xi\omega_0}{\gamma}\left[\frac{2\xi\omega_0(1 - 1/\gamma)s + s^2}{\omega_0^2 + 2\xi\omega_0 s + s^2}\right], \tag{9.35}$$

where

$$\omega_0^2 = -M_w V + M_q Z_w, \quad 2\xi\omega_0 = -M_q - M_{\dot{w}}V - Z_w, \tag{9.36}$$

and

$$\frac{2\xi\omega_0}{\gamma} = \frac{V}{\mu c}, \quad \mu = \frac{2m}{\rho S c C_{L_\alpha}}, \tag{9.37}$$

where m is the mass of the aircraft, c is the mean aerodynamic chord, S is the reference area, ρ is the air density, C_{L_α} is the lift curve slope, and μ is the mass parameter.

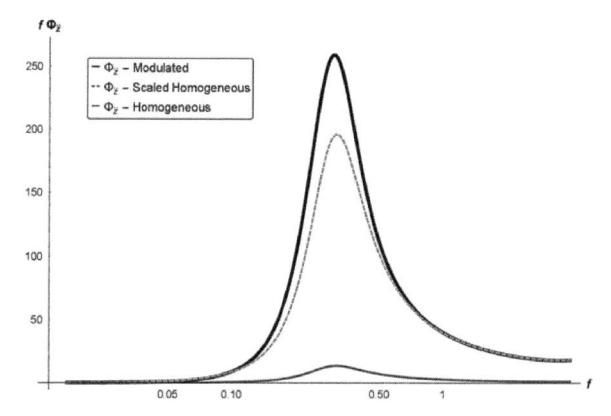

Figure 9.1: Vertical acceleration spectra for the NASA B757 aircraft flying through severe turbulence. The solid gray curve is for homogeneous von Kármán turbulence, while the solid black and the dashed gray curves are for modulated and scaled homogeneous von Kármán turbulence, respectively. Figure taken from Cornman [202].

According to Eq. (9.24), the power spectrum of the vertical acceleration for homogeneous and modulated homogeneous turbulence is the product of the square modulus of the aircraft vertical acceleration frequency response function and the von Kármán lateral velocity spectrum. Figure 9.1 shows a log-linear plot of the vertical acceleration spectrum for the case of the NASA B757 aircraft flying through an area of severe CAT. The three curves correspond to the homogeneous (von Kármán) input wind field (solid gray), the scaled homogeneous (von Kármán) turbulence (dashed gray), and the modulated homogeneous turbulence (solid black). The log-linear scaling comes from the relation $f\Phi(f)d[\log(f)] = \Phi(f)df$, where the power spectrum is multiplied by the linear frequency f and plotted against the logarithm of f. At these flight conditions, the spectra for the scaled and modulated homogeneous turbulence overlap at large frequencies ($f < 0.5$ Hz). In addition, it is evident from Fig. 9.1 that the maximum acceleration power for both the homogeneous and modulated turbulence occurs around the undamped natural frequency for the aircraft (≈ 0.3 Hz). A description of the simulation method and results for homogeneous turbulence is given by Frehlich et al. [303]. A brief account of the results is also provided by Cornman [202].

9.2.2 Airborne Doppler interferometry

The microphysical description of a cloud relies on the details of its droplet properties, such as the size distribution, velocity, and liquid water content. *In-situ* airborne measurements of these parameters represent the best way to explore the structure of clouds at the scale of droplet sizes. As was reviewed in Chapter 7, traditional optical techniques for the determination of droplet sizes were primarily

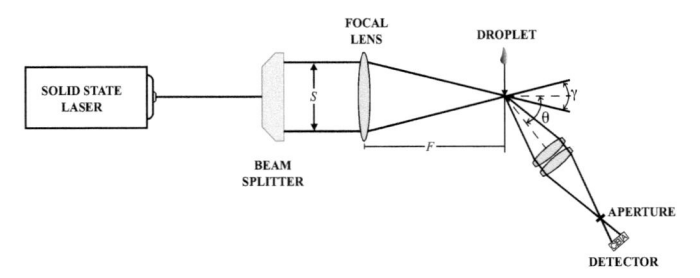

Figure 9.2: Schematic picture showing the optical system of the airborne phase Doppler Interferometer (PDI) for cloud microphysical measurements. Figure adapted from Chuang et al. [178].

based on light-scattering measurements. In particular, over the past 20-30 years, a standard instrument that has been used to measure cloud droplet sizes in the diameter range between 5 and 50 μm is the PMS Forward Scattering Spectrometer Probe (FSSP). While the FSSP suffered from sizing errors and artificial broadening of the observed size distributions, it has been modified over the years, giving rise to the Fast-FSSP, with improved electronics and minimized dead-time effects [115], and later on to the M-Fast FSSP, which minimizes beam-inhomogeneities [795]. A more recent device is the 2D-S instrument, which measures cloud drop sizes in the diameter range between 10 and 1000 μm [536]. This instrument improves the effective size of the sample volume and measures particle shadowgraphs using two crossing laser beams with two linear photodiode arrays.

A new probe for cloud drop size measurements is based on phase Doppler interferometry (PDI) [48, 50, 224]. This new airborne instrument called the *Arthium Flight-PDI* or F/PDI[2] has been described by Chuang et al. [178]. Unlike most existing optical probes, which perform measurements using the intensity of scattered light, the Arthium F/DPI utilizes the wavelength of light as the scale of measurement. Figure 9.2 shows a schematic picture of the F/DPI optical system, which is similar to the laser Doppler velocimetry (LDV) system (see Fig. 7.5). A single polarized laser beam is separated into two equal-intensity beams by a beamsplitter, which are brought together at an angle γ using a frontal focal lens. The measured volume is established at the intersection of the two focused beams. If the angle γ is small, the intersection volume resembles a cylinder of diameter D_{beam} and height L_{aperture}. For small γ, the distance between successive fringes depends on the laser wavelength by means of the relation [50]

$$\delta x = \frac{\lambda}{2\sin(\gamma/2)} \approx \frac{\lambda F}{s}, \qquad (9.38)$$

which is also approximately equal to the product between the wavelength and the focal length, F, over the beam separation before reaching the front focal lens,

---

[2]This instrument was designed and constructed by Arthium Technologies, Inc. of Sunnyvale, CA.

s. Light is scattered into the surrounding space by any liquid droplet passing through the intersection volume. The scattered light is then collected at an off-axis angle θ by a receiver and imaged onto three detectors. Each detector records a time-varying signal, having a Gaussian intensity profile on which a higher frequency sinusoid is superimposed. These signals, or Doppler bursts, represent the raw collected data from which the drop size and velocity are derived. The signals produced with this interferometric method have a distinctive waveform and therefore the use of digital detection techniques permit to distinguish the signal from noise even in cases when the signal-to-noise ratio (SNR) is low and discriminates between spherical and non-spherical particles. Uncertainties in the calculated value of δx affect the accuracy of the measured drop size and velocity, and depend on the ability to properly measure *s* (which is usually measured within less than about 1%).

The drop velocity can be calculated from the fringe separation distance and the Doppler frequency, f_D, of each signal using the simple relation

$$v_{\mathrm{drop}} = f_D \delta x, \tag{9.39}$$

where v_{drop} effectively measures the drop velocity component normal to the fringe plane. The uncertainty in the drop velocity is determined by the uncertainty in δx, which in turn is governed by the uncertainty in *s*. In addition, the drop size can be determined from the phase shift, ϕ, between the two signals, which is linearly related to the droplet diameter [50]. For example, small drops yield a small phase shift, while a larger drop produces a larger phase shift. At very small droplet sizes, light diffraction dominates over refraction and leads to oscillations in the relationship between the phase shift and the drop diameter in the size range below 4 μm and with some bias up to about 8 μm [51]. This results in an uncertainty of ± 0.5 μm in this size range [51]. The PDI technique also presents limitations at the large sizes of natural water drops, which are determined by the drop sphericity. For example, large drops are non-spherical due to the effects of aerodynamic drag forces and therefore their measurable sizes have proportionate uncertainties and their size distribution will be much broader than for small drops. For example, drops smaller than 300 μm are nearly perfectly spherical and have associated uncertainties of $\sim 2\%$, while up to $\sim 10\%$ uncertainties can be found in water drops as large as 2 mm [730].

A further parameter of importance in the calculation of the drop size distribution is the volume of air sampled per unit time, Q_s, which is directly related to the drop concentration (or liquid water content). On an airborne platform, the value of Q_s as a function of the drop diameter is given by

$$Q_s(d) = D_{\mathrm{beam}}(d) L_{\mathrm{aperture}} U, \tag{9.40}$$

where U is the velocity of the instrument relative to the atmosphere and, as was specified above, L_{aperture} is the length of the view volume defined by $L_{\mathrm{aperture}} =$

$W/(m \sin \theta)$, where W is the width of the receiver aperture and m is the magnification. For practical applications of the F/PDI instrument, apertures of width 0.05, 0.1, 0.2, 0.5, and 1 mm are mounted on a motorized mechanism, allowing the operator to choose via a software the aperture size most appropriate for the given sampling conditions. A more detailed description of the sensing of microphysical cloud properties with the F/PDI instrument along with the uncertainties involved and intercomparisons with other instruments can be found in Chuang et al. [178].

9.2.3 Other airborne systems for aerosol sensing

A comprehensive overview of airborne systems for *in situ* measurements of aerosol particles, clouds, and radiation with instruments that have implemented mature and emerging technologies has been reported by Baumgardner et al. [69]. Additional advances in the airborne measurements of microphysical cloud properties has come with the development of sensitive methods for measuring aerosol chemical composition and cloud-forming potential of particles. In particular, the instruments are classified depending on the aerosol characteristics they were designed to sense, i.e., size distributions, the total number and mass concentration, optical properties, and chemical composition. This latter characteristic includes the cloud-forming properties. For example, ground-based instruments for measuring aerosol properties can usually be adapted for operational use inside the aircraft cabin. However, it seems to be more challenging to adapt ground-based instruments for operation outside the aircraft (i.e., attached to the fuselage or wing) due to the many environmental ambient air factors that can affect the measurement as, for example, extreme values of temperature (varying between $-80°C$ and $+40°C$), pressure (between 1000 mb to 50 mb), humidity, air velocities from 50 to 250 m s^{-1}, vibration, CAT, air flow distortion, icing, and eventually bird strikes, to name some of the several factors. Except for the optical spectrometers, nearly all instruments for aerosol measurements in use today are operated inside the aircraft.

Aerosol particle concentrations are measured with airborne Condensation Particle Counters (CPCs). In these devices, such as the continuous flow diffusion CPC which is of current use today in airborne operations [988, 962], the ambient air is sampled from an inlet and is put in contact with the supersaturated vapor of a fluid that condenses rapidly on the particles, which grow in size before traversing a laser beam. The incident light is scattered by the particles and detected by a photodiode. After individual particles are counted, the aerosol concentration is calculated from the known flow rate. These instruments usually detect particles with diameters ranging from a few nanometers to a few micrometers, which are the diameters typically encountered in the atmosphere. More details on the operating principles and developments of CPCs can be found in the overview by McMurry [612]. In addition to the CPCs, a few other airborne instruments for aerosol concentration measurements as a function of particle size, such as the Passive Cavity Aerosol Spectrometer Probe (PCASP) [877, 567], the Ultrahigh Sensitivity Aerosol Spectrometer (UHSAS), and the Model 300 FSSP [70], are mounted outside the aircraft [69]. In particular, the PCASP is sensitive to particle sizes in the range between 0.12 and 10 μm, while the Model 300 FSSP and the UHSAS can detect particle sizes ranging from 0.3 to 20 μm and 0.06 to 1 μm, respectively. However, the FSSP-300 detector differs from the other two in that it measures forward scattered light, while the PCASP and UHSAS collect side scattered light at angles of $\sim 30 - 120°$.

Aerosol optical properties, such as light scattering and absorption coefficients from which extinction, optical depth, and single scattering albedo can be derived, are measured using nephelometry, which is a technique based on the geometrical integration of the angular distribution of the scattered light by gas molecules and aerosol particles in an enclosed finite volume [89]. In this instrument, the incident light is provided by an array of light-emitting diodes (LEDs), a flash lamp, or a quartz-halogen lamp. The emitted light is then measured by a photomultiplier detector arranged in such as a way to avoid stray light detection. Such standard nephelometers have been installed in several commercial transport aircraft [390]. Alternatively to these sensors, the so-called reciprocal nephelometers have been also deployed for use in flight operations [15]. In contrast to standard nephelometers, the light source and the detector position are reversed in reciprocal nephelometers. In addition, standard nephelometers use optical filters to select scattered light in a wavelength band, while in reciprocal nephelometers the laser bandwidth is much narrower and more precisely defined. In fact, these nephelometers have been integrated in the Photoacoustic Absorption Spectrometers (PAS) [34, 639].

On the other hand, the chemical composition of atmospheric aerosols is currently measured and analyzed online either onboard the aircraft or with mass spectrometric methods. Single particle detection and ablation, based on laser methods and thermal vaporization techniques are the most commonly used techniques for aircraft applications [905][3]. The first airborne mass spectrometer was the Particle Analysis by Laser Mass Spectrometer (PALMS). This instrument was installed on the NASA WB-57 aircraft to detect the presence of meteoritic material and organic particulate in the upper troposphere and lower stratosphere [215, 652, 314]. A Z-shaped mass spectrometer was further designed to improve mass resolution in small volumes, which has been coupled to a counter flow virtual impactor to show that biological particles are ice nuclei [720, 721]. The only airborne mass spectrometer based on thermal volatilization of particles is the AMS developed by Jayne et al. [435] and described by Canagaratna et al. [152], which can provide aerosol full mass spectral information as a function of particle size. However, most aircraft equipped with AMS instruments employ

---

[3]Details on the working principle of single particle mass spectrometers can be found in Ref. [651].

either the C-shaped TOFAMS [641] or the high resolution AMS [228], which have sensitivities by factors of 10–30 greater than the quadrupole systems used in the first airborne measurements performed with an AMS [54]. Measurements of refractory black carbon and its size distribution have been performed online using a single particle soot photometer [869, 810]. This instrument was first used on a NASA DC-8 aircraft to investigate the black carbon abundance in the Arctic lower stratosphere [72]. Other applications include the investigation of urban plumes [643], measurements of the vertical distribution of black carbon in global models [810], and further sensings of the black carbon abundance in the upper troposphere and lower stratosphere [392]. Other technologies have been implemented for airborne measurements of cloud condensation nuclei (CCN) based either on the use of static thermal gradient diffusion chambers [864, 916] or continuous flow techniques [839, 419, 316]. A controlled, supersaturated environment is created with both techniques in which CCN will grow. The light scattering signal from the growing CCNs by condensation droplets [518, 854] or from counting droplets in photographs [232] is batch processed in the static chambers. In modern applications, the most frequently used technique is the continuous flow diffusion chamber, which allows cloud and ice formation under controlled temperature and humidity conditions during continuous sampling [768]. This technique has been improved over the years [418, 765, 638] and today it allows continuous and real-time sensing of a wide range of ice nuclei number concentrations and good interfacing with methods for collecting activated ice nuclei for chemical analysis via electron microscopy [508] or single mass particle spectrometry [214]. A newer device for ice nuclei detection has been described by Bundke et al. [139], which uses the Fast Ice Nucleus Chamber (FINCH) where the turbulent mixing of dry air with warm humidified air in a close loop system produces a supersaturated environment. Discrimination of ice nuclei from droplets occurs once the former has grown to diameters above 4 μm during about 10 s residence time by their individual circular depolarization properties, which are then counted in an optical detector.

9.2.4 *Turbulence sensing in clouds and over mountains*

Reliable measurements of CAT have been a longstanding and challenging problem for the aviation industry. For example, clouds are in general turbulent due to strong buoyancy and shear associated with convective motions, latent heat release, and radiation. However, at high aircraft speeds, traditional measurements based on sonic anemometry or differential pressure suffer from low spatial resolution, while the method of hotwire anemometry, which has higher resolution, is not being currently used in airborne operations because of the presence of water droplets and other kind of particles in the atmosphere [834]. Available highly-resolved cloud measurements indicate that turbulence in clouds is very well described by the energy cascade scaling for velocity [835]. In particular, it has been recognized that F/PDI provides a high-resolution technique that cannot only be used to measure cloud droplet size distributions and concentrations, but also to obtain fine-scale turbulent measurements in clouds [178]. Under many cloud conditions, PDI instruments can be configured to simultaneously measure the incoming velocity of each cloud droplet, which corresponds to the velocity component normal to the optical axis and in the plane of the crossed beams (see Fig. 9.2). Compared to typical differential-pressure measurements, PDI methods provide much higher spatial and temporal resolution, thereby allowing for extending the range of eddy sizes that can be resolved within the turbulent energy cascade. The PDI measurements of the droplet incoming-velocity component result in time series that can be statistically analyzed to provide power spectral densities and turbulent EDRs.

In typical clouds, turbulence is fed by the injection of kinetic energy at scales of order 10 to 100 m, which is then transferred to progressively smaller scales by a cascading process (i.e., the inertial range) through nonlinear interactions. Over most of the inertial range, viscous forces are negligible and therefore the input kinetic energy feeds smaller and smaller eddies without significant dissipation until they enter the Kolmogorov microscale (of the order of 1 mm for typical cloud conditions), where significant viscous dissipation occurs. Therefore, velocity fluctuations characteristic of turbulent motions is always present in the inertial scale. Such fluctuations (or eddies) span from ~ 4 to 5 orders of magnitude decrease in size before dissipating in the Kolmogorov microscale. As described by Eq. (1.63), typical velocity differences, u_l, within the inertial range are proportional to $\varepsilon^{1/3} l$. In typical clouds, the EDRs take values in the range $10^{-4} \leq \varepsilon \leq 10^{-2}$ W kg^{-1}, corresponding to velocity fluctuations in the interval $2 \leq u_l \leq 10$ cm s^{-1} for $l = 10$ cm [178]. An estimate of the velocity of each sampled droplet is obtained with the PDI instrument and the corresponding Doppler difference burst frequency is measured using Eq. (9.39), where the frequency estimation is made via a discrete Fourier transform. The uncertainty in the frequency measurement can be estimated through a theoretical (Cramer-Rao) lower bound given by [20, 178]

$$\Delta f_D^2 = \frac{6}{A^2 \pi^2} \frac{\sigma_n^2}{N(N^2 - 1)} f_s^2, \tag{9.41}$$

where A is the amplitude of the Doppler burst signal, σ_n is the measurement noise, N is the number of data points sampled by the instrument within each burst, and f_s is the sampling frequency. This relation provides a measure of the precision in the droplet velocity measurement. Other velocity biases could also come from small uncertainties present in the optical instrument parameters that yield the fringe spacing δx. The processor parameters are set up using the Nyquist criterion to adjust the sampling frequency to correspond to the dynamic range of the velocity measurement. Therefore, the ratio $2\Delta f_D / f_s$ is an estimate of the relative error in the velocity measurement concerning the velocity dynamic range. Instrument settings can in general be optimized by obtaining this relative error within 0.1% or less. The lower bound given by Eq. (9.41) can be used under the assumption that the noise is sampled with a white spectrum. In terms of f_D, δx, the signal-to-noise ratio (SNR), and the beam diameter D_{beam}, Eq. (9.41) can be rewritten as [178]

$$\left(\frac{\Delta f_D}{f_D}\right)^2 = \frac{6}{\pi^2 SNR}\left(\frac{f_D}{f_s}\right)\left(\frac{\delta x}{D_{\text{beam}}}\right)^3, \tag{9.42}$$

where it has been used $N = t_r f_s$ with $t_r = D_{\text{beam}}/v_{\text{drop}}$ being the transit time and $N^2 \gg 1$. Equation (9.42) implies that the relative error in the velocity measurements decays with increasing SNR and f_s, or by decreasing the fringe spacing. However, none of these parameters can be indefinitely increased or decreased. For example, the increase of the sampling frequency is limited by the number of samples, N, the non-constancy of the droplet velocity transiting through the probe volume, and by the fact that the Cramer-Rao error estimate goes as

$$\left(\frac{\Delta f_D}{f_D}\right)^2 \sim \frac{1}{N}, \tag{9.43}$$

which is consistent with the averaging of independent samples. On the other hand, real noise processes will have a finite correlation time τ_{corr} so that they will not be perfectly white. In particular, when $f_s > \tau_{\text{corr}}^{-1}$, the samples are no longer independent of each other and the white noise assumption made for Eq. (9.42) is therefore no longer valid.

As was described by Chuang et al. [178], the PDI instrument can potentially resolve a significant portion of the inertial range of the turbulent energy cascade in clouds under the assumption that the Kolmogorov dissipation range begins at scales of ≈ 1 cm. For airborne applications, the velocity uncertainty scales with the flight velocity, V, as

$$\Delta v_{\text{drop}} \propto \left(\frac{\delta x}{D_{\text{beam}}}\right)^{3/2} V, \tag{9.44}$$

where it has been assumed that V is much greater than the root-mean-square drop velocity so that $f_D \approx V/\delta x$ and the fact that $\Delta v_{\text{drop}} = \Delta f_D \delta x$. To obtain maximum resolution in turbulence velocity statistics, it is advisable to use the minimum possible flight velocity. However, reliable turbulence statistics from PDI instruments can be obtained only from velocity data pertaining to sufficiently small droplets to ensure that they are closely following the airflow pathlines. This is because the motion of large drops may significantly deviate from that of the background air due to differences in the mass density of both phases [49], and therefore measurements of their individual velocities will not provide a good estimate of the intensity of turbulence. The frequency response of a drop in an oscillating background fluid will result in a slip velocity given by

$$\tilde{v}_s = \frac{v_{\text{fluid}} - v_{\text{drop}}}{v_{\text{fluid}}}, \tag{9.45}$$

where v_{fluid} and v_{drop} denote the fluid and drop velocities, respectively. In clouds, where the density of water drops, ρ_{drop}, is much greater than the density of the background fluid, ρ_{fluid}, the drop slip velocity is

$$1 - \tilde{v}_s = \left[1 + (2\pi f \tau_{\text{drop}})^2\right]^{-1/2}, \tag{9.46}$$

where

$$\tau_{\text{drop}} = \frac{1}{18} \frac{\rho_{\text{drop}} d^2}{v \rho_{\text{drop}}}, \tag{9.47}$$

d is the drop diameter, v is the fluid kinematic viscosity, and f is the fluid oscillation frequency, which has to be estimated as the inverse of the Lagrangian velocity correlation time in the drop frame of reference [49]. However, this remains a pending problem in turbulence analysis which has been partly remedied using a frequency derived from the inverse of an Eulerian velocity correlation time, i.e., $f = \tau_l^{-1} \approx (\varepsilon/l^2)^{1/3}$.

Accurate *in situ* measurements of turbulence require that the ambient flow not be perturbed by the instrument housing. Applications of PDI in laboratory and engineering flows are performed with the flow confined to a chamber and the optics and detectors positioned outside it. However, in applications of cloud measurements, the instrument is typically placed inside the flow. In particular, while this puts constraints on the instrument housing design, it permits the benefit of being deployed on different platforms. For example, the F/PDI has been used in various research aircraft, on a helicopter-borne instrument payload (The Airborne Cloud-Turbulence Observation System, ACTOS) [833], and in large research wind tunnels. The F/PDI instrument has been designed to minimize flow distortions for typical speeds in wind tunnels and ground-based experiments with typical aircraft cruise speeds. These experimental tests have demonstrated that as the mean speed is increased from the speeds in wind tunnels to typical aircraft cruise speeds, the irrotational flow pattern around the instrument housing does not change significantly.

Atmospheric turbulence generated in flow over mountains using airborne *in situ* and single-Doppler radar measurements have been conducted over the Medicine Bow Mountains (MBM) in southeast Wyoming [878]. A field campaign took place during the *NASA Orographic Clouds Experiment* (NASA06) in January and February 2006, where two complex mountain flow cases were documented by the University of Wyoming King Air (UWKA) research aircraft carrying the *Wyoming Cloud Radar*. A series of turbulent measurements were performed during the UWKA flights by crossing the highest elevations of the MBM in and against the mean wind direction. Stable upstream conditions and favorable mesoscale dynamics forcing on 26 January and 5 February 2006 has led to a good response to the underlying mountain topography, including large-amplitude lee waves, gravity-wave breaking, and strong low-level turbulence. A detailed account of the NASA06 experiments can be found in French et al. [309] and Grubišić et al. [363], where a comprehensive characterization of the aircraft *in situ* instrumentation and the cloud radar is provided. In particular, two NASA06 events illustrative of large-amplitude mountain waves, midtropospheric wave breaking, and rotor circulations were described by Strauss et al. [878]. One of these events was performed using the UWKA research aircraft, which was equipped with a five-hole gust probe located on an extended nose boom (for measurements of the three components of the wind speed and other turbulence parameters, such as the air temperature and the static pressure) and the Universal Indicated Turbulence System (UITS), also referred to as the Mac-Cready Turbulence Meter [580, 581] (for the estimation of the EDR), which was used as a real-time, on-flight indicator of turbulence [290]. Horizontal along-track and vertical cross-track components of the wind velocity were measured with a precision of ± 0.1 m s^{-1} and ± 0.05 m s^{-1}, respectively. The other event was performed with the Wyoming Cloud Radar (WCR), which consisted of a 95 GHz pulsed, fixed multi-antenna Doppler radar. In order to provide high-resolution measurements of the cloud structure and dynamics, the WCR was operated in three-antenna mode, with beams pointing to the zenith, the nadir, and the 30° down-forward directions when the aircraft was flying straight and level. Since the mountain generated turbulence is highly inhomogeneous and forced from aloft by the presence of gravity waves, instrument resolution is an important factor. In the vertical direction, the WCR resolution was 30 m and at 1 km the radar pulse volume was approximately $10 \times 10 \times 30$ m^{-3}, while in the horizontal direction the sampling range was approximately 3 m at a mean true airspeed of ≈ 100 m s^{-1}. Moreover, the WCR backscattered reflectivity was predominantly from clouds consisting mostly of aggregates of spherical ice particles.

The variance of the vertical wind velocity, σ^2_{wind}, was employed to measure the structure of turbulence, while its intensity was described in terms of the EDR. A thorough analysis of the uncertainties in the measurements of the vertical Doppler velocity showed that 25% accuracy or better can be obtained in regions where turbulence is from moderate to severe in the lee of the mountains. For the two NASA06 events documented by Strauss et al. [878], moderate turbulence was encountered in a wave-breaking region with $\sigma^2_{\text{wind}} \approx 4.8$ m^2 s^{-2} and EDR$= 0.25$ m$^{2/3}$ s^{-1}, while severe turbulence was detected within the rotor circulations with values of σ^2_{wind} and EDR in the ranges from 7.8 to 16.4 m^2 s^{-2} and from 0.50 to 0.77 m$^{2/3}$ s^{-1}, respectively. These experimental studies have demonstrated that combining aircraft *in situ* and Doppler radar measurements can describe mountain-induced turbulence with reasonable accuracy.

9.2.5 *Emerging airborne in situ technologies and methods*

In the last fifty years, significant progress was achieved in the field of laser technology and, as a result, important practical applications have started to emerge over the years. This effort has led to a variety of systems, including airborne laser sensors for both civil and military avionic applications, laboratory devices used for the study of nonlinear optical emissions and propagation, and compact laser-ranging binoculars, among many others. In particular, airborne laser systems have focused on Laser Rangefinders (LRF) and Target Designators (LTD), laser radars (LIDAR), Laser Communication Systems (LCS), and Detected Energy Weapons (DEW). Areas of direct interest in civil aviation include LIDAR, LTD, and LCS systems, which will be discussed separately in the next sections. In this section the most recently emerged turbulence airborne detection technologies will be first considered.

9.2.5.1 Airborne lasers in multi-sensor systems

It is well known that the range of LIDAR systems is degraded by adverse weather and due to their narrow beam widths, they do not provide wide coverage as conventional radars will effectively do [786]. However, these deficiencies are improved when LIDAR is combined with a multi-sensor system. For instance, when LIDAR is made to operate with a Forward Looking Infra-Red (FLIR) sensor, its night vision with terrain flying and obstacle avoidance capabilities are greatly improved. As an example, Fig. 9.3 shows a schematic drawing of a LIDAR combined with a multi-sensor system for electronic surveillance. In particular, such airborne sensors can be used for passive listening over a wide range of frequencies, laser obstacle avoidance and terrain following for covert operations, bad weather detection, and real-time passive night vision. Very complex computer processing responding to different operational situations is often required to control these functions. To improve measurement capabilities from various sensors, including combinations of LIDAR and FLIR, data fusion algorithms have been developed using Extended Kalman Filter (EKF), Unscented Kalman Filter (UKF), and particle filter [35]. On the other hand, EKF-based data fusion involving information from LIDAR, visual sensors, FLIR, and acoustic sensors

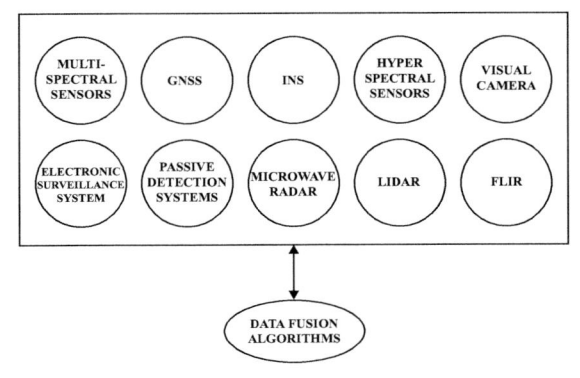

Figure 9.3: Block diagram showing a LIDAR combined with a multi-sensor system for electronic surveillance. Figure adapted from Sabatini et al. [786].

have been developed to provide two- and three-dimensional imagery and allow flying aircraft to autonomously detect and judge targets in uncertain adverse environments and weather [442]. In the recent past, in-flight fusion of hyperspectral imaging and wave from LIDAR for three-dimensional studies of ecosystems have also been developed [37].

9.2.5.2 The multi-purpose airborne sensor Carrier MASC-3

Measurements of the wind speed and direction are of fundamental importance for understanding the processes that take place in the atmospheric boundary layer. In general, measurements of turbulent wind velocities from research aircraft are performed with the aid of a multi-hole probe combined with an inertial navigation system (INS) [741]. Although this technique was originally developed for manned research aircraft [549], it has been more recently adapted to unmanned aircraft systems (UASs) for the measurement of turbulence in the atmospheric boundary layer [920, 993]. In particular, remotely piloted aircraft can be equipped with accurate sensors to measure atmospheric turbulence. Such vehicles can also be equipped with similar detection systems to those used in manned aircraft. However, their use was often limited by the cost and the smaller sizes of UASs. However, during the last ten years, there was a considerable amount of research and design activity to produce low-cost sensors and airborne instrumentation. Most of these systems were based on fixed-wing airframes, such as, for instance, M2AV [861], SUMO [755, 756], MANTA [904], Samartsonde [98], MASC [976], ALADINA [25, 61], Pilatus [226], and BLUECAT5 [993]. Among the capabilities of UASs, they are used for research in atmospheric physics and chemistry [236, 706, 807] and in applications to boundary layer meteorology [405, 598, 921, 97, 977, 950, 110, 506] and wind-energy meteorology [881, 974]. The preferred choice for measurements related to boundary- and surface-layer profiling has been the use of rotary-wing multi-copter systems because of their ability to hover and to slowly ascend and descend vertically [119, 685, 506]. However, these systems have rather limited use for turbulence measurements.

To improve wind and turbulence measurements, a new sensor – the Multi-Purpose Airborne Sensor Carrier (MASC-3) – was developed by the environment-physics group of the Center for Applied Geo-Science (ZAG) at the University of Tübingen in Germany [976, 742]. This device was designed to increase the accuracy of the wind measurement, gain endurance, have more flexibility in implementing further sensors, and allow operations under a much wider range of environmental conditions than previously developed sensors. A detailed description of MASC-3 is provided by Rauntenberg et al. [743]. The sensor is positioned in the front of the UAS fuselage. In contrast to the previous version of MASC, the engine is positioned in the tail of the aircraft (just behind the V-tail) in order to minimize potential vibrational and magnetic influences on the measurements. High maneuverability and allowed cruising speeds from 14 m s^{-1} to 30 m s^{-1} provides high-resolution sampling as well as operational function under high wind speeds and extreme turbulence. As a new feature, MASC-3 is equipped with position and strobe lights, which follow the conventions of manned aircraft and allow take-off and landing during nighttime. In particular, since the lightning system of MASC-3 fulfills the requirements of the SERA 923/2012 regulation, the aircraft is allowed by the local civil flight authorities for UAS to operate during night and beyond the visual line of sight (BVLOS). Moreover, the autopilot system of MASC-3, consisting of a Pixhawk 2.1 "Cube" autopilot using a Here+ RTK GPS and magnetometer for the position, velocity, and heading along with a mRobotics MS5525 digital airspeed sensor connected to a pilot-static tube for airspeed measurement, the airframe, the sensor system, and the data acquisition and post-processing software provide the basis for flight experiments for wind speed and direction measurements as well as turbulence measurements. In particular, the flight experiments described by Rautenberg et al. [743] have demonstrated that the statistical measured properties of turbulence as well as the measurements of the turbulent kinetic energy and the integral length scale all agree well with measurements obtained from a meteorological tower. In this sense, MASC-3 can be used to supplement measurements from meteorological towers and sodar systems and also to provide reliable measurements of stable boundary layers with high temporal resolution [743].

9.2.5.3 *The optical particle counters (OPCs) and optical array probes (OAPs)*

The aerosol-cloud and aerosol-radiation interactions still remain among the largest uncertainties in climate predictions [873]. It has been realized that the size distribution of cloud and aerosol particles plays a fundamental role in such interactions [775, 729]. In particular, if the fraction of cloud coarse particles increases, the radiative direct forcing of desert dust from cooling and warming can be modified [495] and the number of ice-nucleating particles can significantly increase [233]. Airborne *in situ* measurements and observations continue to provide an important avenue for improving the present knowledge of the distributions of cloud coarse and aerosol particles. Coarse particles are currently measured using open-path or passive-inlet optical particle counters (OPCs) and optical array probes (OAPs). Both types of instruments record the number of particles passing through the sampling area within a time interval, and the particle flux is later converted into a particle concentration. However, the calculated particle and hydrometeor concentrations can be largely affected by errors in the airflow speed. For example, if the calculated airflow speed is too low, the conversion will lead to a higher derived particle concentration. On the other hand, airborne *in situ* measurements remain challenging because both the surrounding air and the flow speed measurements are likely to be affected by the aircraft itself [451], which leads to a flow distortion compared to free stream conditions as caused by the aircraft fuselage and wings. Such distortion affects the airflow velocity as well as the air temperature, pressure, and density, and as a consequence, the aerosol and cloud measurements are also affected. Modification of these parameters may lead to several shattering artifacts [961, 502, 887, 925, 356], whose degree also depends on the flight conditions and mounting position of the instrument in the aircraft.

A comprehensive study of flow-induced errors in airborne instruments using CFD simulations is provided by Spanu et al. [860]. In particular, a correction strategy valid for different aircraft configurations and passive-inlet instruments is proposed in this investigation. These numerical results were compared with *in situ* measurements collected with a *Cloud and Aerosol Spectrometer with Depolarization Detection* (CAS-DPOL) [71] and a second-generation *Cloud, Aerosol, and Precipitation Spectrometer* (CAPS). In addition to analyzing flow changes around wing-mounted instruments and their effects on the calculated particle concentrations, these authors have developed a method that provides a corrected particle speed for OAP measurements. The heart of this strategy consisted of expressing the sampling efficiency, f_{eff}, as a function of a modified Stokes number[4] (Stk) defined according to [432]

$$\text{Stk} = \frac{1}{18} \frac{\rho_{\text{p}} U_{\text{air}} d_{\text{p}}^2}{L \mu_{\text{air}}} \psi(\text{Re}_{\text{p}}), \tag{9.48}$$

and a parameter α that describes the difference between the probe and the free stream conditions, given by

$$\alpha = \frac{p_{\text{s,probe}}}{p_{\text{s,free}}} \frac{T_{\text{free}}}{T_{\text{probe}}} \frac{\text{PAS}}{\text{TAS}}, \tag{9.49}$$

so that the sampling efficiency is approximated with the sigmoid equation

$$f_{\text{eff}}(\alpha, \text{Stk}) = k_0 + \frac{100\% - k_0}{1 + (\text{Stk})^{k_2} \exp(k_1)}. \tag{9.50}$$

In the above expressions ρ_{p} is the particle density (where $\rho_{\text{p}} = 2.5$ g cm^{-3} for mineral dust and $\rho_{\text{p}} = 1$ g cm^{-3} for water droplets), U_{air} is the airflow speed in the free stream, d_{p} is the particle diameter, L is a characteristic fluid length (fixed to 1 m), μ_{air} is the dynamic viscosity of air, ψ is a correction factor defined in terms of the particle Reynolds number

$$\text{Re}_{\text{p}} = \frac{p_{\text{s}} U_{\text{air}} d_{\text{p}}}{T \mu_{\text{air}} R_{\text{s}}}, \tag{9.51}$$

---

[4]The Stokes number, Stk, is defined as the ratio of the particle response time to the characteristic fluid timescale. For example, particles moving with small Stokes numbers react spontaneously to sudden flow changes and follow the streamlines. This is the case of atmospheric submicron-sized particles.

where p_s is the static pressure as measured by a Pitot-static tube located at the wing-mounted instrument, T is the air temperature measured at flight conditions, and R_s is the specific gas constant of air (taken to be 287.1 J kg^{-1} K^{-1}), $p_{s,free}$, T_{free} and $p_{s,probe}$, T_{probe} are, respectively, the static pressure and air temperature at free stream conditions and in the vicinity of the probe, PAS is the air speed at the location of the probe, and TAS is the air speed in the free stream (U_{air} = TAS in Eqs. 9.48 and 9.51). For different flight conditions and different distances from the probe, the coefficients k_0, k_1, and k_2 entering in Eq. (9.50) were fitted to be

$$
\begin{aligned}
k_0 &= (83.7\%)\alpha + 14.6\%, \\
k_1 &= 1.86\alpha - 3.66, \\
k_2 &= -0.87.
\end{aligned}
\tag{9.52}
$$

The application of Eq. (9.50) performs corrections of the particle concentrations in terms of the modified Stokes number and the flight conditions. The first step in the correction procedure consists of estimating for each particle diameter the corresponding modified Stokes number from Eqs. (9.48) and (9.51) using free stream conditions (p_s, T, TAS) and a range of particle densities. Using these values, the sampling efficiency is calculated from Eq. (9.50) after evaluation of the parameter α given by Eq. (9.49) and the coefficients defined by relations (9.52). The ambient number concentration, N_i, in each diameter bin i covering the interval from $d_{p,i}$ to $d_{p,i+1}$ is finally determined by the ratio between the number of detected particles in bin i and the product between the size of the sampling area, the value of TAS, the measurement duration, and the sampling efficiency. The benefits of this correction strategy were demonstrated using the dataset collected during the SALTRACE campaign [963] for probes under an aircraft wing. The airflow speed at the probe location (PAS) was shown to be on average 30% smaller than the airflow speed in the free stream (TAS) [860]. For particles of diameters $\lesssim 5\ \mu$m, it is beneficial to employ wing-mounted instruments within this size range using probe conditions measurements (i.e., PAS, $p_{s,probe}$, T_{probe}) because particle concentrations can be determined with low errors (of less than about 2%).

9.2.5.4 *The W-band radar*

The ability of a W-band radar to measure turbulence structure as a function of height throughout the drizzling marine boundary layer was recently reported by Lothon et al. [571]. The analysis was performed using the fine-structure of the Doppler velocity field obtained during the second *Dynamics and Chemistry of Marine Stratocumulus Experiment* (DYCOMS-II) that took place off the coast of southern California in July 2001 [870], using the NCAR C-130 aircraft. This experiment combined airborne radar with the drop counter probes for *in situ* measurements of the dynamics and microphysics of marine stratocumulus. Therefore, it is well-suited to explore the potential of using radar to measure turbulence in the stratocumulus-topped boundary layer (STBL). The NCAR C-130 aircraft flew performing quasi-Lagrangian circle legs of 60 km diameter at different levels within the STBL and in the overlying free troposphere. During this project, the 95 GHz Doppler WCR was mounted on the research aircraft with two downward-looking beams. One antenna looked straight down, while the other backward, 57.5° from the horizontal. The radar emitted a 250 ns pulse with a 1.6 kW peak power and 0.7° beamwidth. In addition, the Doppler velocity was estimated using pulse pair processing. The *in situ* wind velocity components in the horizontal and vertical directions were calculated from the navigation and air motion sensors at 25 s^{-1} in the coordinate reference system defined by the aircraft heading and the horizontal plane, while profiles of the Doppler velocities along the nadir and trailing beams were measured by the WCR at a varying sampling rate of $\sim$ 20 s^{-1}. This way the WCR was able to observe the underlying cloudy atmosphere at high resolution. With this provision, fluctuations of the volume-averaged Doppler velocities were used to obtain turbulence characteristics in stratocumulus.

It was found that the radar-measured drizzle fall velocity contributed to a rather small velocity standard deviation in the range between 0.05 and 0.1 m s^{-1} and that the air vertical velocity and the drizzle fall velocity exhibited a negative correlation in the upper part of the cloud. Although this had an impact on the turbulent energy and dissipation rate deduced from the Doppler velocity, it did not affect the integral scales, which were found to be in good agreement with the *in situ* measurements. Profiles of this key variable were obtained through the entire cloud-capped boundary layer, while estimates of the dissipation in the lower 2/3 of the boundary layer were in excellent agreement with the *in situ* measurements and consistent with the dominant production terms in the turbulent kinetic energy budget [571].

9.2.5.5 *The Stemme S10-VT motor glider for mountain meteorology*

Measurement technologies that emerged during the past two decades have enabled the development and construction of small airborne sensors and devices for measuring basic meteorological quantities like, for instance, wind velocity vectors, temperature, pressure, and humidity. Today, these systems are operated onboard engine-powered research aircraft to systematically probe the atmosphere in much the same way as was previously done in some recent field campaigns devoted to mountain meteorology research, such as the *Mesoscale Alpine Programme* (MAP) [104], the *Terrain-Induced Rotor Experiment* (T-REX) [362], or the *Deep Propagating Gravity Wave Experiment* (DEEPWAVE) [313]. Flow probes installed on small aircraft have also been of current use for atmospheric research. Good examples of such devices are the "Best Aircraft Turbulence Probe" (BAT-probe) [208, 372] as well as a variety of existing sensors (including flow probes for wind measurements) that have been installed on motor gliders, such as, for example, the "Dimona" from MetAir [658]. Such motor gliders have been proved to be valuable tools for mountain meteorology because they can fly at low altitudes and reach valleys that would not be accessible to larger aircraft. Other examples of small aircraft that have been equipped with turbulence probes are the ultralight aircraft [620]. On the other hand, the use of small microelectromechanical system (MEMS) sensors has allowed the construction of lightweight instrumentation that can be installed even in small unmanned aircraft of a few kilograms, which are mainly employed in boundary layer meteorology [743].

The technological advances have also allowed unmanned aircraft to collect thermodynamic data of the same quality as larger research aircraft with the benefit of minimizing the ambient flow perturbation by the instrumentation itself. A measurement system similar to those employed by unmanned aircraft on a motor glider – the "Stemme S10-VT" motor glider – was designed, built, and deployed by Wildmann et al. [975] for three-dimensional, high-resolution *in situ* wind and turbulence measurements during soaring flights over mountain barriers. Similar instrumentation was previously deployed in the Perlan2 glider in 2019, which was developed within the Perlan project and equipped with a turbulence probe for measurements of stratospheric mountain waves [111]. In particular, the Stemme S10-VT motor glider was equipped with a newly developed sensor suite consisting of a five-hole probe (for the measurement of the wind velocity), an inertial navigation and global navigation satellite system (for the measurement of the aircraft motion vector), two temperature sensors (of which one is a semiconductor-based sensor and the other is a PT100 sensor for temperatures below $-40°C$), and a capacitive humidity sensor. A detailed description of the components of the system, as well as its calibration and performance, can be found in Ref. [975]. The Stemme S10-VT glider is an aircraft with a wingspan of 23 m and is fed by a motor with a retractable propeller so that it is essentially a sailplane. In particular, the inertial navigation and global navigation satellite system was chosen because of its suitability in airborne applications [365].

The onboard instrumentation can measure turbulent fluctuations of the three-dimensional wind vector up to frequencies of 100 Hz. However, only data up to 10 Hz can effectively be analyzed because the length of the flow tube (of about 1 m) affect the small-scale fluctuations. However, if the largest eddies have sizes $\gtrsim 100$ m and the flight speeds are less than 100 m s^{-1}, the measured data will correspond to a significant portion of the inertial subrange and therefore they will allow retrieving the EDR from second-order structure function fits. This operation consists of fitting the theoretical second-order structure function (1.69), written as

$$S_k(\tau) = C_k \varepsilon^{2/3} \left(\langle u \rangle \tau \right)^{2/3} \tag{9.53}$$

to the measured structure function given by

$$S_m(\tau) = \langle [u(t) - u(t+\tau)]^2 \rangle, \tag{9.54}$$

in the range from $\tau_{\min}$ to $\tau_{\max}$. In the above relations, $C_k \approx 2$ is the Kolmogorov constant, $u(t)$ is the horizontal wind velocity in the main wind direction, and the brackets indicate averages. In addition, the turbulent kinetic energy is calculated from three wind components using Eq. (1.40). The overall accuracy of the wind velocity vector measurements is estimated to be on the order of ~ 0.5 m s^{-1}. Within the framework of the *Southern Hemisphere Transport, Dynamics, and Chemistry* (SouthTRAC) field campaign, 30 research flights were performed with the Stemme S10-VT motor glider aircraft from September 2019 to January 2020 along the Andes mountain range. Wildmann et al. [975] reported the statistical analysis of the observations by discriminating measurements performed in flights with engine support from those without engine support (i.e., soaring flights) in the lee waves of the Andes. It was found that except for the case of low overall turbulence in wave soaring flights, the energy spectrum is not affected by engine backwash and/or vibrations. The time series of all flights show that during wave soaring, turbulence is generally weak with short outbursts of severe turbulence. In particular, the observed vertical air speeds suggest that 40.6% of the soaring flights are in weak, 14.1% in moderate, and only 0.4% in strong mountain waves. In the quasi laminar flow of the wave lift, the EDRs appeared to be very low ($\sim 10^{-5}$ m^2 s^{-3}) and the turbulent kinetic energy was always below 0.5 m^2 s^{-2}. A case study where the measurements on 11 September 2019 were compared with a high-resolution numerical weather prediction model, showed that data collected from soaring flights can be used for model validation on the mesoscale and within the troposphere.

9.3 LIDAR systems

The advancements in electronics and the advent of Global Navigation Satellite Systems for georeferencing have led to numerous airborne LIDAR applications. In particular, LIDAR sensors can be classified into three main groups: discrete return, full waveform, and profiling sensors. Profiling sensors are the simplest ones because they can record only one return pulse, while discrete return sensors are the more advanced because they can record multiple returns. On the other hand, the full waveform sensors record a digitized profile of the full return pulse. Advanced LIDAR systems not only capture the reflected signal, but also provide georeferencing of the three-dimensional coordinates of the laser returns [732]. The basic principle of operation of a conventional LIDAR has been elucidated in Section 7.6, while the concepts of monostatic and bistatic laser radars have been introduced in Section 7.8.

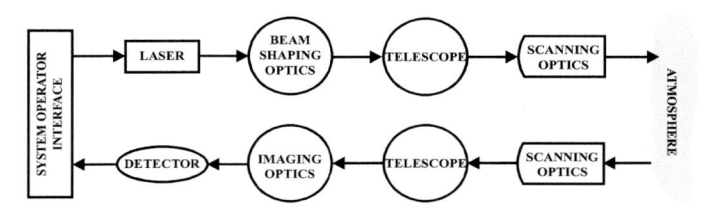

Figure 9.4: Block diagram showing the operation of a direct detection LIDAR. Figure adapted from Sabatini et al. [786].

Typical LIDAR systems operate with the following types of lasers: CO_2 $(9.2 \leq \lambda \leq 11.2 \ \mu m)$, Er:YAG[5] $(\lambda = 2 \ \mu m)$, Nd:YAG[6] $(\lambda = 1.06 \ \mu m)$, Raman Shifted Nd:YAG $(\lambda = 1.54 \ \mu m)$, Nd:YAG $(\lambda = 1.06 \ \mu m)$, GaAlAs[7] $(0.8 \leq \lambda \leq 0.904 \ \mu m)$, HeNe[8] $(\lambda = 0.63 \ \mu m)$, and Frequency Doubled Nd:YAG $(\lambda = 0.53 \ \mu m)$ [452]. According to its receiver detection technique, a LIDAR system can be either of *direct* or *coherent* detection type. Direct detection LIDARs are analogous to conventional passive optical receivers in that the inbound radiation is focused onto a photosensitive element generating a current that is proportional to the incident energy [786]. Figure 9.4 depicts a block diagram showing the operation principle of a direct detection LIDAR. Direct detection is based on molecular or Rayleigh backscattering. A drawback of this technique is the 2 to 3 GHz signal broadening of the detected signal, which is caused by rapid Brownian motion of molecules [796]. Modern approaches employ either molecular absorption cells or fringe images from interferograms combined with image processing to determine the Doppler shift from changes in the fringe radius. For example, in this case, aerosol and molecular signals can be identified by the width of the fringes, while air density, temperature, pressure, and humidity can be derived from molecular scattering only. Among the promising approaches, there are those designed to exploit either the shape of recorded Rayleigh or Raman scattering signals.

In coherent detection LIDARs the interferometer can be either heterodyne or homodyne. As it is well known, coherent detection exploits Doppler-shifted backscatter from aerosol particles whose sizes are bigger than or comparable to the laser wavelength (Mie backscattering). For comparison with Fig. 9.4, Fig. 9.5 shows a block diagram of a heterodyne coherent detection LIDAR. In this case, the laser emitter generates a signal, which in a monostatic system enters

---

[5]This is the abbreviation for Erbium-doped Yttrium Aluminum Garnet laser (Er:$Y_3Al_5O_{12}$), which typically emits infrared light with $\lambda = 2940$ nm.

[6]ND:YAG stands for Neodymmium-doped Yttrium Aluminum Garnet, which is a crystal used as a laser medium for solid-state lasers.

[7]GaAlAs stands for Gallium-Aluminum-Arsenide, which is a semiconductor alloy employed as the light confinement layer in both single- and double-heterostructure diode lasers.

[8]HeNe lasers are gas lasers that use a mixture of Helium and Neon as the gain medium excited by an electrical discharge.

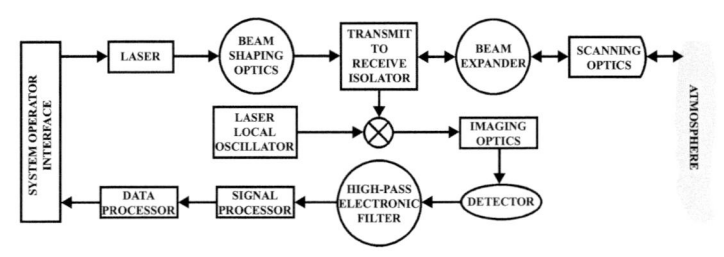

Figure 9.5: Block diagram showing the operation of a coherent detection LIDAR. Figure adapted from Sabatini et al. [786].

a Transmit-to-Receive (T/R) switch. This permits the transmitter and receiver to operate through a common optical aperture. As the optical signal enters the beam expander (or output telescope) and the scanning optics, it is directed to the target. In a monostatic system (see Section 7.8), the incoming radiation that is reflected by the target is now collected by the scanning optics and the beam expander before reaching the T/R switch, which conveys the received signal to an optical mixer where it is combined with a reference signal generated by a local oscillator. The combined signal is then focused onto a photosensitive detector by the imaging optics, where an electrical signal is generated and then passed through a high-pass electronic filter to remove any low-frequency components. As a final step, the target information contained in the high-frequency components is extracted from the electrical signal by signal and data processing. It is important to recall that bistatic systems operate without a T/R switch and that in contrast to heterodyne receivers, homodyne receivers use the transmitter source as the local oscillator rather than requiring a separate laser source. In addition to conventional homodyne receivers, offset homodyne receivers have also been developed. In this case, the local oscillator beam portion is frequency shifted from the transmitter beam [786]. Although coherent detection is very simple and sensitive, it depends on the concentration of aerosol particles in the surrounding airflow, which, in turn, depends on the meteorological conditions and altitude from the ground level. Since the aerosol particle concentration decreases rapidly with altitude, continuous coherent detection is well-suited only for flights at low altitudes and near to the ground, as is the case for helicopter flight altitudes, and therefore it is not applicable for cruising aircraft altitudes [796].

Laser signals possess a high radiance and collimation that have allowed their use to determine target distances with high accuracy. In response to this, a number of Laser Range Finders (LRF) and Laser Target Designators (LTD) have been developed over the last 20 years or so. For example, a state-of-the-art laser range finder is the Nd:YAG LRF ($\lambda = 1064$ nm), which is based on the principle of pulse time-of-flight measurement. Other LRF systems in current use are based on Er:fiber, Raman-shifted Nd:YAG, and CO_2 lasers. Details of the operation principle and architecture of a typical LRF transmitter and receiver system can be found in Refs. [785, 786]. The benefits of using LIDAR as part of a multi-sensor system for airborne applications have already been discussed in Section 9.2.5.1.

9.3.1 *LIDAR remote sensors*

A variety of LIDAR technologies have been developed in the last years to provide atmospheric and surface properties [965, 460, 955]. In particular, the LIDAR operations in the framework of NASA's Cloud Aerosol LIDAR and Infrared Pathfinder Satellite Observations (CALIPSO [991]) and Ice, Cloud, and land Elevation Satellite (ICESat [593]) programs and more recently the ESA's Aeolus wind satellite [454] have opened a new era of LIDAR developments and applications [955]. For example, Doppler LIDARs with different measurement capabilities are widely adopted by the wind energy industry [509, 102]. On the other hand, new areas in scientific research and applications with cheaper ground-based instruments and ultra-light, affordable drones are being highlighted by the progressive miniaturization of sensors and the advent of photon-counting devices [170]. In many cases, three-dimensional mappings and LIDAR images are required to better understand the land-atmosphere interactions [188, 282]. Although there is still room for further development of new LIDAR technologies, applications are progressively occupying most of the LIDAR activities today. For example, airborne LIDAR systems are increasingly used in remote sensing applications because they can operate at all times and during all-weather conditions. Airborne LIDAR systems are employed in most regions of the world for the remote sensing of forest areas [393] and offer great flexibility in data capture, high speed of data delivery, better elevation accuracy in difficult terrain, ease of Geographic Information Systems (GIS) integration, high sample densities, and a high degree of automation [407].

The first successful attempt to remotely detect CAT in the free troposphere using an incoherent Rayleigh LIDAR system was recently reported by Hauchecorne et al. [387]. For this purpose, a ground-based Rayleigh LIDAR was designed and implemented to perform remote detection and characterization of the atmospheric variability induced by CAT on vertical scales. For aeronautical applications, the classification of CAT depends on the vertical acceleration experienced by the aircraft, which is mainly due to the vertical wind and also to fluctuations in atmospheric density, temperature, and horizontal wind. A Rayleigh-Mie LIDAR can measure the vertical profile of light emitted by a pulsed visible or ultraviolet laser and backscattered by atmospheric molecules and particles. The LIDAR equation in elastic scattering mode provides the expression for the backscattered measured signal in the altitude interval $(z, z + dz)$ and reads as follows [387]

$$S(z, z + dz) = E_0 T_{\text{atm}}^2(0, z) Q_{\text{LID}} \frac{A_{\text{tel}}}{z^2} [\beta_r(z) n_r(z) + \beta_m(z) n_m(z)], \qquad (9.55)$$

where E_0 denotes the number of photons emitted by the laser, T_{atm} is the atmospheric transmission between the ground level ($z = 0$) and the altitude z, Q_{LID} is the optical efficiency of the LIDAR, A_{tel} is the telescope area, $\beta_r(z)$ and $\beta_m(z)$ are the backscattering cross sections of air molecules and aerosol particles at height z, while $n_r(z)$ and $n_m(z)$ are the concentrations of air molecules and aerosols, respectively. These systems have been traditionally used to infer the atmospheric vertical density profile at altitudes above 30 km where the air is depleted of aerosols and the vertical temperature profile can be determined from the inferred vertical density profile [386]. However, a more recent application of the Rayleigh-Mie LIDAR has been the detection of atmospheric density fluctuations caused by turbulence as was reported by Hauchecorne et al. [387]. To this purpose only data for which the Mie signal is $< 5\%$ compared to the Rayleigh signal will be of practical use, which typically occurs in the upper troposphere at heights between 5 to 10 km where the formation of clouds is prevented by the presence of dry air. Under the assumption of isotropic turbulence with the scale of the largest eddies being $\gtrsim 100$ m and the hypothesis that the turbulent structures are advected with a mean wind velocity of 10 m s^{-1}, the LIDAR signal should be recorded with at least a vertical resolution of 50 m and a time integration of 5 s. While this is feasible with modern LIDAR data acquisition systems, Hauchecorne et al. [387] proposed to average the square amplitude of the density fluctuations (FQ) instead of averaging the signal itself to circumvent the problem of having too low signal-to-noise ratios for a single measurement. From their listed instrumental and atmospheric parameters, the number of photo-electrons counted at such spatio-temporal resolution was $N_{ph} = 36000$ at 10 km of altitude from the ground level. If only Poisson noise is considered, the accuracy in relative value of the Rayleigh LIDAR measured as $\Delta\rho/\rho = 1/\sqrt{N_{ph}}$ will then be of about 0.5%. It was possible to detect in 1 min turbulent layers larger than 100 m in the stratosphere and 200 m in the troposphere, yielding some confidence in the feasibility of CAT detection using this technique.

The instrumental setup and data processing are described in detail in Ref. [387]. A schematic diagram of the optical layout of the Rayleigh LIDAR is shown in Fig. 9.6. The signal consists of a frequency doubled Nd:YAG laser emitted 6 m off-axis of the receiving telescope. This provision is necessary to avoid excessive illumination of the Photo-Multiplier Tubes by low altitude signal backscattering. The laser beam with a 532 nm wavelength is sent to the sky with a pulse frequency of 50 Hz, duration of 7 ns, and mean power of 15 W. The backscattered light is collected by a Newton telescope with an aperture of 530 nm and conveyed to a narrow-band interference filter. The filtered signal is passed through a polarizer where it is split into two component channels. Approximately 99% of the orthogonal component is detected by the photomultiplier tube R7205-01 PMT2 and 95% of the parallel component by the photomultiplier PMT1. Due to the high intensity of the backscattered signal both photomultipliers are used in analog mode. The signal from the two channels is filtered and

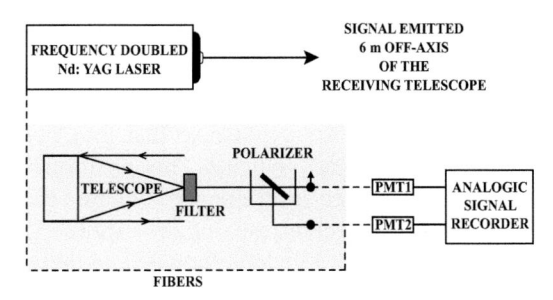

Figure 9.6: Schematic diagram showing the operation of the transmitter and receiver of a Rayleigh LIDAR. Figure adapted from Hauchecorne et al. [387].

digitized pulse to pulse by a dual channel 10 MHz comprised of a 16-bit analog-to-digital converter [387]. Field measurements with this system were performed at the Observatoire de Haute-Provence in France between December 8, 2008, and June 23, 2009. The observed FQ in bidimensional LIDAR signal showed excess compared to the estimated contribution of the instrumental noise that was attributed to the presence of CAT with a 95% confidence level. The turbulence parameters C_n^2 and C_T^2 were found to be consistent with those published in the literature using stratosphere-troposphere radars [238, 239, 989]. The Kolmogorov two-third law was found to apply to scales corresponding to the thickness of the detected turbulent layers, typically at altitudes of 100 to 200 m in the troposphere [239]. The implementation of a high spectral resolution filter discriminating between molecular and aerosol scattering along with the use of a laser with a higher repetition rate in the kHz range will lead to improvements in CAT detection by a Rayleigh LIDAR [387].

9.3.2 Modern Doppler LIDAR systems

It was outlined in Chapter 7 that atmospheric phenomena such as those involving the dynamics of convective clouds, the kinematics of the boundary layer, and atmospheric turbulence have been studied for decades using conventional Doppler radars. However, since most meteorological radars operate at wavelengths in the range between 3 and 10 cm, they are useful in the detection of small particles, like water droplets and hailstones, and therefore they are only effective to detect severe storms and of little use in the detection of CAT [786]. In contrast, Doppler LIDARs have been proved to be more versatile remote sensors in the sense that they enable the measurement of clear-air wind velocities [248, 695]. Measurement techniques based on Doppler LIDAR methods assume that aerosol particles follow the airflow streamlines and therefore their velocities provide a good estimate of the actual airflow speed. By Laser Doppler Velocimetry (LDV)[9] very low wind velocities can be sensed by means of heterodyne detection.

Modern Doppler LIDAR systems are designed for use onboard the aircraft. They are fitted outside into the aircraft nose so that they perform frontal view measurements of the clear-air wind velocity. This position is optimal because the measurements obtained are free from the aerodynamic blockage due to flow past the fuselage or the wings. Recent research on airborne LIDAR systems focused on improving the detection of wind velocities and their variations in order to reduce the pilots' workload by enhancing the autopilot and the autothrottle control logic. In particular, such devices have been designed to improve safety during delicate flight phases, including take-offs and landings. During take-offs and landings air safety can be seriously compromised by strong wind shear or, by even more severe events, such as microbursts, which have caused several aircraft accidents. For example, microbursts typically last for about 15 minutes and span a distance from three to four kilometers. When flying through a wind shear during take-off or landing, the aircraft may experience sudden changes in its dynamical state and the danger lies in the lack of altitude for a safe recovery. In addition, when encountering a microburst during approach, the aircraft experiences a strong head wind, increasing its airspeed and lifting it above the correct path. To correct the flight profile, pilots and autopilots must reduce the engine thrust. However, within a few seconds the aircraft encounters a sudden downdraft and strong tail winds, causing a decrease in its airspeed and a sudden increase in the descent rate and safe recovery by increasing the engine thrust may not occur with sufficient rapidity to avoid the plane crash.

9.3.3 Airborne ultraviolet (UV) Doppler LIDAR

Most research on CAT detection during air navigation has focused on short range detection between 50 to 300 m ahead, allowing the pilots to modify the flight controls just before a CAT encounter [391, 797]. If CAT detection at cruise speeds occurs at least 2 min before the encounter, which corresponds to a frontal distance of about 30 km from the turbulent area, provisions can be taken for complete protection of passengers and crew as there would be enough time for the pilot to warn the passengers to return to their seats and fasten their seat belts. However, a much safer strategy during air navigation would be to detect the CAT area at longer ranges. In general, the vertical velocity component is the most important one within the CAT area since it has a direct impact on the angle of attack, the lift coefficient, and the acceleration experienced by the flying aircraft. Therefore, a long-range detection would require performing a direct measurement of the vertical wind fluctuations. However, this cannot be accomplished

---

[9]LIDARs provide spectral and geometrical laser beams of very narrow widths so that sufficient backscattered radiation is provided, thereby allowing measurement of very low velocities using heterodyne detection techniques.

by conventional LIDAR systems because of their reduced sounded area due to the weak projection of the line of sight over the vertical axis [289]. An alternative technology for long-range CAT detection uses a direct-detection ultraviolet (UV) LIDAR, which can perform long-range detection of density fluctuations associated with the vertical velocity component of the airflow. This is possible because UV LIDAR systems provide intense backscattering from air molecules due to the λ^{-4} dependence of the Rayleigh molecular backscattering. A good theoretical evaluation of the UV LIDAR detection characteristics is provided by Feneyrou et al. [289] and most of the discussion that follows is based on their description.

Random vertical velocity fields are generated using a von Kármán spectrum, where the relative proportion of spectral power at high frequencies is more accurately represented [151]. The generated random vertical velocity fields are then transformed into either density or temperature fluctuations, which are inverted to provide an estimation of the vertical component of the airflow velocity, $\hat{v}_z$. The detection performance of the UV LIDAR is obtained by comparing this estimated component with the initial turbulent field. A maximum detection range can be defined from the correlation between the input vertical velocity and that derived from the computed LIDAR signal contaminated with the photon noise. In the von Kármán model, the velocity spectrum along the vertical coordinate is given by [289]

$$\Phi_{v_z}(\omega) = \frac{3\sigma_{v_z}^2 L_{v_z}}{\pi v_a} \frac{3 + 8\left(1.339 L_{v_z}\omega/v_a\right)^2}{\left[1 + \left(1.339 L_{v_z}\omega/v_a\right)^2\right]^{11/6}}, \tag{9.56}$$

where σ_{v_z} is the rms intensity of v_z, L_{v_z} is the outer scale of turbulence, and v_a is the plane speed. At this outer scale, turbulence is fueled by the kinetic energy. Since the spectrum calculation is performed for altitudes higher than 2000 ft, the length of the outer scale is assumed to be constant at $L_{v_z} = 2500$ ft (762 m). The turbulence intensity is determined as a function of altitude and the turbulence probability of occurrence. For example, as specified in Table 1 of Ref. [289], a probability of 10^{-3} at altitudes between 3 and 15 km from the ground level corresponds to rms turbulence intensities between 2.9 and 1.0 m s^{-1} and is defined as moderate. This corresponds on average to one CAT encounter for every few intercontinental flights. Severe CAT is encountered at a lower probability of 10^{-4}, corresponding to turbulence intensities ranging from 4.2 to 2.5 m s^{-1} in the same range of altitudes as before. In both cases, as the altitude increases, the turbulence intensity decreases.

The turbulence velocity component is generated by passing the band-limited white noise through a properly designed filter $H_{v_z}(\omega)$ defined as $\Phi_{v_z}(\omega) = H_{v_z}(\omega)\hat{H}_{v_z}(\omega)$, where $\hat{H}_{v_z}(\omega)$ is the complex conjugate of $H_{v_z}(\omega)$. Since $\Phi_{v_z}(\omega)$ is not a rational fraction, the shape of the filter is approximated by fitting the numerator of Eq. (9.56) with a second-degree polynomial and the denominator by a third-degree polynomial for the ratio $x_{v_z} = L_{v_z}\omega/v_a$ so that the following expression is obtained for values of x_{v_z} up to 50 [289]

$$H_{v_z}(\omega) = \left(\frac{\sigma_{v_z}^2 L_{v_z}}{\pi v_a}\right)^{1/2} \frac{1 + 2.7478ix_{v_z} - 0.3398x_{v_z}^2}{1 + 2.9958ix_{v_z} - 1.9754x_{v_z}^2 - 0.1539ix_{v_z}^3}, \tag{9.57}$$

where $i = \sqrt{-1}$. The turbulent vertical velocity is obtained as the Fourier transform of the product of the random vertical velocity and the shaping filter. When a fluid particle undergoes a vertical displacement δz with respect to the surrounding fluid, it experiences a relative density and temperature change given by

$$\frac{\delta\rho(x)}{\rho} = -\frac{\delta T(x)}{T} = \frac{N_{BV}^2}{g}\delta z, \tag{9.58}$$

where $N_{BV} = \sqrt{(g/\theta)(\partial\theta/\partial z)}$ is the Brundt-Väisälä frequency, θ is the potential temperature, and g is the gravitational acceleration. In the presence of gravity waves and turbulence, the rms fluctuations of density and temperature are statistically related to the rms fluctuations of the vertical velocity component by the expression

$$\frac{\delta\rho(x)}{\rho} = -\frac{\delta T(x)}{T} = \frac{N_{BV}^2}{g}v_z(x). \tag{9.59}$$

In the troposphere $N_{BV} \sim 0.01$ rad s^{-1}, while in the stratosphere $N_{BV} \sim 0.02$ rad s^{-1}. For wave frequencies lower than N_{BV}, the vertical velocity component will be higher than the expected value at constant density fluctuation. This will also be the case at constant LIDAR signal fluctuation, where the estimated vertical velocity will be higher than the value deduced from the Brunt-Väisälä frequency. At constant pressure and altitude, the molecular scattering coefficient at altitude z is given by

$$\alpha_m(x) = N_0\sigma_R\frac{T_0 p(z)}{p_0[T(z) + \delta T(x)]}, \tag{9.60}$$

where N_0, p_0, and T_0 are the ground molecular density, pressure, and temperature, respectively, σ_R is the molecular scattering cross section, $p(z)$ and $T(z)$ are the pressure and temperature at height z, respectively, and $\delta T(x)$ is the temperature fluctuation associated with turbulence.

The performance evaluation of a UV LIDAR for long-range CAT detection was reported by Feneyrou et al. [289]. In this case, the UV LIDAR system was operating in a backscatter measurement configuration and consisted of a seeded frequency-tripled Nd:YAG laser at 355 nm. As illustrated in Fig. 9.7, a small aperture telescope expands the emitted beam into the atmosphere and a larger aperture telescope is used to collect the scattered beam, which is then filtered when passing through an optical bandpass filter and focused onto a single pixel detector. The fluctuation in the LIDAR signal is assumed to come solely from

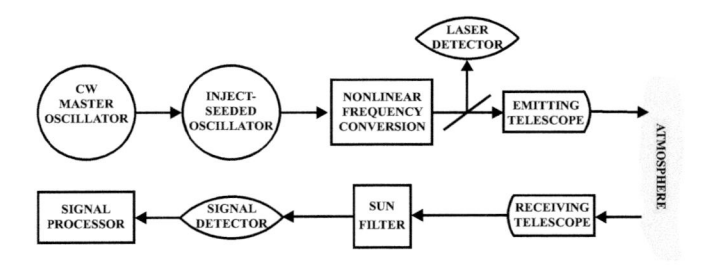

Figure 9.7: Block diagram showing the operation of a direct-detection UV LIDAR system. Figure adapted from Feneyrou et al. [289].

the air density fluctuations, which are the only contributions related to CAT. The LIDAR signal can be written as

$$S(x) \approx \frac{P t_{\text{int}} \lambda T_1 T_2 \tau \rho}{h} \sum_k \frac{\pi R^2}{2(x - k v_a/f)^2} \left[K_a \alpha_a + \frac{3}{8} \alpha_m \left(x - k \frac{v_a}{f} \right) \right]$$

$$\times \ \exp\left[-2 \int_0^{x - k v_a/f} \left(\alpha_a + \alpha_m^0 \right) dz \right] + B, \tag{9.61}$$

where P is the laser power, t_{int} is the integration time, λ is the wavelength, T_1 and T_2 are the transmissions of the emission and collection optics, respectively, R is the radius of the collection optics, x is the range, τ is the pulse duration, α_a is the aerosol scattering coefficient, α_m is the molecular extinction coefficient, α_m^0 is the mean molecular extinction coefficient, ρ is the detector quantum efficiency, K_a is the backscattering to the extinction ratio for aerosols, and B is the photon noise. The form of α_m^0 is known from the standard atmosphere to be

$$\alpha_m^0 = N_0 \sigma_R \frac{p(z)}{p_0} \frac{T_0}{T(z)}. \tag{9.62}$$

The LIDAR signal obtained for $\alpha_m = \alpha_m^0$ is denoted by $S^0(x)$ and obeys a similar expression to Eq. (9.61) with the term $(3/8)\alpha_m(x - k v_a/f)$ replaced by $3\alpha_m^0/(8\pi)$ and $B = 0$. The molecular extinction coefficient is therefore determined by the expression

$$\hat{\alpha}_m = \alpha_m^0 \frac{S(x)}{S^0(x)} + K_a \alpha_a \left(\frac{S(x)}{S^0(x)} - 1 \right) \approx \alpha_m^0 \frac{S(x)}{S^0(x)}. \tag{9.63}$$

From Eqs. (9.59), (9.60), and (9.63), the rms of the vertical velocity component is given by

$$\hat{v}_z(x) = -\frac{g}{N_{\text{BV}}} \frac{\delta \hat{T}(x)}{T} = \frac{g}{N_{\text{BV}}} \left(1 - \frac{\alpha_m^0}{\hat{\alpha}_m(x)} \right) \approx \frac{g}{N_{\text{BV}}} \left(1 - \frac{S(x)}{S^0(x)} \right), \tag{9.64}$$

where by definition $\delta T = 0$ for $\alpha_m = \alpha_m^0$. As a final step, a Wiener filter is applied to $\hat{v}_z$ as computed from Eq. (9.64) and its correlation with the input vertical velocity is calculated. For the LIDAR characteristics listed in Table 2 of Ref. [289], a signal-to-noise ratio of 100 was found for a 15 km detection range, which for the same LIDAR characteristics corresponds to a mean correlation of 0.5 between the input and computed vertical velocity component. The results of the Feneyrou et al.'s [289] analysis showed that moderate CAT can be detected at a distance of 15 km with a probability of a missed alarm of 4% and a false alarm rate of 0.18 per flight hour using a 2 W laser at 355 nm. In addition, it was concluded that turbulence stronger than moderate could be fairly well detected with the same methodology.

The design of a newer airborne, UV direct-detection Doppler wind LIDAR was more recently reported by Herbst and Vrancken [394]. This LIDAR system was proved to be useful for the short-range detection of wake vortices and for implementing turbulence mitigation strategies. It consists of a monolithic, field-widened, fringe-imaging Michelson interferometer, which uses the same fringe-imaging principle described by Cézard et al. [157][10]. However, this new design provides fast and range-resolved airborne measurements of wind speeds in the near field (i.e., 50–300 m) in front of an aircraft. A succinct description of the theoretical performance of a fringe-imaging Michelson interferometer can be found in Ref. [394]. A telescope with a large field-of-view is required for range-resolved detection in the near field. This will allow for full overlap at all ranges to maximize the received signal. However, a large telescopic field-of-view causes an angular distribution of the incident light that has to be compensated for by field widening[11]. In particular, a monolithic fringe-imaging Michelson interferometer is designed to produce linear fringes. This demands that the mirrors of the two arms be inclined to each other. The width of the illuminating beam, d_w, is 10 mm, while the laser wavelength is 354.84 nm and the net inclination, θ, between the mirrors is 17.74 μrad. On the other hand, field widening makes the optical path length difference to be almost independent of the incident angle, α_0, which implies that the refractive indices of arm 1 (n_1) must be different from that of arm 2 (n_2). In the present interferometer case, the refractive index difference, Δn, was maximized by choosing arm 1 to be made of air and the other of glass. This allows for maximization of the lengths of the arms and reduction of the temperature sensitivity [907]. In terms of θ and the angles of refraction (α_1 in air and α_2 in glass), the optical path difference can be written as

$$\Delta L_{op} = 2n_1 l_1 \cos \alpha_1 - n_2 l_2 \left[\cos \alpha_2 + \cos \left(\alpha_2 \pm 2\theta\right)\right], \tag{9.65}$$

---

[10]In particular, these authors considered a dual fringe-imaging Michelson interferometer with inclined mirrors for the measurement of the wind speed as well as the air density and temperature, scattering ratio, etc.

[11]Field widening is the ability that has an interferometer to accept angularly distributed incident light without a reduction of the fringe contrast.

where l_1 and l_2 are the lengths of the arms. Using Snell's law (i.e., $n_0 \sin \alpha_0 = n_1 \sin \alpha_1 = n_2 \sin \alpha_2$, where n_0 is the refractive index of air) and noting that $\cos(2\theta) \sim 10^{-10}$ (i.e., $\theta \to 0$), the above expression becomes

$$\Delta L_{\text{op}}(\alpha_0) = 2n_1 l_1 \left(1 - \frac{\sin^2 \alpha_0}{n_1^2}\right)^{1/2} - 2n_2 l_2 \left(1 - \frac{\sin^2 \alpha_0}{n_2^2}\right)^{1/2}. \tag{9.66}$$

For small α_0, the above form can be approximated to be

$$\Delta L_{\text{op}} \approx 2 \left(n_1 l_1 - n_2 l_2\right) - \sin^2 \alpha_0 \left(\frac{l_1}{n_1} - \frac{l_2}{n_2}\right) - O(\alpha_0^4), \tag{9.67}$$

where the first term is the fixed optical path difference when the tilt angle to the incident light, θ_t, is identically zero. The field widening condition, namely

$$w = \frac{l_1}{n_1} - \frac{l_2}{n_2} = 0, \tag{9.68}$$

implies that for field widening the second term in Eq. (9.67) vanishes. For a nonvanishing tilt angle in the Michelson interferometer, the expressions for the fixed optical path difference, ΔL_{op}^0, and w will be [171]

$$\Delta L_{\text{op}}^0(\theta_t) = 2n_1 l_1 \left(1 - \frac{\sin^2 \theta_t}{n_1^2}\right)^{1/2} - 2n_2 l_2 \left(1 - \frac{\sin^2 \theta_t}{n_2^2}\right)^{1/2}, \tag{9.69}$$

$$w(\theta_t) = \frac{l_1}{\sqrt{n_1^2 - \sin^2 \theta_t}} - \frac{l_2}{\sqrt{n_2^2 - \sin^2 \theta_t}}, \tag{9.70}$$

which must be solved simultaneously to determine the optimal arm lengths for field widening under the condition that $w(\theta_t) = 0$ in Eq. (9.70). Evaluation of the optimal arm lengths into Eq. (9.69) provides the optimal optical path difference. Optimization of the path difference was accomplished by setting $\theta_t = 2°$, allowing for the option of a second detector in back reflection, which will nearly double the efficiency of the interferometer. Fused silica was employed as the glass medium because of its high transmission in the UV. The refractive index of the fused silica, n_g, at wavelength λ in μm relative to air at the reference temperature T_0 and pressure p_0 is calculated using the Sellmeyer equation [425]

$$n_{g_r}(\lambda) = \left(1 + B_1 \frac{\lambda^2}{\lambda^2 - C_1} + B_2 \frac{\lambda^2}{\lambda^2 - C_2} + B_3 \frac{\lambda^2}{\lambda^2 - C_3}\right)^{1/2}, \tag{9.71}$$

where B_1, B_2, B_3, C_1, C_2, and C_3 are the Sellmeyer coefficients. In Eqs. (9.69) and (9.70) n_2 is equal to the absolute refractive index of the silica glass $n_{g_a}(\lambda) = n_a n_{g_r}(\lambda)$, while n_1 is equal to the absolute refractive index of air, n_a, at $T_0 = 22°$C. The fixed optical path difference as a function of λ can be calculated using Eq.

(9.71) into Eq. (9.69) and put into the cosine shaped Michelson monochromatic transmission function given by

$$I(x,y,v) = \mathrm{FI_0}\left[1 + \mathcal{V}\cos\phi\right], \tag{9.72}$$

where the linear interference fringes are aligned perpendicular to the x-axis and parallel to the y-axis, $\mathcal{V}$ is the instrumental interference contrast, and $\phi = 2\pi v/c(\Delta L_{\mathrm{op}}^0 - 2\theta x)$ is the fringe phase. The optimal values of l_1 and l_2 are derived by means of an iterative optimization procedure and the optimal path difference is varied until one fringe period exactly spans a frequency of 10.7 GHz. Moreover, thermal compensation is obtained when the following condition is satisfied

$$
\begin{aligned}
\frac{\partial \Delta L_{\mathrm{op}}^0(\theta_t)}{\partial T}
&= 0 \\
&= 2\gamma_1 l_1 \left(n_1^2 - \sin^2\theta_t\right)^{1/2} + 2\beta_1 n_1 l_1 \left(n_1^2 - \sin^2\theta_t\right)^{-1/2} \\
&\quad - 2\gamma_2 l_2 \left(n_2^2 - \sin^2\theta_t\right)^{1/2} - 2\beta_2 n_2 l_2 \left(n_2^2 - \sin^2\theta_t\right)^{-1/2}, \tag{9.73}
\end{aligned}
$$

where $\gamma_i = (1/l_i)\partial l_i/\partial T$ is the coefficient of the linear thermal expansion of material $i = (1,2)$ and $\beta_i = \partial n_i/\partial T$ is the thermal coefficient of the refractive index. The thermal coefficient of difused silica, β_2, is calculated as follows

$$\frac{dn_{g_a}(\lambda,T)}{dT} = \frac{n_2^2(\lambda,T_0) - 1}{2n_2(\lambda,T_0)}\left(D_0 + 2D_1\Delta T + 3D_2\Delta T^2 + \frac{E_0 + 2E_1\Delta T}{\lambda^2 - \lambda_{TK}^2}\right), \tag{9.74}$$

where $n_2(\lambda,T_0)$ is the refractive index of the glass at the laser wavelength λ in μm and at the reference temperature $T_0 = 22°\mathrm{C}$, $\Delta T = T - T_0$, T is the temperature in $°\mathrm{C}$, D_0, D_1, D_2, E_0, E_1, and λ_{TK} are thermal dispersion coefficients of the glass. Integration of Eq. (9.74) yields the change of the absolute refractive index so that the absolute refractive index of glass at temperature T is

$$n_{g_a}(\lambda,T) = n_{g_a}(\lambda,T_0) + \Delta n_{g_a}(\lambda,\Delta T). \tag{9.75}$$

In addition, the thermal coefficient of air, β_1, and the refractive index of air $n_a(\lambda,T,p)$ are calculated according to

$$\beta_1 = \frac{dn_a(\lambda,T)}{dT} = -0.00367\frac{n_a(\lambda,T,p) - 1}{1 + 0.00367(T/°\mathrm{C})}, \tag{9.76}$$

$$n_a(\lambda,T,p) = 1 + \frac{n_a(\lambda,15°\mathrm{C},p_0) - 1}{1 + 3.4785 \times 10^{-3}(T/°\mathrm{C} - 15)}\frac{p}{p_0}\lambda^2, \tag{9.77}$$

where $p_0 = 0.101325 \times 10^6$ Pa is the standard pressure at 20°C, and $n_a(\lambda,T,p)$ is the absolute refractive index of air at pressure p and temperature T (in °C). Temperature tuning can be done either at constant density (i.e., $\beta_1 = 0$), as in the case of isochoric heating, or at constant pressure, when the fringe-imaging Michelson interferometer is enclosed in a sealed container [394]. When inserting the field-widening equation into the temperature compensation condition, the optimized values of the coefficient of thermal expansion of the spacers for constant density and constant pressure are, respectively

$$\gamma_{1,CD} = n_{g_r}^2 \left(\gamma_2 + \frac{1}{n_{g_r}} \right),$$ (9.78)

and

$$\gamma_{1,CP} = \frac{n_{g_a}(T_{op})}{n_a^2(T_{op})} \beta_2 + \frac{n_{g_a}^2(T_{op})}{n_a^2(T_{op})} \gamma_2 - \frac{1}{n_1} \beta_1.$$ (9.79)

When the interferometer is used for measurements, it will be heated up from the reference (fabrication) temperature $T_0 = 22°C$ to the operation temperature of $40°C$ at a constant density. In the case of temperature tuning at constant density, $\beta_1 = 0$ and $\gamma_1 = 15.5$ ppm K^{-1}. For $\Delta T = 18$ K, the change of length of arm 1 is $\Delta l_1 = 3.1$ μm, while for arm 2 is $\Delta l_2 = 0.15$ μm. Such changes are compensated for by making the initial air arm length, l_1, 3 μm shorter. Reports on the performance of the setup indicate that field widening is an important concept for near-field detection and a large required field of view. Range-resolved line-of-sight Doppler wind measurements with standard deviations of the wind speed of $\lesssim 1$ m s^{-1} were obtained independently of the atmospheric conditions at distances between 60 and 120 m [394]. Such low standard deviations are very important for wake vortex impact alleviation purposes [268].

9.4 The DELICAT airborne LIDAR

In the framework of the EC FP7 project DELICAT (which stands for *DEmonstration of LIdar based Clear Air Turbulence detection*), a UV LIDAR system was designed and manufactured for airborne medium-range (i.e., from 10 to 30 km) CAT detection by a European consortium consisting of industrial partners, research institutes, and universities. This project started in 2003 and its main objective was to validate LIDAR as an advanced and new technology for medium-range CAT detection aimed at ensuring enough protection for passengers and crew against such turbulence encounters [932]. Part of this protection consists of taking actions inside the cabin, such as fixing objects and warning passengers to fasten their seat belts.

The physical principles of UV LIDAR systems for turbulence detection have already been described in Section 9.3.3 based on Refs. [289, 394] and therefore here we shall mainly focus on the DELICAT LIDAR design and functionality. As was previously outlined, a LIDAR system is composed of two main parts, namely a laser transmitter that emits light into the atmosphere and a receiver system that collects and detects the backscattered radiation. Similar to other UV LIDAR setups, the DELICAT LIDAR sounds the atmosphere in front of the aircraft by analyzing the backscattered signals due to local air density fluctuations. To this purpose, DELICAT LASER has been equipped with a special beam steering system that directs the laser beam exactly into the flight path. To achieve molecular density fluctuations of about 1% level, as is appropriate for typical CAT, the DELICAT LIDAR operates by sending UV laser pulses into the atmosphere via a transmitter based on a monolithic, high-power Nd:YAG ring laser of the MOPA design issued from the DLR water vapor, differential absorption DIAL LIDAR WALES [992]. The laser emits infrared pulses at 1064 nm at a rate of 4 kHz and a pulse length of 7.7 ns (FWHM), with a resulting energy per pulse greater than 400 mJ. The infrared signal is then transformed into UV radiation by non-linear frequency conversion. Using a potassium titanyl phosphate crystal, part of the infrared radiation is converted into visible/green radiation at 532 nm. This and the residual infrared radiation are then guided into a Beta barium borate crystal for sum-frequency generation, resulting in 355 nm laser pulses of ∼ 80 mJ. A set of dichroic mirrors is used to separate the residual infrared and green radiations from the UV light so that only UV radiation is directed to the atmosphere. The beam emitted by the DLR LIDAR transmitter is expanded to a diameter of ∼ 15 mm with a divergence of 200 μrad [932].

A telescope of Newtonian architecture for efficient light collection, front optics for filtering and beam forming, and back end optics for detections are the modular subsystems composing the DELICAT LIDAR receiver. The telescope has a window aperture of 140 mm and allows for a small secondary, flat mirror. The telescope focus defines a 1 mrad field of view and a collimation optic generates a parallel beam of a few mm diameter. The latter is fed through a narrow interference filter for blocking the solar background [932]. The backscattered radiation with perpendicular polarization to the emitted signal is separated from the parallel one by the back-end receiver, which are then both fed onto photomultiplier tube detector modules. The synopsis of the DELICAT LIDAR system is similar to that shown in Fig. 9.7. DLR WALES detection modules are integrated into the two polarization modules. The photomultiplier tube digitized signals have a sample rate of 30 MHz, resulting in a LIDAR range resolution of 5 m. The DLR data acquisition stores one LIDAR signal with a 15 km range per laser shot at a rate of 100 Hz, which results in 10 Gbytes per hour of stored data.

The beam steering system is an important integrated component of the DELICAT LIDAR whose main function is to keep the transmit/receive beam parallel to the flight path direction as the aircraft changes its attitude, for example, due to residual movements of the autopilot, variations of the angle of attack due to decreasing weight, and encounters with light turbulence with an identified bandwidth. Invariance of the received beam against such movements is achieved by a couple of two-axes movable mirrors, which are controlled in real time by a unit that gathers its data from the aircraft ARINC stream containing the appropriate attitude and flow angles. Outside the aircraft and inside the protecting fairing there is a third, fixed mirror. Thus, the laser beam after having traveled over the two movable mirrors and through the UV-quartz window converges to this third mirror, where it is bent in a forward direction. For measuring tests the DELICAT LIDAR equipment was installed inside the cabin of the Cessna Citation II aircraft (PH-LAB)[12] of the *Nationaal Lucht- en Ruimtevaartlaboratorium* (NLR, the Netherlands), where the LIDAR was housed in a rigid rack mounted upon the seat rails at both sides of the cabin to reduce loads. Testing the airborne performance of the LIDAR system as a sensitive turbulence sensor will require the beam to point right into the aircraft's heading direction with an accuracy $\lesssim 0.1°$. This requirement is supported by equipping the two movable mirrors on top of the LIDAR rack with actuators that enable the LIDAR to react to local variations in wind speed, short-term altitude changes due to turbulence encounters, and/or lift variations due to fuel consumption [932, 945]. A comprehensive simulator was developed to evaluate the CAT detection performance of the DELICAT system, which uses atmospheric parametric information on localized turbulent events and aircraft trajectory realization, including air density and temperature, Mie and Rayleigh atmospheric backscatter coefficients, and signal processing for turbulence signal recovery.

The DELICAT flight test campaign under airborne conditions started to be active from July 17th, 2013 to August 13th, 2013. During this period a total of 11 flights, including a shake-down flight, were made for a total airborne duration of 31h20m. However, since these flights were made in the summer time, only a few moderate CAT encounters were recorded. Although the CAT collected data were quite limited, they provided interesting and successful results. The results indicate that air density fluctuations can be measured on the percent level on typical aircraft cruise flights, thereby allowing the detection of CAT with moderate to severe strength. The DELICAT project represents an important step toward future integrated CAT sensors for commercial aircraft.

9.5 The JAXA airborne LIDAR

The prevention of turbulence-induced aircraft accidents, which have been the reason for about half of all aviation accidents in Japan between 2008 and 2017, has motivated the Japan Aerospace Exploration Agency (JAXA) in 2011 to develop a Doppler LIDAR for onboard CAT detection [426, 600]. The development of JAXA's airborne Doppler LIDAR has been performed in incremental steps. In 2001 a 1 nautical mile (1NM) class LIDAR was first developed as a low-output prototype airborne Doppler LIDAR [453]. This system was devoted to prove basic functions and was composed of (1) a laser transceiver, which included a master laser oscillator, a commercial optical communication fiber amplifier, and a heterodyne detector, (2) an optical antenna, which was connected to the laser transceiver by a 10 m long flexible optical fiber cable, and (3) a desktop com-

---

[12]It is a turbofan engine aircraft with a maximum ceiling of 43000 ft and maximum cruise speed of 200 m s^{-1}.

puter for signal processing, control, and analysis. The transmitted laser light is deflected by $\pm 10°$ using a double wedge prism installed in the scanner and a steering mirror is mounted on the top of the aircraft fuselage to convey the signal along the aircraft path. A series of flight tests carried on in 2002 have demonstrated that wind speed measurements can be performed remotely with an accuracy of less than 0.7 m s^{-1} [427].

A second development step was accomplished in 2006 with the construction of a 3NM class prototype airborne Doppler LIDAR, which differed from the 1NM class LIDAR in that the laser output was increased to about ten times, thereby allowing for larger detection ranges. To this end, a high output amplifier using a short 9.1 μm optical fiber with a high erbium doping density was developed [29]. The laser pulse repetition frequency was 4 kHz with several incoherent integrations of 4000, providing a measurement rate of 1 Hz. This upgraded LIDAR was installed in the JAXA Beechcraft Model 65 research aircraft and was proved to give encouraging results for the remote detection of CAT. A third step consisted of increasing about three times the laser output of the 3NM class LIDAR, giving rise to a 5NM class airborne Doppler LIDAR. This was achieved by developing and installing a high output amplifier using a 25 μm core diameter optical fiber that was operating as a second amplifier behind the 3NM class LIDAR amplifier [788]. With this improved technology, the new LIDAR was enabled to measure wind speeds at a maximum range of $\approx$ 9 km in clear weather. As a fourth development step, the design of the 5NM class LIDAR was improved to allow CAT observations at cruising altitudes. This high-altitude LIDAR used a Waveguide Amplifier (WGA) that was developed in 2008 [788]. In this newer device, the signal light generated by the transceiver is amplified by a pump light and emitted from an optical antenna with an effective diameter of 150 mm. The backscattered light is then condensed into the antenna and its Doppler shift is measured in the transceiver before being converted to a wind speed signature by a signal processing unit.

The performance index of the Doppler LIDAR output is quantified in terms of a Figure of Merit (FOM), which is defined as [426]

$$FOM = EN^{1/2}, \qquad (9.80)$$

where E is the optical energy per pulse of the transmission laser light and N is the number of integrations performed during each one-second interval, which has the same value as the pulse repetition frequency. For example, the 1NM class LIDAR had a FOM of 1.006 mJ, compared to 3.668 mJ, 11.32 mJ, and 121.7 mJ for the 3NM, 5NM, and high-altitude class LIDARs, respectively [426]. This latter model was installed in a Gulfstream II jet aircraft, while the optical antenna was mounted in a fairing on the bottom of the fuselage. Several airborne CAT measurements were performed over the land and sea around the Chubu area in central Japan. During this campaign moderate CAT was detected in front of the aircraft at altitudes of 10500 ft in clear weather and 30 s before the aircraft started to be affected. Data was also obtained with a confirmed detection range of up 9 km at a height of 40000 ft from the ground, which was the highest test altitude [426]. This flight test has demonstrated that the 2012 JAXA high-altitude LIDAR can be used with some confidence for CAT detection at high altitudes where there are fewer aerosol particles to scatter the incident laser signal. In fact, in a 2013 JAXA note, it was claimed that during a flight experiment in February 2012, CAT was successfully detected at cruise altitudes for the first time in the world[13].

As part of the JAXA ongoing development plans, a new Doppler LIDAR equipped with a more powerful laser output was jointly developed with the Mitsubishi Electric Corporation (MELCO) capable to detect CAT over 10 km ahead of the aircraft. The LIDAR was designed to be small enough to be installed in small aircraft [428]. Its weight has been reduced by a factor of about one-tenth compared to the previous LIDAR class having the same laser power thanks to its advanced optical laser amplifier and signal processing for noise suppression. With the use of this LIDAR, JAXA has also developed an airborne turbulence information system, named "SafeAvio", to make pilots aware of preceding turbulent areas or windshear in clear sky conditions. An advisory function developed in a joint effort with the University of Tokyo was incorporated into the SafeAvio system to further inform pilots of airspeed changes in advance of tens of seconds from an analysis of the LIDAR-measured headwind information [424]. A block diagram illustrating how the SafeAvio information system works is depicted in Fig. 9.8. SaveAvio estimates the headwind component at every 300 m up to about 30 km in front of the aircraft by subtracting the aircraft speed from the LIDAR-prediction value compared to the conventional speed trend vector. The target airspeed calculation logic of the LIDAR-based target airspeed indication (L-TSPD) is embodied in the following equation [600]

$$v_a(t + \Delta t_i) = v_a(t) + a \min(\Delta t_i, 10) + (u_j - u_{j-1}), \tag{9.81}$$

where $j = \max(\text{round}(v_g(t)\Delta t_i/\Delta x), 1)$, v_a is the aircraft speed at time t, Δt_i is the ith prediction time, a is the aircraft longitudinal acceleration, j identifies the LIDAR range bin number, u_j is headwind at the jth range bin, Δx is the LIDAR range bin interval, and v_g is the aircraft ground speed. Minimization of the airspeed deviation from the reference value will demand the L-TSPD to calculate the target airspeed by simply feedbacking the airspeed change predicted by the LIDAR-based predictive airspeed indication (L-PSPD). When the predicted airspeed exceeds the predefined airspeed limits, the L-TSPD is activated according to [600]

$$v_{tg} = v_{ref} - \sum_i \left\{ k_i \left[\Delta v_{a_i}(t) + \frac{\Delta v_{a_i}(t)}{|\Delta v_{a_i}(t)|} k_{sig} \sigma v_a(t + \Delta t_i) \right] \right\}, \tag{9.82}$$

---

[13]https://www.aero.jaxa.jp/eng/research/star/safeavio/.

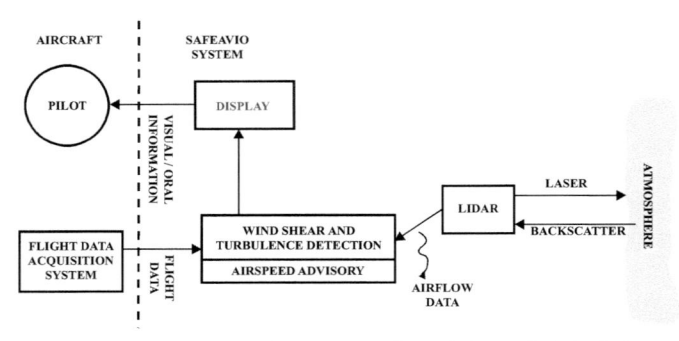

Figure 9.8: Block diagram showing the functionality of the SafeAvio information system developed by the Japanese Aerospace Exploration Agency (JAXA) and the Mitsubishi Electric Corporation (MELCO). Figure adapted from Matayoshi et al. [600].

where $\Delta v_{a_i}(t) = v_a(t + \Delta t_i) - v_a(t)$, v_{tg} is the aircraft target airspeed, v_{ref} is the aircraft reference speed, σv_a is the aircraft speed variation, k_i is the ith feedback gain for airspeed change, and k_{sig} is the feedback gain for airspeed variation.

The new developed system was tested in a series of 16 flight experiments carried out with the Gulfstream II jet from January to February 2017 as part of the JAXA project and on a larger Boeing 777 airplane as part of the Boeing ecoDemonstrator 2018 program. During the 2017 campaign, the SafeAvio system was able to detect windshear in 11 cases out of 13 encounters, corresponding to a detection rate of $\sim 85\%$. Since the deviation between the laser beam direction and the flight path became large for distances farther than 10 km, the windshear detection range often remained close to 10 km [600]. Although the evaluation of turbulence detection was planned at cruising altitude, turbulence was actually encountered at lower altitudes. The SaveAvio system detected turbulence in 10 cases out of 20 encounters from over 5.6 km away, yielding a detection rate of 50%. The measurement range observed in the 2017 flight tests averaged over all altitudes (between 2000 and 40000 ft) was 17.5 km, which exceeded the target of 14 km. This surely corresponds to the world's longest detection range. The measurements indicated that the range was strongly correlated to the aerosol distance, implying that the range decreases at higher altitudes where the atmosphere is almost depleted of aerosol particles. In particular, detection ranges up to over 20 km were achieved below 5000 ft from the ground, while at cruising altitudes of 40000 ft they drop to less than 10 km. Information feedback from the Boeing ecoDemonstrator 2018 program will certainly point toward a standardization of the onboard CAT detection and advisory system to make the JAXA LIDAR available for practical use in commercial aviation.

9.6 The AWIATOR airborne LIDAR

In addition to DELCAT's and JAXA's sensors, the development of a direct detection short pulse UV Doppler LIDAR for the measurement of gusts, turbulence, and potentially wake vortices has been partly supported by the European Commission in the framework of the AWIATOR project[14]. Unlike coherent detection LIDARs, whose principle is suitable for long-range warning, the AWIATOR system was mainly designed for safety-critical applications such as, for example, feed-forward flight control. To this end, the system must provide (1) a good longitudinal spatial resolution from 10 to 15 m, (2) a temporal resolution greater than 10 Hz, (3) a forward-looking measuring distance of ~ 50 to 150 m to ensure that the airflow that is being sensed is the one affecting the aircraft aerodynamics, which is particularly important in the case of wake vortices, and (4) an accuracy of 2 to 4 m s^{-1} normal to the flight path. In addition to the above criteria, the sensor must enable measurements at different angles off the aircraft axis and produce reliable signals in the absence of aerosol, implying that it must be a molecular Rayleigh backscattering system [797]. However, the system must also be capable of measuring aerosol Mie scattering. An interferometer (e.g., an etalon) is employed to measure the wavelength shift, while the required wind speed accuracy of ≈ 1 m s^{-1} in the line-of-sight implies a frequency shift of about 6 MHz for the selected UV wavelength. The thermal broadening of the Rayleigh spectrum is produced by speeds of the air molecules of ≈ 600 m s^{-1}.

The schematic drawing of Fig. 9.9 shows the architecture of the sensor system. It consists of a single-frequency, third harmonics Nd:YAG laser operating at 355 nm, emitting pulses of ~ 10 ns width at kHz rates. The laser beam is multiplexed in 4 directions by a scanner. The backscattered photons are received by a telescope and focused into a 4 UV-fibers. The signal is then de-multiplexed with a rotating mirror into one fiber that is connected to the receiver box. Circular interferograms are generated by a Fabry-Perot etalon, whose radii are proportional to the wavelength change due to the Doppler shift, corresponding to the change in the airflow relative speed. To exclude vibrations, the signal is actively adjusted. A modified DiCAM Pro CCD camera with a UV sensitive microchannel plate makes up the imaging system. The camera data is recorded and transferred to a processing unit by an optical transceiver/receiver interface technology. In particular, the recording and processing units of the AWIATOR system are based on Xilinx Virtex II FPGAs for highly parallel throughput and image processing [797].

The circular interference pattern that forms when the backscattered signal passes through the etalon is described by the well-known Airy function [797]

$$I(\lambda, \theta) = \frac{T^2}{1 - R^2} \frac{1 - R}{(1 + 3R)\sin^2(\phi/2)}, \tag{9.83}$$

---

[14]http://www.awiator.net/project.html, 2004.

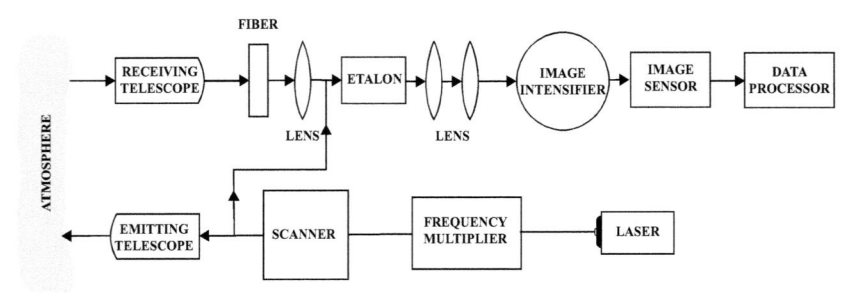

Figure 9.9: Block diagram showing the architecture of the AWIATOR airborne LIDAR turbulent sensor. Figure adapted from Schmitt et al. [797].

where

$$\frac{\phi}{2} = \frac{2\pi n_{\text{gap}} h}{\lambda} \cos \theta, \tag{9.84}$$

T is the transmission coefficient of the etalon, R is its reflection coefficient, h is the gap between the etalon plates, θ is the angle of incidence, and λ the wavelength to be determined. Because of the thermal broadening due to the Doppler shift and caused by Brownian motion, the intensity $I(\lambda, \theta)$ must be convolved with the Gaussian distribution of the one-dimensional speeds of the scattering particles, resulting in a much more complex function. The backscattered wavelengths must be determined with sub-pixel accuracy. This implies that the modified Airy function with unknown parameters must be fitted to the data of the image plane. A calibration signal is used to determine the system parameters that are fairly constant over time. A dedicated software is used to optimize the parameter estimation process, where different circular interferograms could be generated.

Flight tests with this sensor were conducted by installing it in the cabin of the DLR ATTAS VFW 614 test aircraft. Several flights at different height levels up to 24000 ft and speeds between 120 and 240 knots were carried out with a duration of approximately 10 hours using the AWIATOR sensor in conventional mode, i.e., without forward feeding. The experiments were performed under different weather conditions, including rain, clear air, and dense clouds, and as a result, a total of 250 Gbyte data were recorded. At altitudes close to 24000 ft, where the aerosol backscatter signal is weak, the sensor proved to work well in a molecular environment. The sensor was also able to detect signals in dense clouds in conditions where the visibility was reduced by less than 5 m. It is important to mention that during this first stage of tests a standard deviation of the line-of-sight air velocity of 8 to 12 m s^{-1} in one flight and 13 to 20 m s^{-1} in another flight was obtained in clear air conditions, while standard deviations of 3 to 4 m s^{-1} and 5 to 8 m s^{-1} for the same flights were obtained in hazy layers and of 1.8 to 3 m s^{-1} and 3.5 to 6 m s^{-1} in clouds. After using a more powerful laser source along with an optimization of the fiber relay system, the multiplexer optics, and the entrance filter and improved transmission, the sensor was expected to produce CAT signals at a flight altitude of 42000 ft and range of 50 m, resulting in improved accuracy by a factor of 5 in the airspeed measurements, with expected lower values of the standard deviations compared to the values above. As the technology involved in the design and construction of forward-looking optical sensors for measuring airflow, CAT, and wake vortices ahead of an aircraft improves, with the capability of these sensors to feed forward the data into the aircraft flight control, the pilot-assistance systems with automatic control will also improve, leading to a reduction of the pilot commands.

9.7   The KMW method for CAT forecasting

Over many years CAT forecast was founded on empirical rules and diagnostics based on forecasters' observations of the atmosphere and PIREPs on turbulent events. Improvements in the forecast skills commenced with the work of Sharman et al. [826], who developed the *Graphical Turbulence Guidance* (GTG) method. CAT forecasting using this method consisted of finding the best statistical weighting of previously used diagnostics associated with atmospheric observations and PIREPs, and then using this weighting in a forecast model. In spite of the combination of several diagnostics, the GTG method was not able to provide much insight into the physical causes of atmospheric turbulence. A step forward in improving CAT forecasting skills has been provided by the KMW method, where the acronym KMW stands for Knox-McCann-Williams, which was introduced by these authors in 2008 [491] as a method for CAT forecast. The method is based on McCann's forecast technique [605] and relies on the idea that turbulence is triggered by gravity waves. It has been hypothesized that just above the boundary layer these waves can modify the stability and wind shear by the amount necessary to develop turbulent motions. The KMW technique combines the environmental Richardson number (whose components are the vertical wind shear and the vertical temperature stability) and the dimensionless gravity wave amplitude. However, while the estimation of the Richardson number is generally easy to obtain from either observational or numerical data, some difficulties arise when dealing with the dimensionless amplitude of a gravity wave. The problem of estimating the dimensionless amplitudes of gravity waves was solved using the diagnostic from a numerical forecast model of the Lighthill-Ford theory. With this information the KWM method can be used to perform CAT forecasts at cruise flight levels. A brief description of the KMW method is provided below closely following the concise and comprehensive exposition given by McCann et al. [606].

In general, the turbulent kinetic energy is dissipated over characteristic time and space scales that far exceed the life and size of individual eddies. For atmospheric turbulence, such scales are of the order of hours and tens of kilometers, respectively. This way the turbulent kinetic energy dissipation can be written as [326]

$$\varepsilon = K_m \left(\frac{\partial \mathbf{v}}{\partial z}\right)^2 - K_h \frac{g}{\theta_v} \frac{\partial \theta_v}{\partial z}, \tag{9.85}$$

where K_m is the eddy viscosity, K_h is the eddy thermal diffusivity, $\mathbf{v}$ is the wind velocity vector, g is the gravitational acceleration, and θ_v denotes the virtual potential temperature. In CAT, the largest eddies are on average as large as 100 m and it is assumed that the kinetic energy contained by them cascades through the inertial subrange all the way to the molecular dissipation scale. In the KMW method it is assumed that the vertical component of gravity waves can locally modify the environmental stability and wind shear, which in mathematical form can be expressed as [255]

$$\left(\frac{\partial \mathbf{v}}{\partial z}\right)_{\text{loc}} = \left(\frac{\partial \mathbf{v}}{\partial z}\right)_{\text{env}} \left(1 + \text{Ri}_{\text{env}}^{1/2} \hat{a} \sin \phi\right), \tag{9.86}$$

and

$$N_{\text{loc}}^2 = N_{\text{env}}^2 \left(1 + \hat{a} \cos \phi\right), \tag{9.87}$$

respectively, where the subscripts "loc" and "env" stand for local (modified) and environmental conditions, respectively, Ri_{env} is the environmental Richardson number defined as the ratio of $N^2 = (g/\theta_v)(\partial \theta_v/\partial z)$ to the wind shear squared, $(\partial \mathbf{v}/\partial z)^2$, $\hat{a}$ is the dimensionless amplitude of the gravity wave defined as

$$\hat{a} = \frac{a N_{\text{env}}}{|\mathbf{v} - \mathbf{c}|}, \tag{9.88}$$

where a is the dimensional wave acceleration and $|\mathbf{v} - \mathbf{c}|$ is the Doppler adjusted wind speed for the wave phase velocity $\mathbf{c}$, and ϕ is the phase angle of the gravity wave. Equations (9.86) and (9.87) can be combined to produce the locally-modified Richardson number by the gravity wave [605]

$$\text{Ri}_{\text{loc}} = \text{Ri}_{\text{env}} \frac{(1 + \hat{a} \cos \phi)}{\left(1 + \text{Ri}_{\text{env}}^{1/2} \hat{a} \sin \phi\right)^2}. \tag{9.89}$$

When $\text{Ri} < 1/4$, at least a portion of the fluid flow can become turbulent. If in Eq. (9.89) $\phi = \pi$ and $\hat{a} > 1$, then $\text{Ri}_{\text{loc}} < 0$ and turbulence is expected to occur. For $\text{Ri}_{\text{loc}} < 1/4$ it is necessary that $\phi = \pi/2$ and $\hat{a} > (2 - \text{Ri}_{\text{env}}^{-1/2})$. Thus, lower values of Ri_{loc} means that weaker gravity waves would be enough to initiate turbulence. Replacing Eq. (9.86) into Eq. (9.85) for the modified wind shear and finding the maximum produces the form

$$\varepsilon_{\text{w}} = K_m \left(\frac{\partial v}{\partial z}\right)_{\text{loc}}^2 \left(1 + \hat{a} \text{Ri}_{\text{loc}}^{1/2}\right)^2, \tag{9.90}$$

for the estimate of the local kinetic energy dissipation. Similarly, an estimate of the local turbulent kinetic energy dissipation due to buoyancy can be obtained by replacing Eq. (9.87) into Eq. (9.85) and finding the maximum such that

$$\varepsilon_b = K_h(\hat{a} - 1)N_{loc}^2. \tag{9.91}$$

The maximum values of ε_w and ε_b are used in the analyses with the KMW method. Although gravity waves may unfetter turbulence, little is known about their locations, amplitudes, and phase velocities. In particular, based on the observations of the laboratory experiments carried out by Williams et al. [984, 985] on the inertia-gravity waves emitted in rotating two-layer flows, where these waves were seen to best fit their calculations of the Lighthill-Ford radiation, the KMW method adopted the Lighthill-Ford radiation equation written as

$$R = f\mathbf{v} \cdot \nabla\zeta + 2Df\zeta - f\mathbf{k} \cdot \mathbf{v} \times \nabla D - 2\frac{\partial J(u, v)}{\partial t}, \tag{9.92}$$

to produce numerical forecast models, where R is the Lighthill-Ford radiation source, f is the Coriolis parameter, D is the divergence, ζ is the vorticity, u and v are components of the wind vector $\mathbf{v}$, and $J(u, v)$ is the Jacobian of these components. The amplitude of gravity waves are proportional to $R^{1/2}$ in environments typical of the synoptic-scale flows for which the Rossby number is small. Therefore, the Lighthill-Ford radiation source as numerically determined using Eq. (9.92) from gridded atmospheric data, can be used to estimate the wave acceleration $\hat{a}$, which in turn can be used via Eqs. (9.90) and (9.91) to estimate the turbulent kinetic energy dissipation.

9.8 The ULTURB algorithm

The Upper Level TURBulence (ULTURB) algorithm is an improved version of the KMW method [606]. It was created to provide turbulence forecasting at any level above the boundary layer. As a modification of the KMW method, and unlike the GTG and its second version GTG2, the ULTURB algorithm is based on a self-consistent dynamical theory. It differs from KMW in that (1) the altitude range was extended downward to the top of the boundary layer, (2) the algorithm allows the inclusion of the turbulent kinetic energy produced by the environment to the locally produced turbulent kinetic energy, and (3) the eddy dissipation rate (EDR) is used as a metric in place of the turbulent kinetic energy dissipation, according to the worldwide standard.

It is well known that above 20000 ft (≈ 6.1 km) from the ground and near jet streams, the airflow speed increases with height to the tropopause, leading to large vertical wind shear and therefore to large values of the Lighthill-Ford radiation from Eq. (9.92). Below 20000 ft and near the ground surface the large temperature variations with height induce moderate wind shears so that the boundary layer is often characterized by turbulent airflows. However, turbulent encounters at mid level altitudes, i.e., between the top of the boundary layer and 20000 ft are much less probable [826]. On the other hand, applications of the KMW method use a PIREP database consisting of 3996 turbulence PIREPs below 20000 ft that were gathered each day at 1500 UTC ± 1 hour in the period from November 3, 2005 to March 26, 2006. Heidke Skill Scores of light for greater and moderate or greater turbulence at various thresholds stratified by altitude using the turbulent kinetic energy dissipation KMW method showed that below 20000 ft the skill scores were greater than at 20000 ft and above, while there was not enough PIREPs of severe turbulence to provide a stratification with altitude [606].

In addition, above 40000 ft from the ground level, there were only so few CAT PIREPs that it was impossible to determine by analysis the level of turbulent kinetic energy dissipation above such altitudes. However, since just above the tropopause (i.e., between ~ 10 and 20 km from the ground), the vertical temperature structure looks very stable (see Fig. 3.1), it was inferred that this would suppress the possibility for high dissipation of the kinetic energy at such altitudes. This aspect was clarified by looking at Eq. (9.86). At the phase angle $\phi = \pi$, local buoyancy dominates the turbulent kinetic energy dissipation, and from Eq. (9.86) the environmental wind shear is still positive since $\sin \phi = 0$. However, at the phase angle $\phi = \pi/2$, the production of local wind shear reaches a maximum since $\sin \phi = 1$. Moreover, the production of buoyancy is almost always negative and when taking into account this the production of local wind shear is suppressed. In the KMW method, the turbulent kinetic energy dissipation is accounted for by the greater values of ε_w and ε_b calculated from Eqs. (9.90) and (9.91). Therefore, a full account of the gravity wave-modified turbulent kinetic energy equation has the net effect of increasing the production/dissipation of turbulent kinetic energy when the local buoyancy modification is greatest and reducing it when the local wind shear production is largest. This solved the problem in that the apparent high bias in the turbulent kinetic energy dissipation forecasts above the tropopause was effectively reduced.

A further important modification to the KMW method leading to the ULTURB algorithm consisted in transforming the turbulent kinetic energy dissipation per unit mass into the EDR by just taking the cubic root of ε. A powerful reason for making this change was that expressing the turbulence intensity in terms of the EDR complies with the ICAO's standard reporting turbulence metric [204]. At the time the KMW method was developed there was no standard mapping of EDR into subjective turbulence intensity. However, since then there have been enough collected atmospheric EDR data to provide a standard mapping, thereby allowing the transformation of the turbulent kinetic energy dissipation employed in KMW into EDR for ULTURB. In particular, the turbulence of light intensity corresponds to an EDR of 0.15 m$^{2/3}$ s^{-1}, while moderate and severe intensities are mapped with EDRs corresponding to 0.35 and 0.55 m$^{2/3}$ s^{-1} [700], respectively.

Although the core of the ULTURB algorithm is identical to the KMW method, the modifications of the KMW technique that have led to its development provided a logical progression to improve CAT forecasting at all flight levels above the top of the boundary layer. Also, ULTURB has proved to be a superior forecast method to GTG2, especially for severe CAT at jet cruising altitudes. In addition, ULTURB can also successfully forecast CAT at altitudes below 10000 ft from the ground level. The ULTURB algorithm can still be improved because the Lighthill-Ford radiation may not be the only source of gravity waves. For example, other sources may involve the sum of vorticity advection and the Coriolis parameter times divergence, the vertical gradients of the nonlinear balance equation, and the Laplacian of temperature advection [709]. On the other hand, gravity waves can propagate large distances from their source point and therefore there would be a wave-mean flow which may also contribute to wave amplification/dissipation along with the resulting effects of wave-to-wave interactions [83, 302] and wave-turbulence interactions [315]. Regions above mountainous areas would certainly be more susceptible to CAT activity than other regions [995]. Under these circumstances, it is somewhat difficult and to some extent tricky to distinguish between mountain-wave induced turbulence and CAT. Since both are gravity waves it is not difficult to foresee that they may eventually amplify via a wave-to-wave or a wave-turbulence interaction mechanism, possibly causing a much stronger event than either could be able to produce individually.

9.9 The NCAR/NEXRAD turbulence detection algorithm

The NCAR/NEXRAD turbulence detection algorithm (NTDA) was designed to measure turbulence intensity in and around clouds as well as under precipitation using data collected by ground-based weather radars. In particular, convectively induced turbulence usually appears within dynamically evolving storms and can last a few minutes and extend for a few kilometers. Unfortunately, this kind of turbulence, which has been pointed out many times as responsible for the majority of turbulence-related aircraft accidents [203, 457], is known to be difficult to diagnose and/or forecast [524]. The NTDA algorithm has been developed to assist pilots, dispatchers, and air traffic controllers in identifying hazardous areas of convective turbulence and proposing safe flight routes. In particular, this technology permits pinpointing three-dimensional turbulence and providing the data to aviation users in time to take routing and cabin management decisions, and therefore it has been adopted in commercial aircraft. Compared to airborne LIDAR and SODAR units, it is lighter and less expensive. In addition, it does not suffer from severe attenuation in areas with high concentrations of water vapor or hydrometeors. Conversely, ground-based Doppler weather radars typically operate at the longer C- and S-band wavelengths and therefore they can deeply penetrate storms and clouds. As was already outlined in Chapter 7, the approaching and receding wind radial velocity within the radar's measurement volume is represented by the Doppler spectrum width. This can be used as an atmospheric turbulence metric by converting it into EDR independently of range from the radar. It must be mentioned that measurements of the Doppler spectrum width are highly sensitive to low signal-to-noise ratios, overlaid echoes, and contamination by non-atmospheric scatterers. To perform quality control on each spectrum width measurement and compute appropriately weighted averages to extract high-quality turbulence measurements, the NTDA includes a fuzzy-logic procedure as an essential component.

A comprehensive description of the NTDA algorithm is provided by Williams and Meymaris [980]. In particular, they give full details on the above-mentioned quality control procedure, the conversion of the spectrum width into EDR as a metric for measuring turbulence intensity, the NTDA mosaicking algorithm, and its evaluation. The flight testing campaign of NTDA commenced in the spring of 2002, when 11 research flights were performed using the NASA Langley Boeing 757 aircraft [201]. High-rate wind data were obtained and recorded by maneuvering the aircraft into developing thunderstorms and their associated turbulence. As a result, the NTDA analysis yielded a probability of detection of 94% with a false alarm rate of 16%. When extracting the median NTDA EDR value from a disc of radius 2 km around each aircraft location, as projected onto each nearby radar sweep, these data showed that the NTDA achieved a probability of detection of 80% for turbulence exceeding 0.15, 0.35, and 0.55 $m^{2/3}$ s^{-1} for light, moderate, and severe turbulence, respectively. Accompanying false alarm rates were 24%, 23%, and 16%, respectively [979]. An operational demonstration between 2005 and 2007 was further sponsored by the *FAA Aviation Weather Program*. The demonstration consisted of providing real-time NTDA data to the United Airlines dispatchers and en-route pilots. The results from this demonstration indicated that the NTDA turbulence information successfully improved the pilot's situational awareness. Since its development, the NTDA and mosaic software have been optimized. In particular, they have been processing data from over 140 NEXRADs and producing a CONUS mosaic every 5 min with minimal latency. In 2014, the NTDA have been adopted in Taiwan as part of *Taiwan Civil Aeronautics Administration's Advances Operational Aviation Weather System* to provide in-cloud turbulence detection. In addition to its skill for detecting hazardous turbulence, the NTDA output can be combined with satellite (*in situ*) and numerical weather prediction model data to identify and forecast areas of convectively induced turbulence in and near storms [981, 978], which could improve air traffic flow and aviation safety.

9.10 Optical airflow meters

In-flight optical airflow meters are a class of devices that can be used for remote sensing of wind shears, mountain rotors, wake vortices, and ultimately CAT. These sensors are essentially anemometers that operate using the light scattered from inhomogeneities in the refractive index of atmospheric air. Although a great number of different anemometers have been developed, their basic layout consists of a light (laser) source, a scattering medium (the atmosphere), optical elements, a detector, and signal processing electronics [437]. An example of how the different components are interconnected in a typical airborne airflow meter is shown by the block diagram in Fig. 9.10. The thick arrows indicate optical links between the system components, while the thin ones interconnect those components that are linked electronically. In a typical anemometer a focused, coherent light beam is emitted into the atmosphere. Although most optical flow meters use the light scattered by aerosol particles within the measurement volume, they are also sensitive to scattered light by air molecules but with lower power. Before conversion into an electrical signal, the scattered light is passed through a pre-detection signal processing unit and transferred to a detector, whose efficiency depends on the wavelength range of the collected light. Modern optical airflow meters use compact detectors made up of semiconductor materials. The detected signal is then transferred to a post-detection signal processing unit where the signal is Fourier transformed to extract information on the airflow velocity. Alternatively, information on the airflow velocity can also be derived by measuring the time between scattering events. When pulsed lasers or gated detection is applied, the coordination of the process timing is performed using a control unit. This uses the time between the transmitted pulse and the received backscattered signal to control the range from which the measurements are obtained.

The reference beam anemometer is the optical airflow meter most frequently applied for flight purposes. The airflow velocity component in the direction of the light beam is measured by first measuring the Doppler frequency shift between the emitted and scattered light within the measurement volume. Since the

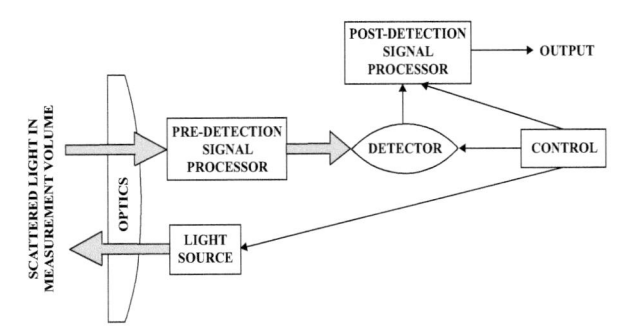

Figure 9.10: Block diagram of a typical in-flight optical airflow meter. Figure adapted from Jentink and Bogue [437].

frequency shift, δf, is small compared to the light frequency, f, it is measured by mixing the scattered light with light in a reference beam with the original light frequency. In this case, the reference beam is said to be in a backscatter configuration. This corresponds to a coherent detection technique, which results in an electrical signal on the detector with a Doppler frequency shift equal to the frequency difference between the scattered and reference beam light

$$\delta f = 2f\frac{v}{c}, \tag{9.93}$$

where v is the velocity component in the direction of the air beam and c is the speed of light in vacuum. This configuration permits airflow velocity measurements at large distances from the anemometer (usually several kilometers). In particular, it can be applied to altitudes of 25000 ft since adequate measurements can be obtained at relatively low densities of scattering particles because of the large measurement volumes and the efficient collection of the scattered light [437].

An alternative configuration to the reference beam anemometer is the so-called dual-beam laser Doppler (or fringe) configuration, which is mainly employed for ground-based applications. These fringe anemometers display well-defined and smaller measurement volumes so that airflow velocities can only be measured at more limited distances from the anemometer. In this case, a beam splitter placed in front of the laser unit is used to divide the emitted signal into two coherent light beams, which are then made to converge by a lens. The convergent beams cross at a certain distance from the lens to produce a measurement volume at the point of beam intersection. Particles crossing the measurement volume will then scatter light from both beams, which is then deflected by an inclined mirror and transmitted to a detector. The detector measures the intensity fluctuations of the scattered light whose frequency is proportional to the velocity component of the scatterer particle in the intersecting beam plane and independent of the detection angle. An interference pattern consisting of parallel light and dark planes (fringes) is formed by the two beams within the measurement volume. The frequency of the intensity fluctuations resulting on the detector when a particle passes through the interference pattern is given by

$$f_D = \frac{2v}{\lambda}\sin\left(\frac{\alpha}{2}\right), \tag{9.94}$$

where v is the component of the particle velocity in the plane of the two intersecting beams and perpendicular to the bisector of the beam directions, α is the angle formed by the intersecting beams, and λ is the light wavelength.

A good overview of the different flight applications of optical airflow meters is given by Jentink and Bogue [437]. For example, these devices have been used to measure the airspeed of aircraft [846, 610] and helicopters. Conventional air data sensors are installed on low-speed helicopters in the flow field generated by the rotor. While this requires substantial flow corrections, the same is not true when the optical measurements are made by installing the sensors outside the influence of the rotor wash. An innovative air data system was designed, built, and tested in the framework of the *New Standby Lidar Instrument* (NESLIE) and *Demonstration of ANemometry InstrumEnt based LAser* (DANIELA) projects supported by the European Commission [936]. This system uses the LIDAR technique to measure the true airspeed, angle of attack, and side slip angle of an aircraft, resulting in a test system of future generation standby instruments for commercial aircraft. In contrast to more traditional pitot-static air data systems, this system has no probes protruding in the air and has different failure modes which work toward increasing flight safety. The flight tests were performed from April to May 2009 and March to May 2011 using the NLR Cessna Citation II research aircraft. The test campaign was conducted in 46 flights covering 100 flight hours from the North Pole to the Equator. A comprehensive overview of these test operational activities and analysis of the recorded data can be found in an unclassified NLR technical report by Verbeek and Jentink [936].

In-flight calibration of air data systems often requires the use of laser anemometers in that errors in conventional systems can be determined by comparing the measured data with those obtained with a laser anemometer. In fact, such devices have been operated in prototype commercial aircraft for this purpose [588, 24, 156]. The data recovered from these experimental flights have also contributed to expanding flight envelopes. However, one of the most important applications is for remote hazard detection at considerable distances and for the ability to operate in clean air. These systems have been under development over the last thirty years or so. Most design studies in this line of research have been addressed to the detection of wind shear [892], CAT, and wake vortices [377, 464]. The feasibility of wake vortex detection from the at-risk aircraft during landing and take-off using laser anemometry technology has been amply demonstrated as has been partly described in Section 7.2. The development of new technologies will probably consider the design of multi-purpose systems, i.e., a single system capable of detecting all hazards, which will certainly be more profitable for commercial aviation applications. The feasibility of this challenging development rests on the continuous advances in optical technology, which have provided incremental improvements in both system performance and data quality. A good example of this is the increased power of lasers that have allowed longer range measurements. On the other hand, the advent of new optical technology applicable to the communication industry is allowing the construction of optical measurement devices around 1.5 μm wavelength components, which would result in much cheaper systems.

In response to these technological developments, the Airborne Coherent Lidar for Advanced In-flight Measurements (ACLAIM) was developed as part of the NASA Airborne Science Program for advanced CAT detection with applications to supersonic inlet control, mitigation of aircraft gust response, and crew and passenger warning for improved seat belt utilization. This sensor is a 2-μm LIDAR operating at 100 pulses per second and using an expanded beam of 8 cm diameter. It was installed on the NASA DC-8 aircraft for testing in mountain rotor conditions to assess its efficiency for the remote detection of turbulence [897]. The results of these tests demonstrated that the ACLAIM system operating at the 2-μm wavelength was capable of reliably detecting turbulence at a range of 2 km. On the other hand, a high-power, pulsed laser source was developed at *Coherent Technologies Inc.* that uses Raman technology and operates at $\lambda = 1.5$ μm with a pulse repetition rate of 20 Hz and an output pulse energy of $\sim$ 200 mJ [633]. With this instrument, the range of 7 km is predicted and with further improvements, could be extended to 10 km. On the other hand, the *Molecular Optical Air Data System* (MOADS) was developed at the Michigan Aerospace Corporation [957]. This is a compact instrument developed to acquire airflow velocities in three dimensions and measure the density and temperature of the air surrounding the aircraft. Other data provided by MOADS include the altitude, the angle of attack, the angle of side-slip, and the Mach number. The instrument is based on direct-detection LIDAR technology that performs measurements from fringe images derived from a Fabry-Perot etalon. The system was installed in a side window of a Beechcraft King Air for flight testing. This was the first attempt to test a reference beam anemometer using Doppler-shift molecular backscatter to provide airspeed measurements [437]. Under the I-Wake project supported by the European Commission, a further anemometer was primarily developed for windshear and wake vortices. However, it can be applied to CAT and volcanic ash detection. The optical system includes a water-cooled solid-state laser emitting light at around 2 μm. With respect to the aircraft body axis, the I-Wake system scans the volume from 800 to 2400 m with a horizontal scan angle from 3° to 15° to the right and with a vertical scan angle from 7.5° to 10.5° down. In particular, this prototype LIDAR was flight tested in 2004 onboard the NLR's Cessna Citation II research aircraft following an Airbus A340-600 prototype.

9.11 Small unmanned aircraft systems

An emerging technology is provided by the so-called small unmanned aircraft system (SUAS) [280]. An unmanned aircraft system (UAS) refers to a general class of remotely piloted aircraft ranging in weight from grams to thousands of kilograms, while SUAS is a class of UAS weighing less than 25 kg. These systems have gained popularity because they require small sensor payloads for their multiple applications, including airborne wind sensing. Compared to manned aircraft, the SUASs involve lower operation costs and require low acquisition. Due to their smaller size they can be operated away from runways and in confined spaces with much more limited risks. SUASs are known in two different configurations, namely fixed wing and rotorcraft.

In particular, the multi-rotor UAS (MUAS) is a class of rotorcraft systems that have a wide range of applicability because of their ability to hover, their mechanical simplicity, ease of use, and great maneuverability [728]. They are particularly well-suited for sensing the atmospheric boundary layer where the winds are heavily influenced by the Earth's surface [960]. In fact, SUASs have been replacing weather balloons using Doppler radar, satellites, and even ground-based SODARs and LIDARs for measurements above the boundary layer (i.e., at altitudes higher than about 1 km) to improve the spatial resolution offered by the former instruments [953]. SUASs are also playing a good job to measure winds at altitudes below 100 m, a task which has been typically performed by anemometers mounted to tall masts at point locations. In the altitude range of the boundary layer the relatively high frequency fluctuations with durations < 1 hour fall within the flight endurance capabilities of SUASs. Because of their high endurance and ability to travel relatively great distances, fixed-wing SUASs are particularly adequate to conduct wind profile measurements in the upper regions of the boundary layer. For example, these systems have been used to measure vertical wind profiles at low altitudes [292] and the air quality with a focus to track pollutants, fires, and water/vapor dispersion [164, 937]. Known SUASs as the *Meteorological Mini Unmanned Vehicle* (M²AV), the *Small Unmanned Meteorological Observer* (SUMO), and the *Multipurpose Airborne Sensor Carrier* (MASC) have been quite successful to perform measurements of high-frequency turbulence fluctuations at altitudes up to 1500 m and flight durations of up to 1 hour [598, 340, 920, 755, 974].

Unlike fixed-wing SUASs, rotorcraft SUASs configurations can hover and maneuver in all directions as well as land and take-off vertically. Although their effectiveness depends on many factors, they can be employed to reliably perform wind measurements at specific points within the altitude range of the boundary layer, including altitudes close to urban structures. However, compared to fixed-wing configurations, rotorcraft have an endurance of half for the same weight and a moderate lift/drag ratio of four [358]. Most small rotorcraft are battery powered. However, batteries must be carefully selected to optimize the endurance and payload capacity requirements because they represent an important mass that must be supported during operations [47]. Attempts have been made to design hybrid systems using fossil fuels to improve the endurance of purely electric systems. On the other hand, unmanned helicopters have been used for many years, and can be scaled to a wide range of sizes. Those systems that are powered by internal combustion engines are particularly suitable for carrying large payloads for extended durations [340].

MUASs have gained increased popularity over other SUASs because of their robustness, affordability, and compactness. Regarding hovering and vertical landing and take-off they share the same benefits as helicopters for wind measurements. To provide pitch, roll, and yaw control they require at least 4 fixed-pitch rotors (quadrotors). A deeper discussion on technical aspects of MUAS is provided by Prudden et al. [728] and references therein. A very important consideration for wind measurements when using SUAS platforms is the integration of wind sensors. Among the methods that have been investigated are: (1) the inertial and power-based methods and (2) the onboard flow sensors. In the former case, the wind vector can be derived by tracking the movement of the aircraft relative to the ground. In this method, no additional sensors, other than those used in the autopilot standard flight, are required for atmospheric flow measurements. Research on this method was first addressed using fixed-wing, inertial-based UASs. For example, early experiments employed a large delta-wing UAS together with a numerical estimation method to calculate the two-dimensional wind direction with reasonably good convergence to ground-based measurements [514]. More recent investigations relied on the use of multiple fixed-wing UASs in formation flight, rather than an individual aircraft, to resolve the three components of the wind velocity after corrections for wake interference between leading and following aircraft [526]. Studies involving the use of quadrotors have also been performed [659, 645, 660, 1001, 857]. However, the first measurements with these systems were limited to two dimensions.

Refinements of the method to allow for three-dimensional wind measurements were developed by several authors [660, 591, 908]. The accuracy of this method was seen to be strongly influenced by the horizontal wind estimation and aerodynamic model, and has therefore been considered to be more complementary than stand-alone. In the second case, the use of onboard flow sensors for wind sensing can effectively reduce payload requirements. However, the use of such technology for turbulence measurements, particularly small scale fluctuating flows, can still pose considerable problems. In response to these difficulties, several sensors and different mounting solutions have been developed for real-world applications at low altitudes, for instance, those encountered in urban places. For high frequency response, the more frequently used techniques for fixed position measurements are hot-wire anemometry (HWA), laser-Doppler anemometry (LDA), particle image velocimetry (PIV), and the multi-hole pressure probes (MHPP). The accurate determination of the airflow velocity will depend on the size, weight, and power requirements for SUAS-mounted sensors. However, the advent of MEMS-based sensors and GPS has improved the accuracy in the determination of the vehicle attitude, heading, position, and speed. If the measurements are accurate enough, the transformed measurements from the aircraft-fixed coordinate frame to the Earth-fixed coordinate system can be used for wind engineering purposes. Not all wind measurement methods can be of practical use for SUAS-based sensing. The most common pressure-based sensor that has been used for UAS-based flow measurements is the Pitot tube, which however cannot solve for components of the airflow speed normal to the flight path. For SUAS applications the MHPP has been revealed to be a highly robust method for measuring wind vectors in the sense that it is less fragile than hot-wire anemometers and more suited to outdoor sensing than PIV and LDA [329].

Fast-response MHPPs, such as those produced by *Aeroprobe and Turbulent Flow Instrumentation* are commercially available and can measure fine-scale turbulence at up to 2 kHz[15]. A limitation of these probes is that they lose accuracy at velocities < 3 m s^{-1}, which can be a concern for aircraft measuring low-velocity wind while hovering. Although the MHPPs are accurate and simple to use, their cost is excessive and they are often exposed to damage during routine operations. As an alternative approach, the so-called flush airdata systems (FADS) have been implemented [531], which use pressure port flush with the aircraft surface. The optimal locations of the pressure ports on the airframe were determined using computational fluid dynamics simulations and calibration of the FADS was performed using a MHPP as a reference instrument. Flight testing results using a SUAS platform have shown that the system is well suited for wind measurement applications.

---

[15] https://www.turbulentflow.com.au.

Chapter 10

The Weather Accident Prevention Project, Pilot Perspectives, and Other Issues

10.1 Introduction

According to both the FAA and the NTSB, nearly 35% of all aviation accidents are due to encounters with nasty weather conditions, while about 75% of those accidents involve fatalities. Therefore, it is clear that both the aviation industry and commercial airline pilots must take steps to reduce the risk of having such unfortunate encounters. Today, part of the risk is mitigated because commercial airlines employ technically advanced aircraft and because a significant emphasis have been placed on meteorology, with the capability of recognizing severe weather, CAT, icing, and thunderstorms. From the pilot's perspective, mitigation of weather hazards must begin with refresher training in seasonal weather conditions and sufficiently good preflight planning, which can be performed with the aid of presently available weather apps. In particular, the FAA's Advisory Circular 91-74B covers icing hazards and proper flight planning. On the other hand, flight service station briefers can provide the pilots a timely and much better outlook than can be obtained using onboard equipment, such as radar or NEXRAD systems, which are designed for avoidance purposes and not for weather navigation. Moreover, technology also points toward mitigating weather risks by reducing the pilot workload. In this sense, graphical information will be much easier to understand and visualize. For example, some apps can provide the pilot with graphics showing the arrival runway with clear indications of the crosswind and tailwind vectors. Weather information and safety can also be improved by the usual PIREP files, which can offer a three-dimensional look at the atmosphere and gauge the accuracy of forecasts by comparing them against METAR and other weather information sources. To guarantee continued growth in this arena, the improvement of flight safety becomes a critical aspect. In 1997, the President of the United States called for an 80% reduction in the rate of fatal accidents by 2007 and a 90% reduction by 2017 [287, 359]. In response to this, the NASA Aeronautics Safety Investment Strategy Team (ASIST) formulated technical objectives for an Aviation Safety Program (AvSP), in partnership with the industry and the FAA.

In this chapter, the salient features of the *Weather Accident Prevention Project of the NASA Aviation Safety Program* (WxAP) will be revisited together with central aspects dealing with the professional pilot perspectives. Inadequacies of present weather information systems will also be discussed along with established standard procedures and checklists for a range of typical scenarios to ease decision-making through cognitive task analysis. The chapter closes with a short description of what is meant by weather avoidance doctrine.

10.2   The Weather Accident Prevention (WxAP) project

In 1997, the Government of the United States, under recommendations of the *Presidential Commission for Aviation Safety and Security*, established as an objective of national priority to reduce by 80% the rate of fatal air accidents in a period of 10 years. The implementation of this plan has required the coordination of several areas of research and development, including the implementation of new technologies and operations, and demanded the participation of the government itself in collaboration with the industry, research institutes, and universities. In response to this, the NASA Aerospace Technology Enterprise developed the AvSP to concentrate resources for the conduction of research and developing technologies that will allow improving the aviation security [831]. To facilitate the development and demonstration of technologies that will meet its main objectives, the AvSP was organized into the following six research areas:

- *Aviation System Modeling and Monitoring* (ASMM). The goal of this research area is to provide decision makers of the NAS with regular, accurate, and insightful measures of health, performance, and safety.

- *System Wide Accident Prevention* (SWAP). In this case, the goal consists of addressing aviation safety issues associated with human error and procedural non-compliance. This aspect is crucially important since 60 to 80% of aviation accidents have been estimated as attributable to human error.

- *Single Aircraft Accident Prevention* (SAAP). The goal of this area is to develop and implement new technologies to help reduce the occurrence of fatal accidents in accordance with the program objective.

- *Weather Accident Prevention* (WxAP). The goal of this area is to develop and implement new technologies to help reduce the occurrence of fatal accidents due to weather hazards and turbulence-related injuries.

- *Accident Mitigation* (AM). Its main goal is to develop, enable, and promote the implementation of emerging technologies to increase the rate of human survival in aviation accidents and prevent in-flight fires.

- *Synthetic Vision Project* (SVP). This project is an important part of the AvSP aimed at developing practical and certifiable cockpit systems capable of utilizing high-resolution digital terrain data and augmented, if necessary, by onboard imaging sensors.

Since adverse weather is responsible for about one-third of commercial aviation accidents, the NASA ASIST has proposed a list of priority investment areas for weather accident reduction, and the WxAP project addresses six of these recommended areas, namely (a) data dissemination, (b) presentation and decision aids, (c) weather product generation, (d) advanced aviation meteorology, (e) turbulence hazard solutions, and (f) near-term tactical sensor/system. To do so, the WxAP project has been organized into three research elements: (1) Aviation Weather Information (AWIN), (2) Weather Information Communications (WINCOMM), and (3) Turbulence Prediction and Warning Systems (TPAWS) [359]. For instance, in many of the "loss of control" accidents, crew errors have been attributed to low visibility. Therefore, the program focused on implementing new techniques to provide complete meteorological information for the support of pilots and ground operators to guarantee the safety of the aircraft. In this sense, the project relies on three main approaches. The first one concerns the development of weather products, displays, and decision support tools addressed to the needs of the flight crew and cockpit environment. A requirement for both commercial transport and/or the general aviation aircraft is that decisions regarding, for instance, change of the flight path are made in collaboration with air traffic management organizations. However, the characterization of weather hazards poses some technical challenges because the physics governing many atmospheric phenomena is so complex that they are poorly understood and difficult to model. In addition, the difficulty increases as different aircraft respond differently to particular weather conditions. Thus, the development of sensors capable of providing accurate measurements of the atmospheric conditions and aircraft response represents a further challenge. On the other hand, the timely and easy-to-process presentation of the weather information to the flight crew has also become a challenge because this demands the crew's response to the graphical weather information supplied to the cockpit and the corresponding decision processes. For example, currently available real-time datalink communication systems are not economically feasible and therefore the aircraft-to-ground, ground-to-aircraft, and ground-to-ground exchange of graphical weather images and atmospheric information is generally affected by coverage and costs.

The characterization of weather-related threats has been partly achieved by the development of sensor systems that have converted the aircraft into an airborne weather data collector. These data include CAT measurements as well as measurement of other atmospheric variables at altitudes below 25000 ft, which are collected by the sensors and then automatically downlinked to forecast modelers and weather product developers. Such sensors are aimed at reducing the workload of the flight crew since it does not intervene in the process of transmitting the information to the ground. Together with the use of technically advanced sensors, the timely dissemination of the weather information must be accompanied by the development of digital datalink technologies and architectures, which in turn must be performed with an awareness of the NAS communication infrastructure and related issues [287, 359].

The AWIN research element was intended to develop new technologies to provide intuitive, timely, and accurate information during the en-route phases of flight to the flight deck, allowing detection and avoidance of atmospheric mishaps. The resulting new products serve to complement existing weather sources with *in situ* and remote sensing capability whenever required, thus providing information with sufficient spatial and temporal resolution for both tactical and strategic decision making for aviation users. AWIN is also dedicated to enhancing weather presentations to minimize interpretation and training, improve situational awareness and engagement, and reduce the crew workload. On the other hand, decision making is also improved by developing aids and including collaborative processes. In particular, these will facilitate the identification of training needs and identify guidelines to support the use of emerging weather information technologies in the cockpit.

The WINCOMM research element was dedicated to the development, assessment, and recommendation of datalinks for high quality and timely transmission of the weather information from and to the flight deck during the flight phases. This aspect is critical for the reduction of aviation accidents and mishaps. Addressing limited bandwidth and coverage will increase the datalink capabilities, thereby allowing enhanced performance in the production of graphical weather images in the en-route service area. Moreover, the need for improved air-to-ground and air-to-air datalink capabilities for both coverage and capacity is in turn driven by further improvements to sensor technology and weather algorithms. In particular, advanced aviation weather communication systems must ensure national access and connectivity, address cross-segment operations, with emphasis on commercial transport and general aviation operators, develop advanced communication systems, which are supported by network system modeling and comply with communication standards and protocols, increase the data rate for weather information between air and ground as well as the communication system capacity, and improve user connectivity and coverage from 50 to 40000 ft above the ground level.

The TPAWS approach is devoted to the development of methods and technologies capable of providing pilots with information on dangerous CAT locations with a 90% level of confidence and at least 2 min prior to encounter during flight. Such technologies exist today with a clear impact on the reduction of accidents and related passenger injuries. This research was intended to provide advisory and warning to the aircraft crew ahead on the flight path. Basic devices in use involve the aircraft weather radar, LIDAR technology components for enhanced radar CAT detection, wake vortex detection, and autopilot redesign for turbulence loads alleviation coupled with advanced TPAWS interfaces, involving communication datalinks with other nearby flying aircraft and communication of turbulence data to ground-based weather stations using the existing datalinks onboard the aircraft. These facilities result in improved weather reports and forecasts, thus providing more efficient flight operations. The TPAWS products include the use of color schema, symbology, hazard levels, and display format.

The WxAP project has given rise to an operational system for use during flight. For example, weather information, including NEXRAD data, is collected and processed on the ground and transmitted via datalink to properly equipped aircraft, which is then displayed as text and graphics in the cockpit. WxAP uses the latest technology datalink and cockpit graphical display to inform the pilot of the weather conditions. The WxAP system updates are transmitted to the cockpit as the flight progresses, providing recent datalink weather information with a data latency factor [287]. As better technologies emerge, this factor will reduce to the point that the displayed information can be used for strategic decisions. Modern commercial aircraft are equipped with additional sensors for real-time CAT detection and warning along with other devices for measuring atmospheric variables. As the destination airport is approached, the pilot performs the transition to the approach phase of flight aided by onboard decision support systems. For example, within the FAA NAS system, a pilot proceeding on instrument flight rules will be informed from the ground of any weather hazard at the destination terminal via radio transmission. The pilot receives this information directly from the tower, the Air Traffic Control (ATC), the Airport Traffic Information System (ATIS), the Hazardous In-flight Weather Advisory Service (HIWAS), the Automated Weather Observing System (AWOS), or the Automated Surface Observing System (ASOS). For pilots proceeding on visual flight rules weather information will be provided by the onboard system and ground radio transmissions.

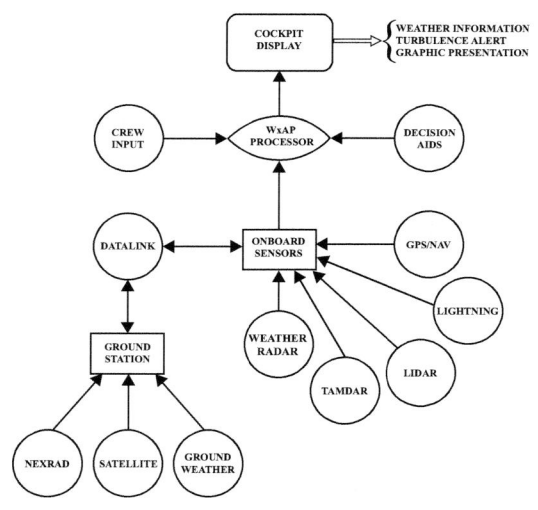

Figure 10.1: Block diagram showing the WxAP system architecture. Figure adapted from Feinberg and Tauss [287].

The basic architecture of the WxAP system consists of both ground-based and airborne components that exchange information over one or more digital datalinks. Figure 10.1 shows a block diagram illustrating the WxAP architecture. The ground-based station processes ground weather, satellite, and NEXRAD information, formats the information for distribution, and manages the distribution network. The airborne components (onboard sensors) display the data received from both ground and onboard systems. The information from ground stations (via datalinks) and onboard sensors converge to the WxAP processor, which is then transmitted to the cockpit. This consists of weather information, turbulence alert, and graphical displays. In addition to weather information, the WxAP systems can provide advanced decision-aiding tools by using the information on aircraft capabilities, operator capabilities, and data on flight path-relevant terrain, obstacles, airspace, and traffic. In particular, aircraft weight and performance, sensor complements, gust alleviation, ice protection, and vision enhancement are part of the aircraft capabilities, while the operator capabilities include experience, training, currency, and human information-processing characteristics. On the other hand, onboard databases store data from terrain, obstacles, and airspace for display on a GPS-based moving map. Near real-time air traffic data can also be available in aircraft equipped with advanced implementations. For commercial air carriers and business aircraft, the WxAP system has been designed to provide weather information on a worldwide basis at normal jetway cruise altitudes. Although WxAP solutions are available and affordable for any class of aircraft, new technologies for both retrofit and fully integrated WxAP implementations will continue to be developed to provide the flight crew with even more accurate weather information, resulting in safer and more efficient actions.

10.3 Professional pilot perspectives

Pilots are qualified professionals in charge of operating aircraft and therefore they must be able to fly jets, passenger planes, helicopters, and rescue planes, among other kinds of airplanes. As commercial airline pilots, they have a great responsibility because they have to make decisions to keep the passengers and crew safe. These skills demand rigorous and intense high level training to man transport aircraft because they are put in charge of passengers' lives. As a consequence, they are relatively well paid and there exists a competitive job market where only the top candidates will make the cut.

In general, pilots are responsible for conducting inspections of fuel, equipment, and navigational systems before and after flights; for operating the aircraft safely and maintaining a reasonable degree of professionalism at all times; for monitoring the weather conditions and maintaining communication with the air traffic control; for liaising with the co-pilot and flight crew; for updating and reassuring passengers and crew during emergencies; and for determining the safest routes and analyzing the flight plans before take-off [243]. In addition, and in accordance with the technological advances developed by the aviation industry, professional pilots have a perspective to be updated with the advances in aviation. Examples of such advances are the modern information and communication systems set up in the cockpit for which many applications based on concepts of optics and communication theory are now available electronically. Pilots must be aware of programs like the Aviation Weather Services Program (AwSP) and projects like NextGen, which were designed to increase the safety of passengers and crew during flights. In addition to the basic flight maneuvers, pilots are trained to acquire knowledge of meteorology and communication systems. In particular, beginning pilots require a lot of practice and flight instructors have the duty and the challenge to teach them about attitude awareness, which requires understanding the movements experienced by an aircraft in flight.

The extreme conditions under which pilots and crews are subject during flights may have a direct impact on their health. In fact, they are frequently exposed to atmospheric effects, changes of altitude, and large doses of cosmic ionizing radiation[1] penetrating the atmosphere, UV radiation, and high-energy solar particles[2] because at normal cruising altitudes the shielding of the Earth's atmosphere against this radiation is weaker than at the surface [310]. However, the different kinds of radiation fields that interact with an aircraft flying at cruise altitudes are extremely complex [145]. The *National Council on Radiation Protection and Measurements* (NCRPM) reported that aircrew have the largest average

---

[1]Cosmic rays consist of high-energy subatomic particles, mainly protons, α-particles, and heavier nuclei, coming from outside the solar system. These particles collide with and disrupt nitrogen, oxygen, and other atomic elements in the atmosphere, producing photons and additional subatomic particles.

[2]At cruise flight altitudes, these energetic particles have qualitatively the same properties as cosmic rays [671].

annual effective dose (3.07 mSv^3) of all radiation-exposed workers in the United States. In fact, flight crews face more radiation exposure than radiology workers or nuclear power plant engineers, according to the NCRPM [296, 771, 52]. Other estimates of annual aircrew cosmic radiation exposure range from 0.2 to 5 mSv per year. As most airline carriers connect farther-flung cities and use polar routes to save time and fuel, pilots and cabin crew are exposed to hazardous UV and cosmic radiation for many hours and therefore they have raised risks of melanoma and other skin cancers [364].

Epidemiological studies about cancer risks in aircrews have been performed since 1990, whose preliminary results indicated high risks. However, these studies involve several challenges since (1) they must take into account synergistic effects from both ionizing and nonionizing radiation coupled to circadian dysrhythmia, reduced atmospheric pressure, mild hypoxia, low humidity, and exposure to sound and vibration; (2) the health effects among aircrews are small, their detection may be difficult; (3) the access to a large population of airline pilots and crews has been traditionally difficult to secure; (4) the physical dosimetry of cosmic radiation is not yet on solid grounds, which may result in an underestimation of doses; and (5) any discussion of potential ionizing radiation risk and/or cancer incidence assessment must include biological markers of past neutron exposure [145]. In particular, Whelan [971] reported an increased cancer incidence and mortality among airline pilots due to prolonged exposure to cosmic ionizing rays. This study revealed that pilots are at increased risk of malignant melanoma, nonmelanoma skin cancer, and possibly acute myeloid leukemia, while cabin crew, made up mostly of women, are at increased risk of breast cancer and malignant melanoma. Other studies have corroborated these findings [561, 613].

Since cruise flights usually exceed 7620 m from the sea level, pilots and cabin crews can also suffer from physiological effects. In fact, tests are required to complete ground training in high-altitude physiology, including hypoxia training. However, the present regulations do not require altitude chamber training (ACT) [373]. It was suggested that professional pilots perceiving that training should include introductory hypoxia training, recurrent hypoxia training, and ACT as well.

10.4 Inadequacies of current weather information systems

The structure and communication systems of modern aircraft are designed to guarantee a high level of safety in ground and flight operations. Despite these technological advances, there are situations in which serious meteorological phenomena at cruise altitudes, such as CAT, continues to represent a threat and are

---

^{3}The symbol mSv stands for milliSievert, where the Sievert (Sv) is a derived unit of ionizing radiation dose in the International System of Units (SI). It is used as a measure of the health effect of low levels of ionizing radiation on the human body.

the cause of aviation incidents and accidents. Since these events are still a matter of concern for the aeronautical industry, in 2018 the EASA has formulated a project called *Weather Information to Pilots (WIP) Strategy Paper: An Outcome of the All Weather Operations Project* to improve the meteorological information provided to pilots pre-flight and the information that is available in-flight [18]. The existing means to attenuate weather effects, including the use of onboard weather radar and other onboard sensors as well as available means to provide the information to pilots, such as the communication systems and the emerging availability of Electronic Flight Bag (EFB) solutions have also been reviewed in this strategy paper. In particular, the following elements have been recognized: (1) although a range of weather information is available at the high spatial and temporal resolution, it is not always utilized; (2) there is a continuous development to improve onboard weather radars and sensors; (3) there are available communications systems for commercial airlines that could support the uplink and downlink of great volumes of meteorological information; and (4) advanced technology already exists that allows displaying the meteorological information in intuitive and interactive ways and exploits the use of colors and symbols for easier and more efficient assimilation of the information [18]. Although these elements can be employed to enhance the situational awareness to pilots and crews prior to and during the flight phases, there are still some challenges.

Before flight pilots are provided with weather information for the flight route, including en-route and destination alternates. This information is provided onboard either in paper or in EFB formats. However, during the flight phases updates to the original briefing material are available subject to variations and the information is often limited. This is the case of updates provided by the *Aircraft Communications Addressing and Reporting System* (ACARS) or the VOLMET[4] broadcasts on the high frequency radio, which are not only subject to limitations of availability, but also of content and format. In particular, this problem is a concern for longer-haul flights, where, for example, a rapidly developing convective disturbance may render useless the meteorological information contained in both the original version and previous updates. Updated meteorological information during flight can reach the pilot either from external sources or from airborne sensors. In the latter case, pilots make their decisions based on the indications of the onboard weather radar, which is the primary source of information available to the pilot ahead of the aircraft and with almost no latency. Since the ATC has no access to the displayed images from the weather radar, they must coordinate their flight routes with ATC. Very frequently, pilots have to build a mental representation of the weather conditions to support their decisions, based on information

---

[4]VOLMET is an abbreviation standing for the french words *vol* (flight) and *météo* (weather report) and means "meteorological information for aircraft in flight". It is a worldwide network of radio stations that broadcast TAF, SIGMET, and METAR reports on short-wave frequencies and in some countries on VHF too. VOLMET provides pilots information on the weather conditions at the surrounding main airports of destination so that they will have time to divert the aircraft to a nearby airport in case of bad weather.

from the airborne weather radar, the pre-flight briefing, the local weather information provided by the Air Traffic Services, and the visual observation from the flight deck, which includes instrument displays and in-flight updates when available.

Sometimes pilots have the perception that they are not always provided with the most relevant up-to-date weather information from the pre-flight stage until they land at the destination airport [480]. This may be due occasionally to several factors, for example, limitations of available meteorological observations and forecasts at the pre-flight stage and the variability of communication capability. However, the analysis of accidents has in some cases revealed that the correct information was available, but the overall situational awareness was compromised because the pilots did not operate the detection system correctly and/or misunderstood the system limitations. The WIP strategy paper has proposed the following elements to enhance situational awareness in the cockpit [18]:

- Derive wind profiles directly from the automatic downlink of weather data;

- Provide weather radar and satellite images;

- Provide lightning location display;

- Perform short term forecasts (nowcasts);

- Provide three-dimensional displays (e.g., radar and volcanic ashes);

- For tailored approaches provide weather information at the destination terminal.

On the other hand, it must be mentioned that sometimes the weather information for flight planning may involve long time intervals between an update and its dissemination, which may be comparable to or even longer than the evolution of the atmospheric event. For example, after several hours of flight forecasts of severe weather hazards, such as thunderstorms, CAT, or icing, the information may be outdated and therefore provide only a rough estimate of the phenomenon. It may also occur that weather events, for example, areas of CAT escape prediction in the planning phase, supersede weather information available at the original briefing. Moreover, for long-haul flights, available synoptic charts and/or satellite images that were included in the pre-flight documentation may become outdated and therefore their use may incur erroneous decisions.

The basic method employed by pilots to receive weather information during flight is voice communication. This is a one-way communication since pilots can only listen. In principle, the information is provided on demand by the operators via ACARS, or via a push-service for SIGMET. Although weather information can be provided via ACARS in alphanumeric form, its global geographic coverage cannot be guaranteed. The transmission of high volumes of information via ACARS is also challenging because of bandwidth/performance limitations and costs. However, emerging datalink technologies could allow to uplink more information to the cockpit than is currently possible via ACARS. For instance, some airline operators use Internet Protocol (IP) satellite communications that have greater bandwidth, guaranteeing connectivity of near-global coverage mainly to support cockpit, cabin, and passenger applications.

EASA's approach to enhancing weather awareness in the cockpit includes: (1) onboard weather radar and basic weather training for pilots; (2) graphical representation of meteorological information at the pre-flight briefing and in-flight updates in a format that facilitates the intuitive and graphical display of the information, the use of color schemes to indicate the severity and/or intensity of weather hazards, the information overlay with other meteorological and/or non-meteorological information (as, for example, the aircraft route), the display of the validity of forecasts and the time of observations, including observations obtained by remote sensing, and the use of clear legends for the description of the meaning of the various symbols, colors, and abbreviations; (3) meteorological information below FL100; (4) access to in-flight updates to meteorological information on EFBs; (5) connectivity solutions supporting uplink weather information; and (6) provision of meteorological information. This latter element is of fundamental importance because it will ensure the availability of weather applications from multiple sources and providers so that pilots and aviation operators will have access to a great amount of meteorological information.

As an airborne detection instrument, weather radars are not entirely reliable for weather forecasting as they are not able to detect wind in dry atmospheric environments. In addition, the use of weather radars requires some level of expertise when analyzing the data because they occupy a huge volume. This implies that a considerable amount of time would be required to complete the analysis. As a consequence, the information cannot be processed in real time. Since the atmosphere behaves in a complex manner and meteorological phenomena are changing all the time, any delay in the information may result in useless data or noise that must be filtered. Under these conditions, the estimates obtained from weather radars may contain uncertainties, which will add important factors that may impact the final decision making. Also, weather radars can experience interference from various aspects of weather, including humidity, wind, and perhaps electromagnetic field interaction, producing low quality data and therefore their analysis can yield unreliable results. However, in the near future, all or part of these disadvantages will certainly improve because today radar technology keeps growing at a really fast pace.

10.5 Cognitive task analysis

Air traffic controllers must take decisions almost all the time concerning the strategic flow management and the tactical solution to aircraft conflicts. Decision-making requires a lot of cognitive effort to repair the situations created by poor human performance. As a realization of these circumstances, the FAA and Eurocontrol have established standard procedures and checklists for a range of typical scenarios to simplify everyday decision-making [500]. For example, the performance of business activities where complex tasks are required may take a lot of time unless the integrated knowledge of controlled (conscious and conceptual) and automated (unconscious, procedural, or strategic) procedures are used in the analysis. Therefore, analysts have employed an instrument called "cognitive task analysis" (CTA) to capture accurate descriptions and complete cognitive processes by using a variety of interview and observation strategies to perform complex tasks. This concept has been developed during the end of the last century and a comprehensive review on the state of CTA in research and practice is given by Clark et al. [181]. This review reports over 100 types of CTA that were used at that time [199]. Originally CTA was associated with behavioral task analysis, with most early work focusing specifically on computer system interfaces and military applications – each with its demands, uses, and research base. To apply CTA, three broad families of techniques have been identified [199], namely (1) observation and interviews, (2) process tracing, and (3) conceptual techniques, which differ in terms of their specificity and formality. However, due to the existing variety of CTA methods, analysts generally use a five-stage process to perform the task, which consists of the following elements: (1) collection of preliminary knowledge, (2) identification of knowledge representations, (3) application of focused knowledge elicitation methods, (4) analysis and verification of the acquired data, and (5) formatting of results for the intended application.

In the field of civil aviation, a CTA was performed to analyze knowledge structures, mental models, skills, and strategies of en route controllers to provide an understanding of the key cognitive components of the air traffic controller's job [814]. This analysis was first used by the FAA to redesign the entire curriculum for the training of en route air traffic controllers [815]. This task was performed in a two-phase procedure, where the first phase focused on the development of a mental model with its associated tasks and cognitive strategies. The second phase was one of validation, where the first phase tasks and strategies were extended to include a greater group of en route controllers. Further applications in the aviation environment suggest that CTA will have a brilliant future. For example, in the operational aviation environment, CTA has produced improved results in applications to training, system design, and human resources management. Other applications include diverse aviation settings for part-task training, crew resource management in the cockpit, the design of systems for improving the human-computer interaction, training in avionics troubleshooting, and even in the training of airport baggage security personnel. These efforts have translated into greater security, reduced operator errors, more efficient and cost-effective training, and fairer and more reliable testing procedures. In addition, CTA methods are helping aviation operators to better understand and prepare for the changing nature of work resulting from system automation and the ever increasing cognitive nature of aviation job tasks. These methods are also being used to provide critical data for the development of effective intelligent aviation interfaces and tutoring systems.

10.6   Weather avoidance doctrine

Despite available modern technologies, adverse weather continues to pose challenges to aviation. In particular, thunderstorms and severe convective weather are more dangerous for small planes than for big ones. For example, according to the 24th Joseph T. Nall Report [9], more than 76% of fatal general aviation accidents are related to adverse weather, while it was suggested that some of these accidents could be attributed to aircraft not equipped to face heavy weather or to pilots with little experience in such adversity [68, 96]. Today, to reduce the risk of accidents pilots are provided with the NEXRAD capability in the cockpit. One of the most important aspects of this technology is that it facilitates the interpretation that pilots can make of the information by simply looking at the color-coded meteorological risk screens of the devices installed in the cockpit, thereby allowing them to visualize the dynamical behavior of the weather [471]. This information is received by a connection to an XM Satellite Radio[5] and displayed in the cockpit of an aircraft equipped with a satellite weather receiver. Since the communication cannot be established continuously in real time, NEXRAD images are updated approximately every 5 minutes. Such time interval limits the pilot sensation of having images moving continuously like a movie on the screen. However, a fast playback of approximately one hour of monitored data, at intervals of 5 minutes, will be enough to reproduce a more or less continuous movie. Therefore, a sensation of apparent movement is transmitted [966], which, for instance, will allow to perceive the direction of motion of a storm. The information retrieved this way can be used to predict the location of both the aircraft and the hazardous weather in the near future [490].

The NEXRAD technology provides pilots with the necessary tools to estimate the minimum distance between the airplane and the storms. However, the estimation is only approximate in the sense that the calculation from the images on the screen have associated errors, which translate into uncertainties in the

---

[5]The XM Satellite Radio is operated by Sirius XM Holdings and was one of the three satellites radio and online radio services in the United States and Canada. Among many other products, it offers weather information for pilots and weather spotters through its *Sirius XM Weather & Emergency Datacasting Service*. Over the years the service has been adopted by many commercial airlines on their aircraft.

estimation of the minimum distances to storms. In addition to instrumentation factors, human errors can also be at the origin of the uncertainties to determine a precise calculation of the distance. A substantial proportion of aviation-related NEXRAD studies have revealed that pilots tend to overestimate the actual minimum separation from bad weather, in most cases getting too close to the storm, and in no case did pilots consistently maintain the 20-nautical-mile separation from bad weather as recommended by the FAA advisory circular 00-24C [471]. To address this problem, Bootsma and Oudejans [101] proposed a mathematical model within a psychophysical context. Their model allowed to verify two issues: first, the presence of detectable information in an object moving towards a designated position and second, the ability of observers to detect that information (albeit imperfectly). The model relies on a simple geometrical exercise, which consists of identifying the parallelism between two lines. Although this is relatively easy to visualize, it must be taken into account that for a flying aircraft the reference system of the observer is moving and therefore the precision of the visual observations is reduced, considering that the NEXRAD loop is included in the operation [335].

The problem of the closest approach point to a storm was also studied by Knecht [471] based on the visual information present in the NEXRAD loop. To illustrate the method consider an aircraft moving along a straight line during a time t with a constant velocity $\mathbf{v}_{plane}$ in the northwest direction as shown in the left picture of Fig. 10.2. On the NEXRAD screen, the airplane icon can be identified as a single point on the nose. In the picture, the airplane icon is approaching the edge of a storm which is moving with constant velocity $\mathbf{v}_{storm}$ in the northeast direction. The problem of avoiding a single point on the edge of such a storm represents possibly the simplest case of "weather avoidance" because in reality the storm does not behave like a rigid body and its morphology is constantly changing. The mathematical model consists of defining the instantaneous range

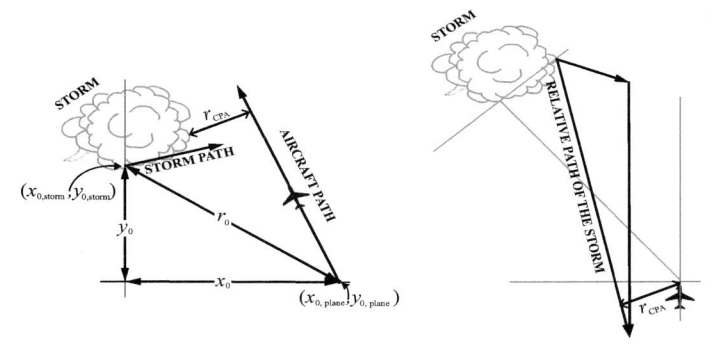

Figure 10.2: Geometry of an aircraft closest approach to a storm cell with initial separation r_0 between the tip of the aircraft and the storm (left). The picture on the right shows the same situation in a frame where the aircraft is at rest and centered. In this case the resulting relative motion of the storm is shown. Figure adapted from Knecht [471].

r_t between the tip of the aircraft icon and the moving point of the storm at time t as

$$r_t = \sqrt{(x_0 + v_x t)^2 + (y_0 + v_y t)^2}, \tag{10.1}$$

where $x_0 = x_{0,\text{plane}} - x_{0,\text{storm}}$, $y_0 = y_{0,\text{plane}} - y_{0,\text{storm}}$, $v_x = v_{x,\text{plane}} - v_{x,\text{storm}}$, and $v_y = v_{y,\text{plane}} - v_{y,\text{storm}}$ are respectively, the initial relative separation distances and the relative velocity components. This time-based parametric form of r_t is also called gap function, or simply gap between the tip of the aircraft icon and that single moving point at the boundary of the storm [471]. The quantities involved in Eq. (10.1) can be easily estimated by comparing at least two views separated by a known time interval. The picture on the right in Fig. 10.2 shows the same geometrical construction of the picture on the left but now in a reference frame where the aircraft is at rest and centered. In this case the moving-map display shows the resulting relative motion of the storm. The point of closest approach to the storm boundary corresponds to the point where the function r_t in Eq. (10.1) reaches a minimum, i.e., for which the condition

$$\frac{dr_t}{dt} = 0, \tag{10.2}$$

holds [470]. Taking the time derivative of r_t and equating the result to zero yields

$$P_{\text{CPA}} = \sqrt{\frac{(x_0 v_y - y_0 v_x)^2}{v_x^2 + v_y^2}}, \tag{10.3}$$

which gives for the simple case illustrated here the closest point of approach to the edge of the storm. The plot of the gap function (typically in units of nautical miles) against time (in minutes) produces a curve of "rounded-V" shape having zero slope at the point of minimum approach to the storm cell.

Although the problem looks simple, in reality it involves many implicit complications due to the impossibility to integrate all the information contained in Eq. (10.1). One way to solve this problem is to look for some clever feature of the situation that will allow us to sidestep complicated calculations, which is called Gibson's direct perception [339]. For deeper details on how pilots' direct perception works the reader is referred to Knecht's [490] paper. An alternative way is to adopt Lee's [543] idea in which an observer perceives a gap of changing size with time. In particular, how the gap changes can serve as a triggering stimulus for actions such as an avoidance maneuver. The information that gives substance to this triggering stimulus is traditionally called ρ_t and is defined by the relation [490]

$$\rho_t = \frac{1}{r_t} \frac{dr_t}{dt}, \tag{10.4}$$

which is the ratio between the instantaneous change in gap size and the instantaneous gap size, or in other words, the ratio between the slope of the gap function at time t and the size of the gap at the same time. The parameter ρ_t defined this way is essentially the inverse of the time-to-contact τ, which forms the basis of the General Tau Theory [544]. An inspection of Eq. (10.4) shows that if the gap shrinks and does it at a fast rate, then the numerator may become bigger than the denominator and ρ_t increases. Equivalently, the same occurs if the denominator becomes smaller due to a smaller gap. A plot of ρ_t versus time t shows that in an on-screen conflict situation such as looping NEXRAD, at the point of closest approach ρ_t will grow large enough to exceed a threshold value and trigger a neural circuit sufficiently far ahead of time to cover reaction and maneuver times [490]. In general, ρ_t may be a key element in hazard avoidance because in current looping NEXRAD, it never reaches the perceptible threshold until too late to facilitate a 20-nautical-miles separation from heavy weather. Further studies focused on analyzing how pilots can visually align the plane with the runway during the landing stage, which is determined by the relative angle between the runway plane and the horizon, and on taking into account how fast the runway expands as the aircraft is approaching it [73].

Chapter 11

Recent Investigations on CAT and Major Challenges

11.1 Introduction

The forecasting and detection of CAT in the free troposphere and stratosphere are still challenging. Although research on CAT focused on the development of new technologies for CAT detection continues, most recent studies on CAT have been devoted to its global response to climate change. These studies have revealed that the busiest international airspace will experience in the future more severe CAT than ever, thereby highlighting the need for improved operational CAT forecasts and their more effective use in flight planning. Many of these studies consist of future model projections of how the CAT strength will be changing across the globe year-round and at different altitudes. For example, multiple computer climate models have shown that the rise of the global temperature will lead to more frequent and powerful air currents in the jet streams and predictions say that severe turbulence on transatlantic flights may go up by 180% [875]. Other case studies have corroborated that a main source of CAT at cruise altitudes is the enhanced shear and flow deformation in the vicinity of a tropopause fold. In particular, understanding the generation mechanisms of CAT in the upper troposphere and lower stratosphere is of fundamental importance for the development of new forecasting algorithms that will help reduce hazardous encounters and enhance aviation safety. On the other hand, microscale measurements would be needed to ensure adequate CAT resolution, which are not regularly available via the present surface and upper-air weather observation networks [767]. This chapter deals with an overview of the most recent research on CAT and turbulence forecasting along with future research needs and major challenges in turbulence prediction and avoidance.

11.2 Recent CAT research studies

Recent CAT research can be divided into four main areas, namely the effects of climate change on CAT, the generation mechanisms of CAT in the upper troposphere and lower stratosphere, the development of new algorithms for the generation of skillful CAT forecasts along with the use of indices for turbulence diagnostics, and the development of new technologies for CAT detection and avoidance. Since the latter area has been amply discussed in previous chapters, in this section, we shall only focus on the first three areas.

11.2.1 CAT response to climate change

Studies of the impact of climate change on CAT have come to light for no more than five to six years [731]. Since the climate is changing not just on the Earth's surface, but also at altitudes of 30000 to 40000 ft, it has important consequences for aviation. Not only are the airport operations affected, but also the flight routes and travel times, and ultimately, what is more important, the safety of passengers and crews. For example, the increase in global temperature causes the rising of the mean sea level and the occurrence of more frequent and intense storms, which may affect the operations in coastal airports [140]. Moreover, the increasing frequency of lightning strikes may induce flight disruptions and delays [772], while at cruising altitudes the occurrence of stronger winds may constrain pilots to modify optimal flight paths with a consequent increase of the journey times [459, 430, 982], and stronger wind shears across the jet streams may cause more severe CAT [987]. It must be recalled that the response of CAT to climate change has already been briefly commented on at the end of Chapter 3 in an attempt to answer the question if the strength and frequency of CAT are effectively increasing. However, here we will go a little more deeply onto the subject.

As a result of enhanced temperature gradients from pole to equator due to increased CO_2 concentrations, through the combined effects of tropospheric warming and stratospheric cooling, the mid-latitude jet streams are expected to strengthen at cruise altitudes in both the northern and southern hemispheres [856, 349]. The consequences of climate change on light, moderate, and severe CAT on the North Atlantic flight corridor in winter was first studied by Williams [983] using climate model simulations. When converting the diagnostics into EDRs, this study concluded that the ensemble-average airspace volume containing light CAT will increase by 59%, while those containing light-to-moderate, moderate, moderate-to-severe, and severe CAT will increase by 75%, 94%, 127%, and 149%, respectively. In a further investigation, Storer et al. [875]

studied the impact that climate change would have on global CAT in the period between 2050 and 2080. They used a current-generation climate model to calculate for the first time the response projections of the various CAT categories to climate change in all four seasons in different regions and flight levels. In these case the climate simulations were performed with the *Met Office Hadley Centre HadGEM2-ES* model [444], which formed part of the *5th Coupled Model Intercomparison Project* (CMIP5) ensemble [894]. Six-hourly snapshots, resolving the diurnal cycle on a suitable set of upper-tropospheric and lower-stratospheric pressure levels provided a better representation of wind shears than the daily mean CMIP3 fields employed by Williams [983]. However, this study corroborated Williams' [983] findings by concluding that the intensification of CAT as was calculated by considering only transatlantic flights in wintertime at altitudes around 39000 ft has a more general application to other flight levels and seasons. As Storer et al. [875] claimed, these predictions will have implications for the aviation industry in that many of the aircraft that will be operating in the second half of the present century will be designed to account for a more turbulent atmosphere. However, it is expected that further technological advances in remote sensing using onboard light detection and ranging as well as improvements in turbulence forecasting may well help to mitigate the adverse effects of a more turbulent atmosphere.

A different approach was recently taken by Lee et al. [546], who analyzed the change in wind speed with height (vertical shear) rather than the wind speed. Using three different reanalysis datasets [450, 229, 492], they found that due to the response of the thermal wind to the enhanced upper-level meridional temperature, the vertical shear in the North Atlantic polar jet stream at 250 hPa has increased between 11 to 17% in contrast to the zonal wind speed which has not changed since the beginning of the observational satellite era in 1979. Evidently, this result shows that the increased vertical wind shear is consistent with previous predictions of the strengthening of shear-driven CAT due to climate change and that it is affecting the North Atlantic jet stream more than has been previously thought. As the temperature difference between the North Pole and the Equator increases, the jet stream will blow stronger from west to east across the North Atlantic, steering and energizing storms and storm systems along with it. On the other hand, in contrast to the cooling of the polar lower stratosphere and the warming of the tropical upper troposphere (which intensify the upper-level jet stream [570]), in the lower atmosphere, the Arctic amplification of global warming is weakening the meridional temperature gradient, which in turn causes a weakening of the upper-level jet stream. This together with the threat posed by the future CAT increase due to climate change has received a lot of media attention in the past few years. An analysis of the future changes to CAT frequency and severity over a pan-Arctic domain was recently reported by Atrill et al. [46], using five-member ensembles of transient regional climate model simulations, corresponding to Representative Concentration Pathways 4.5 and 8.5

for a period from 1980 to 2099. These authors found that the projected increase of severe CAT is likely to double in most cases than for light CAT. Also, the frequency of moderate, moderate-to-severe, and severe CAT over the Arctic circle is projected to decrease by up to 50% in the mid-century for representative concentration pathways 4.5, while by the late century the projection points to an increase in the frequency from 25 to 50%. The results from these studies are providing solid scientific evidence of the strong link existing between climate change and CAT.

11.2.2 Tropopause folding

Over the years there has been a lot of research effort to identify the areas most frequently affected by CAT. An exhaustive analysis reported by Mazon et al. [602] revealed that turbulence was responsible for 66% of the weather-related accidents at cruising altitudes during the period from 1967 to 2010, while 56% of these accidents occurred during descent. In particular, CAT was found to be at the origin of 13% of these cruise flight accidents and 7% of the accidents during descent. A recent review on the impact of weather on aviation meteorology has been presented by Gultepe et al. [368]. A previous study by Kaplan et al. [457] indicated that severe turbulence-induced accidents prevalently occur within the entrance region of the polar or subtropical jet stream, at the synoptic scale, showing that the upstream curvature of the synoptic-scale flow is the most consistent predictor for turbulence. This is in support of the scenario that CAT at cruising altitudes may be related to the formation of tropopause folds [867, 767], which occurs when stratospheric air penetrates the troposphere near an upper-level frontal zone beneath the polar and subtropical jet streams. Although the dynamics of the processes involved in the tropopause folding remains unclear, the global climate simulations presented by Storer et al. [875] support this idea in that CAT events were found to be more pronounced at mid-latitudes in both hemispheres, in response to greenhouse gas forcing. In this respect, a mass exchange of trace species between the lower stratosphere and upper troposphere, including water vapor and aerosols, as well as other chemical species, such as CO, H_2O, N_2O, and O_3 (see Rodriguez Imazio et al. [767] and references therein) can be at the base of the CAT generation process.

Although studies dealing with the generation of CAT in the context of tropopause folding dates from the mid-1990s [519, 411, 175], sustained research in this topic has continued until the present days [748, 494, 994, 867, 855, 767]. However, studies aiming at detecting CAT areas related to tropopause folds have been rather rare, except for the recent works by Stefan et al. [867] and Rodriguez Imazio et al. [767]. In particular, Stefan et al. [867] used turbulence data from PIREPs in the Romanian airspace between June 2017 and December 2018, in conjunction with meteorological data, satellite imagery, and vertical profiles. Diagnostics of CAT related to tropopause folds were obtained by computing a set of indices, including Ellrod, horizontal temperature gradient, Dutton, and Brown indices. An analysis of these indices was provided to explain the physical mechanisms underlying the generation of severe CAT. From a total of 420 PIREPs, only 80 corresponded to cases of severe turbulence of which 13 events were associated with tropopause folding. In addition, it was found that the horizontal temperature gradient, $\nabla_H T$, and Brown indices were the most useful to identify CAT originated from tropopause folding. Since in all cases turbulence was reported just above the tropopause, where other than stability there, the horizontal temperature gradients and the vertical shear induced in the lower stratosphere by tropopause folding (or tropopause sloping in frontal zones) may provide the necessary means for triggering the KH instability, which combined with vorticity and/or deformation flow fields, increase the risk of CAT.

As argued by Rodriguez Imazio et al. [767], aircraft have been in general the most frequently used platforms for CAT measurements. However, accurate measurements are still technically challenging and hence sparse because of the small scales involved. In response to these resolution requirements new research aircraft campaigns have been emerging, which are capable to perform high-resolution *in situ* and remote measurements (not only ahead along the flight path, but also above and below the aircraft) from meteorological parameters and trace gas species at flight level [62, 738]. In particular, Rodriguez Imazio et al. [767] analyzed a CAT event generated by tropopause folding that was recorded on a research flight of the German *High-Altitude Long-Range* (HALO) aircraft as part of the SOUTHTRAC campaign in November 2019. In this case, moderate-to-severe CAT was measured using flight-level observations of wind and temperature with a frequency of 10 Hz. In addition to these measurements, numerical prediction models were employed to describe the synoptic-scale flow and determine the time, location, and large-scale flow patterns characteristic of the CAT event. Results from the synoptic analysis using the ERA5 dataset suggest that advection of a stratospheric air intrusion into the upper troposphere associated with a perturbation at 300 hPa northwest of the Antarctic Peninsula was favored by a polar low originated at the southwest of the Peninsula. The large magnitudes of the turbulent kinetic energy and EDR derived from the HALO measurements, when flying at about 4 km above the folding, were indicative of the presence of severe CAT. What is probably more important, this study has demonstrated for the first time that both CAT and the physical flow properties associated with tropopause folding can be characterized by structure functions and other tools provided by the theory of turbulence. The Ellrod 1 and the Brown indices along with the Ri number were computed from the ERA5 data for a CAT diagnostic at pressure levels of 150, 200, and 250 hPa. All these indices were found to be consistent with strong temperature gradients related to the upper-level front, the tropopause fold, and the curved jet stream exit. Values of Ri in the interval $(0.5, 1.0)$ were diagnosed. This approach provides compatible results, for example, with those obtained from the power spectrum of velocity fluctuations and

CAT indices, thereby opening the possibility for studying intermittency in CAT given its patchy nature. For example, the appearance of these patches remains a longstanding open problem because it involves a constraint to energy dissipation, which has not been completely understood yet [767]. In particular, second-order velocity structure functions comply with atmospheric turbulence following a Kolmogorov scaling with stratification-generated anisotropies, as prescribed by the horizontal and vertical energy spectra, while a third-order structure function supports the evidence of a forward energy cascade and values of the mean EDR close to those derived from spectral analysis. More complete studies of atmospheric turbulence, and in particular CAT, would necessarily require the detection of more turbulent events around the tropopause folds, which will translate into a larger number of flight reports within the lower stratosphere and upper troposphere.

11.2.3 Multi-index prediction of CAT based on machine learning

Numerical weather prediction (NWP) is used to describe the spatio-temporal distribution of turbulence within at least 6 hours from the occurrence of the phenomenon (i.e., within the time frame of nowcasting). This is obtained by numerical integration forward in time of the differential equations governing the state and evolution of the atmosphere [449]. Since in general such NWP models are not sufficient to resolve all atmospheric features leading to turbulence in real time, it has been necessary to develop solutions in the form of turbulence diagnostics, or indices, which are physical quantities derived from the NWP forecast variables that represent the large-scale atmospheric environment. Different indices are used to diagnose different mechanisms and processes, leading to many different formulations of CAT diagnostics [412]. Therefore, the question of whether or how a combination of these indices, as derived from NWP forecasts, would produce good prediction skills might require the use of machine learning techniques to get the optimal index combination. Existing applications of these techniques to the general problem of turbulence deal with numerical solutions of the Navier-Stokes equations for predictive or analytic purposes (e.g., see Refs. [838, 265, 31]). Machine learning algorithms work as a driven-data approach to accelerate, or improve the accuracy of, turbulence simulations [1000, 257]. In the recent past, machine learning within the field of meteorology has been mostly employed to either speed up or improve certain components inside an NWP forecast model [327, 116, 674]. However, studies where machine learning is used to generate skillful CAT forecasts through the optimal combination of multiple conventional CAT indices as produced by an NWP model are missing. An exception of this rule has recently been provided by Hon et al. [412], who applied the XGBoost (Extreme Gradient Boosting) algorithm [168] to generate and validate the multi-index consensus (MIC) turbulence forecast over the Asia-Pacific region. XGBoost is a computationally efficient artificial intelligence algorithm implemented in Python and based on iterative ensembles of regression trees with gradient boosting, which is well-suited for supervised learning involving large datasets [412].

The methodology consists of expressing the MIC turbulence forecast as the weighted sum [412]

$$\text{MIC} = \sum_{i=1}^{19} w_i D_i, \qquad (11.1)$$

where D_i denotes the ith rescaled turbulence index (out of 19 conventional indices as generated using the output of the Weather Research and Forecast (WRF) model software of the *Asian Aviation Meteorological Centre* (AAMC) [840]) and w_i is the corresponding weight obeying the normalization condition

$$\sum_{i}^{19} w_i = 1. \qquad (11.2)$$

The 19 turbulence indices entering in Eq. (11.1) together with their abbreviations, physical units, and significance are listed in Table 1 of Ref. [412]. The MIC values as calculated using Eq. (11.1) vary between 0 and 1, where a value close to 1 indicates a potentially high risk of CAT occurrence.

To evaluate the effectiveness of the present methodology to perform accurate predictions of CAT events a contingency table for the relative operating characteristics (ROC) analysis is constructed, where in a positive forecast the word "HIT" means a positive observation, while "FA" means a false alarm, i.e., a negative observation. In a negative forecast, the word "MISS" means a positive observation and "TN" refers to a negative observation, i.e., a true negative. According to Hon et al. [412], a three-dimensional box of 1° latitude × 1° longitude of horizontal extent and 2000 ft of vertical extent and centered on the observation point is considered for each observation of turbulence. If the maximum MIC value within the box exceeds a pre-determined threshold the forecast is considered to be positive, and negative otherwise. The probability of detection (POD) is calculated according to

$$\text{POD} = \frac{\text{HIT}}{\text{HIT} + \text{MISS}}, \qquad (11.3)$$

while that of false detection (POFD) is given by

$$\text{POFD} = \frac{\text{FA}}{\text{FA} + \text{TN}}. \qquad (11.4)$$

A ROC diagram is constructed by plotting the POD values (ordinate) versus the POFD values (abscissa) so that a curve is generated by displaying a range of thresholds (from 0 to 1 for the case of MIC). This creates a set of data points in the (POFD, POD)-plane. The skill of the prediction method is then reflected by the ROC curve. A positive skill corresponds to points in the top left of the diagram (i.e., large POD and low POFD meaning a high HIT rate with few FA). The optimal weights, w_i, for construction of the MIC turbulence forecast are determined with the aid of the XGBoost algorithm. The study conducted by Hon et al. [412] covered a 1-year period starting from April 2018 and ending in March 2019, used over 16000 PIREPs of moderate-to-severe CAT encounters, and 100 aircraft-derived moderate-to-severe CAT observations, which is one of the largest datasets covering the Asian-Pacific region. The results demonstrated a superior performance of the MIC model forecast, with median gains in skill score (measured as the area under the curve score in the ROC diagram) between 3 to 17% compared to the case when the constituent indices were taken individually. This effectively shows that the real-world problem of CAT prediction can be greatly improved by applying machine learning techniques.

11.2.4 Experimenting a CAT index from the IFS system

Today there is an ever increasing demand in aviation forecasting for reliable CAT estimates because of the hazardous risks imposed by CAT encounters at cruising altitudes. Therefore, it is important to develop observable turbulence diagnostics based on parameters other than the usual diffusion coefficients for heat and momentum and the local Richardson number. For this purpose, a calibrated CAT diagnostic has been developed in the ECMWF Integrated Forecast System (IFS) using the latest upgrade to cycle 47r1 (Cy47r1)[1] and the new moist physics for Cy48r1[2] [76]. The methodology employed is mostly based on the method proposed by Sharman and Pearson [828]. In this method, a log-normal distribution is fitted to the distribution of each CAT index, which is then projected onto a climatological distribution of EDR. High-resolution and ensemble data are used on model levels, with a grid spacing of 300 m across the upper troposphere and lower stratosphere, and on a reduced output grid at $0.1°$ for the probability density functions (pdfs) and $0.3°$ for verification. Preliminary experiments were carried out with only a limited number of CAT indices to construct the CAT product [76], namely (1) the Ellrod 1 index [275] (or TI1, see Section 4.2 where this index is defined); (2) the three-dimensional frontogenesis function (F3D) [409],

---

[1]Cy47r1 is an upgrading cycle of the ECMWF's IFS, which produces the global forecasting system of the *Copernicus Atmosphere Monitoring Service* (CAMS).

[2]This cycle represents a major upgrade in the atmospheric physics, land cryosphere, dynamics, and data assimilation. It is foreseen at the end of 2022, or beginning 2023.

defined as

$$
\begin{aligned}
\text{F3D} \;=\; & \frac{1}{|\nabla\theta|}\left[\frac{\partial\theta}{\partial x}\left(-\frac{\partial v_x}{\partial x}\frac{\partial\theta}{\partial x}-\frac{\partial v_y}{\partial x}\frac{\partial\theta}{\partial y}-\frac{\partial v_z}{\partial x}\frac{\partial\theta}{\partial z}\right)\right. \\
& +\frac{\partial\theta}{\partial y}\left(-\frac{\partial v_x}{\partial y}\frac{\partial\theta}{\partial x}-\frac{\partial v_y}{\partial y}\frac{\partial\theta}{\partial y}-\frac{\partial v_z}{\partial y}\frac{\partial\theta}{\partial z}\right) \\
& \left.+\frac{\partial\theta}{\partial z}\left(-\frac{\partial v_x}{\partial z}\frac{\partial\theta}{\partial x}-\frac{\partial v_y}{\partial z}\frac{\partial\theta}{\partial y}-\frac{\partial v_z}{\partial z}\frac{\partial\theta}{\partial z}\right)\right],
\end{aligned}
\tag{11.5}
$$

where v_x, v_y, and v_z denote the wind velocity components in the zonal (x), meridional (y), and vertical (height z) direction, respectively, and θ is the potential temperature; (3) the mountain wave turbulence index (MWT3), which is defined as the product between the F3D index, the lower tropospheric wind speed, U_{850}, and the orographic elevation H (capped to 2750 m from the ground), that is

$$
\text{MWT3} = \text{F3D} \times U_{850}\min(2750, H),
\tag{11.6}
$$

where U_{850} is taken to be the maximum wind speed in the 500–1500 m layer above the surface; (4) the dissipation from convectively generated gravity waves (GWD), which is defined by scaling the dissipation from the non-orographic gravity wave scheme (see Orr et al. [677]) with the integral convective heating between 500 hPa and the convective top. This is the only way the heating from convective motions can contribute to turbulence. Under this provisions, the GWD index can be written as

$$
\text{GWD} = \left[\left|\left(v_x\frac{\partial v_x}{\partial t}\right)_{\text{gwd}}+\left(v_y\frac{\partial v_y}{\partial t}\right)_{\text{gwd}}\right|\hat{T}_{\text{conv}}\right]^{1/3},
\tag{11.7}
$$

where

$$
\hat{T}_{\text{conv}} = -\frac{c_p}{\hat{T}_0}\int_{500\,\text{hPa}}^{\text{Top}}\frac{1}{g}\left(\frac{\partial T}{\partial t}\right)_{\text{conv}}dp,
\tag{11.8}
$$

T denotes temperature, the subscripts "gwd" and "conv" mean, respectively, gravity wave drag and the temperature tendency from the convection parameterization, g is the acceleration of gravity, c_p is the specific heat at constant pressure, and $\hat{T}_0$ is a normalization factor taken to be 1 W m^{-2} so that GWD and EDR have the same dimensions; (5) the local Richardson number defined by

$$
\text{Ri} = \frac{g}{T_v}\frac{\partial T_v/\partial z+(g/c_p)(T_v/T)}{(\partial v_x/\partial z)^2+(\partial v_y/\partial z)^2},
\tag{11.9}
$$

where T_v is the virtual temperature; and (6) the total dissipation rate index (DISS) given by

$$
\begin{aligned}
\text{DISS} \;=\; & \left|\left(v_x\frac{\partial v_x}{\partial t}\right)_{\text{diff}}+\left(v_y\frac{\partial v_y}{\partial t}\right)_{\text{diff}}\right|^{1/3} \\
& +\left|\left(v_x\frac{\partial v_x}{\partial t}\right)_{\text{conv}}+\left(v_y\frac{\partial v_y}{\partial t}\right)_{\text{conv}}\right|^{1/3}+\text{GWD},
\end{aligned}
\tag{11.10}
$$

where the first term on the right-hand side is the contribution of the turbulent dissipation rate (which contains contributions due to turbulent mixing, orographic wave drag, and orographic blocking and is derived from the vertical diffusion tendencies for momentum in the IFS [81]), the second term is the contribution to dissipation from the convective momentum transport, and the third term accounts for dissipation from the convective gravity wave drag, as defined by Eqs. (11.7) and (11.8). Of the three terms, the turbulent dissipation rate is the most dominant. In the above expressions the horizontal gradients of v_z are set to zero, i.e., $\partial v_z/\partial x = \partial v_z/\partial y = 0$, and

$$\frac{\partial v_x}{\partial y} = \frac{\partial v_y}{\partial x} - \zeta, \quad \frac{\partial v_y}{\partial y} = -\frac{\partial v_x}{\partial x} + \delta, \quad \frac{\partial v_z}{\partial z} = -\delta, \tag{11.11}$$

where ζ and δ are, respectively, the curl (vorticity) and divergence of the wind velocity vector.

Experiments with different combinations of the above indices were performed to choose the CAT product that produces the best matching with the observational trends as measured in terms of correlations or mean absolute errors. The best combinations resulted in the following CAT definitions [76]

$$
\begin{aligned}
\text{CAT}_1 &= 0.6\text{TI1}^\star + \text{GWD}^\star, \\
\text{CAT}_2 &= 0.85\text{TI1}^\star \min(1.0, 0.5\text{Ri}^{-1}) + \text{GWD}^\star, \\
\text{CAT}_3 &= 0.6\text{DISS}^\star, \\
\text{CAT}_4 &= 0.5(\text{CAT}_2 + \text{CAT}_3),
\end{aligned}
\tag{11.12}
$$

where the superscript $\star$ means index values after projection onto a climatological EDR[3]. A graphical comparison of the four index combinations given by Eqs. (11.12) with EDR observations obtained from the NOAA's Meteorological Assimilation Data Ingest System (MADIS) dataset in the height range of 5 to 12 km from the ground level is displayed in Fig. 2 of Bechtold et al. [76]. For a total of 196000 observations with EDR values above 0.005 $\text{m}^{2/3}$ s^{-1}, the combination defined by CAT_4 performed better, approximately following the observed EDR pdf. Although all four CAT definitions were able to predict the turbulence areas in different regions, it was observed large differences between the different CAT products in predicting severe turbulence with EDR > 0.3 $\text{m}^{2/3}$ s^{-1}. All CAT products show a tendency to overpredict CAT. However, CAT_4 was again observed to perform better than the other combinations.

---

[3]After having performed a log-normal fit to each index distribution, it is then projected onto the EDR climatological distribution according to $\ln y^\star = a + b\ln x$, where x stands for the CAT index, $y^\star$ is its transformed value into EDR, and $b = c_2/\sigma$, $a = c_1 - b\mu$, $c_1 = -2.57$, $c_2 = 0.51$, and the parameters μ and σ represent the mean and square root of the variance (i.e., the standard deviation from the mean) of the log-normal distribution.

11.3   Recent improvements in turbulence forecasting

The question of what can pilots and airlines do to avoid CAT while in flight can be answered today in two different ways. One way relies on the use of modern airborne detection systems and the other on the use of sufficiently good predictive forecasting. For example, pilots could spot CAT using UV LIDAR equipment installed at the nose of their aircraft. If the amount of backscattered light is such that the scanned patch of air has wildly varying densities, they could have enough time to either avoid the CAT encounter or, in the worst case, warn the passengers and crew to fasten their seat belts before the shaking starts. However, airlines are not currently using this technology yet because of its relatively high cost, which in most cases would certainly exceed eventual losses of money caused by CAT incidents. Therefore, at present times CAT is mainly avoided through predictive forecasting from national and private weather departments and PIREPs from nearby flying aircraft.

11.3.1   The World Area Forecast System (WAFS)

The WAFS is a forecast system of the *World Area Forecast Centers* (WAFCs) established in 1982 by the ICAO in conjunction with the WMO. There are currently two WAFCs, one based in Washington, D.C., and the other in London, United Kingdom. In support of the ICAO's Aviation System Block Upgrade (ASBU), these centers continue to provide on a 6-hourly basis gridded diagnostics for turbulence, icing, and convection above 800 hPa (6000 ft) across the world, including CAT [342]. This information can be used by airlines for strategic flight planning and is particularly valuable for long-haul flights. For turbulence, the current gridded WAFS forecasts use global NWP model outputs to perform turbulence estimation on a global domain based solely on the Ellrod empirical index as a diagnostic tool. In 2018, as part of the ICAO Global Air Navigation Plan (GANP) aimed at matching the capabilities of modern aircraft toward an improved air traffic management, the ASBU plan started with block 0 (ASBU 0), which consisted of developing new algorithms for upgrading the WAFS turbulence product, providing the EDR as a metric for multiple sources of turbulence, and producing probabilistic forecasts with better spatio-temporal resolution [484]. During the period 2019–2024, the ASBU 1 (block 1) will be aimed at providing calibrated probabilistic forecasts for turbulence, icing, and convection along with the implementation of finer grid resolution for WAFS. Among the other tasks foreseen, there will be the implementation of forecasts for three types of atmospheric turbulence, namely CAT, convectively induced turbulence, and mountain wave turbulence and allowance of the WAFS data to be available via the System Wide Information Management (SWIM). A second block (ASBU 2) from 2025 to 2030 will be devoted to increasing en route weather datasets for performing improved flight planning and supporting decision-making. Finally, beyond 2030 the third block (ASBU 3) plans to provide fully integrated multimember ensemble hazard forecasts, implement high spatio-temporal resolution models, and provide 1-hourly fully automated gridded and significant weather forecast (SIGWX) on a global basis [484].

The sizes of the largest eddies typically encountered in the upper troposphere and lower stratosphere range from a few hundred meters to a few kilometers in horizontal direction, and therefore they are much smaller than the grid spacing of current global NWP models. However, since the kinetic energy flows down from the largest eddies to smaller scales [174], turbulence at the aircraft scale can be predicted using NWP model-based turbulence diagnostics [342, 482, 485, 828]. For instance, NCAR developed the Graphical Turbulence Guidance (GTG) system[4] [828], which has been used mainly for operations in the United States. It provides multiple pieces of information for CAT and mountain wave turbulence (MWT) and turbulence intensities measured in terms of EDR. Among other features, GTG works with stand-alone algorithms that can be applied to any global NWP model and offers the capability of providing probabilistic forecasts based on calibrated EDR thresholds and treating the diagnostics as an ensemble. The domestic operational GTG algorithm has been expanded into a global GTG (G-GTG) system as part of the WAFS ASBU 0 and ASBU 1 upgrades by the NOAA/AWC (WAFC Washington) in collaboration with the Met Office in the United Kingdom (WAFC London) and NCAR. For the G-GTG evaluations, Kim et al. [484] used 15 individual CAT indices and 15 individual MWT diagnostics. The 15 CAT indices were selected from the suite of the GTG, version 3, component diagnostics [828], which had higher skill scores than other indices when verified against long periods of the upper level *in situ* EDR observational data over the conterminous United States (CONUS) and outside them [828]. Similarly, the list of MWT diagnostics includes the 14 diagnostics described by Sharman and Pearson [828] over the CONUS based on a combination of two-dimensional surface mountain-wave parameters (mws) with conventional three-dimensional CAT diagnostics. These were shown to have good forecasting skills against PIREPs data. In addition to these diagnostics, the MWT indices also included the orographic gravity wave drag (OGWD) parameterization method [684]. Tables 11.1 and 11.2 list the 15 CAT and 15 MWT diagnostics used in the G-GTG evaluations, respectively. For a complete description of these turbulence indices, the interested reader is referred to Sharman and Pearson [828] and references therein. The information displayed in Tables 11.1 and 11.2 has been extracted from Kim et al. [484].

After calculation of the individual turbulence diagnostics within the G-GTG from the NWP output, they must be remapped into an EDR scale. Apart from being the standard metric for turbulence reports from commercial aircraft according to the ICAO, it is independent of aircraft type and size. In addition, EDR forecasts are more easily compared with *in situ* EDR observations from aircraft

---

[4]GTG is available on the web page of NOAA/AWC: www.aviationweather.gov/turbulence.

Table 11.1: CAT indices used in the G-GTG evaluations.

Index	Units	Description
Ellrod 3	s^{-2}	Vertical wind shear times the total wind deformation plus divergence tendency [276]
Fth/Ri	$m^2\, s^{-3}$	Normalized two-dimensional frontogenesis function on isentropic surfaces over the Richardson number ratio [828]
NGM1	$m\, s^{-2}$	Wind speed times the total wind deformation [745]
DEFSQ	s^{-2}	Square of the total wind deformation [826]
|DIV|/Ri	s^{-1}	Absolute value of the horizontal wind divergence over the Richardson number ratio [828]
IAWIND	$m\, s^{-2}$	Inertial advective wind [605]
UBF/Ri	s^{-2}	Unbalance flow over the Richardson number ratio [493]
EDRAVG	$m^{4/3}\, s^{-2}$	Estimation of $\varepsilon^{2/3}$ from the average of longitudinal and transverse second-order structure functions for the horizontal wind component [304]
EDRLL	$m^{4/3}\, s^{-2}$	Estimation of $\varepsilon^{2/3}$ from the second-order structure function for the zonal wind component [304]
|TEMPG|/Ri	$K\, m^{-1}$	Absolute value of the horizontal temperature gradient over the Richardson number ratio [828]
NCSU2/Ri	s^{-3}	Vertical vorticity gradients times the gradient of the Montgomery streamfunction over the Richardson number ratio [458]
EDRLUN	$m^{2/3}\, s^{-1}$	Estimation of $\varepsilon^{1/3}$ from $d\mathrm{Ri}/dt$ [342]
WSQ/Ri	$m^2\, s^{-2}$	Square of the vertical wind velocity over the Richardson number [828]
SIGWAVG/Ri	$m^2\, s^{-2}$	Estimation of the variance of the vertical wind velocity from the average of the longitudinal and latitudinal second-order structure function over the Richardson number ratio [306]
RTKE	$m^2\, s^{-2}$	Resolved turbulence kinetic energy [879]

and are more useful for verification purposes [826, 482, 828]. Remapping raw turbulence potential values into an EDR scale is accomplished by assuming that the predicted turbulence indices follow a log-normal distribution [828]. This assumption is supported by previous experimental observations showing that turbulence in the upper troposphere and lower stratosphere actually follows an approximate log-normal distribution [654, 174]. Calculation of the probability density functions for the 15 CAT and 15 MWT diagnostics derived from a 6-month period (from October 2015 to March 2016) of the 18-h forecast of the Global Forecast System (GFS; [816]) model outputs issued at 00.00 UTC have shown that severe CAT corresponding to EDR = 0.34 $m^{2/3}$ has a very small relative

Table 11.2: Mountain wave turbulence indices used in the G-GTG evaluations.

Index	Units	Description
MWT1	$m^3 s^{-3}$	mws[a] times WSQ/Ri
MWT2	$K^2 m^{1/3} s^{-1}$	mws times CTSQ[b]
MWT3	$m^2 s^{-2}$	mws times F3D[c]
MWT4	$m^2 s^{-2}$	mws times the horizontal wind velocity
MWT5	$m s^{-2}$	mws times |DIV|
MWT6	$m s^{-1}$	mws times NGM1
MWT7	$m^3 s^{-2}$	mws times SIGWAVG
MWT8	$m^2 s^{-1}$	mws times LHFK/Ri[d]
MWT9	$m^3 s^{-2}$	mws times IAWIND
MWT10	$m^{7/3} s^{-3}$	mws times EDRAVG
MWT11	$m^{5/3} s^{-2}$	mws times SCHGW[e]
MWT12	$K s^{-1}$	mws times |TEMPG|
MWT13	$m s^{-2}$	mws times DEFSQ
MWT14	$m^{7/3} s^{-3}$	mws times EDRLL
TKE-GWB	$N m^{-2}$	Single-column model-based orographic gravity wave drag [684]

[a] The abbreviation "mws" means gridpoint height (in m) times the horizontal gradient of the height (in m km^{-1}) [828].

[b] Temperature structure constant estimated from the temperature zonal and meridional structure functions [305].

[c] Three-dimensional frontogenesis function on constant z surfaces [828].

[d] Lighthill-Ford radiation source term divided by the Richardson number [491].

[e] $\varepsilon^{1/3}$ derived from the variance of the vertical wind velocity [806].

frequency ($< 0.01\%$), which is consistent with the probability density functions of the observed *in situ* EDR data [824, 484].

Once the CAT and MWT indices are converted into an EDR scale, a multi-diagnostic approach is obtained by averaging the individual CAT or MWT diagnostics according to [484]

$$\text{GTG}(x,y,z) = \frac{1}{N} \sum_{i=1}^{N} \text{EDR}(x,y,z), \qquad (11.13)$$

where N is the number of CAT or MWT diagnostics. This forms an ensemble method and therefore the deterministic G-GTG combination defined by Eq. (11.13) is referred to as the ensemble mean of the individual turbulence indices. The maximum gridpoint-by-gridpoint EDR value of the CAT and MWT ensemble mean products provides the final deterministic forecast. Following this method, it was demonstrated that the higher EDR values occur along the mid-latitude jet stream over the North Atlantic Ocean and northwestern Pacific Ocean in the Northern Hemisphere and in the southern Indian Ocean in the Southern

Hemisphere (see Fig. 6 of Ref. [484]). In particular, these higher EDR forecasts were found to agree well with observations.

11.3.2 ECMWF IFS forecast of severe CAT

In Section 11.2.4, the development of a calibrated CAT parameter for the ECMWF IFS has been described. This method consists of calculating a number of CAT predictors and then projecting them onto the observed climatological distribution of the EDR. A suitable CAT parameter is then finally defined by a combination of the projected indices as shown by Eqs. (11.12). Recently a measure of the EDR has been performed using three predictors: the Ellrod 1, the GWD, and the DISS indices [77]. Expressions for these last two indices are given by Eqs. (11.7) and (11.10), respectively. To obtain the desired horizontal wind gradients additional inverse spectral transforms of the vorticity and divergence fields are involved in the calculation of the Ellrod 1 index, implying an increase in the computational cost of an IFS forecast from 4 to 5%. Climatological distributions of the natural logarithm of these indices, as they were obtained from daily 24-h hindcasts with the IFS during the first half of 2019, were found to closely follow a log-normal law, with the distributions of GWD and DISS being reasonably close to the climatological distribution of the EDR. The actual forecasting of the EDR consists of computing for each forecast the three indices and then project their values onto the climatological EDR [76]. EDR measurements for this experimental study were retrieved from the peak EDR of 12 individual EDR measurements over a 1-minute interval from the U.S. NOAA MADIS public archive for aircraft data.

The EDR algorithm uses either the measured vertical aircraft accelerations, in response to eddies with wavelengths of 10 m to 1 km, or the aircraft's vertical winds as determined using the aircraft's angles of attack and cruising speed. A Fourier transform is then performed on the 8 Hz sampled vertical velocity time series and a von Kármán spectrum is fitted to the retrieved vertical velocity spectrum in the inertial turbulence subrange. When a turbulent event is detected by the algorithm, a report is generated and downlinked at a 1-minute interval. The following combinations of the projected indices were considered to define the CAT products

$$
\begin{aligned}
\mathrm{CAT}_1 &= 0.7\mathrm{TI}1^\star + \mathrm{GWD}^\star, \\
\mathrm{CAT}_2 &= 0.66\mathrm{DISS}^\star + \mathrm{GWD}^\star, \\
\mathrm{CAT}_{12} &= 0.5(\mathrm{CAT}_1 + \mathrm{CAT}_2),
\end{aligned}
\tag{11.14}
$$

where, as before, the asterisks denote EDR projected values of the indices. Since GWD is generally small, CAT_1 is essentially based on the Ellrod 1 (or TI1) index, while CAT_2 represents the total dissipation rate of the IFS. When comparing these two CAT products with observations performed on daily 24-h forecasts from January to March 2019, it was found that both CAT_1 and CAT_2 underestimate the occurrence of weak turbulence intensities (with EDR < 0.05 m$^{2/3}$ s^{-1}). On the other hand, CAT$_2$ was seen to overestimate the observations for moderate-to-severe CAT, while CAT$_1$ underestimates the occurrence of severe-to-extreme CAT (i.e., with EDR > 0.3 m$^{2/3}$ s^{-1}). The linear combination defined by CAT$_{12}$ produced the best overall match with the observations [77].

The intermittent character of turbulence demands an ensemble approach to obtain reliable estimates of CAT. This was also tested by evaluating ensemble forecasts of CAT, comprising a 15-member ensemble at a resolution of ~ 18 km that was run daily during the first two weeks of January 2019. When comparing the point correlations and mean absolute errors (MAEs) against the observations of the forecasts for January to March 2019, the ensemble was found to perform significantly better, with mean correlations achieving values of 0.37 for CAT$_2$ and up to 0.40 for CAT$_{12}$ against values of 0.32 for CAT$_2$ and 0.36 for CAT$_{12}$ for the period from the first to 14 January 2019. Similar mean correlations were also attained for the high-resolution forecasts during the 3-month period from January to March 2019. In terms of the MAE (the ensemble forecasts are actually evaluated using the continuous ranked probability score (CRPS), which compares directly to the MAE of the high-resolution forecasts), the CRPSs of the ensemble forecasts for the three CAT products were around 0.03 m$^{2/3}$ s^{-1} compared to around 0.05 m$^{2/3}$ s^{-1} for the MAEs of the high-resolution forecasts. These results demonstrate that the ensemble forecasts are of reasonable reliability, given that turbulence intensity is typically classified in EDR intervals of 0.1 m$^{2/3}$ s^{-1}.

11.3.3 *Machine learning CAT forecasting based on regression trees*

Since its inception, the original GTG algorithm together with its most recent developments have represented modern references for turbulence forecasting [828, 646]. In particular, GTG algorithms have demonstrated good skills in providing EDR-based diagnostics from regional to global coverage. However, the method is not free of limitations. For example, the method assumes that turbulence indices comply with a log-normal distribution, while there is evidence that only a few diagnostics may actually follow a log-normal distribution. In general, for most turbulence diagnostics departures from the climatological log-normal distribution are observed toward the tails of the distribution [647]. In addition to this, the GTG method uses a forward-selection optimization technique to maximize the skill of the ensemble prediction for a given statistical metric rather than selecting different turbulence indices depending on the atmospheric conditions that may ultimately lead to more skillful and flexible CAT forecasts.

To develop an enhanced GTG-like algorithm and overcome some of the weaknesses of the original method, Muñoz Esparza et al. [647] have implemented a machine learning model based on regression trees for turbulence forecasting at upper levels from 20000 to 45000 ft of altitude. In particular, regression trees are frequently used as a learning method and, in addition to being able to fit complex datasets, it requires less data preparation compared to other machine learning algorithms, particularly those based on neural networks. In one word, regression trees are powerful algorithms that are built by progressively splitting the source dataset according to certain criteria, resulting in the formation of branches and leaves.

The model proposed by Muñoz Esparza et al. [647] includes random forests techniques to reduce model overfitting [114]. Hence, instead of a single tree, the prediction outcome of the machine learning algorithm consists of n trees, which can be used to derive a probability. A series of GTG-derived raw turbulence indices along with prognostic variables from the NWP model forecast were used to train the random forest algorithm. A database of 2.42 million model-observation pairs was created, as the result of 1089 3-km High-Resolution Rapid Refresh (HRRR)-based GTG 6-h forecasts, which were randomly split into independent training (75% of the total) and hold-out testing (25% of the total). Sensitive analysis of the number of trees and the tree depth have shown that an algorithm with 100 trees and 30 layers of depth provides sufficient complexity for this machine learning method. In particular, Muñoz Esparza et al. [647] reported that the algorithm based on forest trees can significantly reduce forecast errors for null and light CAT (i.e., EDR < 0.1 m$^{2/3}$ s^{-1}), increase the probability of detection, and reduce the number of false alarms. These results are in contrast with the high-resolution GTG algorithm which tends to overpredict turbulence intensity. The use of gradient-boosted regression trees along with two feature reduction methods based on relative importance identification by impurity (RFri) and by permutation (RFrp) were found to significantly reduce the complexity of the baseline random forest model and consequently, to speed up the execution of the algorithm, while producing a similar skill to that of the baseline random forest model. On average, the machine learning models show higher skills in discriminating the EDR forecast for light, moderate, and severe CAT compared to current GTG models. Moreover, the analysis conducted by Muñoz Esparza et al. [647] have shown that the random forest models can be used to enhance the GTG predictive skill as well as to reduce the number of false alarms in null and light CAT conditions and simplify the current GTG framework. In the words of these authors, the results obtained with this machine learning model were so promising that this type of random forest algorithm is being considered to be incorporated within the GTG framework for its operational use. As was outlined above in Section 11.2.3 machine learning and artificial intelligence represent emerging techniques providing alternative algorithms for more accurate CAT forecasts. In fact, the popularity of these methods is growing to the point that they have recently found room for application to a large variety of meteorological problems [611], including real-time prediction of convective winds [517], extreme precipitation [396], wind power [503], and severe weather [401], among other applications.

11.3.4 The T^2-Net turbulence forecasting

Sections 11.3.1 and 11.3.2 have described two examples of improved turbulence forecasting based on the use of empirical indices. However, turbulence indices may not be accurate enough to handle complex weather conditions. On the other hand, the development of accurate and effective CAT forecasting systems has been long prevented by the complex spatio-temporal character of turbulence together with the fact that only very limited turbulence labels can actually be obtained. For example, the scarcity of CAT labels is because in addition to being a rare and anomalous event, relevant CAT data can be obtained through rather sparse PIREPs following a CAT encounter. In a attempt to address such impediments, Zhang et al. [1006] have introduced a novel method based on a unified semi-supervised deep learning framework, which they called T^2-Net. This framework consists of two modules: a turbulence forecasting network and a turbulence detection network, which are co-trained in a semi-supervised manner. For example, the turbulence forecasting model works as a sequence-to-sequence multiclass classification system, where from weather features[5] in grid-based three-dimensional regions at previous time slots, turbulence levels are predicted at the next few time slots for all grids in these three-dimensional regions. Similarly, for features forecasted in a three-dimensional region based on a NWP model at a given time slot, the turbulence detection task consists of predicting turbulence conditions at the same time slot for all grids in that region.

The turbulence forecasting module is built upon a convolutional LSTM network (ConvLSTM) [830] to model the complex spatio-temporal correlations among different grids within a region. In particular, it consists of two ConvLSTMs, which serve as the encoder and decoder, respectively. The encoder takes as input the historical turbulence three-dimensional grids at the sequence of time slots $1, 2, \ldots, n$, while the decoder takes the last hidden state of the encoder as the initial hidden state, and applies teacher forcing [351] to generate a sequence of features that correspond to the forecasting time slots $n+1, n+2, \ldots, n+p$. On the other hand, the turbulence detection network uses as input a given NWP forecasted feature of a three-dimensional grid at a given time slot and produces as output the detected turbulence level of the same grid at the same time slot. To extract the spatial correlations and detect the turbulence levels, the detection module employs a Convolutional Neural Network. The problem of label scarcity is tackled using a Dual Label Guessing [542, 88], which serves to generate high-quality pseudo-labels for those unlabeled grids. A more detailed description of the methodology employed by T^2-Net can be found in Zhang et al. [1006]. Extensive experiments conducted by these authors on a real-world dataset have shown that T^2-Net reduces both the problem of spatio-temporal features characteristic of CAT and the scarcity of turbulence labels, thereby providing a superior tur-

---

[5]These may be the wind velocity, temperature, relative humidity, and the pressure integrated with some turbulence indices to enhance the model capacity.

bulence forecasting technique. In recent years there has been a growing interest in applying machine learning techniques to a large variety of spatio-temporal phenomena [954, 1005]. While most of these techniques are based on ConvL-STM to model the spatio-temporal correlations, they do not face the issue of label scarcity. In particular, the T^2-Net approach proposed by Zhang et al. [1006] is one of the few if not the only semi-supervised deep learning approach that takes into account CAT events recorded by PIREPs as ground truth labels and complementary turbulence signals for training a statistical forecasting model.

11.4 Future research and needs

Despite recent progress in aviation turbulence many challenges and needs remain from a research and a development point of view. Much of the difficulty giving rise to such challenges and needs is largely because the nature of turbulent motion is not yet fully understood. This can be summed up in the following words from Turner [914]: *Patches of turbulence in the ocean or atmosphere can arise as a result of the superposition of motions from many sources and on many scales. A completely deterministic theory is therefore unlikely, and detailed forecasting of clear air turbulence will always be difficult, ...* In fact, turbulence is at the base of the outstanding difficulty in hydrodynamics [884], with important implications to the aviation industry.

The scales of aircraft turbulence through which energy production is dissipated by viscous effects range from a few meters to about one kilometer. Therefore, better and more comprehensive observations of such scales are needed through much better *in situ* CAT estimates and the continued development of both ground-based and airborne remote sensing techniques, including improved satellite-based technologies. In parallel to this, the tactical avoidance of CAT patches that were not properly forecasted would certainly need better nowcastings, which may be provided by human-over-the-loop checks. The development of new and more outperforming forecasting products would be a primary necessity to improve passengers and crew safety at cruising flight levels. Research must also focus on better understanding the generation of CAT as well as its advection and propagation mechanisms. In general, this can be accomplished by direct analysis of CAT data gathered in field programs and campaigns. On the other hand, accurate DNS hydrodynamics simulations using high spatial resolution would be imperative to produce useful case studies that will help to improve our understanding of turbulence generation, advection, and propagation at different flight levels. All this information can in turn be used to formulate improved turbulence forecast algorithms. In particular, deep machine learning techniques are emerging as powerful tools for improved CAT forecasting. Ultimately, all these needs will translate into providing more accurate and comprehensive information to the pilots and crew. Evidently, such an effort can only be accomplished by collaborations between universities, research institutes, government laboratories, and industry.

Observation enhancements will need more *in situ* EDR data, provide access to onboard turbulence detection systems, and develop and implement new airborne LIDAR-based detection devices. In addition, detection capabilities will also require the use of high-resolution rawinsondes for upper air observation consisting of direct evaluation of the wind speed and direction, temperature, pressure, and relative humidity aloft. While this has traditionally been done by means of balloon-borne radiosondes tracked by a radar or radio direction finder, nowadays some rawinsondes use GPSs instead. Moreover, errors associated with NWP models and post-processing algorithms are actual limitations for better CAT forecasts. A reduction of the NWP model errors will certainly benefit from enhancing the spatial resolution by incorporating grid nestings along both the horizontal and vertical directions as well as to merge regional models into global ones. In addition, lessening of the NWP model errors will also need refinement of the turbulent kinetic energy subgrid parameterizations for the free atmosphere and sensitivity studies of the NWP models to resolution, model configurations, and parameterizations. However, the minimization of errors associated with post-processing algorithms will probably entail more fundamental research. For example, elements that must be improved include a better understanding of the linkage between the large scales of turbulence represented in the NWP models and the smaller scales, the autotuning of post-processing algorithms when underlying NWP model changes, and the use of ensembles of diagnostics combined with NWP diagnostics. In particular, the latter will provide confidence values, or uncalibrated probabilities.

Other needs are oriented to increase our understanding of the physical processes involved in atmospheric turbulence. Apart from needing a combination of theoretical studies, field programs, and high-resolution numerical simulations, this would demand considering case studies based on reported incidents and accidents through the cooperation of commercial airlines and investigating more deeply the importance of gravity waves and gravity wave breaking because the former are generated in the free atmosphere when air is being displaced vertically due to the presence of mountains, fronts, and rapidly growing convection, among other processes, while gravity wave breaking and steepening may represent dynamical mechanisms conducive to CAT. Another important aspect is to unveil the actual relation between in-cloud and out-of-cloud turbulence, which is necessary to devise optimal avoidance strategies. To better quantify turbulence processes in the atmosphere, more high-resolution observations will be mandatory, which would involve multiple data collection campaigns with high-rate measurements, forward-looking scanning Doppler LIDAR, radiometer, and upward-looking radar capabilities. The need for field programs will also rely on a more intense use of GTG forecasts to identify areas and periods conducive to CAT.

11.5 Major challenges in aviation turbulence

A thorough description of the main challenges in aviation turbulence is not an easy task because it entails a degree of overlap with other grand challenges in the climate and environmental Earth sciences. Regarding the overlapping with climate sciences, one important aspect concerns the effects of climate change on atmospheric turbulence. In this respect, it is worth keeping in mind that climate processes occur over much longer periods than atmospheric processes. In addition to shaping the weather conditions, the latter processes are of central importance in the configuration of the climate as well as in many of the forcings and feedbacks that ultimately determine the magnitude of the climate change and its short- and long-term effects on weather. Improvement of our understanding of the climate and atmospheric processes will demand advances not only in the observational and technological approaches, but also in the conceptual point of view. Of the many different kinds of atmospheric turbulence, the most problematic to predict and avoid is CAT, which is typically encountered at cruising altitudes. Therefore, the prediction and avoidance of CAT remain one of the grand challenges for the aviation industry. Part of the difficulty lies in the fact that current radar technology used by aircraft cannot detect turbulence in dry regions where hydrometeors are absent. While the present LIDAR technology can detect CAT in dry regions and warn pilots in real time of the presence of CAT ahead, its operational use in commercial aircraft has implied, even today, excessive installation costs [461, 874]. Reduction of these costs for operational use in commercial aircraft remains an important challenge.

Challenges in CAT detection and avoidance include observations, data assimilation, forecasting, and technological issues. The description given here is not intended to be an exhaustive one since several other aspects concerning interactions with the Earth's environment may involve other disciplines and processes. One important challenge concerns the scale of turbulence, which is too small to be accurately predicted. For example, forecasts usually rely on numerical models to represent the continuous atmosphere with typical grid spacings of tens of kilometers, while only turbulent eddies of about 100 m in size impact aviation [826]. However, such eddies cannot be explicitly simulated because of the coarse spatial resolution employed by the numerical models, and therefore parameterizations must be used to account for these subgrid scale turbulence processes. Alternatively, data post-processing within the framework of operational NWP models can be used to diagnose turbulence potential by implicitly assuming a downscale cascade. Improving the resolution to achieve better turbulence forecasts remains an important challenge given the projections of an increase in strength and frequency of CAT in the upcoming years, especially at mid-latitudes around the jet stream. On the other hand, the use of DNSs for fluid mechanics problems with finer grids will require increased computational power and a new generation of parameterization schemes for NWP and climate modeling. In addition, meteorologists and weather forecasters must also face a challenging increase in data processing owing to the greater availability of both professional and crowd-sourced observations.

Today, routine ground-based observations are too sparse to allow satisfactory verification, while in most cases PIREPs are nonuniformly distributed in space and time and too subjective in discerning the type and intensity of CAT mainly because it is aircraft dependent. Often the position of the turbulent patches and the times of the reports suffer from inaccuracies, and this is made worse by the fact that pilots usually do not report information about clouds. Therefore, increasing the density of observations is a further challenge that must be met in the near future to enhance our understanding of small scale processes in the atmosphere, including aircraft-sized vortices. As the number of observations will increase in the following years, challenges in data assimilation will relate to technical and general thematic aspects. For example, addressing this challenge will include improvements in weather forecasting, reanalyses, and detection systems. Technical challenges may include performing atmospheric data assimilation at mesoscale or even finer scales and a better representation of forecast errors, model errors, and online bias corrections. Efforts in this direction will demand consolidating and integrating the collaboration between meteorological and space agencies regarding research and operational activities, including *in situ* and satellite observational platforms. For example, large scale turbulence forecasts are susceptible to inaccuracies and errors, which in turn increase with lead time. Reduction of these errors represents a major challenge that must be addressed to improve the safety of passengers and crew, particularly during long-haul flights. However, using ensembles of diagnostics combined with NWP ensembles gives users some confidence in the results. Better combination strategies can be achieved with the use of machine learning and artificial intelligence techniques. Another challenge associated with the difficulty to collect sufficiently dense information concerns the rarity of moderate-to-severe and severe CAT encounters. Based on in situ EDR estimates, light CAT encounters have a probability density function of 0.982 (corresponding to more than 96%), while moderate and severe CAT encounters have probability density functions of $\lesssim 10^{-3}$ and $\lesssim 10^{-4}$, respectively [824].

Weather prediction remains a difficult task because the underlying equations governing the laws of fluid motion are highly nonlinear, and therefore the many interactions involved as well as the multiple types of feedback effects are often hard to understand. Weather prediction together with the establishment of reliable forecasting in the 2- to 7-day range represents two of the major bodies of research in atmospheric sciences. Although weather prediction and turbulence forecasting have traditionally suffered from problems derived from data collection and utilization, new data from satellite as well as ground-based and airborne remote sensing have drastically alleviated the problem. However, the physical challenges continue to be the same as those of two decades ago, for example, the dynamics of deep convection, the role played by the tropopause dynamics in triggering CAT, and the parameterizations used in the wave-based models of weather and climate. In particular, the implications of orographic gravity waves for CAT have not been properly quantified yet. Most operational CAT forecasts today are still based on empirical diagnostics, which have no link to orographic gravity waves, while it is well known that directional wind shear can induce gravity wave breaking, which is expected to be an important source of CAT.

Remote sensing has pioneered the major advances in understanding weather and climate systems, including their changes. Of particular interest for the investigation of CAT is the influence of climate change on it. The intensive use of satellite imagery together with ground-based and airborne sensing has provided valuable information on the spatio-temporal character of CAT. However, the time series concerned are too short to permit the capture of long-term trends of climate variables. Therefore, one important challenge is to extend the duration of the time series to perform more exact predictions of the influence of climate change on CAT in the long term. The advent of next-generation satellites will be of great help to improve CAT nowcasting because their implied higher spatio-temporal resolution will aid avoidance on the short time scales as convection and gravity waves will be better resolved. It is indisputable that one of the highest priorities of the commercial aviation industry in the future is the improvement of CAT forecast systems. In fact, it must be prepared to face the impact that climate change may have on aviation and therefore it must make sure that future aircraft will be up to the challenge [874].

References

[1] Representing WAFS significant weather (SIGWX) data in bufr, November 2013.

[2] Forecast charts, section 8, 2016.

[3] *METAR observation code: Criteria for SPECI.* moratech.com, 2021.

[4] ICAO 2018. A coordinated, risk-based approach to improving global aviation safety. Technical report, International Civil Aviation Organization, Montréal, Canada, 2018.

[5] EASA 2020. Annual safety review 2020. Technical report, European Union Aviation Safety Agency, Koln, Germany, 2020.

[6] FAA 2020. Fact sheet – turbulence, February, 4 2020.

[7] ICAO 2020. A coordinated, risk-based approach to improving global aviation safety. Technical report, International Civil Aviation Organization, Montréal, Canada, 2020.

[8] BTS 2021. On-time performance – reporting operating carrier flight delays at a glance. Technical report, Bureau of Transportation Statistics, U.S. Department of Transportation, Washington, DC, 2021.

[9] 24th Joseph T. Nall Report. General aviation accidents in 2012. Technical report, AOPA Air Safety Institute, 2015.

[10] J. Aarons. Low-angle scintillation of discrete sources. In J. Aarons, editor, *Radio Astronomical and Satellite Studies of the Atmosphere*, pages 65–90. North-Holland Publishing Company, 1962.

[11] J. Aarons. 50 years of radio-scintillation observations. *IEEE Antennas and Propagation Magazine*, 39(6):7–12, 1997.

[12] J. Aarons and R.S. Allen. Scintillation boundary during quiet and disturbed magnetic conditions. *Journal of Geophysical Research*, 76(1):170–177, 1971.

[13] J. Aarons, H.E. Whitney, E. MacKenzie, and S. Basu. Microwave equatorial scintillation intensity during solar maximum. *Radio Science*, 16(5):939–945, 1981.

[14] H.D.I. Abarbanel, D.D. Holm, J.E. Marsden, and T. Ratiu. Richardson number criterion for the nonlinear stability of three-dimensional stratified flow. *Physical Review Letters*, 52:2352–2355, 1984.

[15] A. Abu-Rahmah, W.P. Arnott, and H. Moosmüller. Integrating nephelometer with a low truncation angle and an extended calibration scheme. *Measurement and Science Technology*, 17(7):1723–1732, 2006.

[16] D.J. Acheson. *Elementary Fluid Dynamics*. Clarendon Press, 1990.

[17] A.A. Adedoyin, D.K. Walters, and S. Bhushan. Investigation of turbulence model and numerical scheme combinations for practical finite-volume large eddy simulations. *Engineering Applications of Computational Fluid Mechanics*, 9:324–342, 2015.

[18] European Aviation Safety Agency. Weather information to pilots strategy paper: An outcome of the all weather operations project, January 2018.

[19] C.D. Ahrens and R. Henson. *Meteorology Today: An Introduction to Weather, Climate, and the Environment*. Cengage Learning, 2019.

[20] H.-E. Albrecht, N. Damaschke, M. Borys, and C. Tropea. *Laser Doppler and Phase Doppler Measurement Techniques*. Springer Verlag, Berlin, 2003.

[21] A.A. Aldama. *Filtering Techniques for Turbulent Flow Simulation*. Springer, 2013.

[22] F. Alfonso, J. Vale, É. Oliveira, F. Lau, and A. Suleman. A review on nonlinear aeroelasticity of high aspect-ratio wings. *Progress in Aerospace Sciences*, 89:40–57, 2017.

[23] M. Alldritt, R. Jones, C.J. Oliver, and J.M. Vaughan. The processing of digital signals by a surface acoustic wave spectrum analyser. *Journal of Physics E: Scientific Instruments*, 11(2):116–119, 1978.

[24] F. Alonso. Anemometric calibration test methods for airbus a330/340. In *Proceedings ETTC '91*, pages 2–7, June 1991.

[25] B. Altstädter, A. Platis, B. Wehner, A. Scholtz, N. Wildmann, M. Hermann, R. Käthner, H. Baars, J. Bange, and A. Lampert. Aladina – an unmanned research aircraft for observing vertical and horizontal distributions of ultrafine particles within the atmospheric boundary layer. *Atmospheric Measurement Techniques*, 8(4):1627–1639, 2015.

[26] J. Amezcua, S.P. Näsholm, E.M. Blixt, and A.J. Charlton-Perez. Assimilation of atmospheric infrasound data to constrain tropospheric and stratospheric winds. *Quarterly Journal of the Royal Meteorological Society*, 146(731):2634–2653, 2020.

[27] J.A. Anderson. Astronomical seeing. *Journal of the Optical Society of America*, 25(5):152–155, 1935.

[28] R.K. Anderson, J.J. Gurka, and S.J. Steinmetz. Application of vas multispectral imagery to aviation forecasting. In *Ninth Conference on Weather Forecasting and Analysis*, pages 227–234. American Meteorological Society, 1982.

[29] T. Ando, S. Kameyama, T. Sakimura, K. Asaka, and Y. Hirano. Long range all-fiber coherent doppler lidar (cdl) system for wind sensing. In *Proceedings of the 14th Coherent Laser Radar Conference (CLRC XIV)*, pages 31–34. Universities Space Research Association (USRA), July 2007.

[30] A. Ansmann, M. Riebesell, U. Wandinger, C. Weitkamp, E. Voss, W. Lahmann, and W. Michaelis. Combined raman elastic-backscatter lidar for vertical profiling of moisture, aerosol extinction, backscatter, and lidar ratio. *Applied Physics B*, 55(1):18–28, 1992.

[31] D. Anwar, K. Sembiring, K. Tarigan, M. Sinambela, A.R. Abubar, and D. Hasibuan. Machine learning approach for turbulence forecasting using support vector machine. *IOP Conference Series: Materials Science and Engineering*, 725:012087, 2020.

[32] P.S. Argall and R.J. Sica. Lidar — atmospheric sounding introduction. In J.R. Holton, J.A. Curry, and J.A. Pyle, editors, *Encyclopedia of Atmospheric Sciences*, pages 1169–1176. Academic Press, 2003.

[33] L. Armijo. A theory for the determination of wind and precipitation velocities with doppler radars. *Journal of the Atmospheric Sciences*, 26(3):570–573, 1969.

[34] W.P. Arnott, H. Moosmüller, C.F. Rogers, T. Jin, and R. Bruch. Photoacoustic spectrometer for measuring light absorption by aerosols: instrument description. *Atmospheric Environment*, 33(17):2845–2852, 1999.

[35] K.O. Arras, N. Tomatis, and R. Siegwart. Multisensor on-the-fly localization using laser and vision. In *Proceedings 2000 IEEE/RJS International Conference on Intelligent Robots and Systems (IROS 2000)*, volume 1, pages 462–467. IEEE, 2000.

[36] M. Asch, M. Bocquet, and M. Nodet. *Data Assimilation: Methods, Algorithms, and Applications*. SIAM, 2016.

[37] G.P. Asner, D.E. Knapp, T. Kennedy-Bowdoin, M.O. Jones, R.E. Martin, J.W. Boardman, and C.B. Field. Carnegie airborne observatory: in-flight fusion of hyperspectral imaging and waveform light detection and ranging for three-dimensional studies of ecosystems. *Journal of Applied Remote Sensing*, 1(1):013536, 2007.

[38] J.D. Assink, R. Waxler, W.G. Frazier, and J. Lonzaga. The estimation of upper atmospheric wind model updates from infrasound data. *Journal of Geophysics Research: Atmospheres*, 118(19):10707–10724, 2013.

[39] R.W. Astheimer. The remote detection of clear air turbulence by infrared radiation. *Applied Optics*, 9(8):1789–1797, 1970.

[40] D. Atlas, K.R. Hardy, K.M. Glover, I. Katz, and T.G. Konrad. Tropopause detected by radar. *Science*, 153(3740):1110–1112, 1966.

[41] D. Atlas, K.R. Hardy, and K. Naito. Optimizing the radar detection of clear air turbulence. *Journal of Applied Meteorology and Climatology*, 5(4):450–460, 1966.

[42] D. Atlas, J.I. Metcalf, J.H. Richter, and E.E. Gossard. The birth of "CAT" and microscale turbulence. *Journal of the Atmospheric Sciences*, 27(6):903–913, 1970.

[43] D. Atlas, K. Naito, and R.E. Carbone. Bistatic microwave probing of a refractively perturbed clear atmosphere. *Journal of the Atmospheric Sciences*, 25(2):257–268, 1968.

[44] D. Atlas, R.C. Srivastava, R.E. Carbone, and D.H. Sargeant. Doppler crosswind relations in radio tropo-scatter beam swinging for a thin scatter layer. *Journal of the Atmospheric Sciences*, 26(5):1104–1117, 1969.

[45] D. Atlas and W.S. Srivastava, R.C. Marker. The influence of specular reflections on bistatic tropospheric radio scatter from turbulent strata. *Journal of the Atmospheric Sciences*, 26(5):1118–1121, 1969.

[46] J. Atrill, L. Sushama, and B. Teufel. Clear-air turbulence in a changing climate and its impact on polar aviation. *Safety in Extreme Environments*, 3:103–124, 2021.

[47] G. Avanzini, E.L. de Angelis, and F. Giulietti. Optimal performance and sizing of a battery-powered aircaft. *Aerospace Science and Technology*, 59:132–144, 2016.

[48] W.D. Bachalo. A method for measuring the size and velocity of spheres by dual-beam light-scatter interferometry. *Applied Optics*, 19(3):363–370, 1980.

[49] W.D. Bachalo. Measurement techniques for turbulent two-phase flow research. In *International Symposium on Multiphase Fluid, Non-Newtonian Fluid and Physicochemical Fluid Flows (ISMNP)*. Beijing, China, October 1997.

[50] W.D. Bachalo and M.J. Houser. Phase/doppler spray analyzer for simultaneous measurements of drop size and velocity distributions. *Optical Engineering*, 23(5):235583, 1984.

[51] W.D. Bachalo and S.V. Sankar. *Phase Doppler Particle Analyzer*, pages 37.1–37.19. CRC, Idaho Falls, 1996.

[52] J. Bachman. Super long-haul trips expose flight crews and passengers to cosmic radiation. *Skift Research*, March 2018.

[53] J.T. Bacmeister, P.A. Newman, B.L. Gary, and K.R. Chan. An algorithm for forecasting mountain wave-related turbulence in the stratosphere. *Weather and Forecasting*, 9(2):241–253, 1994.

[54] R. Bahreini, J.L. Jimenez, J. Wang, R.C. Flagan, J.H. Seinfeld, J.T. Jayne, and D.R. Worsnop. Aircraft-based aerosol size and composition measurements during ace-asia using an aerodyne aerosol mass spectrometer. *Journal of Geophysical Research: Atmospheres*, 108(D23):8645, 2003.

[55] F.Z. Bai, Y. Miao, H. Guo, X.J. Gao, and X.Q. Wang. Simple quasi-common path point diffraction interferometer with adjustable fringe contrast and carrier frequency. *Journal of the European Optical Society—Rapid Publications*, 10:15042–1–6, 2015.

[56] H.A. Baluch and M. van Tooren. Multidisciplinary design of flexible aircraft. In R. Curran, S.Y. Chou, and A. Trappey, editors, *Collaborative Product and Service Life Cycle Management for a Sustainable World*, pages 375–385. Springer, 2008.

[57] P.R. Bandyopadhyay and R. Balasubramanian. Vortex reynols number in turbulent boundary layers. *Theoretical and Computational Fluid Dynamics*, 7(2):101–117, 1995.

[58] J. Barat. Étude experimentale de la structure du champ de la turbulence dans la moyenne stratosphère. *Comptes Rendus de l'Académie des Sciences, Série B*, 280(22):691–693, 1975.

[59] J. Barat. Initial results from the use of ionic anemometers under stratospheric balloons: Application to the high resolution analysis of stratospheric motions. *Journal of Applied Meteorology and Climatology*, 21(10):1489–1496, 1982.

[60] J. Barat. Some characteristics of clear-air turbulence in the middle stratosphere. *Journal of the Atmospheric Sciences*, 39(11):2553–2564, 1982.

[61] K. Bärfuss, F. Pätzold, B. Altstädter, E. Kathe, S. Nowak, L. Bretschneider, U. Bestmann, and A. Lampert. New setup of the uas aladina for measuring boundary layer properties, atmospheric particles and solar radiation. *Atmosphere*, 9(1):28, 2018.

[62] I. Bartolome Garcia, R. Spang, J. Ungermann, S. Griessbach, M. Krämer, M. Höpfner, and M. Riese. Observation of cirrus clouds with gloria during the wise campaign: detection methods and cirrus characterization. *Atmospheric Measurement Tecniques*, 14(4):3153–3168, 2021.

[63] E.J. Bass. Towards a pilot-centered turbulence assessment and monitoring system. In *Gateway to the New Millennium. Proceedings of the AIAA/IEEE 18th Digital Avionics Systems Conference*, volume 2, pages 6.D.3.1–6.D.3.8. IEEE, 1999.

[64] S. Basu and A.D. Gupta. Scintillations of satellite signals by ionospheric irregularities with sharp boundary. *Journal of Geophysical Research*, 74(5):1294–1300, 1969.

[65] G.K. Batchelor. Kolmogoroff's theory of locally isotropic turbulence. *Mathematical Proceedings of the Cambridge Philosophical Society*, 43:533–559, 1947.

[66] G.K. Batchelor. *The Theory of Homogeneous Turbulence*. Cambridge University Press, 1971.

[67] G.K. Batchelor and A.A. Townsend. The nature of turbulent motion at large wave numbers. *Proceedings of the Royal Society of London. Series A, Mathematical and Physical Sciences*, 199:238–255, 1949.

[68] R. Batt and D. O'Hare. General aviation pilot behaviours in the face of adverse weather. Technical Report B2005/0127, Australian Transport Safety Bureau, June 2005.

[69] D. Baumgardner, J.L. Brenguier, A. Bucholtz, H. Coe, P. DeMott, T.J. Garrett, J.F. Gayet, M. Hermann, A. Heymsfield, A. Korolev, M. Krämer, A. Petzold, W. Strapp, P. Pilewskie, J. Taylor, C. Twohy, M. Wendisch, W. Bachalo, and P. Chuang. Airborne instruments to measure atmospheric aerosol particles, clouds and radiation: A cook's tour of mature and emerging technology. *Atmospheric Research*, 102(1-2):10–29, 2011.

[70] D. Baumgardner, J.E. Dye, B.W. Gandrud, and R.G. Knollenberg. Interpretation of measurements made by the forward scattering spectrometer probe (fssp-300) during the airborne arctic stratospheric expedition. *Journal of Geophysical Research: Atmospheres*, 97(D8):8035–8046, 1992.

[71] D. Baumgardner, H. Jonsson, W. Dawson, D. O'Connor, and R. Newton. The cloud, aerosol and precipitation spectrometer: a new instrument for cloud investigations. *Atmospheric Research*, 59-60:251–264, 2001.

[72] D. Baumgardner, G. Kok, and G. Raga. Warming of the arctic lower stratosphere by light absorbing particles. *Geophysics Research Letters*, 31(6):L06117, 2004.

[73] A.C. Beall and J.M. Loomis. Optic flow and visual analysis of the base-to-final turn. *The International Journal of Aviation Psychology*, 7(3):201–223, 1997.

[74] B.R. Bean and E.J. Dutton. *Radio Meteorology, NBS Monograph 92*. US Government Printing Office, 1966.

[75] B.R. Bean, A.S. Frisch, L.G. McAllister, and J.R. Pollard. Planetary boundary-layer turbulence studies from acoustic echo sounder and in-situ measurements. *Boundary-Layer Meteorology*, 4(1-4):449–474, 1973.

[76] P. Bechtold, M. Bramberger, A. Dörnbrack, L. Isaksen, and M. Leutbecher. Experimenting with a clear air turbulence (CAT) index from the IFS. Technical Report 874, European Centre for Medium-Range Weather Forecasts (ECMWF), Shinfield Park, Reading, UK, January 2021.

[77] P. Bechtold, M. Bramberger, A. Dörnbrack, M. Leutbecher, and L. Isaksen. Forecasting clear-air turbulence. Technical Report 168, European Centre for Medium-Range Weather Forecasts (ECMWF), Shinfield Park, Reading, UK, July 2021.

[78] R.A. Becker. Effects of atmospheric turbulence on optical instrumentation. *IRE Transactions on Military Electronics*, MIL-5(4):352–356, 1961.

[79] A.J. Bedard. Aviation weather hazards. In G.R. North, J. Pyle, and F. Zhang, editors, *Encyclopedia of Atmospheric Sciences (Second Edition)*, pages 166–176. Elsevier, 2015.

[80] R.R. Beland. Propagation through atmospheric optical turbulence. In F.G. Smith, editor, *Atmospheric Propagation of Radiation*, pages 157–232. SPIE Press, Bellingham, WA, 1993.

[81] A.C.M. Beljaars, A.R. Brown, and N. Wood. A new parametrization of turbulent orographic form drag. *Quarterly Journal of the Royal Meteorological Society*, 130(599):1327–1347, 2004.

[82] J.S. Bendat and A.G. Piersol. *Random Data. Wiley Series in Probability and Statistics*. Wiley, Hoboken, NJ, fourth edition, 2011.

[83] D.J. Benney. A general theory for interactions between short and long waves. *Studies in Applied Mathematics*, 56(1):81–94, 1977.

[84] R. Benzi, S. Ciliberto, R. Tripiccione, C. Baudet, F. Massaioli, and S. Succi. Extended self-similarity in turbulent flows. *Physical Review E*, 48:R29–R32, 1993.

[85] D.W. Beran. Acoustics: A new approach for monitoring the environment near airports. *Journal of Aircraft*, 8(11):934–936, 1971.

[86] D.W. Beran, W.H. Hooke, and S.F. Clifford. Acoustic echo-sounding techniques and their application to gravity-wave, turbulence, and stability studies. *Boundary-Layer Meteorology*, 4(1-4):133–153, 1973.

[87] D.W. Beran, C.G. Little, and B.C. Willmarth. Acoustic doppler measurements of vertical velocities in the atmosphere. *Nature*, 230:160–162, 1971.

[88] D. Berthelot, N. Carlini, I. Goodfellow, A. Oliver, N. Papernot, and C. Raffel. Mixmatch: A holistic approach to semi-supervised learning. In *NIPS 2019: Proceedings of the 33rd Conference on Neural Information Processing Systems (NeurIPS)*, number 454, pages 5049–5059. Currant Associates Inc., Red Hook, NY, USA, December 2019.

[89] R.G. Beuttell and A.W. Brewer. Instruments for the measurement of the visual range. *Journal of Scientific Instruments*, 26(11):357–359, 1949.

[90] P.E. Bieringer, B. Martin, B. Collins, and J. Shaw. Commercial aviation enconters with severe low altitude turbulence. In *Proceedings of the 11th Conference on Aviation, Range and Aerospace Meteorology*, pages 1–8. American Meteorology Society, October 2004.

[91] J.W. Bilbro. Atmospheric laser doppler velocimetry: An overview. *Optical Engineering*, 19(4):194533, 1980.

[92] J. Bjerknes. On the structure of moving cyclones. *Geofysiske Plublikasjoner*, 1(2):1–8, 1919.

[93] J. Bjerknes and H. Solberg. Life cycle of cyclones and the polar front theory of atmospheric circulation. *Geofysiske Plublikasjoner*, 3(1):3–18, 1922.

[94] E. Blanc, L. Ceranna, A. Hauchecorne, A. Charlton-Perez, E. Marchetti, L.G. Evers, T. Kvaerna, J. Lastovicka, L. Eliasson, N.B. Crosby, P. Blanc-Benon, A. Le Pichon, N. Brachet, C. Pilger, P. Keckhut, J.D. Assink, P. S.M. Smets, C.F. Lee, J. Kero, T. Sindelarova, N. Kämpfer, R. Rüfenacht, T. Farges, C. Millet, S.P. Näsholm, S.J. Gibbons, P.J. Espy, R.E. Hibbins, P. Heinrich, M. Ripepe, S. Khaykin, N. Mze, and J. Chum. Toward an improved representation of middle atmospheric dynamics thanks to the arise project. *Surveys in Geophysics*, 39(2):171–225, 2018.

[95] E.M. Blixt, S.P. Näsholm, S.J. Gibbons, L.G. Evers, A.J. Charlton-Perez, Y.J. Orsolini, and T. Kværna. Estimating tropospheric and stratospheric winds using infrasound from explosions. *The Journal of the Acoustical Society of America*, 146(2):973–982, 2019.

[96] National Transport Safety Board. Risk factors associated with weather-related general aviation accidents. Technical Report NTSB/SS-05/01, September 2005.

[97] T. Bonin, P. Chilson, B. Zielke, and E. Fedorovich. Observations of the early evening boundary-layer transition using a small unmanned aerial system. *Boundary-Layer Meteorology*, 146:119–132, 2013.

[98] T.A. Bonin, D.C. Goines, A.K. Scott, C.E. Wainwright, J.A. Gibbs, and P.B. Chilson. Measurements of the temperature structure-function parameters with a small unmanned aerial system compared with a sodar. *Boundary-Layer Meteorology*, 155:417–434, 2015.

[99] H.G. Booker. The use of radio stars to study irregular refraction of radio waves in the ionosphere. *Proceedings of the IRE*, 46(1):298–314, 1958.

[100] H.G. Booker and W.E. Gordon. A theory of radio scattering in the troposphere. *Proceedings of the IRE*, 38(4):401–412, 1950.

[101] R.J. Bootsma and R.D. Oudejans. Visual information about time-to-collision between two objects. *Journal of Experimental Psychology, Human Perception and Performance*, 19(5):1041–1052, 1993.

[102] R. Bos, A. Giyanani, and W. Bierbooms. Assessing the severity of wind gusts with lidar. *Remote Sensing*, 8(9):758, 2016.

[103] E. Bou-Zeid, C. Meneveau, and Parlange M. A scale-dependent lagrangian dynamic model for large eddy simulation of complex turbulent flows. *Physics of Fluids*, 17(2):025105, 2005.

[104] P. Bougeault, P. Binder, A. Buzzi, R. Dirks, R. Houze, J. Kuettner, R.B. Smith, R. Steinacker, and H. Volkert. The map special observing period. *Bulletin of the American Meteorological Society*, 82(3):433–462, 2001.

[105] D.A. Bowdle, J. Rothermel, J.M. Vaughan, D.W. Brown, and M.J. Post. Aerosol backscatter measurements at 10.6 micrometers with airborne and ground-based CO_2 Doppler lidars over the Colorado High Plains: 1. Lidar intercomparison. *Journal of Geophysical Research: Atmospheres*, 96(D3):5327–5335, 1991.

[106] D.A. Bowdle, J. Rothermel, J.M. Vaughan, and M.J. Post. Aerosol backscatter measurements at 10.6 micrometers with airborne and ground-based CO_2 Doppler lidars over the Colorado High Plains: 2. Backscatter structure. *Journal of Geophysical Research: Atmospheres*, 96(D3):5337–5344, 1991.

[107] S.A. Bowhill. Statistics of a radio wave diffracted by a random ionosphere. *Journal of Research of the National Bureau of Standards*, 65D(3):275–292, 1961.

[108] S.A. Bowhill and S. Gnanalingam. Gravity waves in severe weather. In *Proceedings of the Third Workshop on Technical and Scientific Aspects of MST Radar*, volume 80. International Council of Scientific Unions, Middle Atmosphere Program. Handbook for MAP, June 1986.

[109] N.E. Bowler, A. Arribas, K.R. Mylne, K.B. Robertson, and S.E. Beare. The mogreps short-range ensemble prediction system. *Quarterly Journal of the Royal Meteorological Society*, 134(632):703–722, 2008.

[110] M. Braam, F. Beyrich, J. Bange, A. Platis, S. Martin, B. Maronga, and A.F. Moene. On the discrepancy in simultaneous observations of the structure parameter of temperature using scintillometers and unmanned aircraft. *Boundary-Layer Meteorology*, 158:257–283, 2016.

[111] C. Brahic. The perlan project: flying on the thinnest air. *New Scientist*, 213(2846):34–37, 2012.

[112] D.B. Brayton and W.H. Goethert. New velocity measuring technique using dual-scatter laser. Technical Report AD876458, Arnold Engineering Development Center, Arnold Air Force Station, Tennessee, USA, 1970.

[113] J.G. Breiland. Vertical distribution of atmospheric ozone and its relation to synoptic meteorological conditions. *Journal of Geophysical Research*, 69(18):3801–3808, 1964.

[114] L. Breiman. Random forests. *Machine Learning*, 45:5–32, 2001.

[115] J.-L. Brenguier, T. Bourrianne, A. Araujo Coelho, J. Isbert, R. Peytavi, D. Trevarin, and P. Weschler. Improvements of droplet size distribution measurements with the fast-fssp (forward scattering spectrometer probe). *Journal of Atmospheric and Oceanic Technology*, 15(5):1077–1090, 1998.

[116] N.D. Brenowitz and C.S. Bretherton. Prognostic validation of a neural network unified physics parameterization. *Geophysical Research Letters*, 45(12):6289–6298, 2018.

[117] B.H. Briggs and I.A. Parkin. On the variation of radio star and satellite scintillations with zenith angle. *Journal of Atmospheric and Terrestrial Physics*, 25(6):339–366, 1963.

[118] W.A.R. Brinkmann. What is a foehn? *Weather*, 26(6):230–240, 1971.

[119] R.L. Brouwer, M.A. de Schipper, P.F. Rynne, F.J. Graham, A.J.H.M. Reniers, and J.M. MacMahan. Surfzone monitoring using rotary wing unmanned aerial vehicles. *Journal of Atmospheric and Oceanic Technology*, 32(4):855–863, 2015.

[120] E.V. Browell, S. Ismail, and W.B. Grant. Lidar — dial. In J.R. Holton, editor, *Encyclopedia of Atmospheric Sciences*, pages 1183–1194. Elsevier, 2003.

[121] R. Brown. New indices to locate clear-air turbulence. *The Meteorological Magazine*, 102:347–361, 1973.

[122] R.A. Brown. On the inflection point instability of a stratified Ekman boundary layer. *Journal of the Atmospheric Sciences*, 29(5):851–859, 1972.

[123] R.A. Brown. *Analytical Methods in Planetary Boundary-Layer Modelling*. Wiley, 1974.

[124] R.A. Brown. Longitudinal instabilities and secondary flows in the planetary boundary layer: A review. *Reviews of Geophysics*, 18(3):683–697, 1980.

[125] R.A. Brown. Doppler weather radar. In P.T. Bobrowsky, editor, *Encyclopedia of Natural Hazards. Encyclopedia of Earth Sciences Series*, pages 188–188, Dordrecht, Netherlands, 2013. Springer.

[126] K.A. Browning. Structure of the atmosphere in the vicinity of large-amplitude Kelvin-Helmholtz billows. *Quarterly Journal of the Royal Meteorological Society*, 97(413):283–299, 1971.

[127] K.A. Browning, J.R. Starr, and A.J. Whyman. The structure of an inversion above a convective layer as observed using high-power pulsed doppler radar. *Boundary-Layer Meteorology*, 4(1-4):91–111, 1973.

[128] K.A. Browning and C.D. Watkins. Observations of clear air turbulence by high power radar. *Nature*, 227:260–263, 1970.

[129] K.A. Browning, C.D. Watkins, J.R. Starr, and A. McPherson. Simultaneous measurements of clear air turbulence at the tropopause by high-power radar and instrumented aircraft. *Nature*, 228:1065–1067, 1970.

[130] K.A. Brucker and S. Sarkar. A comparative study of self-propelled and towed wakes in a stratified fluid. *Journal of Fluid Mechanics*, 652:373–404, 2010.

[131] P. Buchanan. Aviation turbulence ensemble techniques. In R. Sharman and T. Lane, editors, *Aviation Turbulence*, pages 285–296. Springer, 2016.

[132] A. Bucharles and P. Vacher. Flexible aircraft model identification for control law design. *Aerospace Science and Technology*, 6(8):591–598, 2002.

[133] K.G. Budden. The amplitude fluctuations of the radio wave scattered from a thick ionospheric layer with weak irregularities. *Journal of Atmospheric and Terrestrial Physics*, 27(2):155–172, 1965.

[134] K.G. Budden. The theory of the correlation of amplitude fluctuations of radio signals at two frequencies, simultaneously scattered by the ionosphere. *Journal of Atmospheric and Terrestrial Physics*, 27(8):883–897, 1965.

[135] W.E. Buehler, C.H. King, and C.D. Lunden. Radar echoes from clear air inhomogeneities. In Y.-H. Pao and A. Goldburg, editors, *Clear Air Turbulence and Its Detection*, pages 425–435. Springer Science, 1969.

[136] J. Bufton and S.H. Genatt. Simultaneous observations of atmospheric turbulence effects on stellar irradiance and phase. *The Astronomical Journal*, 76(4):378–386, 1971.

[137] G.S. Buldovskii, S.A. Bortnikov, and M.V. Rubinsztejn. Forecasting of heavy turbulence zones in the upper troposphere. *Meteorologija i Gidrologija*, 2:9–18, 1976.

[138] K. Bullrich. Scattered radiation in the atmosphere and the natural aerosol. In H.E. Landsberg and J. Van Mieghem, editors, *Advances in Geophysics*, volume 10, pages 99–260. Elsevier, 1964.

[139] U. Bundke, B. Nillius, R. Jaenicke, T. Wetter, H. Klein, and H. Bingemer. The fast ice nucleus chamber finch. *Atmospheric Research*, 90(2-4):180–186, 2008.

[140] R. Burbidge. Adapting european airports to a changing climate. *Transportation Research Procedia*, 14:14–23, 2016.

[141] K.J. Bures. High-resolution imaging of tropospheric radio scatterers. *IEEE Transactions on Aerospace and Electronic Systems*, 11(2):137–146, 1975.

[142] D.W. Burgess. Single doppler radar vortex recognition: Part I—mesocyclone signatures. In *The 17th Conference on Radar Meteorology*, pages 97–103. American Meteorological Society, 1976.

[143] M.A. Burgess and R.P. Thomas. The effect of NEXRAD image looping and national convective weather forecast product on pilot decision making in the use of a cockpit weather information display. Technical Report NASA/CR-2004-212655, NASA, March 2004.

[144] N.E. Busch, H. Tennekes, and H.A. Panofsky. Turbulence structure in the planetary boundary layer. *Boundary-Layer Meteorology*, 4(1-4):211–211, 1973.

[145] G.C. Butler, J. Nicholas, D.T. Lackland, and W. Friedberg. Perspectives of those impacted: Airline pilot's perspective. *Health Physics: The Radiation Safety Journal*, 79(5):602–607, 2000.

[146] H.E. Butler. Observations of stellar scintillation. *Quarterly Journal of the Royal Meteorological Society*, 80(344):241–245, 1954.

[147] J.A.T. Bye. A Richardson number criterion for the air-sea interface. *Boundary-Layer Meteorology*, 44:407–410, 1988.

[148] A. Cahn. An investigation of the free oscillations of a simple current system. *Journal of the Atmospheric Sciences*, 2(2):113–119, 1945.

[149] R.D. Callan, D.A. Huckridge, C.R. Nash, and J.M. Vaughan. Active airborne linescan experiment. In *Airborne Reconnaissance XVIII*, volume 2272, pages 183–191. SPIE, October 1994.

[150] J.C. Camparo. Stellar scintillation and the atmosphere's vertical turbulence profile. Technical Report TR-2000(8555)-6, Space and Missile Systems Center Air Force Materiel Command, Los Angeles Air Force Base, El Segundo, CA 90245, July 2001.

[151] C.W. Campbell. Monte Carlo turbulence simulation using rational approximation to von Karman spectra. *AIAA Journal*, 24(1):62–66, 1986.

[152] M.R. Canagaratna, J.T. Jayne, J.L. Jimenez, J.D. Allan, M.R. Alfarra, Q. Zhang, T.B. Onasch, F. Drewnick, H. Coe, A. Middlebrook, A. Delia, L.R. Williams, A.M. Trimborn, M.J. Northway, P.F. DeCarlo, C.E. Kolb, P. Davidovits, and D.R. Worsnop. Chemical and microphysical characterization of ambient aerosols with the aerodyne aerosol mass spectrometer. *Mass Spectrometry Reviews*, 26(2):185–222, 2007.

[153] V.M. Canuto and M.S. Dubobikov. Turbulence.

[154] D. Carati, S. Ghosal, and P. Moin. On the representation of backscatter in dynamic localization models. *Physics of Fluids*, 7:606–616, 1995.

[155] O. Cardoso, B. Gluckmann, O. Parcollet, and P. Tabeling. Dispersion in a quasi-two-dimensional-turbulent flow: An experimental study. *Physics of Fluids*, 8(1):209–214, 1996.

[156] F. Cattin-Valsecchi and C. Lopez. Utilisation d'un vélocimètre laser triaxe pour calibration anémométrique sur airbus a340. In *AGARD-FMP-Symposium, AGARD-CP-519*, number 26, June 1992.

[157] N. Cézard, A. Dolfi-Bouteyre, J.-P. Huignard, and P.H. Flamant. Performance evaluation of a dual fringe-imaging Michelson interferometer for air parameter measurements with a 355 nm Rayleigh-Mie lidar. *Applied Optics*, 48(12):2321–2332, 2009.

[158] E. Chambers. Clear air turbulence and civil jet operations. *The Aeronautical Journal*, 59(537):613–628, 1955.

[159] P.W. Chan. Atmospheric turbulence in complex terrain: Verifying numerical model results with observations by remote-sensing instruments. *Meteorology and Atmospheric Physics*, 103:145–157, 2009.

[160] H. Chandra and R.G. Rastogi. Scintillations of satellite signals near magnetic equator. *Current Science*, 43(18):567–568, 1974.

[161] S. Chandrasekhar. A statistical basis for the theory of stellar scintillation. *Monthly Notices of the Royal Astronomical Society*, 112(5):475–483, 1952.

[162] S. Chandrasekhar. *Hydrodynamic and Hydromagnetic Stability*. Dover Publications, Inc. New York, USA, 1981.

[163] C.-S. Chang, D.H. Hodges, and M.J. Patil. Flight dynamics of highly flexible aircraft. *Journal of Aircraft*, 45(2):538–545, 2008.

[164] H. Chao and Y. Chen. Surface wind profile measurement using multiple small unmanned aerial vehicles. In *2010 American Control Conference (ACC)*. IEEE, 2010. Paper number 17.

[165] A.J. Charlton and L.M. Polvani. A new look at stratospheric sudden warming. part i: Climatology and modeling benchmarks. *Journal of Climate*, 20(3):449–469, 2007.

[166] S.K. Chatterjee, A.K. Bandyopadhyay, B.K. Guhathakurta, and P. Bandyopadhyay. The equatorial scintillation belt as observed at huancayo, peru. *Annales de Géophysique*, 30:329–337, 1974.

[167] A. Chaudhari, V. Vuorinen, O. Agafonova, A. Hellsten, and J. Hämäläinen. Large-eddy simulations for atmospheric boundary layer flows over complex terrains with applications in wind energy. In E. Oñate, J. Oliver, and A. Huerta, editors, *6th European Conference on Computational Fluid Dynamics (ECFD VI)*, pages 1–12, 2014.

[168] T. Chen and C. Guestrin. Xgboost: A scalable tree boosting system. In *Proceedings of the 22nd ACM SIGKDD International Conference on Knowledge Discovery and Data Mining*, pages 785–794. Association for Computing Machinery (ACM), August 2016.

[169] W. Chen and H.B. Zhou. Lévy-Kolmogorov scaling of turbulence. *eprint arXiv:math-ph/0506039*, 2005.

[170] Z. Chen, R. Fan, X. Li, Z. Dong, Z. Zhou, G. Ye, and D. Chen. Accuracy improvement of imaging lidar based on time-correlated single-photon counting using three laser beams. *Optics Communications*, 429:175–179, 2018.

[171] Z. Cheng, D. Liu, J. Luo, Y. Yang, Y. Zhou, L. Zhang, Y. Duan, L. Su, L. Yang, Y. Shen, K. Wang, and J. Bai. Field-widened Michelson interferometer for spectral discrimination in high-spectral-resolution lidar: theoretical framework. *Optics Express*, 23(9):12117–12134, 2015.

[172] J. Chimot. *Global mapping of atmospheric composition from space – Retrieving aerosol height and tropospheric NO_2 from OMI*. PhD thesis, Delft University of Technology (TU Delft), 2018.

[173] H.J.A. Chivers. The simultaneous observation of radio star scintillations on different radio frequencies. *Journal of Atmospheric and Terrestrial Physics*, 17(3):181–187, 1960.

[174] J.Y.N. Cho and E. Lindborg. Horizontal velocity structure functions in the upper troposphere and lower stratosphere: 1. observations. *Journal of Geophysical Research: Atmospheres*, 106(D10):10223–10232, 2001.

[175] J.Y.N. Cho, R.E. Newell, T.P. Bui, E.V. Browell, M.A. Fenn, M.J. Mahoney, G.L. Gregory, G.W. Sachse, S.A. Vay, T.L. Kucsera, and A.M. Thompson. Observations of convective and dynamical instabilities in tropopause folds and their contribution to stratosphere-troposphere exchange. *Journal of Geophysical Research: Atmospheres*, 104(D17):21549–21568, 1999.

[176] N. Chonacky and R.W. Deuel. Atmospheric optical turbulence profiling by stellar scintillometers: an instrument evaluation experiment. *Applied Optics*, 27(11):2214–2221, 1988.

[177] F.K. Chow and R.L. Street. Evaluation of turbulence closure models for large-eddy simulation over complex terrain: Flow over askervein hill. *Journal of Applied Meteorology and Climatology*, 48:1050–1065, 2009.

[178] P.Y. Chuang, E.W. Saw, J.D. Small, R.A. Shaw, C.M. Sipperley, G.A. Payne, and W.D. Bachalo. Airborne phase doppler interferometry for cloud microphysical measurements. *Aerosol Science and Technology*, 42(8):685–703, 2008.

[179] I. Chunchuzov, S. Kulichkov, V. Perepelkin, O. Popov, P. Firstov, J.D. Assink, and E. Marchetti. Study of the wind velocity-layered structure in the stratosphere, mesosphere, and lower thermosphere by using infrasound probing of the atmosphere. *Journal of Geophysical Research: Atmospheres*, 120(17):8828–8840, 2015.

[180] I.P. Chunchuzov, S.N. Kulichkov, O.E. Popov, V.G. Perepelkin, A.P. Vasil'ev, A.I. Glushkov, and P.P. Firstov. Characteristics of a fine vertical wind-field structure in the stratosphere and lower thermosphere according to infrasound signals in the zone of acoustic shadow. *Izvestiya, Atmospheric and Oceanic Physics*, 51:57–74, 2015.

[181] R.E. Clark, D.F. Feldon, J.J.G. van Merriënboer, K.A. Yates, and S. Early. Cognitive task analysis. In J.M. Spector, M.D. Merrill, J.J.G. van Merriënboer, and M.P. Driscoll, editors, *Handbook of Research on Educational Communications and Technology*, volume 43, pages 577–593. Macmillan/Gate, New York, 2008.

[182] T.L. Clark, W.D. Hall, R.M. Kerr, D. Middleton, L. Radke, F.M. Ralph, P.J. Neiman, and D. Levinson. Origins of aircraft-damaging clear-air turbulence during the 9 december 1992 Colorado downslope windstorm: Numerical simulations and comparison with observations. *Journal of the Atmospheric Sciences*, 57(8):1105–1131, 2000.

[183] W.M. Clark. High altitude daytime sky radiance. *Optical Engineering*, 7(2):070240, 1969.

[184] S.F. Clifford and T.M. Georges. Refractive effects in acoustic-echo sounding of the atmosphere. *The Journal of the Acoustical Society of America*, 54(1):307–307, 1973.

[185] J. Clodman, G.M. Morgan, and J.T. Ball. High level turbulence. Technical Report 158, New York University, 1960.

[186] S.A. Clough and M.J. Iacono. Line-by-line calculation of atmospheric fluxes and cooling rates: 2. Application to carbon dioxide, ozone, methane, nitrous oxide and the halocarbons. *Journal of Geophysical Research: Atmospheres*, 100(D8):16519–16535, 1995.

[187] S.A. Clough, M.J. Iacono, and J.-L. Moncet. Line-by-line calculation of atmospheric fluxes and cooling rates: Application to water vapor. *Journal of Geophysical Research: Atmospheres*, 97(D14):15761–15785, 1992.

[188] J. Colin, R. Faivre, and M. Menenti. Aerodynamic roughness length estimation from very high-resolution imaging LIDAR observations over the Heihe basin in China. *Hydrology and Earth System Sciences Discussions*, 7:3397–3421, 2010.

[189] B.A. Colle and F.M. Clifford. Windstorms along the western side of the washington cascade mountains. Part I: A high-resolution observational and modeling study of the 12 February 1995 event. *Monthly Weather Review*, 126(1):28–52, 1998.

[190] M. Collins, R. Knutti, J. Arblaster, J.-L. Dufresne, T. Fichefet, P. Friedling-stein, X. Gao, W.J. Gutowski, T. Johns, G. Krinner, M. Shongwe, C. Tebaldi, A.J. Weaver, M.F. Wehner, M.R. Allen, T. Andrews, U. Beyerle, C.M. Bitz, S. Bony, and B.B.B. Booth. Long-term climate change: Projections, commitments and irreversibility. In T.F. Stocker, D. Qin, G.-K. Plattner, S.K. Tignor, M.M.B. Allen, J. Boschung, A. Nauels, Y. Xia, V. Bex, and P.M. Midgley, editors, *Climate Change 2013 – The Physical Science Basis: Contribution of Working Group I to the Fifth Assessment Report of the Intergovernmental Panel on Climate Change*, pages 1029–1136. Cambridge University Press, 2013.

[191] R.T.H. Collis. Lidar. *Applied Optics*, 9(8):1782–1788, 1970.

[192] R.T.H. Collis, F.G. Fernald, and M.G.H. Ligda. Laser radar echoes from a stratified clear atmosphere. *Nature*, 203(4951):1274–1275, 1964.

[193] R.T.H. Collis and M.G.H. Ligda. Laser radar echoes from the clear atmosphere. *Nature*, 203(4944):508–508, 1964.

[194] J.R. Colquhoun. A summary of four, five-day clear air turbulence reporting periods. *Autralian Meteorological and Oceanographic Journal*, 15(2):131–132, 1967.

[195] D. Colson and H.A. Panofsky. An index of clear air turbulence. *Quarterly Journal of the Royal Meteorological Society*, 91(390):507–513, 1965.

[196] H. Combe, F. Köpp, and M. Keane. On-board wake vortex detection. Technical report, Industrial & Materials Technologies Programme, Great Malvern, England, 2000.

[197] A. Consortini, F. Cochetti, J.H. Churnside, and R.J. Hill. Inner-scale effect on irradiance variance measured for weak-to-strong atmospheric scintillation. *Journal of the Optical Society of America A*, 10(11):2354–2362, 1993.

[198] L.S. Cook, B. Wood, A. Klein, R. Lee, and B. Memarzadeh. Analyzing the share of individual weather factors affecting NAS performance using the weather impacted traffic index. In *9th AIAA Aviation Technology, Integration, and Operations Conference (ATIO)*. American Institute of Aeronautics and Astronautics, 2009.

[199] N.J. Cooke. Varieties of knowledge elicitation techniques. *International Journal of Human-Computer Studies*, 41(6):801–849, 1994.

[200] G.A. Corby and C.E. Wallington. Air flow over mountains: The lee-wave amplitude. *Quarterly Journal of the Royal Meteorological Society*, 82(353):266–274, 1956.

[201] L.B. Cornamn, J. Williams, G. Meymaris, and B. Chorbajian. Verification of an airborne radar turbulence detection algorithm. In *6th International Symposium on Tropospheric Profiling: Needs and Technologies*, pages 9–12. American Meteorological Society, 2003.

[202] L.B. Cornman. *Airborne in situ measurements of turbulence*, chapter 5. Springer, 2016.

[203] L.B. Cornman and B. Carmichael. Varied research efforts are underway to find means of avoiding air turbulence. *ICAO Journal*, 48:10–15, 1993.

[204] L.B. Cornman, G. Meymaris, J. Prestopnik, and T. Tasset. An update on turbulence reporting from commercial aircraft. In *Proceedings of the 13th Conference on Aviation, Range, and Aerospace Meteorology*. American Meteorological Society, 2008. Paper number 10.1.

[205] L.B. Cornman, C.S. Morse, and G. Cunning. Real-time estimation of atmospheric turbulence severity from in-situ aircraft measurements. *Journal of Aircraft*, 32(1):171–177, 1995.

[206] R. Cowen. Clearing the air about turbulence: A fearful flier's foray. *Science News*, 153(26):408–410, June 27 1998.

[207] R.K. Crane. Ionospheric scintillation. *Proceedings of the IEEE*, 65(2):180–199, 1977.

[208] T.L. Crawford and R.J. Dobosy. A sensitive fast-response probe to measure turbulence and heat flux from any airplane. *Boundary-Layer Meteorology*, 59:257–278, 1992.

[209] J. Croft. Truer pictures of turbulence. *Aerospace America*, 43(4):34–38, 2005.

[210] W.M. Crooks. *Project HICAT, An Investigation of High Altitude Clear Air Turbulence*, volume 1. PN, 1967.

[211] W.M. Crooks, F.M. Hoblit, and F.A. Mitchell. Project hicat: High altitude clear air turbulence measurements and meteorological correlations. Technical Report AFFDL-TR-68-127, Air Force Flight Dynamics Laboratory, Wright-Patterson Air Force Base, Ohio, November 1968.

[212] T.D. Crum and R.L. Alberty. The WSR-88D and the WSR-88D operational support facility. *Bulletin of the American Meteorological Society*, 74(9):1669–1688, 1993.

[213] B. Cushman-Roisin and J.-M. Beckers. *Introduction to Geophysical Fluid Dynamics: Physical and Numerical Aspects*. Elsevier, Amsterdam, The Netherlands, 2011.

[214] D.J. Cziczo, P.J. DeMott, C. Brock, P.K. Hudson, B. Jesse, S.M. Kreidenweis, A.J. Prenni, J. Schreiner, D.S. Thomson, and D.M. Murphy. A method for single particle mass spectrometry of ice nuclei. *Aerosol Science and Technology*, 37(5):460–470, 2003.

[215] D.J. Cziczo, D.S. Thomson, and D.M. Murphy. Ablation, flux, and atmospheric implications of meteors inferred from stratospheric aerosol. *Science*, 291(5509):1772–1775, 2001.

[216] J. Daney and X. Lesceu. Wind shear: An invisible enemy to pilots? *Safety First*, 2015.

[217] T.S. Daniels. Tropospheric airborne meteorological data reporting (TAMDAR) sensor development. *SAE International*, (2002-01-1523), 2002.

[218] T.S. Daniels, J.J. Murray, C.A. Grainger, D.K. Zhou, M.A. Avery, G. Tsoucalas, M.F. Cagle, P.R. Schaffner, and R.T. Neece. Validation of tropospheric airborne meteorological data reporting (TAMDAR) temperature, relative humidity, and wind sensors during 2003 Atlantic thorpex regional campaign and the alliance icing research study (AIRS II). In *11th AMS Conference on Aviation, Range, and Aerospace*. Airbus S.A.S., 2004.

[219] T.S. Daniels, W.L. Smith, and S. Kireev. Recent developments on airborne forward looking interferometer for the detection of wake vortices. In *45th AIAA Atmospheric and Space Environments Conference*, pages 1–7. AIAA, 2012.

[220] T.S. Daniels, G. Tsoucalas, M. Anderson, D.J. Mullaly, W. Moninger, and R. Mamrosh. Tropospheric airborne meteorological data reporting (TAMDAR) sensor development. In *11th AMS Conference on Aviation, Range, and Aerospace*. Airbus S.A.S., 2004.

[221] S.S. Das, A.K. Ghosh, K. Satheesan, A.R. Jain, and K.N. Uma. Characteristics of atmospheric turbulence in terms of background atmospheric parameters inferred using mst radar at gadanki (13.5n,79.2e). *Radio Science*, 45(4):RS4008, 2010.

[222] G. d'Auria. Propagation effects of a variable scatter mechanism. In *AGARD Conference Proceedings No. 70 Tropospheric Radio Wave Propagation. Part II*, number AGARD-CP-70-71, pages 25.1–25.16. Advisory Group for Aerospace Research & Development (AGARD), 1971.

[223] K. Davies, R.B. Fritz, R.N. Grubb, and J.E. Jones. Some early results from the ats-6 radio beacon experiment. *Radio Science*, 10(8-9):785–799, 1975.

[224] E.J. Davis and G. Schweiger. *The Airborne Microparticle: Its Physics, Chemistry, Optics and Transport Phenomena*. Springer Verlag, Heidelberg, 2002.

[225] P.A. Davis. The analysis of lidar signatures of cirrus clouds. *Applied Optics*, 8(10):2099–2102, 1969.

[226] G. de Boer, S. Palo, B. Argrow, G. LoDolce, J. Mack, R.-S. Gao, H. Telg, C. Trussel, J. Fromm, C.N. Long, G. Bland, J. Maslanik, B. Schmid, and T. Hock. The PILATUS unmanned aircraft system for lower atmospheric research. *Atmospheric Measurement Techniques*, 9(4):1845–1857, 2016.

[227] M.P. de Villiers and J. van Heerden. Clear air turbulence over South Africa. *Meteorological Applications*, 8(1):119–126, 2001.

[228] P.F. DeCarlo, E.J. Dunlea, J.R. Kimmel, A.C. Aiken, D. Sueper, J. Crounse, P.O. Wennberg, L. Emmons, Y. Shinozuka, A. Clarke, J. Zhou, J. Tomlinson, D.R. Collins, D. Knapp, A.J. Weinheimer, D.D. Montzka, T. Campos, and J.L. Jimenez. Fast airborne aerosol size and chemistry measurements above Mexico City and Central Mexico during the milagro campaign. *Atmospheric Chemistry and Physics*, 8(14):4027–4048, 2008.

[229] D.P. Dee, S.M. Uppala, A.J. Simmons, P. Berrisford, P. Poll, S. Kobayashi, U. Andrae, M.A. Balmaseda, G. Balsamo, P. Bauer, P. Bechtold, A.C.M. Beljaars, L. van de Berg, J. Bidlot, N. Bormann, C. Delsol, R. Dragani, M. Fuentes, A.J. Geer, L. Haimberger, S.B. Healy, H. Hersbach, E.V. Hólm, L. Isaksen, P. Kållberg, M. Köhler, M. Matricardi, A.P. McNally, B.M. Monge-Sanz, J.-J. Morcrette, B.-K. Park, C. Peubey, P. de Rosnay, C. Tavolato, J.-N. Thépaut, and F. Vitart. The era-interim reanalysis: configuration and performance of the data assimilation system. *Quarterly Journal of the Royal Meteorological Society*, 137(656):553–597, 2011.

[230] D. del Castillo-Negrete, B.A. Carreras, and V.E. Lynch. Non-diffusive transport in plasma turbulence: A fractional diffusion approach. *Physical Review Letters*, 94(6):065003, 2005.

[231] S.C. Delcambre, D.J. Lorenz, D.J. Vimont, and J.E. Martin. Diagnosing northern hemisphere jet portrayal in 17 CMIP3 global climate models: Twenty-first-century projections. *Journal of Climate*, 26(14):4930–4946, 2013.

[232] D.J. Delene, T. Deshler, P. Wechsler, and G.A. Vall. A balloon-borne cloud condensation nuclei counter. *Journal of Geophysical Research: Atmospheres*, 103(D8):8927–8934, 1998.

[233] P.J. DeMott, A.J. Prenni, X. Liu, S.M. Kreidenweis, M.D. Petters, C.H. Twohy, M.S. Richardson, T. Eidhammer, and D.C. Rogers. Predicting global atmospheric ice nuclei distributions and their impacts on climate. *Proceedings of the National Academy of Sciences of the United States of America*, 107(25):11217–11222, 2010.

[234] G. Deskos, A. del Carre, and R. Palacios. Assessment of low-altitude atmospheric turbulence models for aircraft aeroelasticity. *Journal of Fluids and Structures*, 95:102981, 2020.

[235] A.C. DeVoria and K. Mohseni. Vortex sheet roll-up revisited. *Journal of Fluid Mechanics*, 855:299–321, 2018.

[236] J.A. Diaz, D. Pieri, K. Wright, P. Sorensen, R. Kline-Shoder, C.R. Arkin, M. Fladeland, G. Band, M.F. Buongiorno, C. Ramirez, E. Corrales, A. Alan, O. Alegria, D. Diaz, and J. Linick. The PILATUS unmanned aircraft system for lower atmospheric research. *Journal of the American Society for Mass Spectrometry*, 26(2):292–304, 2015.

[237] R. Dmowska, J.R. Holton, and H.T. Rossby. Observational techniques. In C.J. Nappo, editor, *An Introduction to Atmospheric Gravity Waves. International Geophysics Series*, volume 85, pages 181–207. Elsevier, 2002.

[238] J. Dole and R. Wilson. Estimates of turbulent parameters in the lower stratosphere-upper troposphere by radar observations: A novel twist. *Geophysical Research Letters*, 27(17):2625–2628, 2000.

[239] J. Dole, R. Wilson, F. Dalaudier, and C. Sidi. Energetics of small scale turbulence in the lower stratosphere from high resolution radar measurements. *Annales Geophysicae*, 19:945–952, 2001.

[240] R.J. Donaldson. Vortex signature recognition by a doppler radar. *Journal of Applied Meteorology and Climatology*, 9(4):661–670, 1970.

[241] W.L. Donn and D. Rind. Microbaroms and the temperature and wind of the upper atmosphere. *Journal of the Atmospheric Sciences*, 19(1):156–172, 1972.

[242] P.M. Doran. *Bioprocess Engineering Principles*. Academic Press, Massachusetts, USA, 2013.

[243] M. Dosunmu. How to become a professional pilot. *STEM*, November 2020.

[244] R.J. Doviak, D. Burgess, L. Lemon, and D. Sirmans. Doppler velocity and reflectivity structure observed within a tornadic storm. *Journal de Recherches Atmosphériques*, 8(1-2):235–243, 1974.

[245] R.J. Doviak and C.T. Jobson. Dual doppler radar observations of clear air wind perturbations in the planetary boundary layer. *Journal of Geophysical Research: Oceans*, 84(C2):697–702, 1979.

[246] R.J. Doviak and D.S. Zrnic. *Doppler Radar and Weather Observations*. Dover Publications, New York, second edition edition, 2006.

[247] R.J. Doviak, D.S. Zrnic, and D.S. Sirmans. Doppler weather radar. *Proceedings of the IEEE*, 67(11):1522–1553, 1979.

[248] L.E. Drain. *The Laser Doppler Technique*. John Wiley & Sons, New York, 1980.

[249] D. Dravins, L. Lindegren, E. Mezey, and A.T. Young. Atmospheric intensity scintillation of stars. I. statistical distributions and temporal properties. *Publications of the Astronomical Society of the Pacific*, 109(732):173–207, 1997.

[250] P.G. Drazin. Kelvin-Helmholtz instability of finite amplitude. *Journal of Fluid Mechanics*, 42:321–335, 1970.

[251] D.P. Drob, J.T. Emmert, G. Crowley, J.M. Picone, G.G. Shepherd, W. Skinner, P. Hays, R.J. Niciejewski, M. Larsen, C.Y. She, J.W. Meriwether, G. Hernández, M.J. Jarvis, D.P. Sipler, C.A. Tepley, M.S. O'Brien, J.R. Bowman, Q. Wu, Y. Murayama, S. Kawamura, I.M. Reid, and R.A. Vincent. An empirical model of the earth's horizontal wind fields: HWM07. *Journal of Geophysics Research: Space Physics*, 113(A12):A12304, 2008.

[252] D.P. Drob, R.R. Meier, J.M. Picone, and M. Garcés. Inversion of infrasound signals for passive atmospheric remote sensing. In A. Le Pichon, E. Blanc, and A. Hauchecorne, editors, *Infrasound Monitoring for Atmospheric Studies*, pages 701–732. Springer, 2010.

[253] D.P. Drob, J.M. Picone, and M.A. Garcés. The global morphology of infrasound propagation. *Journal of Geophysics Research: Atmospheres*, 108(D21):4680, 2003.

[254] F. du Castel, P. Misme, A. Spizzichino, and J. Voge. Réflexions partielles dans l'atmosphère et propagation à grande distance. quatrième partie: Réflexion spéculaire et réflexion diffuse sur des feuillets atmosphériques. *Annales des Télécommunications*, 15(1-2):38–47, 1960.

[255] T.J. Dunkerton. Shear instability of inertia-gravity waves. *Journal of the Atmospheric Sciences*, 54(12):1628–1641, 1997.

[256] E.F. Duquette and K. Glover. A study of clear air turbulence using sensitive radars. In *Preprints of the 14th Radar Meteorological Conference*, pages 89–94. American Meteorological Society, 1970.

[257] K. Duraisamy, G. Iaccarino, and H. Xiao. Turbulence modeling in the age of data. *Annual Review of Fluid Mechanics*, 51:357–377, 2019.

[258] D.R. Durran. Lee waves and mountain waves. In J.R. Holton, J. Pyle, and J.A. Curry, editors, *Encyclopedia of Atmospheric Sciences*, pages 1161–1169. Academic Press, 2003.

[259] D.R. Durran. Mountain meteorology. lee waves and mountain waves. In G.R. North, J. Pyle, and F. Zhang, editors, *Encyclopedia of Atmospheric Sciences (Second edition)*, pages 95–102. Elsevier, 2015.

[260] J. Dutton. Broadening horizons in prediction of the effects of atmospheric turbulence on aeronautical systems. In *5th Annual Meeting and Technical Display. AIAA Paper No. 68-1065*. American Institute of Aeronautics and Astronautics, 1968.

[261] J.A. Dutton and H.A. Panofsky. Clear air turbulence: A mistery may be unfolding. *Science*, 167(3920):937–944, 1970.

[262] J.A. Dutton, G.J. Thompson, and D.G. Deaven. The probabilistic structure of clear air turbulence – some observational results and implications. In Y.H. Pao and A. Goldburg, editors, *Clear Air Turbulence and its Detection*, pages 183–206. Springer, 1969.

[263] M.J.O. Dutton. Probability forecasts of clear-air turbulence based on numerical model output. *The Meteorological Magazine*, 109(1299):293–310, 1980.

[264] C. Eckart. Vortices and streams caused by sound waves. *Physical Review*, 73:68–76, 1948.

[265] W.N. Edeling, G. Iaccarino, and P. Cinnella. Data-free and data-driven rans predictions with quantified uncertainty. *Flow, Turbulence and Combustion*, 100(3):593–616, 2018.

[266] M. Eghdami, S. Bhushan, and A.P. Barros. Direct numerical simulations to investigate energy transfer between meso- and synoptic scales. *Journal of the Atmospheric Sciences*, 75:1163–1171, 2018.

[267] L.J. Ehernberger. *Atmospheric Conditions Associated with Turbulence Encountered by the XB-70 Airplane Above 40000 Feet Altitude*. NASA, 1968.

[268] J. Ehlers and N. Fezans. Airborne doppler lidar sensor parameter analysis for wake vortex impact alleviation purposes. In J. Bordeneuve-Guibé, A. Drouin, and C. Roos, editors, *Advances in Aerospace Guidance, Navigation and Control: Selected Papers of the Third CEAS Specialist Conference on Guidance, Navigation and Control*, pages 433–453. Springer, 2015.

[269] A. Einstein and R. Fürth. *Investigations on the Theory of Brownian Movement.* Dover Publications, New York, USA, 1956.

[270] A. Eliassen and E. Palm. On the transfer of energy in stationary mountain waves. *Geofysiske Plublikasjoner*, 22(3):1–23, 1961.

[271] M.A. Ellison. Location, size and speed of refractional irregularities causing scintillation. *Quarterly Journal of the Royal Meteorological Society*, 80(344):246–248, 1954.

[272] M.A. Ellison and H. Seddom. Some experiments on the scintillation of stars and planets. *Monthly Notices of the Royal Astronomical Society*, 112(1):73–87, 1952.

[273] G.P. Ellrod. Detection of high level turbulence using satellite imagery and upper air data. Technical Report NOAA NESDIS 10, Satellite Applications Laboratory, Washington, D.C., April 1985.

[274] G.P. Ellrod. An index for clear air turbulence based on horizontal deformation and vertical wind shear. In *Third International Conference on the Aviation Weather System*, pages 339–344. American Meteorological Society, 1989.

[275] G.P. Ellrod and D.I. Knapp. An objective clear-air turbulence forecasting technique: Verification and operational use. *Weather and Forecasting*, 7(1):150–165, 1992.

[276] G.P. Ellrod and J.A. Knox. Improvements to an operational clear-air turbulence diagnostic index by addition of a divergence trend term. *Weather and Forecasting*, 25(2):789–798, 2010.

[277] G.P. Ellrod, J.A. Knox, P.F. Lester, and L.J. Ehernberger. Clear air turbulence. In G.R. North, J. Pyle, and F. Zhang, editors, *Encyclopedia of Atmospheric Sciences*, pages 177–186. Elsevier, 2015.

[278] G.P. Ellrod, P.F. Lester, and L.J. Ehernberger. Clear air turbulence. In J.R. Holton, J. Pyle, and J.A. Curry, editors, *Encyclopedia of Atmospheric Sciences*, pages 393–403. Elsevier, 2003.

[279] N. Engler, R. Latteck, B. Strelnikov, W. Singer, and M. Rapp. Turbulent energy dissipation rates observed by doppler mst radar and by rocket-borne instruments during the midas/macwave campaign 2002. *Annales Geophysicae*, 23(4):1147–1156, 2005.

[280] B. English, J. Kraus, and P. Pillar. Using unmanned aerial systems for aircraft accident investigations. *Journal of the International Society of the Air Safety Investigations*, 48(3):4–9, 2015.

[281] J.G. Esler and L.M. Polvani. Kelvin-Helmholtz instability of potential vorticity layers: A route to mixing. *Journal of the Atmospheric Sciences*, 61(12):1392–1405, 2004.

[282] R. Faivre, J. Colin, and M. Menenti. Evaluation of methods for aerodynamic roughness length retrieval from very high-resolution imaging LIDAR observations over the Heihe basin in China. *Remote Sensing*, 9(1):63, 2017.

[283] M. Farge. Wavelet transforms and their applications to turbulence. *Annual Review of Fluid Mechanics*, 24:395–458, 1992.

[284] A. Favre. Equations des gaz turbulents compressibles. *Journal de Mécanique*, 4:391, 1965.

[285] A.J.A. Favre. Formulation of the statistical equations of turbulent flows with variable density. In T.B. Gatski, C.G. Speziale, and S. Sarkar, editors, *Studies in Turbulence*, pages 324–341. Springer, 1992.

[286] L. Fehlhaber. Methods of distinguishing scatter and partial reflection at tropospheric transhorizon paths. In *AGARD Conference Proceedings No. 70 Tropospheric Radio Wave Propagation. Part II*, number AGARD-CP-70-71, pages 30.1–30.13. Advisory Group for Aerospace Research & Development (AGARD), 1971.

[287] A. Feinberg and J. Tauss. NASA aviation safety program weather accident prevention/weather information communications (WINCOMM). Technical Report NASA/CR–2002-211903, Glenn Research Center, NASA, October 2002.

[288] P.B. Fellgett. On the ultimate sensitivity and practical performance of radiation detectors. *Journal of the Optical Society of America*, 39(11):970–976, 1949.

[289] P. Feneyrou, J.-C. Lehureau, and H. Barny. Performance evaluation for long-range turbulence-detection using ultraviolet lidar. *Applied Optics*, 48(19):3750–3759, 2009.

[290] X. Feng. Evaluation of the maccready turbulence sensor. Master's thesis, Department of Atmospheric Sciences, University of Wyoming, Laramie, Wyoming, December 2001.

[291] S.J. Findlay and N.D. Harrison. Why aircraft fail. *Materials Today*, 5(11):18–25, 2002.

[292] A. Finn, K. Rogers, F. Rice, J. Meade, G. Holland, and P. May. A comparison of vertical atmospheric wind profiles obtained from monostatic sodar and unmanned aerial vehicle based acoustic tomography. *Journal of Atmospheric and Oceanic Technology*, 34(10):2311–2328, 2017.

[293] G. Fiocco and L.D. Smullin. Detection of scattering layers in the upper atmosphere (60–140 km) by optical radar. *Nature*, 199(4900):1275–1276, 1963.

[294] S.M. Flatté, C. Bracher, and G.-Y. Wang. Probability-density functions of irradiance for waves in atmospheric turbulence calculated by numerical simulation. *Journal of the Optical Society of America A*, 11(7):2080–2092, 1994.

[295] S.M. Flatté, G.-Y. Wang, and J. Martin. Irradiance variance of optical waves through atmospheric turbulence by numerical simulation and comparison with experiment. *Journal of the Optical Society of America A*, 10(11):2363–2370, 1993.

[296] T. Foelsche, R.B. Mendell, J.W. Wilson, and R.R. Adams. Measured and calculated neutron spectra and dose equivalent rates at high altitudes; relevance to sst operations and space research. Technical Report NASA/TN D-7715, Langley Research Center, NASA, October 1974.

[297] T. Foken. 50 years of the Monin-Obukhov similarity theory. *Boundary-Layer Meteorology*, 119(3):431–447, 2006.

[298] R. Foord, R. Jones, J.M. Vaughan, and D.V. Willets. Precise comparison of experimental and theoretical SNRs in CO_2 laser heterodyne systems. *Applied Optics*, 22(23):3787–3795, 1983.

[299] NASA & The FAA: A Technology Partnership for the New Millennium. Aviation weather information (AWIN).

[300] R. Ford. Gravity wave radiation from vortex trains in rotating shallow water. *Journal of Fluid Mechanics*, 281:81–118, 1994.

[301] The Flight Safety Foundation. Low level wind shear. SKYbrary.

[302] P.M. Franke and W.A. Robinson. Nonlinear behavior in the propagation of atmospheric gravity waves. *Journal of the Atmospheric Sciences*, 56(17):3010–3027, 1999.

[303] R. Frehlich, L. Cornman, and R. Sharman. Simulation of three-dimensional turbulence velocity fields. *Journal of Applied Meteorology and Climatology*, 40(2):246–258, 2001.

[304] R. Frehlich and R. Sharman. Estimates of turbulence from numerical weather prediction model output with applications to turbulence diagnosis and data assimilation. *Monthly Weather Review*, 132(10):2308–2324, 2004.

[305] R. Frehlich and R. Sharman. Climatology of velocity and temperature turbulence statistics determined from rawinsonde and ACARS/AMDAR data. *Journal of Applied Meteorology and Climatology*, 49(6):1149–1169, 2010.

[306] R.G. Frehlich and R. Sharman. Estimates of upper level turbulence based on second order structure functions derived from numerical weather prediction model output. In *11th Conference on Aviation, Range, and Aerospace*, pages 1–18. American Meteorology Society, 2004.

[307] E.J. Fremouw, H.C. Carlson, T.A. Potemra, P.F. Bythrow, C.L. Rino, J.F. Vickrey, R.L. Livingston, R.E. Huffman, C.-I. Meng, D.A. Hardy, F.J. Rich, R.A. Heelis, W.B. Hanson, and L.A. Wittwer. The hilat satellite mission. *Radio Science*, 20(3):416–424, 1985.

[308] E.J. Fremouw, R.L. Leadabrand, R.C. Livingston, M.D. Cousins, C.L. Rino, B.C. Fair, and R.A. Long. Early results from dna wideband satellite experiment—complex-signal scintillation. *Radio Science*, 13(1):167–187, 1978.

[309] J.R. French, S.J. Haimov, L.D. Oolman, V. Grubišić, S. Serafin, and L. Strauss. Wave-induced boundary layer separation in the lee of the medicine bow mountains. Part I: Observations. *Journal of the Atmospheric Sciences*, 72(12):4845–4863, 2015.

[310] W. Friedberg, D.N. Faulkner, L. Snyder, E.B. Darden, and K. O'Brien. Galactic cosmic radiation exposure and associated health risks for air carrier crewmembers. *Aviation, Space, and Environmental Medicine*, 60(11):1104–1108, 1989.

[311] H.T. Friis, A.B. Crawford, and D.C. Hogg. A reflection theory for propagation beyond the horizon. *The Bell System Technical Journal*, 36(3):627–644, 1957.

[312] U. Frisch. *Turbulence*. Cambridge University Press, 1995.

[313] S. Fritts. How to decode terminal aerodrome forecasts (TAFs). *Think Aviation*, 1, 2016.

[314] K.D. Froyd, D.M. Murphy, T.J. Stanford, D.S. Thomson, J.C. Wilson, L. Pfister, and L. Lait. Aerosol composition of the tropical upper troposphere. *Atmospheric Chemistry and Physics*, 9:4363–4385, 2009.

[315] D. Fua, G. Chimonas, F. Einaudi, and O. Zeman. An analysis of wave-turbulence interactions. *Journal of the Atmospheric Sciences*, 39(11):2450–2463, 1982.

[316] N. Fukuta and V.K. Saxena. A horizontal thermal gradient cloud condensation nucleus spectrometer. *Journal of Applied Meteorology and Climatology*, 18(10):1352–1362, 1979.

[317] C. Fureby. Towards the use of large eddy simulation in engineering. *Progress in Aerospace Sciences*, 44:381–396, 2008.

[318] R. Fürth. Statistical analysis of scintillations of stars. In Z. Kopal, editor, *Astronomical Optics and Related Subjects*, pages 300–309. North-Holland Publishing Company, 1956.

[319] K.S. Gage and B.B. Balsley. Doppler radar probing of the clear atmosphere. *Bulletin of the American Meteorological Society*, 59(9):1074–1094, 1978.

[320] K.S. Gage and W.L. Clark. Mesoscale variability of jet stream winds observed by the sunset VHF doppler radar. *Journal of Applied Meteorology and Climatology*, 17(9):1412–1416, 1978.

[321] K.S. Gage, J.L. Green, and T.E. Van Zandt. Use of doppler radar for the measurement of atmospheric turbulence parameters from the intensity of clear air echoes. *Radio Science*, 15(2):407–416, 1980.

[322] B. Galperin, S. Sukoriansky, and P.S. Anderson. On the critical Richardson number in stably stratified turbulence. *Atmospheric Science Letters*, 8:65–69, 2007.

[323] F. Gao, X. Zhang, N.A. Jacobs, X-Y.. Huang, X. Zhang, and P.P. Childs. Estimation of tamdar observational error and assimilation experiments. *Weather and Forecasting*, 27(4):856–877, 2012.

[324] A.H. Gardiner, H.L. Giclas, F. Gifford, H.L. Johnson, R.L.I. Mitchell, and A.G. Wilson. Optical studies of atmospheric turbulence. Technical Report AFCRC-TR-56-261, Lowell Observatory, Flagstaff, Arizona, USA, 1956.

[325] C.S. Gardner. Sodium resonance fluorescence lidar applications in atmospheric science and astronomy. *Proceedings of the IEEE*, 77(3):408–418, 1989.

[326] J.R. Garratt. *The Atmospheric Boundary Layer*. Cambridge University Press, Cambridge, UK, 1992.

[327] P. Gentine, M. Pritchard, S. Rasp, G. Reinaudi, and G. Yacalis. Could machine learning break the convection parameterization deadlock? *Geophysical Research Letters*, 45(11):5742–5751, 2018.

[328] R. George and J. Yang. A survey for methods of detecting aircraft vortices. In *Proceedings of the ASME 2012: International Design Engineering Technical Conferences & Computers and Information in Engineering Conference (IDETC/CIE)*, number DETC2012-70632, pages 41–50. The American Society of Mechanical Engineers, 2013.

[329] D. Georgiou and K.F. Milidonis. Fabrication and calibration of a subminiature 5-hole probe with embedded pressure sensors for use in extremely confined and complex flow areas in turbomachinery research facilities. *Flow Measurement and Instrumentation*, 39:54–63, 2014.

[330] M. Germano, U. Piomelli, P. Moin, and W.H. Cabot. A dynamic subgrid-scale eddy viscosity model. *Physics of Fluids A: Fluid Dynamics*, 3:1760–1765, 1991.

[331] T. Gerz, F. Holzäpfel, and D. Darracq. Commercial aircraft wake vortices. *Progress in Aerospace Sciences*, 38(3):181–208, 2002.

[332] A.V. Getling and E.A. Spiegel. Rayleigh-Bénard convection: Structure and dynamics. *Physics Today*, 52(9):59–60, 1999.

[333] W.R. Geyer and J.D. Smith. Shear instability in a highly stratified estuary. *Journal of Physical Oceanography*, 17:1668–1679, 1987.

[334] S. Ghosal, T.S. Lund, P. Moin, and K. Akselvoll. A dynamic localization model for large-eddy simulation of turbulent flows. *Journal of Fluid Mechanics*, 286:229–255, 1995.

[335] R. Gibb, R. Gray, and L. Scharff. *Aviation Visual Perception: Research, Misperception and Mishaps (Asgate Studies in Human Factors for Flight Operations)*. Routledge, first edition, 2010.

[336] C.H. Gibson. Turbulence, mixing and microstructure. In B. Le Mahaute and D.M. Hanes, editors, *Ocean Engineering Science: The Sea*, volume 9, pages 631–659. Wiley Interscience, 1990.

[337] C.H. Gibson. Fossil turbulence revisited. *Journal of Marine Systems*, 21(1–4):147–167, 1999.

[338] C.H. Gibson. What is turbulence, what is fossil turbulence, and which ways do they cascade? *arXiv:1203.5819v1*, 2012.

[339] J.J. Gibson. *The Ecological Approach to Visual Perception*. Houghton-Mifflin, New York, 1979.

[340] G. Giebel, U. Schmidt Paulsen, J. Bange, A. La Cour-Harbo, J. Reuder, S. Mayer, A. Van den Kroonenberg, and J. Mølgaard. Autonomous aerial sensors for wind power meteorology – a pre-project. Technical Report Risoe-R No. 1798(EN), Danmarks Tekniske Universitet, Risø Nationallaboratoriet for Bæredygtig Energi, Denmark, January 2012.

[341] A.E. Gill. *Atmosphere-Ocean Dynamics*. Pergamon Press, 1982.

[342] P.G. Gill. Objective verification of world area forecast centre clear air turbulence forecasts. *Meteorological Applications*, 21(1):3–11, 2014.

[343] P.G. Gill. Aviation turbulence forecast verification. In R. Sharman and T. Lane, editors, *Aviation Turbulence*, pages 261–283. Springer, 2016.

[344] P.G. Gill and P. Buchanan. An ensemble based turbulence forecasting system. *Meteorological Applications*, 21(1):12–19, 2014.

[345] G.G. Gimmestad, L.L. West, W.L. Smith, S. Kireev, X. Liu, P.R. Schaffner, and J.J. Murray. Airborne forward looking interferometer turbulence investigation. In *45th AIAA Aerospace Sciences Meeting and Exhibit*. AIAA, 2007.

[346] D.T. Gjessing. On the scattering of electromagnetic waves by nonisotropic inhomogeneities in the atmosphere. *Journal of Geophysical Research*, 67(3):1017–1026, 1962.

[347] D.T. Gjessing, A.G. Kjelaas, and E. Golton. Small-scale atmospheric structure deduced from measurements of temperature, humidity and refractive index. *Boundary-Layer Meteorology*, 4(1-4):475–492, 1973.

[348] S.O. Gladkov. Theoretical description of the temperature distribution in the earth's atmosphere. *Atmospheric and Oceanic Optics*, 22(2):180–185, 2009.

[349] H.F. Goessling and S. Bathiany. Why CO_2 cools the middle atmosphere – a consolidating model perspective. *Earth System Dynamics*, 7(3):697–715, 2016.

[350] W.L. Golding. Turbulence and its impact on commercial aviation. *Journal of Aviation/Aerospace Education & Research*, 11(2):19–29, 2000.

[351] I. Goodfellow, Y. Bengio, and A. Courville. *Deep Learning*. MIT Press, 2016.

[352] E.E. Gossard and L.J. Anderson. The effect of superrefractive layers on 50–5000 mc nonoptical fields. *IRE Transactions on Antennas and Propagation*, 4(2):175–178, 1956.

[353] Y. Gotaas. Mother of pearl clouds over southern Norway, February 21, 1959. *Geofysiske Plublikasjoner*, 22(4):1–13, 1961.

[354] G.G. Goyer and R. Watson. The laser and its application to meteorology. *Bulletin of the American Meteorological Society*, 44(9):564–570, 1963.

[355] M.E. Gracheva, A.S. Gurvich, S.S. Kashkarov, and Vl.V. Pokasov. Similarity relations and their experimental verification for strong intensity fluctuations of laser radiation. In J.W. Strohbehn, editor, *Laser Beam Propagation in the Atmosphere*. *Topics in Applied Physics*, volume 25, pages 107–127. Springer-Verlag, Berlin, 1978.

[356] L. Graig, A. Moharreri, A. Schanot, D.C. Rogers, B. Anderson, and S. Dhaniyala. Characterizations of cloud droplet shatter artifacts in two airborne aerosol inlets. *Aerosol Science and Technology*, 47(6):662–671, 2013.

[357] H.L. Grant, R.W. Stewart, and A. Moillet. Turbulence spectra from a tidal channel. *Journal of Fluid Mechanics*, 12:241–268, 1962.

[358] W.E. Green and P.Y. Oh. A MAV that flies like an airplane and hovers like a helicopter. In *Proceedings of the 2005 IEEE/ASME International Conference on Advanced Intelligent Mechatronics*, pages 24–28. IEEE, September 2005.

[359] W.S. Green, G. Tsoucalas, and T. Tanger. Concept of operations for the NASA weather accident prevention (WxAP) project. Technical Report NASA/TM-2003-212424, Langley Research Center, NASA, September 2003.

[360] J.S. Greenhow, H.K. Sutcliffe, and C.D. Watkins. The electron scattering cross-section in incoherent backscatter. *Journal of Atmospheric and Terrestrial Physics*, 25(4):197–207, 1963.

[361] E. Grenier, N. Adams, and P. Sagaut. *Large Eddy Simulation for Compressible Flows*. Springer, 2009.

[362] V. Grubišić and J.M. Lewis. Sierra wave project revisited: 50 years later. *Bulletin of the American Meteorological Society*, 85(8):1127–1142, 2008.

[363] V. Grubišić, S. Serafin, L. Strauss, S.J. Haimov, J.R. French, and L.D. Oolman. Wave-induced boundary layer separation in the lee of the medicine bow mountains. Part II: Numerical modeling. *Journal of the Atmospheric Sciences*, 72(12):4865–4883, 2015.

[364] E.M. Gudmundsdottir, J. Hrafnkelsson, and V. Rafnsson. Incidence of cancer among licenced commercial pilots flying north atlantic routes. *Environmental Health*, 16(86):1–10, 2017.

[365] A. Guinamard. Ellipse, ekinox and apogee – high performance inertial sensors. Technical Report SBGTECHRM.1.2, SGB Systems, 2018.

[366] I. Gultepe and W.F. Feltz. Aviation meteorology: Observations and models. introduction. *Pure and Applied Geophysics*, 176:1863–1867, 2019.

[367] I. Gultepe, M. Pawgoski, and J. Reid. A satellite-based fog detection scheme using screen air temperature. *Weather and Forecasting*, 22(3):444–456, 2007.

[368] I. Gultepe, R. Sharman, P.D. Williams, B. Zhou, G. Ellrod, P. Minnis, S. Trier, S. Griffin, S.S. Yum, B. Gharabaghi, W. Feltz, M. Temimi, Z. Pu, L.N. Storer, P. Kneringer, M.J. Weston, H. Chuang, L. Thobois, A.P. Dimri, S.J. Dietz, B.F. Gutemberg, M.V. Almeida, and F.L. Albquerque Neto. A review of high impact weather for aviation meteorology. *Pure and Applied Geophysics*, 176:1869–1921, 2019.

[369] A. Gurvitch and A.M. Yaglom. Breakdown of eddies and probability distributions for small-scale turbulence. *Physics of Fluids*, 10(9):S959–S65, 1967.

[370] S. Haase, K. Matthes, M. Latif, and N.-E. Omrani. The importance of a properly represented stratosphere for northern hemisphere surface variability in the atmosphere and the ocean. *Journal of Climate*, 31(20):8481–8497, 2018.

[371] Meteorologist Jeff Haby. The 300/200 mb chart. Webpage.

[372] J.M. Hacker and T. Crawford. The bat-probe: The ultimate tool to measure turbulence from any kind of aircraft (or sailplane). *Technical Soaring: An International Journal*, 23(2):43–46, 1999.

[373] C. Hackworth, L. Peterson, D. Jack, and C. Williams. Altitude training experiences and perspectives: survey of 67 professional pilots. *Aviation, Space, and Environmental Medicine*, 76(4):392–394, 2005.

[374] K.-U. Hahn. Effect of wind shear on flight safety. *Progress in Aerospace Sciences*, 26(3):225–259, 1989.

[375] L.A. Hajkowicz. Equatorwards limits of the southern scintillation oval. *Journal of Atmospheric and Terrestrial Physics*, 44(6):539–545, 1982.

[376] Y. Han, M. Stoellinger, and J. Naughton. Large eddy simulation for atmospheric boundary layer flow over flat and complex terrains. *Journal of Physics: Conference Series*, 753:032044, 2016.

[377] S.M. Hannon and J.A. Thomson. Aircraft wake vortex detection and measurement with pulsed solid-state coherent laser radar. *Journal of Modern Optics*, 41(11):2175–2196, 1994.

[378] A.R. Hansen, G.D. Nastrom, and F.D. Eaton. Seasonal variation of gravity wave activity at 5-20 km observed with VHF radar at white sands missile range, New Mexico. *Journal of Geophysical Research: Atmospheres*, 106(D15):17171–17183, 2001.

[379] A.R. Hansen, G.D. Nastrom, J.A. Otkin, and F.D. Eaton. MST radar observations of gravity waves and turbulence near thunderstorms. *Journal of Geophysical Research: Atmospheres*, 41(3):298–305, 2002.

[380] K.R. Hardy. Wind shear and clear air turbulence. *Journal of Air Law and Commerce*, 42(1):165–183, 1976.

[381] K.R. Hardy, K.M. Glover, and H. Ottersten. Radar investigations of atmospheric structure an cat in the 3 to 20-km region. In Y.H. Pao and A. Goldburg, editors, *Clear Air Turbulence and Its Detection*, pages 402–416. Springer, 1969.

[382] P. Hariharan. *Basics of Interferometry*. Academic Press, 2007.

[383] F.H. Harlow and P.I. Nakayama. Transport of turbulence energy decay rate. Technical Report LA-3854, Los Alamos Scientific Laboratory of the University of California, Los Alamos, New Mexico, January 1968.

[384] L.P. Harrison. Mathematical theory of the graphical evaluation of meteorograph soundings by means of the Stuve (Lindenberg) adiabatic chart. *Monthly Weather Review*, 63(4):123–135, 1935.

[385] G. Hattenberger, G. Cayez, and G. Roberts. Flight tests for meteorological studies with MAV. In *International Micro Air Vehicle Conference and Flight Competition (IMAV2013)*, number HAL00936235, pages 1–8. IMAV, January 2014.

[386] A. Hauchecorne and M.-L. Chanin. Density and temperature profiles obtained by lidar between 35 and 70 km. *Geophysical Research Letters*, 7(8):565–568, 1980.

[387] A. Hauchecorne, C. Cot, F. Dalaudier, J. Porteneuve, T. Gaudo, R. Wilson, C. Cénac, C. Laqui, P. Keckhut, J.-M. Perrin, A. Dolfi, N. Cézard, L. Lombard, and C. Besson. Tentative detection of clear air turbulence using a ground-based Rayleigh lidar. *Applied Optics*, 55(13):3420–3428, 2016.

[388] H. Haverdings and P.W. Chan. Quick access recorder data analysis software for windshear and turbulence studies. *Journal of Aircraft*, 47(4):1443–1447, 2010.

[389] Y. He, S. Zhao, H. Wei, and Y. Li. Noise reduction of air turbulence via the quasi-common-path method. *Applied Optics*, 56(23):6668–6672, 2017.

[390] J. Heintzenberg, A. Wiedensohler, T.M. Tuch, D.S. Covert, P. Sheridan, J.A. Ogren, J. Gras, R. Nessler, C. Kleefeld, N. Kalivitis, V. Aaltonen, R.-T. Wilhelm, and M. Havlicek. Intercomparisons and aerosol calibrations of 12 commercial integrating nephelometers of three manufacturers. *Journal of Atmospheric and Oceanic Technology*, 23(7):902–914, 2006.

[391] S.W. Henderson, P. Gatt, D. Rees, and R.M. Huffaker. *Wind Lidar*, chapter 7, pages 469–722. CRC Press, first edition, 2005.

[392] J. Hendricks, B. Kärcher, A. Döpelheuer, J. Feichter, U. Lohmann, and D. Baumgardner. Simulating the global atmospheric black carbon cycle: a revisit to the contribution of aircraft emissions. *Atmospheric Chemistry and Physics*, 4:2521–2541, 2004.

[393] J.G. Henning and P.J. Radtke. Ground-based laser imaging for assessing three-dimensional forecast canopy structure. *Photogrammetric Engineering & Remote Sensing*, 72(12):1349–1358, 2006.

[394] J. Herbst and P. Vrancken. Design of a monolithic michelson interferometer for fringe imaging in a near-field, UV, direct-detection doppler wind lidar. *Applied Optics*, 55(25):6911–6920, 2016.

[395] J.W. Herbstreit and M.C. Thompson. Measurements of the phase of radio waves received over transmission paths with electrical lengths varying as a result of atmospheric turbulence. *Proceedings of the IRE*, 43(10):1391–1401, 1955.

[396] G.R. Herman and R.S. Schumacher. Money doesn't grow on trees, but forecasts do: Forecasting extreme precipitation with random forests. *Monthly Weather Review*, 146(5):1571–1600, 2018.

[397] S.L. Hess. *Introduction to Theoretical Meteorology*. Krieger Publishing Company, Malabar, 2006.

[398] A. Hewish. The diffraction of galactic radio waves as a method of investigating the irregular structure of the ionosphere. *Proceedings of the Royal Society A: Mathematical, Physical and Engineering Sciences*, 214(1119):494–514, 1952.

[399] J.S. Hey, J.S. Parsons, and J.W. Phillips. Fluctuations in cosmic radiation at radio frequencies. *Nature*, 158(4007):234–234, 1946.

[400] J.J. Hicks, I. Katz, C.R. Landry, and K.R. Hardy. Clear-air turbulence: Simultaneous observations by radar and aircraft. *Science*, 157(3790):808–809, 1967.

[401] A.J. Hill, G.R. Herman, and R.S. Schumacher. Forecasting severe weather with random forests. *Monthly Weather Review*, 148(5):2135–2161, 2020.

[402] R.J. Hill and R.G. Frehlich. Probability distribution of irradiance for the onset of strong scintillation. *Journal of the Optical Society of America A*, 14(7):1530–1540, 1997.

[403] C.O. Hines. Internal atmospheric gravity waves at ionospheric heights. *Canadian Journal of Physics*, 38(11):1441–1481, 1960.

[404] J.O. Hinze. *Turbulence*. McGraw-Hill, New York, second edition, 1975.

[405] S. Hobbs, D. Dyer, D. Courault, A. Olioso, J.-P. Lagouarde, Y. Kerr, J. Mcaneney, and J. Bonnefond. Surface layer profiles of air temperature and humidity measured from unmanned aircraft. *Agronomie*, 22(6):635–640, 2002.

[406] U. Högström. Non-dimensional wind and temperature profiles in the atmospheric surface layer: A re-evaluation. *Boundary-Layer Meteorology*, 42(1-2):55–78, 1988.

[407] J. Holmgren and Å. Persson. Identifying species of individual trees using airborne laser scanner. *Remote Sensing of Environment*, 90(4):415–423, 2004.

[408] J.R. Holton. The role of gravity wave induced drag and diffusion in the momentum budget of the mesosphere. *Journal of Atmospheric Sciences*, 39(4):791–799, 1982.

[409] J.R. Holton. *An Introduction to Dynamic Meteorology*. Academic Press, London, fourth edition, 2004.

[410] J.R. Holton. Dynamical meteorology. overview. In G.R. North, J. Pyle, and F. Zhang, editors, *Encyclopedia of Atmospheric Sciences (Second Edition)*, pages 265–271. Elsevier, 2015.

[411] J.R. Holton, P.H. Haynes, M.E. McIntyre, A.R. Douglass, R.B. Rood, and L. Pfister. Stratosphere-troposphere exchange. *Reviews of Geophysics*, 33(4):403–439, 1995.

[412] K.K. Hon, C.W. Ng, and P.W. Chan. Machine learning based multi-index prediction of aviation turbulence over the Asia-Pacific. *Machine Learning with Applications*, 2:100008, 2020.

[413] R.H. Hopkins. Forecasting techniques of clear-air turbulence including that associated with mountain waves. Technical Report 155, World Meteorological Organization, Geneva, Switzerland, 1977.

[414] B.J. Hoskins. The role of potential vorticity in symmetric stability and instability. *Quarterly Journal of the Royal Meteorological Society*, 100(425):480–482, 1974.

[415] J.C. Houbolt and G.G. Williamson. Spectral gust response for an airplane with vertical motion and pitch. Technical Report AFFDL-TR-75-121, Air Force Flight Dynamics Laboratory, October 1975.

[416] R.A. Houze. Chapter 11 - clouds and precipitation in extratropical cyclones. *International Geophysics*, 104:329–367, 2014.

[417] L.N. Howard. Note on a paper of John W. Miles. *Journal of Fluid Mechanics*, 10:509–512, 1961.

[418] J.G. Hudson. An instantaneous ccn spectrometer. *Journal of Atmospheric and Oceanic Technology*, 6(6):1055–1065, 1989.

[419] J.G. Hudson and P. Squires. An improved continuous flow diffusion cloud chamber. *Journal of Applied Meteorology and Climatology*, 15(7):776–782, 1976.

[420] R.M. Huffaker, A.V. Jelalian, and J.A.L. Thomson. Laser-doppler system for detection of aircraft trailing vortices. *Proceedings of the IEEE*, 58(3):322–326, 1970.

[421] R.E. Hufnagel and N.R. Stanley. Modulation transfer function associated with image transmission through turbulent media. *Journal of the Optical Society of America*, 54(1):52–61, 1964.

[422] M. Hughes, A. Hall, and R.G. Fovell. Blocking in areas of complex topography, and its influence on rainfall distribution. *Journal of the Atmospheric Sciences*, 66(2):508–518, 2009.

[423] W.J. Humphreys. Vertical temperature-gradients of the atmosphere, especially in the region of the upper inversion. *Astrophysical Journal*, 29:14–32, 1909.

[424] T. Iijima, T. Uemura, N. Matayoshi, J.O. Entzinger, J. Matsumoto, S. Ueda, and E. Yoshikawa. Development and evaluation of a new airspeed information system utilizing airborne doppler lidar. In *2017 IEEE/AIAA 36th Digital Avionics Systems Conference (DASC)*, pages 1–10. IEEE, 2017.

[425] SCHOTT Technical Information. Refractive index and dispersion, 2007.

[426] H. Inokuchi, M. Furuta, and T. Inagaki. High altitude turbulence detection using an airborne doppler lidar. In *29th Congress of the International Council of the Aeronautical Sciences (ICAS)*, pages 1–6. ICAS, September 2014. Paper number 0208.

[427] H. Inokuchi, H. Tanaka, and T. Ando. Development of an onboard doppler lidar for flight safety. *Journal of Aircraft*, 46(4):1411–1415, 2009.

[428] H. Inokuchi, H. Tanaka, and T. Ando. Development of a long range airborne doppler lidar. In *Proceedings of the 27th Congress of International Council of the Aeronautical Sciences*, pages 1–7. ICAS, 2010. Paper ICAS 2010-10.4.3.

[429] R. Inoue and W.D. Smyth. Efficiency of mixing forced by unsteady shear flow. *Journal of Physical Oceanography*, 39:1150–1166, 2009.

[430] E.A. Irvine, K.P. Shine, and M.A. Stringer. What are the implications of climate change for trans-atlantic aircraft routing and flight time? *Transportation Research Part D: Transport and Environment*, 47:44–53, 2016.

[431] A. Ishida. Two-wavelength displacement-measuring interferometer using second-harmonic light to eliminate air-turbulence-induced errors. *Japanese Journal of Applied Physics*, 28(3A):L473–L475, 1989.

[432] R. Israel and D.E. Rosner. Use of a generalized Stokes number to determine the aerodynamic capture efficiency of non-Stokesian particles from a compressible gas flow. *Aerosol Science and Technology*, 2(1):45–51, 1982.

[433] P. Jacquinot. The luminosity of spectrometers with prisms, gratings, or fabry-perot etalons. *Journal of the Optical Society of America*, 44(10):761–765, 1954.

[434] E.B. Jaeger and M. Sprenger. A northern hemispheric climatology of indices for clear air turbulence in the tropopause region derived from ERA40 reanalysis data. *Journal of Geophysical Research: Atmospheres*, 112(D20):D20106, 2007.

[435] J.T. Jayne, D.C. Leard, X. Zhang, P. Davidovits, K.A. Smith, C.E. Kolb, and D.R. Worsnop. Development of an aerosol mass spectrometer for size and composition analysis of submicron particles. *Aerosol Science and Technology*, 33(1-2):49–70, 2000.

[436] A. Jelalian. Review of atmospheric turbulence detection sensors. *AIAA Meeting Paper*, pages 417–439, 1971.

[437] H.W. Jentink and R.K. Bogue. Optical air flow measurements for flight tests and flight testing optical air flow meters. In *Flight Test – Sharing Knowledge and Experience. Meeting Proceedings RTO-MP-SCI-162*, pages 11–1–11–14. NATO/OTAN, 2005. Paper number 11.

[438] J. Jiménez. Small scale intermittency in turbulence. *European Journal of Mechanics - B/Fluids*, 17(4):405–419, 1998.

[439] J. Jiménez. Intermittency and cascades. *Journal of Fluid Mechanics*, 409:99–120, 2000.

[440] J. Jiménez. The contributions of A.N. Kolmogorov to the theory of turbulence. *Arbor*, 178:589–606, 2004.

[441] R.E. Jimenez. The use of a vertical scan technique in the radiometric detection of clear air turbulence. In *Final Report: FAA Symposium on Turbulence*, pages 53–53. Federal Aviation Administration, March 1971.

[442] E.N. Johnson, A.J. Calise, A.R. Tannenbaum, A.J. Yezzi, S. Soatto, G. Barbastathis, and N. Hovakimyan. Active-vision control systems for complex adversarial 3-D environments. Technical Report AFRL-SA-AR-TR-10-0338, Airforce Research Laboratory, Arlington, Virginia, March 2009.

[443] M.A. Johnson. A review of tropospheric scatter propagation theory and its application to experiment. *Proceedings of the IEEE—Part B: Radio and Electronic Engineering*, 105(8S):165–176, 1958.

[444] C.D. Jones, J.K. Hughes, N. Bellouin, S.C. Hardiman, G.S. Jones, J. Knight, S. Liddicoat, F.M. O'Connor, R.J. Andres, C. Bell, K.-O. Boo, A. Bozzo, N. Butchart, P. Cadule, K.D. Corbin, M. Doutriaux-Boucher, P. Friedlingstein, J. Gornall, L. Gray, P.R. Halloran, G. Hurtt, W.J. Ingram, J.-F. Lamarque, R.M. Law, M. Meinshausen, S. Osprey, E.J. Palin, L. Parsons Chini, T. Raddatz, A.A. Sanderson, M.G. Sellar, A. Schurer, P. Valdes, N. Wood, S. Woodward, M. Yoshioka, and M. Zerroukat. The HadGEM2-ES implementation of CMIP5 centennial simulations. *Geoscientific Model Development*, 4(3):543–570, 2011.

[445] W.P. Jones and B. Launder. The prediction of laminarization with a two-equation model of turbulence. *International Journal of Heat and Mass Transfer*, 15:301–314, 1972.

[446] P.W. Kadlec. Exploration of the relationship between atmospheric temperature change and clear air turbulence. Technical Report 660181, Society of Automotive Engineering (SAE) International, Washington, DC, February 1966.

[447] J.C. Kaimal and D.A. Haugen. An acoustic doppler sounder for measuring wind profiles in the lower boundary layer. *Journal of Applied Meteorology and Climatology*, 16(12):1298–1305, 1977.

[448] J.C. Kaimal, J.C. Wyngaard, Y. Izumi, and O.R. Coté. Spectral characteristics of surface-layer turbulence. *Quarterly Journal of the Royal Meteorological Society*, 98(417):563–589, 1972.

[449] E. Kalnay. *Atmospheric Modeling, Data Assimilation and Predictability.* Cambridge University Press, Cambridge, UK, 2002.

[450] E. Kalnay, M. Kanamitsu, R. Kistler, W. Collins, D. Deaven, L. Gandin, M. Iredell, S. Saha, G. White, J. Woollen, Y. Zhu, M. Chelliah, W. Ebisuzaki, W. Higgins, J. Janowiak, K.C. Mo, C. Ropelewski, J. Wang, A. Leetmaa, R. Reynolds, R. Jenne, and D. Joseph. The NCEP/NCAR 40-year reanalysis project. *Bulletin of the American Meteorological Society*, 77(3):437–472, 1996.

[451] J.A. Kalogiros and Q. Wang. Aerodynamic effects on wind turbulence measurements with research aircraft. *Journal of Atmospheric and Oceanic Technology*, 19(10):1567–1576, 2002.

[452] G.W. Kamerman. *Laser radar*, volume 6, chapter 1, pages 3–70. Infrared Information Analysis Center and SPIE Optical Engineering Press, 1993.

[453] S. Kameyama, T. Ando, K. Asaka, Y. Hirano, and S. Wadaka. Compact all-fiber pulsed coherent doppler lidar system for wind sensing. *Applied Optics*, 46(11):1953–1962, 2007.

[454] T. Kanitz, A. Ciapponi, A. Mondello, A. D'Ottavi, A. Baselga Mateo, A.-G. Straume, C. Voland, D. Bon, E. Checa, E. Alvarez, I. Bellucci, J. Pereira Do Carmo, J. Brewster, J. Marshall, M. Schillinger, M. Hannington, M. Rennie, O. Reitebuch, O. Lecrenier, P. Bravetti, V. Sacchieri, V. De Sanctis, A. Lefebvre, T. Parrinello, and D. Wernham. ESA's lidar missions aeolus and earthcare. *EPJ Web Conferences*, 237:01006, 2020.

[455] S.-K. Kao and J.B. Gebhard. An analysis of heat-, momentum-transports, and spectra for clear air turbulence in mid-stratosphere. *Pure and Applied Geophysics*, 88(1):180–185, 1971.

[456] L.D. Kaplan. Inference of atmospheric structure from remote radiation measurements. *Journal of the Optical Society of America*, 49(10):1004–1007, 1959.

[457] M.L. Kaplan, A.W. Huffman, K.M. Lux, J. Charney, A.J. Riordan, and Y.-L. Lin. Characterizing the severe turbulence environments associated with commercial aviation accidents. part 1: A 44-case study synoptic observational analysis. *Meteorology and Atmospheric Physics*, 88(3-4):129–152, 2005.

[458] M.L. Kaplan, K.M. Lux, J.D. Cetola, A.W. Huffman, A.J. Riordan, S.W. Slusser, Y.-L. Lin, J.J. Charney, and K.T. Waight. Characterizing the severe turbulence environments associated with commercial aviation accidents: A real-time turbulence model (RTTM) designed for the operational prediction of hazardous aviation turbulence environments. Technical Report NASA/CR-2004-213025, Langley Research Center, Hampton, Virginia 23681-2199, August 2004.

[459] K.B. Karnauskas, J.P. Donnelly, H.C. Barkley, and J.E. Martin. Coupling between air travel and climate. *Nature Climate Change*, 5(12):1068–1073, 2015.

[460] A.G. Kashani, M.J. Olsen, C.E. Parrish, and N. Wilson. A review of lidar radiometric processing: From ad hoc intensity correction to rigorous radiometric calibration. *Sensors*, 15(11):28099–28128, 2015.

[461] P. Kauffmann. The business case for turbulence sensing systems in the US air transport sector. *Journal of Air Transport Management*, 8(2):99–107, 2002.

[462] M.J. Kavaya and R.T. Menzies. Lidar aerosol backscatter measurements: Systematic, modeling, and calibration error considerations. *Applied Optics*, 24(21):3444–3453, 1985.

[463] Y. Kawatani, K. Hamilton, L.J. Gray, S.M. Osprey, S. Watanabe, and Y. Yamashita. The effects of a well-resolved stratosphere on the simulated boreal winter circulation in a climate model. *Journal of the Atmospheric Sciences*, 76(5):1203–1226, 2019.

[464] M. Keane, D. Buckton, M. Redfern, C. Bollig, C. Wedekind, F. Köpp, and F. Berni. Axial detection of aircraft wake vortices using doppler lidar. *Journal of Aircraft*, 39(5):850–861, 2002.

[465] R.J. Keeler, R.J. Serafin, R.L. Schwiesow, D.H. Lenschow, J.M. Vaughan, and A.A. Woodfield. An airborne laser air motion sensing system, Part I: Concept and preliminary experiment. *Journal of Atmospheric and Oceanic Technology*, 4(1):113–127, 1987.

[466] G. Keller. Relation between the structure of stellar shadow band patterns and stellar scintillation. *Journal of the Optical Society of America*, 45(10):845–851, 1955.

[467] G. Keller. Astronomical scintillation and atmospheric turbulence. Comments on several recent papers. *Astronomische Nachrichten*, 283(2-3):85–86, 1956.

[468] G. Keller and R.H. Hardie. Experimental verification of a recently proposed theory of astronomical seeing. *The Astronomical Journal*, 59(3):105–113, 1954.

[469] J.L. Keller. Clear air turbulence as a response to meso- and synoptic-scale dynamic processes. *Monthly Weather Review*, 118(10):2228–2243, 1990.

[470] W. Kencht, L. Murphy, and K. Smith. Sector congestion analytical modeling program (scamp) and the standard index of sector congestion (SISCO). Technical Report KSU HFEEL 99-G-020-4, Federal Aviation Administration (FAA), December 2000.

[471] W.R. Kencht. Current NEXRAD cannot reliably enable safe flight around heavy weather. *The International Journal of Aviation Psychology*, 26(1-2):46–61, 2016.

[472] G.S. Kent. High frequency fading observed on the 40 mc/s wave radiated from artificial satellite 1957α. *Journal of Atmospheric and Terrestrial Physics*, 16(1-2):10–20, 1959.

[473] G.S. Kent, B.R. Clemesha, and R.W. Wright. High altitude atmospheric scattering of light from a laser beam. *Journal of Atmospheric and Terrestrial Physics*, 29(2):169–181, 1967.

[474] G.S. Kent and S.K. Schaffner. Comparison of 1 μm satellite aerosol extinction with CO_2 lidar backscatter. In J.W. Bilbro and C. Werner, editors, *5th Conference on Coherent Laser Radar: Technology and Applications*, volume 1181, pages 252–259. SPIE, 1989.

[475] L. Kersley, D.B. Jenkins, and K.J. Edwards. Relative movements of mid-latitude trough and scintillation boundary. *Nature*, 239(88):11–11, 1972.

[476] L. Kersley, C.D. Russell, and D.L. Rice. Distribution of irregularities in the northern polar region determined from hilat observations. *Radio Science*, 25(2):115–124, 1990.

[477] L. Kersley, C.D. Russell, and D.L. Rice. Phase scintillations and irregularities in the northern polar ionosphere. *Radio Science*, 30(3):619–629, 1995.

[478] D. Keyser. Atmospheric fronts: An observational perspective. In P.S. Ray, editor, *Mesoscale Meteorology and Forecasting*, pages 216–258. American Meteorological Society, 1986.

[479] M. Khalid. Crosswise wind shear represented by a ramped velocity profile impacting a forward-moving aircraft. *International Journal of Aerospace Engineering*, 2019(7594737):1–18, 2019.

[480] R. Khatwa. An analysis of the operation and interpretation of weather radar by flight crews. In *60th Annual International Air Safety Seminar 2007: Sharing Global Safety Knowledge*, volume 2, pages 1121–1182. Flight Safety Foundation, October 2007.

[481] J. Kidston, A.A. Scaife, S.C. Hardiman, D.M. Mitchell, and N. Butchart. Stratospheric influence on tropospheric jet streams, storm tracks and surface weather. *Nature Geoscience*, 8:433–440, 2015.

[482] J.-H. Kim, W.N. Chan, B. Sridhar, and R.D. Sharman. Combined winds and turbulence prediction system for automated air-traffic management applications. *Journal of Applied Meteorology and Climatology*, 54(4):766–784, 2015.

[483] J.-H. Kim and H.-Y. Chun. A numerical study of clear-air turbulence (CAT) encounters over South Korea on 2 april 2007. *Journal of Applied Meteorology and Climatology*, 40(12):2381–2403, 2010.

[484] J.-H. Kim, R. Sharman, M. Strahan, J.W. Scheck, C. Bartholomew, J.C.H. Cheung, P. Buchanan, and N. Gait. Improvements in nonconvective aviation turbulence prediction for the world area forecast system. *Bulletin of the American Meteorological Society*, 99(11):2295–2311, 2018.

[485] S.-H. Kim and H.-Y. Chun. Aviation turbulence encounters detected from aircraft observations: spatiotemporal characteristics and applications to Korean aviation turbulence guidance. *Meteorological Applications*, 23(4):594–604, 2016.

[486] Y.-J. Kim, S.D. Eckermann, and H.-Y. Chun. An overview of the past, present and future of gravity-wave drag parametrization for numerical climate and weather prediction models. *Atmosphere-Ocean*, 41(1):65–98, 2003.

[487] G.P. Klaassen and W.R. Peltier. The evolution of finite-amplitude Kelvin-Helmholtz billows in two spatial dimensions. *Journal of the Atmospheric Sciences*, 227:1321–1339, 1985.

[488] D. Klingle-Wilson and J. Evans. Description of the corridor integrated weather system (CIWS). weather products. Technical Report ATC-317, Lincoln Laboratory, Massachusetts Institute of Technology, Lexington, Massachusetts, August 2005.

[489] J. Klostermeyer and R. Rüster. Further study of a jet stream-generated Kelvin-Helmholtz instability. *Journal of Geophysical Research: Oceans*, 86(C7):6631–6637, 1981.

[490] W.R. Knecht. Is rho the key to hazardous weather avoidance? In *19th International Symposium on Aviation Psychology*, pages 13–18. CORE Scholar, 2017.

[491] J.A. Knox, D.W. McCann, and P.D. Williams. Application of the Lighthill-Ford theory of spontaneous imbalance to clear-air turbulence forecasting. *Journal of the Atmospheric Sciences*, 65(10):3292–3304, 2008.

[492] S. Kobayashi, Y. Ota, Y. Harada, A. Ebita, M. Moriya, H. Onoda, K. Onogi, H. Kamahori, C. Kobayashi, H. Endo, K. Miyaoka, and K. Takahashi. The JRA-55 reanalysis: General specifications and basic characteristics. *Journal of the Meteorological Society of Japan*, 93(1):5–48, 2015.

[493] S.E. Koch and F. Caracena. Predicting clear-air turbulence from diagnosis of unbalanced flow. In *13th Conference of Applied Climatology and 10th Conference on Aviation, Range, and Aerospace Meteorology*, pages 1–5. American Meteorology Society, 2002.

[494] S.E. Koch, B.D. Jamison, C. Lu, T.L. Smith, E.I. Tollerud, C. Girz, N. Wang, T.P. Lane, M.A. Shapiro, D.D. Parrish, and O.R. Cooper. Turbulence and gravity waves within an upper-level front. *Journal of the Atmospheric Sciences*, 62(11):3885–3908, 2005.

[495] J.F. Kok, D.A. Ridley, Q. Zhou, R.L. Miller, C. Zhao, C.L. Heald, D.S. Ward, S. Albani, and K. Haustein. Smaller desert dust cooling effect estimated from analysis of dust size and abundance. *Nature Geoscience*, 10:274–278, 2017.

[496] A.N. Kolmogorov. Dissipation of energy in locally isotropic turbulence. *Doklady Akademii Nauk SSSR*, 32:19–21, 1941.

[497] A.N. Kolmogorov. The local structure of turbulence in incompressible viscous fluids at very large Reynolds' numbers. *Doklady Akademii Nauk SSSR*, 30:301–305, 1941.

[498] A.N. Kolmogorov. On the degeneration of isotropic turbulence in an incompressible viscous liquid. *Doklady Akademii Nauk SSSR*, 31:538–540, 1941.

[499] A.N. Kolmogorov. A refinement of previous hypotheses concerning the local structure of turbulence in a various incompressible fluid at high Reynolds number. *Journal of Fluid Mechanics*, 13:82–285, 1962.

[500] T. Kontogiannis and S. Malakis. *Cognitive Engineering and Safety Organization in Air Traffic Management*. CRC Press, Boca Raton, USA, 2017.

[501] F. Köpp. Doppler lidar investigations of wake vortex transport between closely separated parallel runways. *AIAA Journal*, 32(4):805–812, 1994.

[502] A. Korolev and G.A. Isaac. Shattering during sampling by OAPs and HVPS. Part I: Snow particles. *Journal of Atmospheric and Oceanic Technology*, 22(5):528–542, 2005.

[503] B. Kosović, S.E. Haupt, D. Adriaansen, S. Alessandrini, G. Wiener, L. Delle Monache, Y. Liu, S. Linden, T. Jensen, W. Cheng, M. Politovich, and P. Prestopnik. A comprehensive wind power forecasting system integrating artificial intelligence and numerical weather prediction. *Energies*, 13(6):1372, 2020.

[504] J.R. Koster and L.G. Grimes. Total electron measurements of the ionosphere using beacon satellites BEB(S66) and BEC. Technical Report AD0751517, Defense Technical Information Center, Fort Belvoir, Virginia, USA, June 1967.

[505] J.R. Koster and R.W. Wright. Scintillations, spread F and transequatorial scatter. *Journal of Geophysical Research*, 65(8):2303–2306, 1960.

[506] S.T. Kral, J. Reuder, T. Vihma, I. Suomi, E. O'Connor, R. Kouznetsov, B. Wrenger, A. Rautenberg, G. Urbancic, M.O. Jonassen, L. Båserud, B. Maronga, S. Mayer, T. Lorenz, A.A.M. Holtslag, G.-J. Steeneveld, A. Seidl, M. Müller, C. Lindenberg, C. Langohr, H. Voss, J. Bange, M. Hundhausen, P. Hilsheimer, and M. Schygulla. Innovative strategies for observations in the arctic atmospheric boundary layer (ISOBAR) – the Hailuoto 2017 campaign. *Atmosphere*, 9(7):268, 2018.

[507] R. Krasny. Vortex sheet computations: Roll-up, wakes, separation. In C. Anderson and C. Greengard, editors, *Vortex Dynamics and Vortex Methods*, volume 28, pages 385–402. American Mathematical Society, 1991.

[508] S.M. Kreidenweis, Y. Chen, D.C. Rogers, and P.J. DeMott. Isolating and identifying atmospheric ice-nucleating aerosols: a new technique. *Atmospheric Research*, 46(3-4):263–278, 1998.

[509] R. Krishnamurthy, A. Choukulkar, R. Calhoun, J. Fine, A. Oliver, and K.S. Barr. Coherent doppler lidar for wind farm characterization. *Wind Energy*, 16:189–206, 2012.

[510] G.W. Kronebach. An automated procedure for forecasting clear air turbulence. *Journal of Applied Meteorology and Climatology*, 3(2):119–125, 1964.

[511] P. Kuhn, F. Caracena, and C.M. Gillespie. Clear air turbulence: detection by infrared observations of water vapor. *Science*, 196(4294):1099–1100, 1977.

[512] Y.B. Kumar, V.S. Kumar, A.R. Jain, and P.B. Rao. MST radar and polarization lidar observations of tropical cirrus. *Annales Geophysicae*, 19(8):873–882, 2001.

[513] Y. Kumaran and N. Sumathi. Aviation meteorology. *International Journal of Latest Technology in Engineering, Management & Applied Science*, VI(VIIIS):90–94, 2017.

[514] M. Kumon, I. Mizumoto, Z. Iwai, and Nagata. Wind estimation by unmanned air vehicle with delta wing. In *Proceedings of the 2005 IEEE International Conference on Robotics and Automation*, pages 1896–1901. IEEE, April 2005.

[515] P.K. Kundu and I.M. Cohen. *Fluid Mechanics*. Elsevier Academic Press, Amsterdam, The Netherlands, 2004.

[516] D.F. Kurtulus. Introduction to micro air vehicles: concepts, design and applications. In R. Decuypere and M. Carbonaro, editors, *Recent Developments in Unmanned Aircraft Systems (UAS, Including UAV and MAV)*, pages 219–255. Von Karman Institute for Fluid Dynamics, 2011.

[517] R. Lagerquist, A. McGovern, and T. Smith. Machine learning for real-time prediction of damaging straight-line convective wind. *Weather and Forecasting*, 32(6):2175–2193, 2017.

[518] G.G. Lala and J.E. Jiusto. An automatic light scattering ccn counter. *Journal of Applied Meteorology and Climatology*, 16(4):413–418, 1977.

[519] J.-F. Lamarque and P.G. Hess. Cross-tropopause mass exchange and potential vorticity budget in a simulated tropopause folding. *Journal of the Atmospheric Sciences*, 51(15):2246–2269, 1994.

[520] U.H.W. Lammers. Imaging properties of monostatic and bistatic troposcatter radars. *Radio Science*, 8(2):89–101, 1973.

[521] L.D. Landau and E.M. Lifshitz. *Fluid Mechanics*. Pergamon Press, second edition, 1987.

[522] J.A. Lane. Radar echoes from clear air in relation to refractive-index variations in the troposphere. *Proceedings of the IEEE*, 116(10):1656–1660, 1969.

[523] T.P. Lane, J.D. Doyle, R. Plougonven, M.A. Shapiro, and R. D. Sharman. Observations and numerical simulations of inertia-gravity waves and shearing instabilities in the vicinity of a jet stream. *Journal of the Atmospheric Sciences*, 61(22):2692–2706, 2004.

[524] T.P. Lane, R.D. Sharman, S.B. Trier, R.G. Fovell, and J.K. Williams. Recent advances in the understanding of near-cloud turbulence. *Bulletin of the American Meteorological Society*, 93(4):499–515, 2012.

[525] A. Lapworth and S.R. Osborne. Gravity-wave drag in the stable boundary layer over moderate terrain. *Boundary-Layer Meteorology*, 171:175–189, 2019.

[526] T. Larrabee, H. Chao, M. Rhudy, Y. Gu, and M.R. Napolitano. Wind field estimation in uav formation flight. In *Proceedings of the 2014 American Control Conference*, pages 5408–5413. IEEE, July 2014.

[527] M.F. Larsen and J. Röttger. VHF and UHF doppler radars as tools for synoptic research. *Bulletin of the American Meteorological Society*, 63(9):996–1008, 1982.

[528] K.A. Latorella. Awin overview, 2003.

[529] B.E. Launder, G.J. Reece, and W. Rodi. Progress in the development of a Reynolds-stress turbulence closure. *Journal of Fluid Mechanics*, 68:537–566, 1975.

[530] B.E. Launder and D.B. Spalding. The numerical computation of turbulent flows. *Computer Methods in Applied Mechanics and Engineering*, 3:269–289, 1974.

[531] R.J. Laurence III and B.M. Argrow. Development and flight test results of a small UAS distributed flush airdata system. *Journal of Atmospheric and Oceanic Technology*, 35(5):1127–1140, 2018.

[532] J.-P. Laval, J.C. McWilliams, and B. Dubrulle. Forced stratified turbulence: Successive transitions with Reynolds number. *Physical Review E*, 68:036308, 2007.

[533] R.S. Lawrence, J.L. Jespersen, and R.C. Lamb. Amplitude and angular scintillations of the radio source cygnus-a observed at boulder, colorado. *Journal of Research of the National Bureau of Standards*, 65D(4):333–349, 1961.

[534] R.S. Lawrence, C.G. Little, and H.J.A. Chivers. A survey of ionospheric effects upon earth-space radio propagation. *Proceedings of the IEEE*, 52(1):4–27, 1964.

[535] J.D. Lawrence Jr., M.P. McCormick, S.H. Melfi, and D.P. Woodman. Laser backscatter correlation with turbulent regions of the atmosphere. *Applied Physics Letters*, 12(3):72–73, 1968.

[536] R.P. Lawson, D. O'Connor, P. Amarzly, K. Weaver, B. Baker, Q. Mo, and H. Jonsson. The 2D-S (stereo) probe: Design and preliminary tests of a new airborne, high-speed, high-resolution particle imaging probe. *Journal of Atmospheric and Oceanic Technology*, 23(11):1462–1477, 2006.

[537] A. Le Pichon, E. Blanc, D.P. Drob, S. Lambotte, J.X. Dessa, M. Lardy, P. Bani, and S. Vergniolle. Infrasound monitoring of volcanoes to probe high-altitude winds. *Journal of Geophysics Research: Atmospheres*, 110(D13):D13106, 2005.

[538] A. Le Pichon, E. Blanc, and A. Hauchecorne. *Infrasound Monitoring for Atmospheric Studies: Challenges in Middle Atmosphere Dynamics and Societal Benefits*. Springer, Berlin, 2018.

[539] A. Le Pichon, J. Vergoz, Y. Cansi, L. Ceranna, and D. Drob. Contribution of infrasound monitoring for atmospheric remote sensing. In A. Le Pichon, E. Blanc, and A. Hauchecorne, editors, *Infrasound Monitoring for Atmospheric Studies*, pages 629–646. Springer, 2010.

[540] D. Learmount. Runway excursions most frequent type of landing accident: Flight safety foundation, March, 14 2008.

[541] D.-B. Lee and H.-Y. Chun. A numerical study of aviation turbulence encountered on 13 February 2013 over the yellow sea between China and the Korean peninsula. *Journal of Applied Meteorology and Climatology*, 57(4):1043–1060, 2013.

[542] D.-H. Lee. Pseudo-label: The simple and efficient semi-supervised learning method for deep neural networks. In *ICML 2013 Workshop: Challenges in Representation Learning (WREPL)*, pages 896–901. ICML, 2013.

[543] D.N. Lee. Moving to make contact. *Ecological Psychology*, 26(1-2):47–59, 2014.

[544] D.N. Lee, R.J. Bootsma, M. Land, D. Regan, and R. Gray. General tau theory: Evolution to date. *Perception*, 38(6):837–858, 2009.

[545] L. Lee. *A Climatological Study of Clear Air Turbulence over the North Atlantic*. Department of Earth Sciences, Uppsala University, 2013.

[546] S.H. Lee, P.D. Williams, and T.H.A. Frame. Increased shear in the North Atlantic upper-level jet stream over the past four decades. *Nature*, 572(7771):639–642, 2019.

[547] M.A. LeMone. The structure and dynamics of horizontal vorticities in the planetary boundary layer. *Journal of the Atmospheric Sciences*, 30(6):1077–1091, 1973.

[548] K. Lemos and J. Chamberlain. In-flight weather trending information: Optimal looping characteristics for animated nexrad images. In *The 23rd Digital Avionics Systems Conference (IEEE Cat. No. 04CH37576)*, volume 1, pages 1015–1051. IEEE, 2004.

[549] D.H. Lenschow. Aircraft measurements in the boundary layer. In D.H. Lenschow, editor, *Probing the Atmospheric Boundary Layer*, volume 1, pages 39–53. American Meteorological Society, 1986.

[550] A. Leonard. Energy cascade in large-eddy simulations of turbulent fluid flows. In F.N. Frenkiel and R.E. Munn, editors, *Advances in Geophysics*, volume 18A, pages 237–248. Elsevier, 1974.

[551] M. Lesieur and O. Metais. New trends in large-eddy simulations of turbulence. *Annual Review of Fluid Mechanics*, 28:45–82, 1996.

[552] P.F. Lester. *Some physical and statistical aspects of clear air turbulence*. PhD thesis, Colorado State University, Fort Collins, Colorado, 1970.

[553] H.H. Lettau. Wind and temperature profile prediction for diabatic surface layers including strong inversion cases. *Boundary-Layer Meteorology*, 17(4):443–464, 1979.

[554] P.T. Leung. Coastal microstructure: From active overturn to fossil turbulence. *Journal of Cosmology*, 17(23):7612–7750, 2011.

[555] R. Lhermitte. Dual-doppler radar observation of convective storm circulation. In *Preprints of the 14th Conference on Radar Meteorology*, pages 139–144. American Meteorological Society, 1970.

[556] R.M. Lhermitte and M. Gilet. Dual-doppler radar observation and study of sea breeze convective storm development. *Journal of Applied Meteorology and Climatology*, 14(7):1346–1361, 1975.

[557] M.J. Lighthill. On sound generated aerodynamically I. general theory. *Proceedings of the Royal Society of London. Series A. Mathematical and Physical Sciences*, 211(1107):564–587, 1952.

[558] D.K. Lilly and P.J. Kennedy. Observations of a stationary mountain wave and its associated momentum flux and energy dissipation. *Journal of the Atmospheric Sciences*, 30(6):1135–1152, 1973.

[559] Y.-L. Lin. *Mesoscale Dynamics*. Cambridge University Press, 2007.

[560] R. Linkden and S. Burgmann. Laser-optical methods for transport studies in low temperature fuel cells. In C. Hartnig and C. Roth, editors, *Polymer Electrolyte Membrane and Direct Methanol Fuel Cell Technology: In Situ Characterization Techniques for Low Temperature Fuel Cells*, volume 2 of *Woodhead Publishing Series in Energy*, pages 425–461. Woodhead Publishing, 2012.

[561] A. Linnersjö, N. Hammar, B.-G. Dammström, M. Johansson, and H. Eliasch. Cancer incidence in airline cabin crew: experience from Sweden. *Occupational and Environmental Medicine*, 60(11):810–814, 2003.

[562] Flight Literacy. Cockpit weather systems (part one).

[563] Flight Literacy. Cockpit weather systems (part two).

[564] C.G. Little. A diffraction theory of the scintillation of stars on optical and radio wave-lengths. *Monthly Notices of the Royal Astronomical Society*, 111(3):289–302, 1951.

[565] C.G. Little. Acoustic methods for the remote probing of the lower atmosphere. *Proceedings of the IEEE*, 57(4):571–578, 1969.

[566] C.G. Little, W.M. Rayton, and R.B. Roof. Review of ionospheric effects at VHF and UHF. *Proceedings of the IRE*, 44(8):992–1018, 1956.

[567] P.S.K. Liu, W.R. Leaitch, J.W. Strapp, and M.A. Wasey. The response of the particle measuring systems airborne ASAP and PCASP to NaCl and latex particles. *Aerosol Science and Technology*, 16(2):83–95, 1992.

[568] Y. Liu, A. Mamtimin, W. Huo, X. Yang, X. Liu, F. Yang, and Q. He. Nondimensional wind and temperature profiles in the atmospheric surface layer over hinterland of the Taklimakan desert in China. *Advances in Meteorology*, 2016(9325953):1–8, 2016.

[569] R.L. Livingston. Comparison of multifrequency equatorial scintillation: American and Pacific sectors. *Radio Science*, 15(4):801–814, 1980.

[570] D.J. Lorenz and E.T. DeWeaver. Tropopause height and zonal wind response to global warming in the IPCC scenario integrations. *Journal of Geophysical Research: Atmospheres*, 112(D10):D10119, 2007.

[571] M. Lothon, D.H. Lenschow, D. Leon, and G. Vali. Turbulence measurements in marine stratocumulus with airborne doppler radar. *Quarterly Journal of the Royal Meteorological Society*, 131(609):263–280, 2021.

[572] F. Lott, B. Deremble, and C. Soufflet. Mountain waves produced by a stratified boundary layer flow. Part I: Hydrostatic case. *Journal of the Atmospheric Sciences*, 77(5):1683–1697, 2020.

[573] F. Lott, B. Deremble, and C. Soufflet. Mountain waves produced by a stratified shear flow with a boundary layer. Part II: Form drag, wave drag, and transition from downstream sheltering to upstream blocking. *Journal of the Atmospheric Sciences*, 78(4):1101–1112, 2021.

[574] H. Lu and F. Porté-Agel. Large-eddy simulation of a very large wind farm in a stable atmospheric boundary layer. *Physics of Fluids*, 6:065101, 2011.

[575] H. Luce, T. Mega, M.K. Yamamoto, M. Yamamoto, H. Hashiguchi, S. Fukao, N. Nishi, T. Tajiri, and M. Nakazato. Observations of Kelvin-Helmoholtz instability at a cloud base with the middle and upper atmosphere (mu) and weather radars. *Journal of Geophysical Research: Atmospheres*, 115(D19):D19116, 2010.

[576] F.H. Ludlam. Characteristics of billow clouds and their relation to clear-air turbulence. *Quarterly Journal of the Royal Meteorological Society*, 93(398):419–435, 1967.

[577] J.L. Lumley. Coherent structures in turbulence. In R.E. Meyer, editor, *Transition and Turbulence*, pages 215–242. Academic Press, 1981.

[578] B. Lunnon. Turbulence events interpreted by vortex rolls. In R. Sharman and T. Lane, editors, *Aviation turbulence*, pages 83–94. Springer, 2016.

[579] T. Ma and S. Wang. Rayleigh-Bénard convection: Dynamics and structure in the physical space. *Communications in Mathematical Sciences*, 5(3):553–574, 2007.

[580] P.B. MacCready. The inertial subrange of atmospheric turbulence. *Journal of Geophysical Research*, 67(3):1051–1059, 1962.

[581] P.B. MacCready. Standardization of gustiness values from aircraft. *Journal of Applied Meteorology and Climatology*, 3(4):439–449, 1964.

[582] A. Mahalov, M. Moustaoui, and B. Nichols. Characterization of stratospheric clear air turbulence for air force platforms. In *2006 HPCMP Users Group Conference*, pages 296–302. IEEE, 2006.

[583] A.R. Mahoney, L.G. McAllister, and J.R. Pollard. The remote sensing of wind velocity in the lower troposphere using an acoustic sounder. *Boundary-Layer Meteorology*, 4(1-4):155–167, 1973.

[584] C. Malandri, L. Mantecchini, and M.N. Pastorino. Airport ground access reliability and resilience of transit network: a case study. *Transportation Research Procedia*, 27:1129–1136, 2017.

[585] C. Malandri, L. Mantecchini, and V. Reis. Aircraft turnaround and industrial actions: How ground handlers' strikes affect airport airside operational efficiency. *Journal of Air Transport Management*, 78:23–32, 2019.

[586] S. Mallick and D. Malacara. *Common-path interferometers*, chapter 3. Wiley, third edition, 2006.

[587] R.L. Mancuso and R.M. Endlich. Clear air turbulence frequency as a function of wind shear and deformation. *Monthly Weather Review*, 94(9):581–585, 1966.

[588] J. Mandle and E. Fabacher. Laser anemometry for flight test. In *Proceedings ETTC '91*, pages 3–4, June 1991.

[589] K. Manomaiphiboon and A.G. Russell. Formulation of joint probability density functions of velocity for turbulent flows: An alternative approach. *Atmospheric Environment*, 37:4917–4925, 2003.

[590] D. Marconnet, C. Norden, and L. Vidal. Optimum use of weather radar. *Safety First*, 22:1–23, 2016.

[591] M. Marino, A. Fisher, R. Clothier, S. Watkins, S. Prudden, and C.S. Leung. An evaluation of multi-rotor unmanned aircraft as flying wind sensors. *International Journal of Micro Air Vehicles*, 7(3):285–299, 2015.

[592] W.D. Mark and R.W. Fischer. Investigation of the effects of nonhomogeneous (or nonstationary) behavior on the spectra of atmospheric turbulence. Technical Report NASA-CR-2745, NASA, October 1976.

[593] T. Markus, T. Neumann, A. Martino, W. Abdalati, K. Brunt, B. Csatho, S. Farrell, H. Fricker, A. Gardner, D. Harding, M. Jasinski, R. Kwok, L. Magruder, D. Lubin, S. Luthcke, J. Morison, R. Nelson, A. Neuenschwander, S. Palm, S. Popescu, C.K. Shum, B.E. Schutz, B. Smith, Y. Yang, and J. Zwally. The ice, cloud, and land elevation satellite-2 (ICESAT-2): Science requirements, concept, and implementation. *Remote Sensing of Environment*, 190:260–273, 2017.

[594] S. Marlatt, S. Waggy, and S. Biringen. Direct numerical simulation of the turbulent Ekman layer: Evaluation of closure models. *Journal of the Atmospheric Sciences*, 69:1106–1117, 2012.

[595] J. Marshall, R. Langille, and W.M.K. Palmer. Measurement of rainfall by radar. *Journal of the Atmospheric Sciences*, 4(6):186–192, 1947.

[596] J. Marshall and W.M.K. Palmer. The distribution of raindrops with size. *Journal of the Atmospheric Sciences*, 5(4):165–166, 1948.

[597] J. Marshall and R.A. Plumb. *Atmosphere, Ocean, and Climate Dynamics*. Academic Press, 1982.

[598] S. Martin, J. Bange, and F. Beyrich. Meteorological profiling of the lower troposphere using the research UAV "M^{2}AV Carolo". *Atmospheric Measurement Techniques*, 4(4):705–716, 2011.

[599] J. Maruhashi, P. Serrão, and M. Belo-Perreira. Analysis of mountain wave effects on a hard landing incident in pico aerodrome using the arome model and airborne observations. *Atmosphere*, 10(7):350, 2019.

[600] N. Matayoshi, T. Iijima, E. Yoshikawa, J.O. Entzinger, T. Uemura, T. Akiyama, and H. Inokuchi. Development and flight demonstration of a new lidar-based onboard turbulence information system. In *31st Congress of the International Council of the Aeronautical Sciences (ICAS)*, pages 1–10. ICAS, September 2018. Paper number 0284.

[601] J. Mawdsley and I.R. Richards. Ionospheric scattering of satellite transmissions. *Nature*, 189(4768):806–907, 1961.

[602] J. Mazon, J.I. Rojas, M. Lozano, D. Pino, X. Prats, and M.M. Miglietta. Influence of meteorological phenomena on worldwide aircraft accidents, 1967-2010. *Meteorological Applications*, 25(2):236–245, 2018.

[603] L.G. McAllister, J.R. Pollard, A.R. Mahoney, and P.J.R. Shaw. Acoustic sounding – a new approach to the study of atmospheric structure. *Proceedings of the IEEE*, 57(4):579–587, 1969.

[604] M.S. McBeth and R.F. Jones. A compact interferometric radar for studying clear air turbulence. In *2017 IEEE Radar Conference (RadarConf)*, pages 0233–0237. IEEE, 2017.

[605] D.W. McCann. Gravity waves, unbalanced flows, and clear air turbulence. *National Weather Digest*, 25(1-2):3–14, 2001.

[606] D.W. McCann, J.A. Knox, and P.D. Williams. An improvement in clear-air turbulence forecasting based on spontaneous imbalance theory: the UL-TURB algorithm. *Meteorological Applications*, 19:71–78, 2012.

[607] W.D. McComb. *The Physics of Fluid Turbulence*. Oxford University Press, 1990.

[608] W.D. McComb and M.Q. May. The effect of Kolmogorov (1962) scaling on the universality of turbulence energy spectra. *arXiv:1812.09174v1 [physics.flu-dyn]*, pages 1–11, 2018.

[609] N.A. McFarlane. The effect of orographically excited gravity wave drag on the general circulation of the lower stratosphere and troposphere. *Journal of Atmospheric Sciences*, 44(14):1775–1800, 1987.

[610] R.L. McGann. Flight test results from a low-power doppler optical air data sensor. In *Proceedings of SPIE 2464, Air Traffic Control Technologies*, pages 116–124. SPIE, June 1995.

[611] A. McGovern, K.L. Elmore, D.J. Gagne II, S.E. Haupt, C.D. Karstens, R. Lagerquist, T. Smith, and J.K. Williams. Using artificial intelligence to improve real-time decision-making for high-impact weather. *Bulletin of the American Meteorological Society*, 98(10):2073–2090, 2017.

[612] P.H. McMurry. The history of condensation nucleus counters. *Aerosol Science and Technology*, 33(4):297–322, 2000.

[613] E. McNeely, I. Mordukhovich, S. Staffa, S. Tideman, S. Gale, and B. Coull. Cancer prevalence among flight attendants compared to the general population. *Environmental Health*, 17(49):1–9, 2018.

[614] D. McRuer, I. Ashkenas, and D. Graham. *Aircraft Dynamics and Automatic Control*. Princeton University Press, Princeton, NJ, 1973.

[615] E.C.S. Megaw. Fundamental radio scatter propagation theory. *Proceedings of the IEEE—Part C: Monographs*, 104(6):441–455, 1957.

[616] D. Mehta, Y. Zhang, A. Van Zuijlen, and H. Bijl. Large eddy simulation with energy-conserving schemes and the smagorinsky model: A note on accuracy and computational efficiency. *Energies*, 12:129, 2019.

[617] J.P. Mellado, C.S. Bretherton, B. Stevens, and M.C. Wyant. DNS and LES for simulating stratocumulus: Better together. *Journal of Advances in Modeling Earth Systems*, 10:1421–1438, 2018.

[618] M.Q. Menchaca and D.R. Durran. Mountain waves, downslope winds, and low-level blocking forced by a midlatitude cyclone encountering an isolated ridge. *Journal of the Atmospheric Sciences*, 74(2):617–639, 2017.

[619] J.J. Metz and D.R. Durran. Are finite-amplitude effects important in non-breaking mountain waves? *Quarterly Journal of the Royal Meteorological Society*, 2021. submitted.

[620] S. Metzger, W. Junkermann, K. Butterbach-Bahl, H.P. Schmid, and T. Foken. Corrigendum to "measuring the 3-D wind vector with a weight-shift microlight aircraft" published in atmos. meas. tech., 4, 1421–1444,2011. *Atmospheric Measurement Techniques*, 4(7):1515–1539, 2011.

[621] J.R. Meyer-Arendt and C.B. Emmanuel. Optical scintillation; a survey of the literature. Technical Report 225, U.S. Department of Commerce, National Bureau of Standards (NBS), Gaithersburg, Maryland, USA, April 1965.

[622] R.C. Michelson. Overview of micro air vehicle system design and integration issues. *Encyclopedia of Aerospace Engineering*, pages 1–12, 2010.

[623] G.V. Middleton. Sediment deposition from turbidity currents. *Annual Review of Earth and Planetary Sciences*, 21:89–114, 1993.

[624] G. Mie. Beiträge zur optik trüber medien, speziell kolloidaler metallösungen. *Annalen der Physik*, 330(3):377–445, 1908.

[625] A.H. Mikesell. The scintillation of starlight. *Publications of the United States Naval Observatory*, 17:130–191, 1955.

[626] A.H. Mikesell. Star visibility in daylight at high altitudes. *Journal of the Optical Society of America*, 50(1):85–85, 1960.

[627] A.H. Mikesell, A.A. Hoag, and J.S. Hall. The scintillation of starlight. *Journal of the Optical Society of America*, 41(10):689–695, 1951.

[628] J.W. Miles. Richardson's criterion for the stability of stratified shear flow. *Physics of Fluids*, 29:3470–3471, 1986.

[629] P. Misme, F. du Castel, A. Spizzichino, and J. Voge. Réflexions partielles dans l'atmosphère et propagation à grande distance. Cinquième partie: Réflexion et diffusion atmosphériques des ondes radioélectriques. *Annales des Télécommunications*, 15(5-6):107–121, 1960.

[630] D.M. Mitchell, L.J. Gray, J. Anstey, M.P. Baldwin, and A.J. Charlton-Perez. The influence of stratospheric vortex displacements and splits on surface climate. *Journal of Climate*, 26(8):2668–2682, 2013.

[631] F.A. Mitchell and D.T. Prophet. Meteorological analysis of clear air turbulence in the stratosphere. In Y.-H. Pao and A. Goldburg, editors, *Clear Air Turbulence and its Detection*, pages 144–182. Springer, 1969.

[632] W.F. Moler and D.B. Holden. Tropospheric scatter propagation and atmospheric circulations. *Journal of Research of the National Bureau of Standards*, 64D(1):61–64, 1960.

[633] T.W. Monarski, S.M. Hannon, and P. Gatt. Eye-safe coherent lidar detection using a 1.5-μm raman laser. In G.W. Kammerman, editor, *Laser Radar Technology and Applications VI*, volume 4377, pages 229–236. SPIE, September 2001.

[634] A.S. Monin. Characteristics of the scattering of sound in a turbulent atmosphere. *Soviet Physics Acoustics*, 7(4):370–373, 1962.

[635] W.R. Moninger, S.G. Benjamin, B.D. Jamison, T.W. Schlatter, T.L. Smith, and E.J. Szoke. Evaluation of regional aircraft observations using TAMDAR. *Weather and Forecasting*, 25(2):627–645, 2010.

[636] W.R. Moninger, R.D. Mamrosh, and P.M. Pauley. Automated meteorological reports from commercial aircraft. *Bulletin of the American Meteorological Society*, 84(2):203–216, 2003.

[637] A.J. Montgomery and A. Weigandt. Study of passive optical techniques for detecting clear air turbulence. Technical Report CR-86161, IIT Research Institute, Chicago, Illinois, 1968.

[638] R.H. Moore and A. Nenes. A scanning flow CCN analysis—a method for fast measurements of CCN spectra. *Aerosol Science and Technology*, 43(12):1192–1207, 2009.

[639] H. Moosmüller, R.K. Chakrabarty, and W.P. Arnott. Aerosol light absorption and its instrument: A review. *Journal of Quantitative Spectroscopy & Radiative Transfer*, 110(11):844–878, 2009.

[640] B. Morbieu and J. Mandle. True airspeed measurement with a coherent laser radar. In *3rd International Conference on Advanced Infrared Detectors & Systems*, pages 154–157, London, UK, June 1986.

[641] W.T. Morgan, J.D. Allan, K.N. Bower, G. Capes, J. Crosier, P.I. Williams, and H. Coe. Vertical distribution of sub-micron aerosol chemical composition from north-western Europe and the North-East Atlantic. *Atmospheric Chemistry and Physics*, 9:5389–5401, 2009.

[642] M. Moriche, O. Flores, and M. García-Villalba. On the aerodynamic forces on heaving and pitching airfoils at low Reynolds number. *Journal of Fluid Mechanics*, 828:395–423, 2017.

[643] N. Moteki, Y. Kondo, Y. Miyazaki, N. Takegawa, Y. Komazaki, G. Kurata, T. Shirai, D.R. Blake, T. Miyakawa, and M. Koike. Evolution of mixing state of black carbon particles: Aircraft measurements over the western Pacific in March 2004. *Geophysics Research Letters*, 34(11):L11803, 2007.

[644] J.N. Moum, J.D. Nash, and W.D. Smyth. Narrowband oscillations in the upper equatorial ocean. Part I: Interpretation as shear instabilities. *Journal of Physical Oceanography*, 41:397–411, 2011.

[645] J. Moyano Cano. Quadrotor uav for wind profile characterization, July 2013.

[646] D. Muñoz Esparza and R. Sharman. An improved algorithm for low-level turbulence forecasting. *Journal of Applied Meteorology and Climatology*, 57(6):1249–1263, 2018.

[647] D. Muñoz Esparza, R.D. Sharman, and W. Deierling. Aviation turbulence forecasting at upper levels with machine learning techniques based on regression trees. *Journal of Applied Meteorology and Climatology*, 59(11):1883–1899, 2020.

[648] R.B. Muchmore and A.D. Wheelon. Frequency correlation of line-of-sight signal scintillations. *IEEE Transactions on Antennas and Propagation*, 11(1):46–51, 1963.

[649] S.J. Munchak. Remote sensing of precipitation from airborne and spaceborne radar. In T. Islam, Y. Hu, A. Kokhanovsky, and J. Wang, editors, *Remote Sensing of Aerosols, Clouds, and Precipitation*, chapter 13, pages 267–299. Elsevier, 2018.

[650] R.J. Munick. Turbulent backscatter of light. *Journal of the Optical Society of America*, 55(7):893–893, 1965.

[651] D.M. Murphy. The design of single particle mass spectrometers. *Mass Spectrometry Reviews*, 26(2):150–165, 2007.

[652] D.M. Murphy, D.J. Cziczo, K.D. Froyd, P.K. Hudson, B.M. Matthew, A.M. Middlebrook, R.E. Peltier, A. Sullivan, D.S. Thomson, and R.J. Weber. Single-particle mass spectrometry of tropospheric aerosol particles. *Journal of Geophysical Research: Atmospheres*, 111(D23):D23104, 2006.

[653] G.D. Nastrom and K.S. Gage. A brief climatology of vertical wind variability in the troposphere and stratosphere as seen by the Poker Flat, Alaska, MST radar. *Journal of Applied Meteorology and Climatology*, 23(3):453–460, 1984.

[654] G.D. Nastrom and K.S. Gage. A climatology of atmospheric wavenumber spectra of wind and temperature observed by commercial aircraft. *Journal of the Atmospheric Sciences*, 42(9):950–960, 1985.

[655] G.D. Nastrom, K.S. Gage, and B.B. Balsley. Variability of Cn2 at Poker Flat, Alaska, from mesosphere, stratosphere, troposphere (MST) Doppler radar observations. *Optical Engineering*, 21(2):212347, 1982.

[656] A.H. Nayfeh and W.S. Saric. Nonlinear waves in a Kelvin-Helmholtz flow. *Journal of Fluid Mechanics*, 55:311–327, 1972.

[657] P.J. Neiman, M.A. Shapiro, and L.S. Fedor. The life cycle of an extratropical marine cyclone. Part II: Mesoscale structure and diagnostics. *Monthly Weather Review*, 121(8):2177–2199, 1993.

[658] B. Neininger, W. Fuchs, M. Baeumle, A. Volz-Thomas, A.S.H. Prévôt, and J. Dommen. A small aircraft for more than just ozone: Merair's 'dimona' after ten years of evolving development. In *Proceedings of the 11th Symposium on Meteorological Observations and Instrumentation, 81st AMS Annual Meeting*, pages 1–6. American Meteorological Society, 2001.

[659] P. Neumann, M. Bartholmai, J.H. Schiller, B. Wiggerich, and M. Manolov. Micro-drone for the characterization and self-optimizing search of hazardous gaseous substance sources: A new approach to determine wind speed and direction. In *Proceedings of the 2010 IEEE International Workshop on Robotic and Sensors Environments (ROSE 2010)*, pages 1–6. IEEE, December 2010.

[660] P.P. Neumann and M. Bartholmai. Real-time wind estimation on a micro unmanned aerial vehicle using its inertial measurement unit. *Sensors and Actuators A: Physical*, 235:300–310, 2015.

[661] NextGen. Update to the business case for the next generation air transportation system based on the future of the NAS report, 2016.

[662] Next Generation Air Transportation System (NextGen). This is NextGen, 2021.

[663] E.G. Njoku and O.B. Brown. Sea surface temperature. In R.J. Gurney, J.L. Foster, and C.L. Parkinson, editors, *Atlas of Satellite Observations Related to Global Change*, pages 237–249. Cambridge University Press, 1993.

[664] T.E. Noll, J.M. Brown, M.E. Perez-Davis, S.D. Ishmael, G.G. Tiffany, and M. Gaier. *Investigation of the Helios Prototype Aircraft Mishap*, volume I. Createspace Independent Publishing Platform, 2012. Mishap Report.

[665] R.J. Noriega-Manez and J.C. Gutiérrez-Vega. Rytov theory for Helmholtz-Gauss beams in turbulent atmosphere. *Optics Express*, 15(25):16328–16341, 2007.

[666] K.A. Norton. Carrier-frequency dependence of the basic transmission loss in tropospheric forward scatter propagation. *Journal of Geophysical Research*, 65(7):2029–2045, 1960.

[667] K.A. Norton, J.W. Herbstreit, H.B. Janes, K.O. Hornberg, C.F. Peterson, A.F. Barghausen, W.E. Johnson, P.I. Wells, M.C. Thompson, M.J. Vetter, and A.W. Kirkpatrick. An experimental study of phase variations in line-of-sight microwave transmissions. In *NBS Monograph 33*. U. S. Department of Commerce, National Bureau of Standards, 1961.

[668] C.B. Notess. The effects of atmospheric turbulence upon flight at low altitude and high speed. Technical Report FDM-325, Cornell Aeronautical Laboratory, Inc., Buffalo, New York, October 1961.

[669] D.R. Novak, L.F. Bosart, D. Keyser, and J.S. Waldstreicher. An observational study of cold season-banded precipitation in Northeast U.S. cyclones. *Weather and Forecasting*, 19(6):993–1010, 2004.

[670] W. Nupen. Bibliography on atmospheric aspects of radio astronomy. Technical Report 171, U.S. Department of Commerce, National Bureau of Standards (NBS), Gaithersburg, Maryland, USA, May 1963.

[671] K. O'Brien, W. Friedberg, H.H. Sauer, and D.F. Smart. Atmospheric cosmic rays and solar energetic particles at aircraft altitudes. *Environment International*, 22(Supplement 1):9–44, 1996.

[672] G.R. Ochs, T.-I. Wang, R.S. Lawrence, and S.F. Clifford. Refractive-turbulence profiles measured by one-dimensional spatial filtering of scintillations. *Applied Optics*, 15(10):2504–2510, 1976.

[673] Civil Aviation Authority of New Zealand. Gap: Wake turbulence. Booklet, 2008.

[674] P.A. O'Gorman and J.G. Dwyer. Using machine learning to parameterize moist convection: Potential for modeling of climate, climate change, and extreme events. *Journal of Advances in Modeling Earth Systems*, 10(10):2548–2563, 2018.

[675] L. Ong and J.M. Wallace. Joint probability density analysis of the structure and dynamics of the vorticity field of a turbulent boundary layer. *Journal of Fluid Mechanics*, 367:291–328, 1998.

[676] L. Onsager. The distribution of energy in turbulence. *Physical Review*, 68:281–288, 1945.

[677] A. Orr, P. Bechtold, J. Scinocca, M. Ern, and M. Janiskova. Improved middle atmosphere climate and forecasts in the ecmwf model through a nonorographic gravity wave drag parameterization. *Journal of Climate*, 23(22):5905–5926, 2010.

[678] H. Ottersten. Atmospheric structure and radar backscattering in clear air. *Radio Science*, 4(12):1179–1193, 1969.

[679] H. Ottersten. Radar backscattering from the turbulent clear atmosphere. *Radio Science*, 4(12):1251–1255, 1969.

[680] H. Ottersten. Radar monitoring of turbulence at the tropopause level. Technical Report ERL 243-WPL 21, National Oceanic and Atmospheric Administration, Boulder, Colorado, 1972.

[681] H. Ottersten, K.R. Hardy, and C.G. Little. Radar and sodar probing of waves and turbulence in statically stable clear-air layers. *Boundary-Layer Meteorology*, 4(1-4):47–89, 1973.

[682] A. Overeem. Verification of clear-air turbulence forecasts. Technical Report TR-224, Koninklijk Nederlands Meteorologisch Instituut, KNMI, De Bilt, Utrecht, June 2002.

[683] E. Palmén and C.W. Newton, editors. *Principal tropospheric jet streams*, volume 13 of *Atmospheric Circulation Systems. Their Structure and Physical Interpretation. International Geophysics.*, chapter 8, pages 195–236. Elsevier, 1969.

[684] T.N. Palmer, G.J. Shutts, and R. Swinbank. Alleviation of a systematic westerly bias in general circulation and numerical weather prediction models through an orographic gravity wave drag parameterization. *Quarterly Journal of the Royal Meteorological Society*, 112(474):1001–1039, 1986.

[685] R.T. Palomaki, N.T. Rose, M. van den Bossche, T.J. Sherman, and S.F.J. De Wekker. Wind estimation in the lower atmosphere using multirotor aircraft. *Journal of Atmospheric and Oceanic Technology*, 34(5):1183–1191, 2017.

[686] A. Papoulis. Minimum-bias windows for high-resolution spectral estimates. *IEEE Transactions on Information Theory*, 19(1):9–12, 1973.

[687] I.A. Parkin. The effects of field-aligned ionospheric irregularities on satellite scintillations. *Journal of Atmospheric and Terrestrial Physics*, 30(6):1135–1142, 1968.

[688] E.K. Parks, R.C. Wingrove, R.E. Bach, and R.S. Mehta. Identification of vortex-induced clear air turbulence using airline flight records. *Journal of Aircraft*, 22(2):124–129, 1985.

[689] G. Parry and J.G. Walker. Statistics of stellar scintillation. *Journal of the Optical Society of America*, 70(9):1157–1159, 1980.

[690] G. Parry, J.G. Walker, and R.J. Scaddan. On the statistics of stellar speckle patterns and pupil plane scintillation. *Optica Acta: International Journal of Optics*, 26(5):563–574, 1979.

[691] J.E. Passner, K. Stephen, and T. Jameson. Using real-time weather data from an unmanned aircraft system to support the advanced research version of the weather research and forecast model. Technical Report ARL-TR-5950, U.S. Army Research Laboratory, Adelphi, MD 20783-1197, April 2012.

[692] M.J. Patil, D.H. Hodges, and C.E.S. Cesnik. Limit-cycle oscillations of high-aspect-ratio wings. *Journal of Fluids and Structures*, 15(1):107–132, 2001.

[693] J.M. Pearson and R.D. Sharman. Prediction of energy dissipation rates for aviation turbulence. Part II: Nowcasting convective and nonconvective turbulence. *Journal of Applied Meteorology and Climatology*, 56(2):339–351, 2017.

[694] N. Pedatella, J. Chau, H. Schmidt, L. Goncharenko, C. Stolle, K. Hocke, V. Harvey, B. Funke, and T. Siddiqui. How sudden stratosphere warming affects the whole atmosphere. *Eos*, 99:35–38, 2018.

[695] J. Pelon, G. Vali, G. Ancellet, G. Ehret, P.H. Flamant, S. Haimov, G. Heymsfield, D. Leon, J.B. Mead, A.L. Pazmany, A. Protat, Z. Wang, and M. Wolde. *LIDAR and RADAR observations*, chapter 9, pages 457–526. Wiley Series in Atmospheric Physics and Remote Sensing. Wiley-VCH, 2013.

[696] W.R. Peltier and T.L. Clark. The evolution and stability of finite-amplitude mountain waves. Part II: Surface wave drag and severe downslope windstorms. *Journal of the Atmospheric Sciences*, 36(8):1498–1529, 1979.

[697] J.M. Pernter and F.M. Exner. *Meteorologische Optik*. Salzwasser-Verlag GmbH, 1922.

[698] M. Petitdidier, P.B. Chilson, and C.W. Ulbrich. Observations of gravity waves and turbulence associated with a tropical thunderstorm. In *Proceedings of the 1st SPARC General Assembly*, volume 2, pages 401–404. World Meterorological Organization Publications, 1997.

[699] L. Petricca, P. Ohlckers, and C. Grinde. Micro- and nano-air vehicles: State of the art. *International Journal of Aerospace Engineering*, 2011(214549):1–17, 2011.

[700] B. Pettergrew, A.F. Loughe, J.E. Hart, J.L. Mahoney, J.K. Henderson, and J.E. Hart. On the use of eddy dissipation rate observations in verification. In *Proceedings of the 14th Conference on Aviation, Range, and Aerospace Meteorology*, number Paper P473. American Meteorological Society, 2010.

[701] S. Petterssen. Contribution to the theory of frontogenesis. *Geofysiske Plublikasjoner*, 11(6):1–27, 1936.

[702] H.T. Pham, S. Sarkar, and K.A. Brucker. Dynamics of a stratified shear layer above a region of uniform stratification. *Journal of Fluid Mechanics*, 630:191–223, 2009.

[703] PhisOrg. New instrument could detect hidden aviation hazards, 2009.

[704] N.Z. Pinus, E.R. Reiter, G.N. Shur, and N.K. Vinnichenko. Power spectra of turbulence in the free atmosphere. *Tellus*, 19(2):206–213, 1966.

[705] U. Piomelli and J.R. Chasnov. Large-eddy simulations: Theory and applications. In A.V. Johansson and P.H. Alfredsson, editors, *Turbulence and Transition Modelling. ERCOFTAC Series*, volume 2, pages 269–336. Springer, 1996.

[706] A. Platis, B. Altstädter, B. Wehner, N. Wildmann, A. Lampert, M. Hermann, W. Birmili, and J. Bange. An observational case study on the influence of atmospheric boundary-layer dynamics on new particle formation. *Boundary-Layer Meteorology*, 158:67–92, 2016.

[707] C.M.R. Platt. Lidar — backscatter. In J.R. Holton, J.A. Curry, and J.A. Pyle, editors, *Encyclopedia of Atmospheric Sciences*, pages 1176–1183. Elsevier, 2003.

[708] C.M.R. Platt, S.A. Young, P.J. Manson, G.R. Patterson, S.C. Marsden, R.T. Austin, and J.H. Churnside. The optical properties of equatorial cirrus from observations in the arm pilot radiation observation experiment. *Journal of the Atmospheric Sciences*, 55(11):1977–1996, 1998.

[709] R. Plougonven, C. Snyder, and Zhang F. Comments on "application of the Lighthill-Ford theory of spontaneous imbalance to clear-air turbulence forecasting". *Journal of the Atmospheric Sciences*, 66(8):2506–2510, 2009.

[710] R.W. Pohl. *Einführung in die Optik*. Springer, Berlin, 1948.

[711] S. Polavarapu, T.G. Shepherd, Y. Rochon, and S. Ren. Some challenges of middle atmosphere data assimilation. *Quarterly Journal of the Royal Meteorological Society*, 131(613):3513–3527, 2005.

[712] J. Polívka. Microwave radiometry and applications. *International Journal of Infrared and Millimeter Waves*, 16(9):1593–1672, 1995.

[713] S.B. Pope. Transport equation for the joint probability density function of velocity and scalars in turbulent flow. *The Physics of Fluids*, 24:588–596, 1981.

[714] S.B. Pope. *The statistical description of turbulent flows*. Cambridge University Press, 2000.

[715] S.B. Pope. *Turbulent Flows*. Cambridge University Press, 2000.

[716] S.B. Pope. Ten questions concerning the large eddy simulation of turbulent flows. *New Journal of Physics*, 4:1–24, 2004.

[717] A.L. Porta, G.A. Voth, A.M. Crawford, J. Alexander, and E. Bodenschatz. Fluid particle accelerations in fully developed turbulence. *Nature*, 409(6823):1017–1019, 2001.

[718] F. Porté-Agel, C. Meneveau, and M.B. Parlange. A scale-dependent dynamic model for large-eddy simulation: application to a neutral atmospheric boundary layer. *Journal of Fluid Mechanics*, 415:261–284, 2000.

[719] N. Possiel and J.R. Scoggins. Curvature of the wind profile in the troposphere versus regions of CAT and non-CAT in the stratosphere. *Monthly Weather Review*, 104(1):57–62, 1976.

[720] K.A. Pratt, P.J. DeMott, J.R. French, Z. Wang, D.L. Westphal, A.J. Heymsfield, C.H. Twohy, A.J. Prenni, and D.L. Prather. In situ detection of biological particles in cloud ice-crystals. *Nature Geoscience*, 2:398–401, 2009.

[721] K.A. Pratt, J.E. Mayer, J.C. Holecek, R.C. Moffet, R.O. Sanchez, T.P. Rebotier, H. Furutani, M. Gonin, K. Fuhrer, Y. Su, S. Guazzotti, and K.A. Prather. Development and characterization of an aircraft aerosol time-of-flight mass spectrometer. *Analytical Chemistry*, 81(5):1792–1800, 2009.

[722] J.B. Prince, B.K. Buck, P.A. Robinson, and T. Ryan. *In-Service Evaluation of the Turbulence Auto-PIREP System and Enhanced Turbulence Radar Technologies*. BiblioGov, 2013.

[723] S. Priyadarshi. A review of ionospheric scintillation models. *Surveys in Geophysics*, 36(2):295–324, 2015.

[724] ATR Products. Aerotech research's conops for turbulence avoidance and decision making, 2016.

[725] W.M. Protheroe. Determination of shadow band structure from stellar scintillation measurements. *Journal of the Optical Society of America*, 45(10):851–855, 1955.

[726] W.M. Protheroe. Stellar scintillation. *Science*, 134(3490):1593–1599, 1961.

[727] W.M. Protheroe. The motion and structure of stellar shadow patterns. *Quarterly Journal of the Royal Meteorological Society*, 90(383):27–42, 1964.

[728] S. Prudden, A. Fisher, M. Marino, A. Mohamed, S. Watkins, and G. Wild. Measuring wind with small unmanned aircraft systems. *Journal of Wind Engineering and Industrial Aerodynamics*, 176(3):197–210, 2018.

[729] H. Pruppacher and J. Klett. *Microphysics of Clouds and Precipitation*. Springer, Dordrecht, Netherlands, 2010.

[730] H.R. Pruppacher and J.D. Klett. *Microphysics of Clouds and Precipitation*. Reidel, Dordrecht, The Netherlands, 1997.

[731] H. Puempel and P.D. Williams. The impacts of climate change on aviation: Scientific challenges and adaptation pathways. In *ICAO Environmental Report 2016: On Board A Sustainable Future*, pages 205–207. International Civil Aviation Organization (ICAO), 2016.

[732] S.J. Purkis and J.C. Brock. Lidar overview. In J.A. Goodman, S.J. Purkis, and S.R. Phinn, editors, *Coral Reef Remote Sensing*, pages 115–143. Springer, Dordrecht, 2013.

[733] M. Quante and D.O'C. Starr. Dynamical process in cirrus clouds: A review of observational results. In D.K. Lynch, K. Sassen, and G. Stephens, editors, *Cirrus*, pages 346–374. Oxford University Press, 2002.

[734] P. Queney. The problem of air flow over mountains: A summary of theoretical studies. *Bulletin of the American Meteorological Society*, 29(1):16–26, 1948.

[735] B. Raghavan and M.J. Patil. Flight dynamics of high aspect-ratio flying wings: Effects of large trim deformation. *Journal of Aircraft*, 46(5):1808–1812, 2009.

[736] S. Ragland, R.M. Chambers, R.J. Crosbie, and L. Hitchcock. Simulation and effects of severe turbulence on jet airlines pilots. Technical Report NADCL-ML-6411, U. S. Naval Air Development Center, Johnsville, Pennsylvania, 1964.

[737] S. Rao, editor. *Direct Numerical Simulations. An Introduction and Applications*. IntechOpen, 2021.

[738] M. Rapp, B. Kaifler, A. Dörnbrack, S. Gisinger, T. Mixa, R. Reichert, N. Kaifler, S. Knobloch, R. Eckert, N. Wildmann, A. Giez, L. Krasauskas, P. Preusse, M. Geldenhuys, M. Riese, W. Woiwode, F. Friedl-Vallon, B.-M. Sinnhuber, A. de la Torre, P. Alexander, J.L. Hormaechea, D. Janches, M. Garhammer, J.L. Chau, F. Conte, P. Hoor, and A. Engel. Southtrac-gw: An airborne field campaign to explore gravity wave dynamics at the world's strongest hotspot. *Bulletin of the American Meteorological Society*, 102(4):E871–E894, 2021.

[739] R.G. Rastogi. Nighttime ionospheric scintillations & E- & F-region irregularities at the magnetic equator. *Indian Journal of Radio & Space Physics*, 11:1–14, 1982.

[740] J.A. Ratcliffe. Some aspects of diffraction theory and their application to the ionosphere. *Reports on Progress in Physics*, 19(1):188–267, 1956.

[741] A. Rautenberg, J. Allgeier, S. Jung, and J. Bange. Calibration procedure and accuracy of wind and turbulence measurements with five-hole probes on fixed-wing unmanned aircraft in the atmospheric boundary layer and wind turbine wakes. *Atmosphere*, 10(3):124, 2019.

[742] A. Rautenberg, M.S. Graf, N. Wildmann, A. Platis, and J. Bange. Reviewing wind measurement approaches for fixed-wing unmanned aircraft. *Atmosphere*, 9(11):422, 2018.

[743] A. Rautenberg, M. Schön, K. zum Berge, M. Mauz, P. Manz, A. Platis, B. van Kesteren, I. Suomi, S.T. Kral, and J. Bange. The multi-purpose airborne sensor carrier MASC-3 for wind and turbulence measurements in the atmospheric boundary layer. *Sensors*, 19(10):2292, 2019.

[744] P.S. Ray, D.P. Jorgensen, and S.-I. Wang. Airbone doppler radar observations of a convective storm. *Journal of Climate and Applied Meteorology*, 24(7):687–698, 1985.

[745] R.M. Reap. Probability forecasts of clar-air-turbulence for the contiguous U.S. Technical Procedures Bulletin W/OM21:MEH, National Oceanic and Atmospheric Administration, U.S. Department of Commerce, Silver Spring, Md. 20910, February 1996.

[746] R.J. Reed. A study of the relation of clear air turbulence to the mesoscale structure of the jet stream region. In Y.-H. Pao and A. Goldburg, editors, *Clear Air Turbulence and Its Detection*, pages 288–307. Springer, 1969.

[747] B.P. Reen and R.E. Dumais. Assimilating tropospheric airborne meteorological data reporting (TAMDAR) observations and the relative value of other observation types. Technical Report ARL-TR-7022, U.S. Army Research Laboratory, Adelphi, MD 20783-1197, August 2014.

[748] H.J. Reid and G. Vaughan. Convective mixing in a tropopause fold. *Quarterly Journal of the Royal Meteorological Society*, 130(599):1195–1212, 2004.

[749] S.H. Reiger. Atmospheric turbulence and the scintillation of starlight. Technical Report R-406-PR, The RAND Corporation, Santa Monica, California, USA, September 1962.

[750] S.H. Reiger. Starlight scintillation and atmospheric turbulence. *The Astronomical Journal*, 68(6):395–406, 1963.

[751] E. Reiter. Tropospheric circulation and jet streams. *Paper No. 153*, pages 85–203, 1969.

[752] E.R. Reiter. The nature of clear air turbulence: A review. In Y.-H. Pao and A. Goldburg, editors, *Clear Air Turbulence and Its Detection*, pages 7–33. Springer, 1969.

[753] E.R. Reiter and A. Burns. Atmospheric structure and clean-air turbulence. *Technical Paper No. 65*, pages 1–13, 1965.

[754] E.R. Reiter and A. Nania. Jet-stream structure and clear-air turbulence (CAT). *Journal of Applied Meteorology and Climatology*, 3(3):247–260, 1964.

[755] J. Reuder, P. Brisset, M. Jonassen, M. Müller, and S. Mayer. The small unmanned meteorological observer sumo: A new tool for atmospheric boundary layer research. *Meteorologische Zeitschrift*, 18(2):141–147, 2009.

[756] J. Reuder, M.O. Jonassen, and H. Ólafsson. The small unmanned meteorological observer sumo: Recent developments and applications of a micro-UAS for atmospheric boundary layer research. *Acta Geophysica*, 60:1454–1473, 2012.

[757] R. Reynolds. Synoptic meteorology. Weather maps. In G.R. North, J. Pyle, and F. Zhang, editors, *Encyclopedia of Atmospheric Sciences (Second Edition)*, pages 289–298. Elsevier, 2015.

[758] W.C. Reynolds. The potential and limitations of direct and large eddy simulations. In J. Lumley, editor, *Whither Turbulence? Turbulence at the Crossroads*, volume 2, pages 313–343. Springer-Verlag, 1990.

[759] L.F. Richardson. Atmospheric diffusion shown in a distance-neighbour graph. *Proceedings of the Royal Society A*, 110:709–737, 1926.

[760] R.E. Rinehart. *Radar for Meteorologists*. Radar Met, 2010.

[761] D.L. Ringwalt, W.S. Ament, and F.C. MacDonald. Measurements of 1250-mc scatter propagation as function of meteorology. *IRE Transactions on Antennas and Propagation*, 6(2):208–209, 1958.

[762] C. Rino. *The Theory of Scintillation with Applications in Remote Sensing.* Wiley-IEEE Press, New Jersey, 2011.

[763] C.L. Rino. Some new results on the statistics of radio wave scintillation. 2. Scattering from a random ensemble of locally homogeneous patches. *Journal of Geophysical Research*, 81(13):2059–2064, 1976.

[764] W. Roach and R. Dixon. A note on the paper 'on the influence of synoptic development on the production of high level turbulence'. *Quarterly Journal of the Royal Meteorological Society*, 96(410):758–760, 1970.

[765] G.C. Roberts and A. Nenes. A continuous-flow streamwise thermal-gradient CCN chamber for atmospheric measurements. *Aerosol Science and Technology*, 39(3):206–221, 2005.

[766] R.G. Roble and R.E. Dickinson. How will changes in carbon dioxide and methane modify the mean structure of the mesosphere and thermosphere? *Geophysical Research Letters*, 16(12):1441–1444, 1989.

[767] P. Rodriguez Imazio, A. Dörnbrack, R. Delgado Urzua, N. Rivaben, and A. Godoy. Clear air turbulence observed across a tropopause fold over the drake passage – a case study. *Earth and Space Science Open Archive*, pages 1–37, 2021.

[768] D.C. Rogers, P.J. DeMott, S.M. Kreidenweis, and Y. Chen. A continuous-flow diffusion chamber for airborne measurements of ice nuclei. *Journal of Atmospheric and Oceanic Technology*, 18(5):725–741, 2001.

[769] G.L. Rogers and W. Ireland. Ionospheric holography— II. The analysis of a set of ionospheric holograms. *Journal of Atmospheric and Terrestrial Physics*, 42(4):397–399,401–406, 1980.

[770] P.J. Rogers and P.J. Eccles. The bistatic radial equation for radomly distributed targets. *Proceedings of the IEEE*, 59(6):1019–1021, 1971.

[771] E. Romano, L. Ferrucci, F. Nicolai, V. Derme, and G.F. De Stefano. Increase of chromosomal aberrations induced by ionising radiation in peripheral blood lymphocytes of civil aviation pilots and crew members. *Mutation Research/Fundamental and Molecular Mechanisms of Mutagenesis*, 377:1, June 1997.

[772] D.M. Romps, J.T. Seeley, D. Vollaro, and J. Molinari. Projected increase in lightning strikes in the united states due to global warming. *Science*, 346(6211):851–854, 2014.

[773] J. Rösch. L'agitation des images télescopiques. *Ciel et Terre*, 71:205–222, 1955.

[774] P. Rosenberg. Clear air turbulence detection and warning. *Navigation*, 13(2):162–165, 1966.

[775] D. Rosenfeld and I.M. Lensky. Satellite-based insights into precipitation formation processes in continental and maritime convective clouds. *Bulletin of the American Meteorological Society*, 79(11):2457–2476, 1998.

[776] C.-G. Rossby. On the mutual adjustment of pressure and velocity distributions in certain simple current systems ii. *Journal of Marine Research*, 2(01-02):15–28, 1938.

[777] J. Rothermel, D.A. Bowdle, and J.M. Vaughan. Calculation of aerosol backscatter from airborne continuous wave focused CO_2 doppler lidar measurements: 2. Algorithm performance. *Journal of Geophysical Research: Atmospheres*, 96(D3):5299–5305, 1991.

[778] J. Rothermel, D.A. Bowdle, J.M. Vaughan, D.W. Brown, and A.A. Woodfield. Calculation of aerosol backscatter from airborne continuous wave focused CO_2 doppler lidar measurements: 1. Algorithm description. *Journal of Geophysical Research: Atmospheres*, 96(D3):5293–5298, 1991.

[779] J. Rothermel, D.A. Bowdle, J.M. Vaughan, and M.J. Post. Evidence of a tropospheric aerosol backscatter background mode. *Applied Optics*, 28(6):1040–1042, 1989.

[780] J. Rothermel and W.D. Jones. Ground-based measurements of atmospheric backscatter and absorption using coherent CO_2 lidar. *Applied Optics*, 24(21):3487–3496, 1985.

[781] R. Rüster, G.D. Nastrom, and G. Schmidt. High-resolution VHF radar measurements in the troposphere with a vertically pointing beam. *Journal of Applied Meteorology and Climatology*, 37(11):1522–1529, 1998.

[782] W.D. Ryan. Radio star scintillation near the auroral zone. *Canadian Journal of Physics*, 42(3):458–464, 1964.

[783] M.S. Ryerson, M. Hansen, L. Hao, and M. Seelhorst. Landing on empty: estimating the benefits from reducing fuel uplift in US civil aviation. *Environmental Research Letters*, 10(9):094002, 2015.

[784] E. Ryznar. Visual resolution and optical scintillation in stable stratification over snow. *Journal of Applied Meteorology and Climatology*, 2(4):526–530, 1963.

[785] R. Sabatini and M.A. Richardson. Airborne laser systems testing and analysis. RTO AGARDograph 300, flight test techniques series – volume 26. Technical Report RTO-AG-300-V26, North Atlantic Treaty Organization (NATO), April 2010.

[786] R. Sabatini, M.A. Richardson, A. Gardi, and S. Ramasamy. Airborne laser sensors and integrated systems. *Progress in Aerospace Sciences*, 79:15–63, 2015.

[787] K. Saha. *The Earth's Atmosphere. Its Physics and Dynamics.* Springer-Verlag, 2008.

[788] T. Sakimura, Y. Watanabe, T. Ando, S. Kameyama, K. Asaka, H. Tanaka, T. Yanagisawa, Y. Hirano, and H. Inokuchi. 3.2-mJ, 1.5-μm laser power amplifier using an er, YB:glass planar waveguide for coherent doppler LIDAR. In *Proceedings of the 17th Coherent Laser Radar Conference (CLRC 2013)*, pages 35–38. Universities Space Research Association (USRA), June 2013.

[789] F. Sanders. An investigation of the structure and dynamics of an intense surface frontal zone. *Journal of the Atmospheric Sciences*, 12(6):542–552, 1955.

[790] K. Sassen. Lidar backscatter depolarization technique for cloud and aerosol research. In M.I.J. Mischenko, J.W. Hovenier, and L.D. Travis, editors, *Light Scattering by Nonspherical Particles: Theory, Measurements, and Applications*, pages 393–416. Academic Press, 2000.

[791] K. Sassen, Arnott W.P., D.O'C. Starr, G.G. Mace, Z. Wang, and M.R. Poellet. Midlatitude cirrus clouds derived from hurricane Nora: A case of study with implications for ice crystal nucleation and shape. *Journal of the Atmospheric Sciences*, 60(7):873–891, 2003.

[792] K. Sato, H. Hashiguchi, and S. Fukao. Gravity waves and turbulence associated with cumulus convection observed with the UHF/VHF clear-air doppler radars. *Journal of Geophysical Research: Atmospheres*, 100(D4):7111–7119, 1995.

[793] J.A. Saxton, J.A. Lane, R.W. Meadows, and P.A. Mathews. Layer structure of the troposphere – simultaneous radar and microwave refractometer investigations. *Proceedings of the IEEE*, 111(2):275–283, 1964.

[794] D.K. Schmidt. *Modern Flight Dynamics.* Mc-Graw Hill, New York, 2012.

[795] S. Schmidt, K. Lehmann, and M. Wendisch. Minimizing instrumental broadening of the drop size distribution with the m-fast-fssp. *Journal of Atmospheric and Oceanic Technology*, 21(12):1855–1867, 2004.

[796] N.P. Schmitt. Onboard lidar detects turbulence, volcanic ash near and far. *Photonics Spectra*, 52(2):36–42, 2018.

[797] N.P. Schmitt, W. Rehm, T. Pistner, P. Zeller, H. Diehl, and P. Navé. The awiator airborne lidar turbulence sensor – der bordgetragene awiator lidar turbulenz-sensor. *Aerospace Science and Technology*, 11(7-8):546–552, 2007.

[798] T. Schmugge. Radiometry at infrared and microwave frequencies. In B.J. Choudhury, Y.H. Kerr, E.G. Njoku, and P. Pampaloni, editors, *Passive Microwave Remote Sensing of Land-Atmosphere Interactions*, pages 193–204. De Gruyter, 1994.

[799] K. Schneider and O.V. Vasilyev. Wavelet methods in computational fluid dynamics. *Annual Review of Fluid Mechanics*, 42:473–503, 2010.

[800] J.F. Schubert. Acoustic detection of momentum transfer during the abrupt transition from a laminar to a turbulent atmospheric boundary layer. *Journal of Applied Meteorology and Climatology*, 16(12):1292–1297, 1977.

[801] D.M. Schultz. Perspectives on Fred Sanders' research on cold fronts. *Meteorological Monographs*, 33(55):109–126, 2008.

[802] D.M. Schultz and W. Blumen. Synoptic meteorology. Fronts. In G.R. North, J. Pyle, and F. Zhang, editors, *Encyclopedia of Atmospheric Sciences (Second Edition)*, pages 337–343. Elsevier, 2015.

[803] D.M. Schultz and P.J. Roebber. The fiftieth anniversary of Sanders (1955): A mesoscale model simulation of the cold front of 17-18 april 1953. *Meteorological Monographs*, 33(55):127–143, 2008.

[804] D.M. Schultz and P.N. Schumacher. The use and misuse of conditional symmetric instability. *Monthly Weather Review*, 127(12):2709–2732, 1999.

[805] D.M. Schultz and G. Vaughan. Occluded fronts and the occlusion process: A fresh look at conventional wisdom. *Bulletin of the American Meteorological Society*, 92(4):443–466, 2011.

[806] U. Schumann. A contrail cirrus prediction model. *Geoscientific Model Development*, 5(3):543–580, 2012.

[807] T.J. Schuyler and M.I. Guzman. Unmanned aerial systems for monitoring trace tropospheric gases. *Atmosphere*, 8(10):206, 2017.

[808] C. Schwarz and D. Fischenberg. Wake turbulence hazard analysis for a general aviation accident. *Deutscher Luft- und Raumfahrtkongress*, 340177:1–7, 2014.

[809] C.W. Schwarz, D. Fischenberg, and F.N. Holzäpfel. Wake turbulence evolution and hazard analysis for a general aviation takeoff accident. In *10th AIAA Atmospheric and Space Environments Conference*. American Institute of Aeronautics and Astronautics, 2018.

[810] J.P. Schwarz, R.S. Gao, D.W. Fahey, D.S. Thomson, L.A. Watts, J.C. Wilson, J.M. Reeves, M. Darbeheshti, D.G. Baumgardner, G.L. Kok, S.H. Chung, M. Schulz, J. Hendricks, A. Lauer, B. Kärcher, J.G. Slowik, K.H. Rosenlof, T.L. Thompson, A.O. Langford, M. Loewenstein, and K.C. Alkin. Single-particle measurements of midlatitude black carbon and light-scattering aerosols from the boundary layer to the lower stratosphere. *Journal of Geophysical Research: Atmospheres*, 111(D16):D16207, 2006.

[811] J.R. Scoggins. Meteorological variables vs CAT in the stratosphere: A statistical approach. *Journal of Aircraft*, 12(7):567–571, 1975.

[812] R.S. Scorer. Theory of waves in the lee of mountains. *Quarterly Journal of the Royal Meteorological Society*, 75(323):41–56, 1949.

[813] R.S. Scorer. Lee waves in the atmosphere. *Scientific American*, pages 124–137, 1961.

[814] T.L. Seamster, R.E. Redding, J.R. Cannon, J.M. Ryder, and J.A. Purcell. Cognitive task analysis of expertise in air traffic control. *The International Journal of Aviation Psychology*, 3(4):257–283, 1993.

[815] T.L. Seamster, R.E. Redding, and G.E. Kaempf. *Applied Cognitive Task Analysis in Aviation*. Routledge, 1997.

[816] J. Sela. The derivation of the sigma pressure hybrid coordinate semi-lagrangian model equations for GFS. Office note 462, National Oceanic and Atmospheric Administration, U.S. Department of Commerce, Silver Spring, Md. 20910, 2010.

[817] F.J.J. Serrano and A. Kazda. Airline disruption management: yesterday, today and tomorrow. *Transportation Research Procedia*, 28:3–10, 2017.

[818] National Weather Service. *Aviation Weather: For pilots and flight operations personnel, AC 00-6A*. Aviation Supplies & Academics, 2012.

[819] S. Shamsoddin and F. Porté-Agel. Large-eddy simulation of atmospheric boundary-layer flow through a wind farm sited on topography. *Boundary-Layer Meteorology*, 163:1–17, 2017.

[820] M.A. Shapiro. Further evidence of the mesoscale and turbulent structure of upper level jet stream-frontal zone systems. *Monthly Weather Review*, 106(8):1100–1111, 1978.

[821] M.A. Shapiro. Turbulent mixing within tropopause folds as a mechanism for the exchange of chemical constituents between the stratosphere and troposphere. *Journal of the Atmospheric Sciences*, 37(5):994–1004, 1980.

[822] M.A. Shapiro, T. Hample, and D.W. Van De Kamp. Radar wind profiler observations of fronts and jet streams. *Monthly Weather Review*, 112(6):1263–1266, 1984.

[823] M.A. Shapiro and D. Keyser. Fronts, jet streams and the tropopause. In C.W. Newton and E.O. Holopainen, editors, *Extratropical Cyclones*, pages 167–191. Springer, 1990.

[824] R. Sharman, L.B. Cornman, G. Meymaris, J. Pearson, and T. Farrar. Description and derived climatologies of automated in situ eddy-dissipation-rate reports of atmospheric turbulence. *Journal of Applied Meteorology and Climatology*, 53:1416–1432, 2014.

[825] R. Sharman, T. Lane, and S. Trier. Convectively-generated gravity waves and clear-air turbulence (CAT). In *EGU General Assembly Conference Abstracts*, volume 15, pages EGU2013–6555, 2013.

[826] R. Sharman, C. Tebaldi, G. Wiener, and J. Wolff. An integrated approach to mid- and upper-level turbulence forecasting. *Weather and Forecasting*, 21(3):268–287, 2006.

[827] R.D. Sharman, J.D. Doyle, and M.A. Shapiro. An investigation of a commercial aircraft encounter with severe clear-air turbulence over Western Greenland. *Journal of Applied Meteorology and Climatology*, 51(1):42–53, 2012.

[828] R.D. Sharman and J.M. Pearson. Prediction of energy dissipation rates for aviation turbulence. Part I: Forecasting nonconvective turbulence. *Journal of Applied Meteorology and Climatology*, 56(2):317–337, 2017.

[829] R.D. Sharman, S.B. Trier, T.P. Lane, and J.D. Doyle. Sources and dynamics of turbulence in the upper troposphere and lower stratosphere: A review. *Geophysics Research Letters*, 39:L12803, 2012.

[830] X. Shi, Z. Chen, H. Wang, D.-Y. Yeung, W. Wong, and W. Woo. Convolutional LSTM network: A machine learning approach for precipitation nowcasting. In *NIPS 2015: Proceedings of the 28th International Conference on Neural Information Processing Systems*, volume 1, pages 802–810. MIT Press, Cambridge, MA, USA, December 2015.

[831] J. Shin. The NASA aviation safety program: Overview. Technical Report NASA/TM-2000-209810, Glenn Research Center, NASA, March 2000.

[832] M.F. Shlesinger, B.J. West, and J. Klafter. Lévy dynamics of enhanced diffusion: Application to turbulence. *Physical Review Letters*, 58(11):1100–1103, 1987.

[833] H. Siebert, H. Franke, K. Lehmann, R. Maser, E.W. Saw, D. Schell, R.A. Shaw, and M. Wendisch. Probing fine-scale dynamics and microphysics of clouds with helicopter-borne measurements. *Bulletin of the American Meteorological Society*, 87(12):1727–1738, 2006.

[834] H. Siebert, K. Lehmann, and R.A. Shaw. On the use of hot-wire anemometers for turbulence measurements in clouds. *Journal of Atmospheric and Oceanic Technology*, 24(6):980–993, 2007.

[835] H. Siebert, K. Lehmann, and M. Wendisch. Observations of small-scale turbulence and energy dissipation rates in the cloud boundary layer. *Journal of the Atmospheric Sciences*, 63(5):1451–1466, 2006.

[836] J.E. Simpson. Gravity currents in the laboratory, atmosphere, and ocean. *Annual Review of Fluid Mechanics*, 14:213–234, 1982.

[837] J. Sinclair and R.F. Kelleher. The f-region equatorial irregularity belt as observed from scintillation of satellite transmissions. *Journal of Atmospheric and Terrestrial Physics*, 31(1):201–206, 1969.

[838] A.P. Singh, S. Medida, and K. Duraisamy. Machine-learning-augmented predictive modeling of turbulent separated flows over airfoils. *AIAA Journal*, 55(7):1–13, 2017.

[839] A.M. Sinnarwalla and D.J. Alofs. A cloud nucleus counter with long available growth time. *Journal of Applied Meteorology and Climatology*, 12(5):831–835, 1973.

[840] W.C. Skamarock and J.B. Klemp. A time-split nonhydrostatic atmospheric model for weather research and forecasting applications. *Journal of Computational Physics*, 227(7):3465–3485, 2008.

[841] W.C. Skamarock, J.B. Klemp, J. Dudhia, D.O. Gill, D.M. Barker, M.G. Duda, X.-Y. Huang, W. Wang, and J.G. Powers. A description of the advanced research WRF version 3. Technical Report NCAR/TN-474+STR, National Center for Atmospheric Research, NCAR, Boulder, Colorado, USA, June 2008.

[842] N.J. Skinner, R.F. Kelleher, J.B. Hacking, and C.W. Benson. Scintillation fading of signals in the SHF band. *Nature Physical Science*, 232(27):19–21, 1971.

[843] M.I. Skolnik. *Introduction to Radar Systems*. McGraw Hill: New York, 1962.

[844] Skybrary. Nextgen, 2021.

[845] J. Smagorinsky. General circulation experiments with the primitive equations. I. The basic experiment. *Monthly Weather Review*, 91:99–164, 1963.

[846] A.E. Smart. Optical velocity sensor for air data applications. *Optical Engineering*, 31(1):166–173, 1992.

[847] R.B. Smith. The influence of mountains on the atmosphere. *Advances in Geophysics*, 21:87–230, 1979.

[848] R.B. Smith. Stratified flow over topography. In R. Grimshaw, editor, *Environmental Stratified Flows. Topics in Environmental Fluid Mechanics*, volume 3, pages 119–159. Springer, 2002.

[849] S.A. Smith, G.J. Romick, and K. Jayaweera. Poker flat MST radar observations of shear-induced turbulence. *Journal of Geophysical Research: Oceans*, 88(C9):5265–5271, 1983.

[850] W.L. Smith, S. Kireev, L.L. West, G.G. Gimmestad, L. Cornman, W. Feltz, G. Perram, and T. Daniels. Interferometric radiometer for in-flight detection of aviation hazards. In W.F. Feltz and J.J. Murray, editors, *Remote Sensing Applications for Aviation Weather Hazard Detection and Decision Support*, volume 7088, pages 89–99. International Society for Optics and Photonics, SPIE, 2008.

[851] W.D. Smyth and J.N. Moum. Ocean mixing by Kelvin-Helmholtz instability. *Oceanography*, 25:140–149, 2012.

[852] W.D. Smyth, J.N. Moum, and J.D. Nash. Narrowband oscillations in the upper equatorial ocean. Part II: Properties of shear instabilities. *Journal of Physical Oceanography*, 41:412–428, 2011.

[853] W.D. Smyth and S.A. Thorpe. Glider measurements of overturning in a Kelvin-Helmholtz billow train. *Journal of Marine Research*, 70:119–140, 2012.

[854] J.R. Snider and J.-L. Brenguier. A comparison of cloud condensation nuclei and cloud droplet measurements obtained during ACE-2. *Tellus. Series B: Chemical and Physical Meteorology*, 52(2):828–842, 2000.

[855] J. Söder, C. Zülicke, M. Gerding, and F.-J. Lübken. High-resolution observations of turbulence distributions across tropopause folds. *Journal of Geophysical Research: Atmospheres*, 126(6):e2020JD033857, 2021.

[856] A. Solomon and L.M. Polvani. Highly significant responses to anthropogenic forcings of the midlatitude jet in the southern hemisphere. *Journal of Climate*, 29(9):3463–3470, 2016.

[857] Y. Song, Q.-H. Meng, B. Luo, M. Zeng, S.-G. Ma, and P.-F. Qi. A wind estimation method for quadrotors using inertial measurement units. In *Proceedings of the 2016 IEEE International Conference on Robotics and Biomimetics (ROBIO)*, pages 426–431. IEEE, March 2016.

[858] J. Soria and Willert C. On measuring the joint probability density function of three-dimensional velocity components in turbulent flows. *Measurement Science and Technology*, 23:065301, 2012.

[859] P.R. Spalart. Strategies for turbulence modeling and simulations. *International Journal of Heat and Fluid Flow*, 21:252–263, 2000.

[860] A. Spanu, M. Dollner, J. Gasteiger, T.P. Bui, and B. Weinzierl. Flow-induced errors in airborne in situ measurements of aerosols and clouds. *Atmospheric Measurement Techniques*, 13:1963–1987, 2020.

[861] T. Spiess, J. Bange, M. Buschmann, and P. Vörsmann. First application of the meteorological Mini-UAV 'M2AV'. *Meteorologische Zeitschrift*, 16(2):159–169, 2007.

[862] P.N. Spotts. Detecting turbulence that no one can see. *The Christian Science Monitor*, pages 3–5, December 31 1997.

[863] H.B. Squire. On the stability for three-dimensional disturbances of viscous fluid flow between parallel walls. *Proceedings of the Royal Society of London A*, 142:621–628, 1933.

[864] P. Squires and S. Twomey. A comparison of cloud nucleus measurements over central North America and the Caribbean Sea. *Journal of the Atmospheric Sciences*, 22(4):401–404, 1966.

[865] K.R. Sreenivasan and R.A. Antonia. The phenomenology of small-scale turbulence. *Annual Review of Fluid Mechanics*, 29:435–472, 1997.

[866] B. Stanford and P. Beran. Direct flutter and limit cycle computations of highly flexible wings for efficient analysis and optimization. *Journal of Fluids and Structures*, 36:111–123, 2013.

[867] S. Stefan, B. Antonescu, A.D. Urlea, L. Buzdugan, M.D. Andrei, C. Necula, and S. Voinea. Study of clear air turbulence related to tropopause folding over the romanian airspace. *Atmosphere*, 11(10):1099, 2020.

[868] J.J. Stephens and E.R. Reiter. Estimating refractive index spectra in regions of clear air turbulence. *Journal of Applied Meteorology and Climatology*, 6(5):911–913, 1967.

[869] M. Stephens, N. Turner, and J. Sandberg. Particle identification by laser-induced incandescence in a solid-state laser cavity. *Applied Optics*, 42(19):3726–3736, 2003.

[870] B. Stevens, D.H. Lenschow, G. Vali, H. Gerber, A. Bandy, B. Blomquist, J.-L. Brenguier, C.S. Bretherton, F. Burnet, T. Campos, S. Chai, I. Faloona, D. Friesen, S. Haimov, K. Laursen, D.K. Lilly, S.M. Loehrer, S.P. Malinowski, B. Morley, M.D. Petters, D.C. Rogers, L. Russell, V. Savic-Jovcic, J.R. Snider, D. Straub, M.J. Szumowski, H. Takagi, D.C. Thornton, M. Tschudi, C. Twohy, M. Wetzel, and M.C. van Zanten. Dynamics and chemistry of marine stratocumulus – DYCOMS-II. *Bulletin of the American Meteorological Society*, 84(5):551–578, 2003.

[871] J.J. Stickland. An assessment of two algorithms for automatic measurement and reporting of turbulence from commercial public transport aircraft. Technical report, Bureau of Meteorology. Department of the Environment, Melbourne, Australia, February 1998.

[872] J. Stock and G. Keller. Astronomical seeing. In G.P. Kuiper and B.M. Middlehurst, editors, *Telescopes*, volume 9, pages 138–153. University of Chicago Press, 1960.

[873] T.F. Stocker, D. Qin, G.-K. Plattner, M.M.B. Tignor, S.K. Allen, J. Boschung, A. Nauels, Y. Xia, V. Bex, and P.M. Midgley. *Climate Change 2013 – The Physical Science Basis.* Cambridge University Press, Cambridge, 2014.

[874] L.N. Storer, P.D. Williams, and P.G. Gill. Aviation turbulence: Dynamics, forecasting, and response to climate change. *Pure and Applied Geophysics*, 176(5):2081–2095, 2019.

[875] L.N. Storer, P.D. Williams, and M.M. Joshi. Global response of clear-air turbulence to climate change. *Geophysical Research Letters*, 44(19):9976–9984, 2017.

[876] H.P. Stough, J.F. Watson, and M.A. Jarrell. New technologies for reducing aviation weather-related accidents. In *ICAS 2006 – 25th Congress of the International Council of Aeronautical Sciences*, number 20060048302, pages 1–10, 2006.

[877] J.W. Strapp, W.R. Leaitch, and P.S.K. Liu. Hydrated and dried aerosol-size-distribution measurements from the particle measuring systems FSSP-300 probe and the deiced PCASP-100x probe. *Journal of Atmospheric and Oceanic Technology*, 9(5):548–555, 1992.

[878] L. Strauss, S. Serafin, S. Haimov, and Grubišić. Turbulence in breaking mountain waves and atmospheric rotors estimated from airborne in situ and doppler radar measurements. *Quarterly Journal of the Royal Meteorological Society*, 141(693):3207–3225, 2015.

[879] R.B. Stull. *An Introduction to Boundary Layer Meteorology.* Kluwer Academic Publishers, Dordrecht, The Netherlands, 1988.

[880] R.B. Stull. Reply (to comments on "a convective transport theory for surface fluxes"). *Journal of the Atmospheric Sciences*, 54(4):579–579, 1997.

[881] B. Subramanian, N. Chokani, and R.S. Abhari. Drone-based experimental investigation of three-dimensional flow structure of a multi-megawatt wind turbine in complex terrain. *Journal of Solar Energy Engineering*, 137(5):051007, 2015.

[882] P.P. Sullivan, J.C. McWilliams, and C.-H. Moeng. A subgrid-scale model for large-eddy simulation of planetary boundary-layer flows. *Boundary-Layer Meteorology*, 71:247–276, 1994.

[883] O.G. Sutton. *Micrometeorology: A Study of Physical Processes in the Lowest Layers of the Earth's Atmosphere.* MCGraw-Hill, New York-London, 1953.

[884] O.G. Sutton. *The Challenge of the Atmosphere.* Greenwood Publishing Group, Connecticut, USA, 1969.

[885] National Aircraft System. The future of the NAS, 2016.

[886] Next Generation Air Transport System. Integrated plan, 2004.

[887] M. Szakáll, K. Diehl, S.K. Mitra, and S. Borrmann. A wind tunnel study on the shape, oscillation, and internal circulation of large raindrops with sizes between 2.5 and 7.5 mm. *Journal of the Atmospheric Sciences*, 66(3):755–765, 2009.

[888] H. Takayasu. Stable distribution and Lévy process in fractal turbulence. *Progress of Theoretical Physics*, 72(3):471–479, 1984.

[889] K. Takeuchi. Ozone. In P. Worsfold, A. Townshend, and C. Poole, editors, *Encyclopedia of Analytical Science (Second Edition)*, pages 462–471. Elsevier, 2005.

[890] A. Tang and C. Morgan. Why airplane takeoffs and landings are so dangerous, October, 31 2020.

[891] J.C. Tannehill, R.H. Pletcher, and D.A. Anderson. *Computational Fluid Mechanics and Heat Transfer*. CRC Press, 2012.

[892] R. Targ, M.J. Kavaya, R.M. Huffaker, and R.I. Bowles. Coherent lidar airborne windshear sensor: performance evaluation. *Applied Optics*, 30(15):2013–2026, 1991.

[893] V.I. Tatarski. *Wave Propagation in a Turbulent Medium*. McGraw-Hill Book Company, 1961.

[894] K.E. Taylor, R.J. Stouffer, and G.A. Meehl. An overview of CMIP5 and the experiment design. *Bulletin of the American Meteorological Society*, 93(4):485–498, 2012.

[895] T.S. Taylor. *Introduction to Laser Science and Engineering*. CRC Press, Boca Raton, USA, 2019.

[896] *TC Aeronautical information manual*, 2019.

[897] E.H. Teets, J. Ehernberger, R. Bogue, and C. Ashburn. Turbulence and mountain wave conditions observed with an airborne 2-micron lidar. In *45th AIAA Aerospace Sciences Meeting and Exhibit*, pages 1–8. American Institute of Aeronautics and Astronautics, January 2007.

[898] M.A.C. Teixeira. The physics of orographic gravity wave drag. *Frontiers in Physics*, 2(43):1–24, 2014.

[899] M.A.C. Teixeira, J.L. Argaín, and P.M.A. Miranda. Drag produced by trapped lee waves and propagating mountain waves in a two-layer atmosphere. *Quarterly Journal of the Royal Meteorological Society*, 139(673):964–981, 2013.

[900] M.A.C. Teixeira, J.L. Argaín, and P.M.A. Miranda. Orographic drag associated with lee waves trapped at an inversion. *Journal of the Atmospheric Sciences*, 70(9):2930–2947, 2013.

[901] J. Tenenbaum. Jet stream winds: Comparisons of analyses with independent aircraft data over southwest Asia. *Weather and Forecasting*, 6(3):320–336, 1991.

[902] H. Tennekes and J.L. Lumley. *A First Course in Turbulence*. MIT Press, 1972.

[903] J. Thiel, T. Pfeifer, and M. Hartmann. Interferometric measurement of absolute distances of up to 40 m. *Measurement*, 16(1):1–6, 1995.

[904] R.M. Thomas, K. Lehmann, H. Nguyen, D.L. Jackson, D. Wolfe, and V. Ramanathan. Measurement of turbulent water vapor fluxes using a lightweight unmanned aerial vehicle system. *Atmospheric Measurement Techniques*, 5(1):243–257, 2012.

[905] D.S. Thomson, M.E. Schein, and D.M. Murphy. Particle analysis by laser mass spectrometry WB-57F instrument overview. *Aerosol Science and Technology*, 33(1-2):153–169, 2000.

[906] S.A. Thorpe. The axial coherence of Kelvin-Helmholtz billows. *Quarterly Journal of the Royal Meteorological Society*, 128:1529–1542, 2002.

[907] A.M. Title. *Imaging Michelson interferometers*, chapter 9, pages 349–361. ISSI Scientific Report Series. Springer, New York, 2013.

[908] T. Tomić, K. Schmid, P. Lutz, A. Mathers, and S. Haddadin. The flying anemometer: Unified estimation of wind velocity from aerodynamic power and wrenches. In *Proceedings of the 2016 IEEE/RSJ International Conference on Intelligent Robots and Systems (IROS)*, pages 1637–1644. IEEE, December 2016.

[909] A.A. Townsend. Local isotropy in the turbulent wake of a cylinder. *Australian Journal of Scientific Research, Series A: Physical Sciences*, 1(2):161–174, 1948.

[910] A. Tsinober. Anomalous diffusion in geophysical and laboratory turbulence. *Nonlinear Processes in Geophysics*, 1(2/3):80–94, 1994.

[911] A. Tsiringakis, G.J. Steeneveld, and A.A.M. Holtslag. Small-scale orographic gravity wave drag in stable boundary layers and its impact on synoptic systems and mear-surface meteorology. *Quarterly Journal of the Royal Meteorological Society*, 143(704):1504–1516, 2017.

[912] A. Tsonis. *An Introduction to Atmospheric Thermodynamics*. Cambridge University Press, 2007.

[913] J.C. Tu. Cockpit weather information (CWIN) system. In *Proceedings of the IEEE 1996 National Aerospace and Electronics Conference NAECON 1996*, volume 1, pages 29–32. IEEE, 1996.

[914] J.S. Turner. *Buoyancy Effects in Fluids*. Cambridge University Press, Cambridge, 1973.

[915] A.P. Tvaryanas. Epidemiology of turbulence-related injuries in airline cabin crew, 1992–2001. *Aviation, Space, and Environmental Medicine*, 74(9):970–976, 2003.

[916] S. Twomey and T.A. Wojciechowski. Observations of the geographic variations of cloud nuclei. *Journal of Atmospheric Sciences*, 26(4):684–688, 1969.

[917] L.W. Uccellini and D.R. Johnson. The coupling of upper and lower tropospheric jet streaks and implications for the development of severe convective storms. *Monthly Weather Review*, 107(6):682–703, 1979.

[918] T. Uchida. Numerical investigation of terrain-induced turbulence in complex terrain by large-eddy simulation (les) technique. *Energies*, 11(10):2638, 2018.

[919] S.L. Valley. *Handbook of Geophysics and Space Environments*. McGraw-Hill Book Company, 1965.

[920] A. Van den Kroonenberg, T. Martin, M. Buschmann, J. Bange, and P. Vörsmann. Measuring the wind vector using the autonomous mini aerial vehicle M^2AV. *Journal of Atmospheric and Oceanic Technology*, 25(11):1969–1982, 2008.

[921] A.C. Van den Kroonenberg, S. Martin, F. Beyrich, and J. Bange. Spatially-averaged temperature structure parameter over a heterogeneous surface measured by an unmanned aerial vehicle. *Boundary-Layer Meteorology*, 142:55–77, 2012.

[922] J.A. van Hooft, S. Popinet, C.C. van Heerwaarden, S.J.A. van der Linden, S.R. de Roode, and B.J.H. van de Wiel. Towards adaptive grids for atmospheric boundary-layer simulations. *Boundary-Layer Meteorology*, 167:421–443, 2018.

[923] T.E. Van Zandt. A brief history of the development of wind-profiling or MST radars. *Annales Geophysicae*, 18(7):740–749, 2000.

[924] T.E. Van Zandt, J.L. Green, W.L. Clark, and J.R. Grant. Buoyancy waves in the troposphere: Doppler radar observations and a theoretical model. *Geophysical Research Letters*, 6(6):429–432, 1979.

[925] M. Vargas and A. Feo. Experimental observations on the deformation and breakup of water droplets near the leading edge of an airfoil. In *AIAA Atmospheric and Space Environments Conference*. American Institute of Aeronautics and Astronautics, 2010.

[926] R. Vasaturo, I. Kalkman, B. Blocken, and P.J.V. van Wesemael. Large eddy simulation of the neutral atmospheric boundary layer: Performance evaluation of three inflow methods for terrains with different roughness. *Journal of Wind Engineering and Industrial Aerodynamics*, 173:241–261, 2018.

[927] J.M. Vaughan, D.W. Brown, P.H. Davies, C. Nash, G. Kent, and M.P. McCormick. Comparison of SAGE II solar extinction with airborne measurements of atmospheric backscattering in the troposphere and lower stratosphere. *Nature*, 332(6166):709–711, 1988.

[928] J.M. Vaughan, D.W. Brown, C. Nash, S.B. Alejandro, and G.G. Koenig. Atlantic atmospheric aerosol studies: 2. Compendium of airborne backscatter measurements at 10.6 μm. *Journal of Geophysical Research: Atmospheres*, 100(D1):1043–1065, 1995.

[929] J.M. Vaughan, R.D. Callan, D.A. Bowdle, and J. Rothermel. Spectral analysis, digital integration, and measurement of low backscatter in coherent laser radar. *Applied Optics*, 28(15):3008–3014, 1989.

[930] J.M. Vaughan, K.O. Steinvall, C. Werner, and P.H. Flamant. Coherent laser radar in Europe. *Proceedings of the IEEE*, 84(2):205–226, 1996.

[931] P. Vedula and P.K. Yeung. Similarly scaling of acceleration and pressure statistics in numerical simulations of isotropic turbulence. *Physics of Fluids*, 11(5):1208–1220, 1999.

[932] H.P.J. Veerman, P. Vrancken, and L. Lombard. Flight testing delicat – a promise for medium-range clear air turbulence protection. In *European 46th SETP and 25th SFTE Symposium*, pages 1–23. HAL archives-ouvertes, June 2014.

[933] A.V. Velasco, P. Cheben, M. Florjańczyk, and M.L. Calvo. Spatial heterodyne fourier-transform waveguide spectrometers. In E. Wolf, editor, *Progress in Optics*, chapter 3, pages 159–208. Elsevier, 2014.

[934] T.N. Venkatesh and J. Mathew. The problem of clear air turbulence: Changing perspectives in the understanding of the phenomenon. *Sadhana*, 38(4):707–722, 2013.

[935] I. Vera Rodriguez, S.P. Näsholm, and A. Le Pichon. Atmospheric wind and temperature profiles inversion using infrasound: An ensemble model context. *The Journal of the Acoustical Society of America*, 148(5):2923–2934, 2020.

[936] M.J. Verbeek and H.W. Jentink. Optical air data system flight testing. Technical Report NLR-TP-2012-068, Nationaal Lucht- en Ruimtevaartlaboratorium (NLR), Amsterdam, The Netherlands, March 2012.

[937] T.F. Villa, F. Gonzalez, B. Miljievic, Z.D. Ristovski, and L. Morawska. An overview of small unmanned aerial vehicles for air quality measurements: Present applications and future prospectives. *Sensors*, 16(7):1072, 2016.

[938] F. Villars and V.F. Weisskopf. On the scattering of radio waves by turbulent fluctuations of the atmosphere. *Proceedings of the IRE*, 43(10):1232–1239, 1955.

[939] E. Villermaux. Mixing versus stirring. *Annual Review of Fluid Mechanics*, 51:245–273, 2019.

[940] N.K. Vinnichenko, N.Z. Pinus, and G.N. Shur. Some results of experimental investigations of turbulence in the troposphere. In *International Colloquium on the Fine-Scale Structure of the Atmosphere and Its Relation to Radio-Wave Propagation*, pages 65–75. Nauka, Moscow, 1965.

[941] F.E. Volz and R.M. Goody. The intensity of the twilight and upper atmospheric dust. *Journal of the Atmospheric Sciences*, 19(5):385–406, 1962.

[942] V.I. Vorobyev and Tarakanov G.G. *Introduction to Synoptic Meteorology*. Russian State Hydrometeorological University (RSHU), 2005.

[943] S.B. Vosper. Inversion effects on mountain lee waves. *Quarterly Journal of the Royal Meteorological Society*, 130:1723–1748, 2004.

[944] G.A. Voth, K. Satyanarayan, and E. Bodenschatz. Lagrangian acceleration measurements at large Reynolds numbers. *Physics of Fluids*, 10(9):2268–2280, 1998.

[945] P. Vrancken, M. Wirth, D. Rempel, G. Ehret, A. Dolfi-Bouteyre, L. Lombard, T. Gaudo, D. Rees, H. Barny, and P. Rondeau. Clear air turbulence detection and characterization in the delicat airborne lidar project. In *Proceedings of the 25th International Laser Radar Conference (ILRC)*, pages 301–304. Deutsches Zentrum für Luft- und Raumfahrt (DLR), July 2010.

[946] B. Vreman, B. Geurts, and H. Kuerten. Subgrid-modelling in LES of compressible flow. *Applied Scientific Research*, 54:191–203, 1995.

[947] D.W. Waco. A statistical analysis of wind and temperature variables associated with high altitude clear turbulence (HICAT). *Journal of Applied Meteorology and Climatology*, 9(2):300–309, 1970.

[948] E. Wada, H. Hashiguchi, M.K. Yamamoto, M. Teshiba, and S. Fukao. Simultaneous observations of cirrus clouds with a millimeter-wave radar and the MU radar. *Journal of Applied Meteorology and Climatology*, 44(3):313–323, 2005.

[949] L.S. Wagner. Zenith angle dependence of radio star scintillation. *Journal of Geophysical Research*, 67(11):4187–4194, 1962.

[950] C.E. Wainwright, T.A. Bonin, P.B. Chilson, J.A. Gibbs, E. Fedorovich, and R.D. Palmer. Methods for evaluating the temperature structure-function parameter using unmanned aerial systems and large-eddy simulation. *Boundary-Layer Meteorology*, 155:189–208, 2015.

[951] J.M. Wallace and P.V. Hobb. *Atmospheric Science: An Introductory Survey*. Elsevier, 2006.

[952] J.L. Walmsley. On theoretical wind speed and temperature profiles over the sea with applications to data from Sable Island, Nova Scotia. *Atmosphere-Ocean*, 26(2):203–233, 1988.

[953] R.L. Walterscheid. High-altitude wind prediction and measurement technology assessment. Technical Report ATR-2009(5427)-1, Volpe National Transportation Systems Center, U.S. Department of Transportation, Cambridge, Massachusetts, June 2009.

[954] R. Wang, K. Kashinath, M. Mustafa, A. Albert, and R. Yu. Towards physics-informed deep learning for turbulent flow prediction. In *KDD 2020: Proceedings of the 26th ACM SIGKDD International Conference on Knowledge Discovery & Data Mining*, pages 1457–1466. Association for Computing Machinery, NY, USA, August 2019.

[955] Z. Wang and M. Menenti. Challenges and opportunities in lidar remote sensing. *Frontiers in Remote Sensing*, 2:641723, 2021.

[956] D.O. Wark and H.E. Fleming. Indirect measurements of atmospheric temperature profiles from satellites: I. introduction. *Monthly Weather Review*, 94(6):351–362, 1966.

[957] C.B. Watkins, C.J. Richey, P. Tchoryk, G.A. Ritter, M.T. Dehring, P.B. Hays, C.A. Nardell, and R. Urzi. Molecular optical air data system (MOADS) prototype II. In G.W. Kamerman, editor, *Laser Radar Technology and Applications IX*, volume SPIE 5412, pages 10–20. International Society for Optics and Photonics, SPIE, September 2004.

[958] C.D. Watkins. High-power radar for meteorological studies in clear air. *Proceedings of the IEEE*, 118(3):519–528, 1971.

[959] C.D. Watkins and K.A. Browning. The detection of clear air turbulence by radar. *Physics in Technology*, 4(1):28–61, 1973.

[960] S. Watkins, M. Thompson, and B. Loxton. On low altitude flight through the atmospheric boundary layer. *International Journal of Micro Air Vehicles*, 2(2):55–68, 2010.

[961] R.J. Weber, A.D. Clarke, M. Litchy, J. Li, G. Kok, R.D. Schillawski, and P.H. McMurry. Spurious aerosol measurements when sampling from aircraft in the vicinity of clouds. *Journal of Geophysical Research: Atmospheres*, 103(D21):28337–28346, 1998.

[962] R. Weigel, M. Hermann, J. Curtius, C. Voigt, S. Walter, T. Böttger, B. Lepukhov, G. Belyaev, and S. Borrmann. Experimental characterization of the condensation particle counting system for high altitude aircraft-borne application. *Atmospheric Measurement Techniques*, 2(1):243–258, 2009.

[963] B. Weinzierl, A. Ansmann, J.M. Prospero, D. Althausen, N. Benker, F. Chouza, M. Dollner, D. Farrell, W.K. Fomba, V. Freudenthaler, J. Gasteiger, S. Groß, M. Haarig, B. Heinold, K. Kandler, T.B. Kristensen, T. Mayol-Bracero, O.L. Müller, O. Reitebuch, D. Sauer, A. Schäfler, K. Schepanski, A. Spanu, I. Tegen, C. Toledano, and A. Walser. The saharan aerosol long-range transport and aerosol-cloud-interaction experiment: Overview and selected highlights. *Bulletin of the American Meteorological Society*, 98(7):1427–1451, 2017.

[964] M.A. Weissman. Nonlinear wave packets in the Kelvin-Helmholtz instability. *Philosophical Transactions of the Royal Society A*, 290:639–685, 1979.

[965] C. Weitkamp. *Lidar. Range-Resolved Optical Remote Sensing of the Atmosphere*, volume 102 of *Springer Series in Optical Sciences (SSOS)*. Springer Nature Switzerland AG, 2005.

[966] M. Wertheimer. Experimentelle studien über das sehen von bewengung. *Zeitschrift für Psychologie*, 61:161–265, 1912.

[967] L. West, G. Gimmestad, R. Herkert, W.L. Smith, S. Kireev, P.R. Schaffner, T.S. Daniels, L.B. Cornman, R. Sharman, A. Weekley, G. Perram, K. Gross, G. Smith, W. Feltz, J. Taylor, and E. Olson. Hazard detection analysis for a forward-looking interferometer. Technical Report NASA/TP-2010-216845, NASA, August 2010.

[968] L. West, G. Gimmestad, W. Smith, S. Kireev, L. Cornman, P.R. Schaffner, and G. Tsoucalas. Applications of a forward-looking interferometer for the on-board detection of aviation weather hazards. Technical Report NASA/TP-2008-215536, NASA, October 2008.

[969] A.D. Wheelon. Radio scattering by tropospheric irregularities. *Journal of Atmospheric and Terrestrial Physics*, 15(3-4):185–205, 1959.

[970] A.D. Wheelon. Radio-wave scattering by tropospheric irregularities. *Journal of Research of the National Bureau of Standards*, 63D(2):205–233, 1959.

[971] E.A. Whelan. Cancer incidence in airline cabin crew. *Occupational and Environmental Medicine*, 60(11):805–806, 2003.

[972] F.J.W. Whipple. The high temperature of the upper atmosphere as an explanation of zones of audibility. *Nature*, 111(2780):187–187, 1923.

[973] D.C. Wilcox. *Turbulence Modelling for CFD*. DCW Industries, 1993.

[974] N. Wildmann, S. Bernard, and J. Bange. Measuring the local wind field at an escarpment using small remotely-piloted aircraft. *Renewable Energy*, 103:613–619, 2017.

[975] N. Wildmann, R. Eckert, A. Dörnbrack, S. Gisinger, M. Rapp, K. Ohlmann, and A. van Niekerk. In situ measurements of wind and turbulence by a motor glider in the Andes. *Journal of Atmospheric and Oceanic Technology*, 38(4):921–935, 2021.

[976] N. Wildmann, M. Hofsäß, F. Weimer, A. Joos, and J. Bange. MASC – a small remotely piloted aircraft (RPA) for wind energy research. *Advances in Science and Research*, 11(1):55–61, 2014.

[977] N. Wildmann, G.A. Rau, and J. Bange. Observations of the early morning boundary-layer transition with small remotely-piloted aircraft. *Boundary-Layer Meteorology*, 157:345–373, 2015.

[978] J.K. Williams. Using random forests to diagnose aviation turbulence. *Machine Learning*, 95:51–70, 2014.

[979] J.K. Williams, L. Cornman, J. Yee, S.G. Carson, and A. Cotter. Real-time remote detection of convectively-induced turbulence. In *AMS 32nd Radar Meteorology Conference*. American Meteorological Society, 2005. Paper number P12R.1.

[980] J.K. Williams and G. Meymaris. *Remote turbulence detection using ground-based Doppler weather radar*, chapter 7. Springer, 2016.

[981] J.K. Williams, G. Meymaris, J. Craig, G. Blackburn, W. Deierling, and F. McDonough. Measuring in-cloud turbulence: the nexrad turbulence detection algorithm. In *AMS 15th Conference on Aviation, Range, and Aerospace Algorithm*. American Meteorological Society, 2011. Paper number 2.1.

[982] P.D. Williams. Transatlantic flight times and climate change. *Environmental Research Letters*, 11(2):024008, 2016.

[983] P.D. Williams. Increased light, moderate, and severe clear-air turbulence in response to climate change. *Advances in Atmospheric Sciences*, 34(5):576–586, 2017.

[984] P.D. Williams, T.W.N. Haine, and P.L. Read. On the generation mechanisms of short-scale unbalanced modes in rotating two-layer flows with vertical shear. *Journal of Fluid Mechanics*, 528:1–22, 2005.

[985] P.D. Williams, T.W.N. Haine, and P.L. Read. Inertia-gravity waves emitted from balanced flow: Observations, properties, and consequences. *Journal of the Atmospheric Sciences*, 65(11):3543–3556, 2008.

[986] P.D. Williams and M.J. Joshi. Intensification of winter transatlantic aviation turbulence in response to climate change. *Nature Climate Change*, 3:644–648, 2013.

[987] P.D. Williams and M.M. Joshi. Clear-air turbulence in a changing climate. In R. Sharman and T. Lane, editors, *Aviation Turbulence*, pages 465–480. Springer, 2016.

[988] J.C. Wilson, J.H. Hyun, and E.D. Blackshear. The function and response of an improved stratospheric condensation nucleus counter. *Journal of Geophysical Research: Oceans*, 88(C11):6781–6785, 1983.

[989] R. Wilson, F. Dalaudier, and F. Bertin. Estimates of the turbulent heat flux in the lower stratosphere from high resolution radar meaurements. *Geophysical Research Letters*, 32(21):L21811, 2005.

[990] R. Wilson, F. Dalaudier, and F. Bertin. Estimation of the turbulent fraction in the free atmosphere from mst radar measurements. *Journal of Atmospheric and Oceanic Technology*, 22(9):1326–1339, 2005.

[991] D.M. Winker, J. Pelon, J.A. Coakley, S.A. Ackerman, R.J. Charlson, P.R. Colarco, P. Flamant, Q. Fu, R.M. Hoff, C. Kittaka, T.L. Kubar, H. Le Treut, M.P. McCormick, G. Mégie, L. Poole, K. Powell, C. Trepte, M.A. Vaughan, and B.A. Wielicki. The CALIPSO mission: A global 3D view of aerosols and clouds. *Bulletin of the American Meteorological Society*, 91(9):1211–1230, 2010.

[992] M. Wirth, A. Fix, P. Mahnke, H. Schwarzer, F. Schrandt, and G. Ehret. The airborne multiwavelength water vapor differential absorption lidar Wales: system design and performance. *Applied Physics B: Lasers and Optics*, 96(1):201–213, 2009.

[993] B.M. Witte, R.F. Singler, and S.C.C. Bailey. Development of an unmanned aerial vehicle for the measurement of turbulence in the atmospheric boundary layer. *Atmosphere*, 8(10):195, 2017.

[994] W. Woiwode, A. Dörnbrack, M. Bramberger, F. Friedl-Vallon, F. Haenel, M. Höpfner, S. Johansson, E. Kretschmer, I. Krisch, T. Latzko, H. Oelhaf, J. Orphal, P. Preusse, B.-M. Sinnhuber, and J. Ungermann. Mesoscale fine structure of a tropopause fold over mountains. *Atmospheric Chemistry and Physics*, 18(21):15643–15667, 2018.

[995] J.K. Wolff and R.D. Sharman. Climatology of upper-level turbulence over the contiguous United States. *Journal of Applied Meteorology and Climatology*, 47(8):2198–2214, 2008.

[996] R.W. Wood. *Physical Optics*. MacMillan, New York, third edition, 1959.

[997] A.A. Woodfield and J.M. Vaughan. Airspeed and wind shear measurements with an airborne CO_2 continuous wave laser. Technical Report RAE-TR–83061, Royal Aircraft Establishment, Farnborough, United Kingdom, 1983.

[998] J.D. Woods, V. Högström, P. Misme, H. Ottersten, and O.M. Phillips. Fossil turbulence. *Radio Science*, 4(12):1365–1367, 1969.

[999] R.M. Worthington. MST radar observations of turbulent altocumulus layers. *Atmospheric Science Letters*, 16(4):500–505, 2015.

[1000] J.-L. Wu, H. Xiao, and E. Paterson. Physics-informed machine learning approach for augmenting turbulence models: a comprehensive framework. *Physical Review Fluids*, 3(7):074602, 2018.

[1001] X. Xiang, Z. Wang, Z. Mo, G. Chen, K. Pham, and E. Blasch. Wind field estimation through autonomous quadcopter avionics. In *Proceedings of the 2016 IEEE/AIAA 35th Digital Avionics Systems Conference (DASC)*, pages 1–6. IEEE, December 2016.

[1002] K.C. Yeh and C.-H. Liu. Radio wave scintillations in the ionosphere. *Proceedings of the IEEE*, 70(4):324–360, 1982.

[1003] S.A. Young. Analysis of lidar backscatter profiles in optically thin clouds. *Applied Optics*, 34(30):7019–7031, 1995.

[1004] S.A. Young, C.M.R. Platt, R.T. Austin, and G.R. Patterson. Optical properties and phase of some midlatitude, midlevel clouds in eclips. *Journal of Applied Meteorology and Climatology*, 39(2):135–153, 2000.

[1005] Z. Yuan, H. Liu, Y. Liu, D. Zhang, F. Yi, N. Zhu, and H. Xiong. Spatio-temporal dual graph attention network for query-poi matching. In *SIGIR 2020: Proceedings of the 43rd International ACM SIGIR Conference on Research and Development in Information Retrieval*, pages 629–638. Association for Computing Machinery, NY, USA, July 2020.

[1006] D. Zhang, Y. Liu, W. Cheng, B. Zong, J. Ni, Z. Chen, H. Chen, and H. Xiong. T^2-net: A semi-supervised deep model for turbulence forecasting. In C. Plant, H. Wang, A. Cuzzocrea, C. Zaniolo, and X. Wu, editors, *Proceedings—20th IEEE International Conference on Data Mining (IDM 2020)*, pages 1388–1393. IEEE, November 2020.

[1007] F. Zwicky. Aerial blobs. *Science*, 122(3160):159–160, 1955.

Index